Computational Modeling for Fluid Flow and Interfacial Transport

TRANSPORT PROCESSES IN ENGINEERING SERIES

Advisory Editor: A.S. Mujumdar, Department of Chemical Engineering,
   McGill University, Montreal, Quebec, Canada

Transport Processes in Engineering, 5

Computational Modeling for Fluid Flow and Interfacial Transport

by

Wei Shyy

Department of Aerospace Engineering, Mechanics and Engineering Science, University of Florida, Gainesville, FL 32611, USA

ELSEVIER

Amsterdam — London — New York —Tokyo   1994

ELSEVIER SCIENCE PUBLISHERS B.V.
Sara Burgerhartstraat 25
P.O. Box 211, 1000 AE Amsterdam, The Netherlands

ISBN: 0-444-81760-3

To
My Family and My Students

PREFACE

Transport processes are often characterized by the simultaneous presence of multiple dependent variables, multiple length scales, body forces, free boundaries, thin interfaces, and strong nonlinearities. They may also contain additional physical mechanisms, such as turbulence, combustion, phase change, capillarity, and electromagnetic effects. The various physical and chemical mechanisms cannot be treated simply as modules that can be added or deleted in the course of computation since they are all intrinsically coupled. In addition, it is difficult to completely resolve these mechanisms using the presently available (super)computing capability. Appropriate modeling is often required in order to make the problems computationally tractable. By coupling physical modeling with recent advances in computer hardware, numerical solution algorithms and grid generation schemes, many fluid flow and interfacial transport problems with physical as well as geometrical complexities can now be modelled and solved. There is a clear need for books dealing with computational modeling for handling complex flowfields, containing multiple physicochemical features, and interfacial dynamics for problems of practical importance. This book is a personal account of the subject area covered; it summarizes a substantial portion of my research and teaching activities during the last decade. It presents the various computational elements important for the prediction of complex fluid flows and interfacial transport. Also emphasized are practical applications, presented in the form of illustrations and examples, as well as physical interpretation of the computed results.

This book addresses both macroscopic and microscopic (but still continuum) features. In order to lay down a good foundation to facilitate discussion of more advanced techniques, I have divided this book into three parts. Part I is the outcome of a course that I have offered at the University of Florida for beginning graduate students. It presents the basic concepts of finite difference schemes for solving parabolic, elliptic and hyperbolic partial differential equations. My intention is to emphasize the development of tools for analyzing and assessing finite difference schemes for different equations, rather than compiling all the available methods published in the literature.

Part II deals with issues related to computational modeling for fluid flow and transport phenomena. Existing algorithms to solve the Navier–Stokes equations can be generally classified as density–based methods and pressure–based methods. The density–based methods use the continuity equation to specify the density and extract pressure information using the equation of state. These methods need to be modified to handle low Mach number regimes where the flows are incompressible and hence density has no role to play in determining the pressure field. The pressure–based methods obtain the pressure field via a pressure or a pressure correction equation, which is formulated by manipulating the continuity and momentum equations. In the present book, the pressure–based method is emphasized because, first, this method has not been covered in depth in most texts currently available, and, second, it is suitable for handling many issues encountered in flows containing interfaces and phase change, especially in materials processing and low speed heat transfer applications. Recent efforts to improve the performance of the pressure–based algorithm, both qualitatively and quantitatively, are emphasized, including formulation of the pressure–based algorithm and its generalization to all flow speeds, choice of the coordinate system and primary velocity variables, issues of grid layout, open boundary treatment and the role of global mass conservation, convection treatment, and convergence. Part II also discusses practical engineering applications, including gas–turbine combustor flow, heat transfer and convection in high pressure discharge lamps, thermal management under microgravity, and flow through hydraulic turbines.

Part III discusses the transport processes involving interfacial dynamics. Specifically, those influenced by phase change, gravity, and capillarity are emphasized, and both the macroscopic and morphological (microscopic) scales are presented. Basic concepts of interface, capillarity, and phase change processes are summarized to help clarify physical mechanisms, followed by a discussion of recent developments in computational modeling. Since the existing books on theoretical and computational fluid dynamics usually do not emphasize interfacial transport, the presentation in this area starts by summarizing fundamental concepts and basic analyses, then addresses scaling procedures and complex systems. Computational algorithms and a variety of numerical solutions are also discussed to illustrate the salient features of interfacial dynamics. Solutions obtained by using the schemes discussed in Part II and Part III are also presented in the form of case studies. These cases serve to demonstrate the interplay between the fluid and thermal transport at macroscopic scales and their interaction with the interfacial transport. Topics arising from materials processing and solidification are emphasized in Part III although the computational schemes developed in this part can be applied to other problems as well.

This book is written as a reference for researchers and graduate students with different background but sharing common interests. Topics already extensively reviewed in the literature, such as turbulence and combustion modeling, are not emphasized. Due to the diverse nature of the topics, it is not possible to cover the whole area comprehensively. However, efforts have been made to demonstrate that the algorithms and models discussed in this book can handle problems of practical interest.

The material presented in this book draws heavily from the research conducted by myself and with my collaborators. In particular, I would like to mention the following colleagues at General Electric Company with whom I conducted research in various areas: Mark Braaten (computational algorithms), Sanjay Correa, Mike Drake, Bob Pitz and Dave Burrus (turbulent combustion and gas turbine combustor flows), Peggy Chang and Jim Dakin (lamp modeling), and Thi Vu (hydraulic turbine analyses). Norm Lipstein and Ramani Mani gave me much appreciated trust and encouragement when I first embarked on research in computational fluid dynamics at GE research and Development Center in Schenectady, New York. Support and encouragement from Dan Backman, Yuan Pang and Dan Wei was instrumental for my work in materials processing and interfacial dynamics. The many discussions with and input from Yuan Pang have been particularly helpful. The late Bill Gingrich and I shared a good deal of research interests and friendship; he will be forever remembered. At University of Florida, work conducted with C.–S. Sun, M.–H. Chen, S. Thakur, H.S. Udaykumar, J. Wright, S.–J. Liang, M. Rao, E. Blosch, J. Liu, R. Smith, and R. Mittal has been included. My colleague Richard Fearn has provided many helpful comments on the material presented in Part III. I also appreciate the substantial help given by my students Madhukar Rao, Siddharth Thakur, Shin–Jye Liang, H.S. Udaykumar, Jeff Wright, Jian Liu, Venkat Krishnamurty and Jeff Burke during the course of manuscript preparation.

According to Chinese teaching, family is the root of a person's life; my family has given me much appreciated trust and love over the years. The relationship between a teacher and students is also a special one, similar to that in a family. The close and friendly atmosphere existing among my students and myself is truly satisfying. Hopefully more people interested in the subject area covered in this book can benefit from it. These thoughts have helped me in the course of preparing this work.

Gainesville, Florida
August 1993

CONTENTS

CHAPTER III. ELLIPTIC EQUATIONS

PART II. PRESSURE–BASED ALGORITHMS AND THEIR APPLICATIONS

CHAPTER V. PRESSURE–BASED ALGORITHMS

CHAPTER VI. PRACTICAL APPLICATIONS

PART III. INTERFACIAL TRANSPORT

CHAPTER VII. BASIC CONCEPTS OF THERMODYNAMICS

PART I

BASIC CONCEPTS OF FINITE DIFFERENCE METHODS

This part summarizes the basic concepts of computational methods applicable to a single partial differential equation. It presents the theory and techniques useful for understanding finite difference equations and suitable methods for solving parabolic, elliptic, and hyperbolic equations. Basic error analysis, stability, and consistency properties of different numerical schemes, and the physical realizability of solutions yielded by these schemes are emphasized. The emphasis of this part is on the development of tools for analyzing and assessing finite difference schemes for different type of equations. In this part we provide the foundation for more sophisticated techniques to be developed in Part II and Part III.

CHAPTER I
INTRODUCTION TO FINITE DIFFERENCE METHODS

Due to the advanced capabilities of modern computers the numerical solutions of differential equations is very commonplace and in fact necessary in order to solve a variety of problems of practical interest which defy any analytical treatment. Several techniques to solve the differential equations numerically have become popular such as finite difference methods, finite element methods, spectral methods and boundary element methods. A wealth of literature is available in the form of books and journals on these methods. We confine ourselves to the essential basis of finite difference methods in this part; we start with the basic concepts of finite difference schemes.

1 BASIC CONCEPTS OF FINITE DIFFERENCE SCHEMES

A differential equation defines the variation of a function, say U, with respect to one or many continuous independent variables. In the following we shall restrict ourselves to two independent variables, namely x and t, where x represents, say, the space coordinate in one dimension and t denotes, say, the time coordinate. While the definition of the independent variables depend on the physics of the problems encountered, the application of the finite difference methods are not dictated by them.

1.1 Finite Difference Operators

The basic idea of the finite difference method is to replace U, which, for now, is a function of two continuous independent variables, with a discrete function u, which is a function of the variables x and t at discrete points. Let h and k be positive constants, so that $x_m = mh$ and $t_n = nk$ are, respectively, the spatial grid points and time instants, for arbitrary integers m and n. We define the corresponding discrete function at (x_m, t_n) as u_m^n, i.e, $u_m^n = U(x_m, t_n)$. Thus the essence of the finite difference operators is to replace the original differential operators by finite difference operators, i.e., $\mathcal{L}(u) \cong L(U)$. For example,

$$
\begin{aligned}
\textit{forward difference} \quad &: \quad \Delta u_m = u_{m+1} - u_m \\
\textit{backward difference} \quad &: \quad \nabla u_m = u_m - u_{m-1} \\
\textit{central difference} \quad &: \quad \delta u_m = u_{m+1/2} - u_{m-1/2}
\end{aligned}
\qquad (1.1)
$$

1) Note that for the central difference operator given above, $u_{m \pm 1/2}$ need to be further defined since they are not located at the discrete grid points. This can be achieved via appropriate interpolations, e.g.,

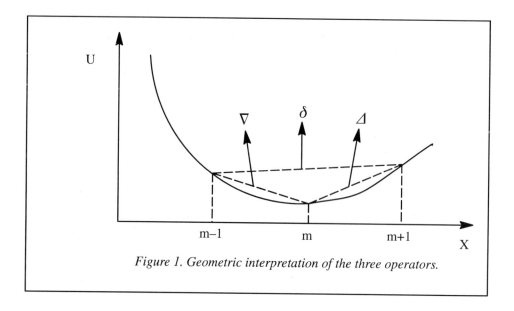

Figure 1. Geometric interpretation of the three operators.

$$u_{m \pm 1/2} = \frac{u_m + u_{m \pm 1}}{2} \qquad (1.2)$$

2) $U_x|_m$ can be approximated as $\Delta u_m/h$, $\nabla u_m/h$ or $\delta u_m/h$.

3) Similarly, $U_t|_n$ can be approximated as $\Delta u^n/k$, $\nabla u^n/k$ or $\delta u^n/k$.

In general, terms such as U_x and U_{xx} can be approximated with any number of the discrete u values at the neighboring grid points. The formal order of accuracy (to be defined later) of these numerical approximations generally increases with the number of grid points involved. A natural question to ask then is the following– does this mean that if we desire a high accuracy finite difference operator to approximate a differential equation, can we simply use more grid points? The involvement of more and more grid points in the approximation is associated with some problems:

(i) The effort of obtaining the numerical solution usually increases with the number of grid points which leads to a need to optimize the numerical accuracy on a per (computer) operation count basis.

(ii) Formulas involving more than two grid points in general need a starting procedure, i.e., near the boundary points.

(iii) More critically, it often turns out that even in the sense of absolute accuracy, i.e., completely ignoring the consideration of the computing cost, more number of grid points in the numerical approximation may NOT yield higher accuracy after all ! Such situations can appear when the solutions do not vary smoothly and modestly with respect to the independent variables. A case in point is the fluid dynamics problem with strong convective effects. In an extreme but frequently encountered case, the solution variable may exhibit a "jump" in its profile; under such a circumstance, a nominally high–order difference scheme is fundamentally unsuitable since only a weak form of the solution can be produced. By weak form we mean that instead of dealing directly with the equation containing the solution discontinuity, we first integrate the equation to eliminate the jump and then proceed with the integral form of the equation. The point is that in situations where discontinuities exist, since the derivatives at the discontinuous locations are not defined, those high–order schemes based on the notion of containing higher–order derivatives in the truncation error term are no longer effective.

(iv) Another problem is related to numerical stability, i.e., if an initial value problem is solved by marching in time, then some schemes will simply fail because error grows with time and becomes unbounded, no matter what is done. For some other schemes, success depends on the way one conducts the computation, i.e., the choice of time and space steps.

For the finite difference methods, there are many choices of schemes, many of them not too difficult to construct. However, one must determine

(a) whether these schemes can be reduced to the original differential operators as, say, h, k $\to 0$,

(b) whether they are "good" approximations to the original differential equation in the sense that the numerical solutions remain stable as t$\to \infty$, where t designates either the physical time in initial value problems or the number of steps in iterative methods often used to solve boundary value problems, and

c) whether they are accurate (with due consideration to computing cost).

The above considerations lead to the following issues associated with any finite difference scheme:

A) **consistency**, i.e, whether a difference operator $L(u)$ can truly represent the differential operator $\mathcal{L}(U)$ in the limiting situation, i.e., $L(u) \to \mathcal{L}(U)$ as h,k $\to 0$ and **convergence**, i.e., whether the finite difference solution approaches the exact differential solution in the limit, i.e., $u_m^n \to U(x_m, t_n)$ as h,k $\to 0$.

B) **stability**, i.e., whether u_m^n is bounded as n$\to \infty$, and

C) **accuracy**, i.e., how closely u_m^n approximates $U(x_m, t_n)$ as h and k vary.

These issues are the central themes of the analysis of finite difference schemes and are all interrelated. The accuracy of a scheme is often characterized by the "order" of accuracy which is formally defined next.

1.2 Order of Accuracy

The order of accuracy of a scheme is defined as the power of the mesh size (both in space and time) with which the truncation error of the scheme tends to zero. For example, consider the operator Δ

$$\frac{\Delta u_m}{h} = \frac{u_{m+1} - u_m}{h} \tag{1.3}$$

$$u_m = U(x_m)$$

$$u_{m+1} = U(x_m + h) = U(x_m) + h\, U_x(x_m) + \frac{h^2}{2} U_{xx}(x_m) + \ldots$$

$$\therefore \quad \frac{\Delta u_m}{h} = U_x(x_m) + \mathcal{O}(h) \tag{1.4}$$

The truncation error of the above scheme goes to zero like the first power in h; therefore, it is a first order scheme. Similarly,

$$\frac{\nabla u_m}{h} = U_x(x_m) + \mathcal{O}(h) \qquad \text{is first order, and} \qquad (1.5)$$

$$\frac{\delta u_m}{h} = U_x(x_m) + \mathcal{O}(h^2) \qquad \text{is second order.} \qquad (1.6)$$

A pertinent question at this point is that what happens to the order of accuracy of the above schemes if the grid spacing h is not a constant. This issue will be addressed later.

As another example, the second derivative U_{xx} can be approximated by repeated applications of the schemes for the first derivatives, as follows:

$$(U_{xx})_m = \left[(U_x)_x \right]_m \qquad (1.7)$$

Approach (i):

$$(U_{xx})_m \cong \frac{(U_x)_{m+1/2} - (U_x)_{m-1/2}}{h} \qquad \text{central difference}$$

$$\cong \frac{\left(\frac{\delta u}{h} \right)_{m+1/2} - \left(\frac{\delta u}{h} \right)_{m-1/2}}{h} \qquad \text{another central difference}$$

$$\cong \frac{u_{m+1} - 2u_m + u_{m-1}}{h^2} \qquad (1.8)$$

$$\text{where} \quad (\delta u)_{m \pm 1/2} = \begin{cases} u_{m+1} - u_m & \text{for } m + 1/2 \\ u_m - u_{m-1} & \text{for } m - 1/2 \end{cases}$$

The leading truncation error here is $\mathcal{O}(h^2)$. Thus it is a second order scheme.
Approach (ii):

$$(U_{xx})_m \cong \frac{(U_x)_{m+1} - (U_x)_m}{h}$$

$$\cong \frac{\left(\frac{\nabla u}{h} \right)_{m+1} - \left(\frac{\nabla u}{h} \right)_m}{h}$$

$$\cong \frac{\left(\frac{u_{m+1} - u_m}{h} \right) - \left(\frac{u_m - u_{m-1}}{h} \right)}{h}$$

$$\cong \frac{u_{m+1} - 2u_m + u_{m-1}}{h^2} \qquad (1.9)$$

Here the leading truncation error is, of course, again $\mathcal{O}(h^2)$. Thus, in this approach, by repeatedly applying two first order schemes, that are of opposite bias, the leading truncation error terms are cancelled out and the resulting scheme becomes second–order accurate. Again, this is possible here due to the constant grid spacing, h.

It must be emphasized that the order of accuracy referred here is called the *"local order of accuracy"* since it does not consider the propagation and accumulation of errors outside the stencil of the grid points directly utilized by the scheme applied at the grid point x_m. In other words, all the values of the discrete variables at the grid points are assumed to be exact; the only issue dealt with here is how good a job those finite difference operators can do under such a circumstance. As expected, the *"global order of accuracy"* is usually lower than the local one, since the inaccuracy of the finite difference operators at the "upstream" locations will propagate "downstream". We will define and discuss the global order of accuracy later.

So far we have discussed the basic issues related to the finite difference concept and looked at the idea of the local order of accuracy of finite difference schemes. Next we look at the finite difference equations obtained from the differential equations by applying finite difference operators to the derivatives, and show how to solve these difference equations.

2 SOLUTION OF FINITE DIFFERENCE EQUATIONS

A finite difference equation of the form

$$f(u_n, u_{n+1}, \cdots, u_{n+k}) = 0 \tag{2.1}$$

which can also be written in the following recursive form

$$u_{n+k} = g(u_n, u_{n+1}, \cdots, u_{n+k-1}) \tag{2.2}$$

is called a kth–order difference equation. The general solution of this equation contains k arbitrary constants. Hence, its solution is uniquely determined when k initial values, namely, $u_0, u_1, \ldots, u_{k-1}$, are given, and these grid locations can be arbitrarily chosen. This, as one may recall, is similar to the case of a kth–order differential equation.

Consider a homogeneous, linear difference equation of kth–order with constant coefficients

$$u_{n+k} + a_1 u_{n+k-1} + \ldots + a_k u_n = 0. \tag{2.3}$$

Its fundamental solution can be obtained by solving the characteristic equation

$$\lambda^{n+k} + a_1 \lambda^{n+k-1} + \ldots + a_k \lambda^n = 0 \tag{2.4}$$

whose general solution is given by

$$u_n = c_1 \lambda_1^n + c_2 \lambda_2^n + \ldots + c_k \lambda_k^n \tag{2.5}$$

where $\lambda_1, \lambda_2, \ldots, \lambda_k$ are the different roots of the characteristic equation. This is, again analogous to the differential equation case where a fundamental solution is $e^{\lambda x}$, instead of λ^n (note that $x_n = nh$).

Example 1 Difference equation:

$$u_{n+2} - 5u_{n+1} + 6u_n = 0$$
$$I.C.s : \quad u_0 = 0, \quad u_1 = 1$$

The characteristic equation is

$$\lambda^2 - 5\lambda + 6 = 0$$

and its roots are $\lambda_1 = 3$, $\lambda_2 = 2$. Thus, the general solution of the difference equation is

$$u_n = c_1 . 3^n + c_2 . 2^n$$

Applying the initial conditions, we get the following:

$$n = 0 : \quad c_1 + c_2 = 0$$
$$n = 1 : \quad 3c_1 + 2c_2 = 1$$

Thus the constants are $c_1 = 1$, $c_2 = -1$ and the final solution to the difference equation is

$$u_n = 3^n - 2^n .$$

A pertinent question at this juncture is the following: what happens when the multiple roots of the difference equation are identical ? To illustrate this case, consider a second–order equation of the form

$$au_n + bu_{n+1} + cu_{n+2} = 0 \tag{2.6}$$

Let the two roots of the characteristic equation be denoted by $\lambda_1, \lambda_2.$ Then the solution is

$$u_n = a\lambda_1^n + \beta\lambda_2^n \tag{2.7}$$

where α, β are determined by any two values of u_n, e.g., u_0 and u_1. Then we have

$$a + \beta = u_0$$

$$a\lambda_1 + \beta\lambda_2 = u_1$$

$$\text{Hence} \quad a = \frac{u_0\lambda_2 - u_1}{\lambda_2 - \lambda_1} \tag{2.8a}$$

$$\beta = \frac{u_1 - u_0\lambda_1}{\lambda_2 - \lambda_1} \tag{2.8b}$$

The above procedure breaks down if $\lambda_1 = \lambda_2$ .Recall that for a differential equation of the form

$$aU + bU' + cU'' = 0$$

with two roots λ_1 and λ_2 of the characteristic equation, the general solution is given as follows:

$$(i) \text{ if } \lambda_1 \neq \lambda_2$$
$$U = ae^{\lambda_1 x} + \beta e^{\lambda_2 x}$$
$$(ii) \text{ if } \lambda_1 = \lambda_2$$
$$U = ae^{\lambda_1 x} + \beta xe^{\lambda_1 x}$$

Here, in the case of $\lambda_1 = \lambda_2$ for the difference equation, we write the first solution as

$$(u_n)_1 = \lambda_1^n \tag{2.9a}$$

and the second solution as

$$(u_n)_2 = y_n \lambda_1^n \tag{2.9b}$$

where y_n is yet to be determined. Substituting $(u_n)_2$ into the original difference equation, we get

$$ay_n + b\lambda_1 y_{n+1} + c\lambda_1^2 y_{n+2} = 0 \tag{2.10}$$

Also from the characteristic equation

$$a + b\lambda_1 + c\lambda_1^2 = 0 \tag{2.11}$$

we know that $a/c = \lambda_1^2$ and $b/c = -2\lambda_1$. Thus Eq. (2.10) becomes

$$c\lambda_1^2 y_n - 2c\lambda_1^2 y_{n+1} + c\lambda_1^2 y_{n+2} = 0$$

$$\Rightarrow \quad y_n - 2y_{n+1} + y_{n+2} = 0$$

$$\Rightarrow \quad y_{n+2} - y_{n+1} = y_{n+1} - y_n \tag{2.12}$$

i.e., any arbitrary arithmetic progression is a solution to y_n , e.g., $y_n=n$ (or $y_n=n+1$, etc.). Hence the general solution is

$$u_n = \alpha \lambda_1^n + \beta n \lambda_1^n \tag{2.13}$$

As already discussed, the first derivative U_x can be approximated by many different finite difference operators, each involving a different number of grid points. Hence the order of the finite difference equation may be greater than the order of the differential equation it approximates. Consequently, the requirement of the initial conditions and the boundary conditions for the difference equation may not be compatible to that for the differential equation. This means that some artificial measures must be taken since there is not enough information available to obtain the general solution to the difference equation. Usually, one may need to resort to a lower order scheme to start a computation, then use this lower order solution to supplement the need for initial conditions for higher–order schemes.

3 ORDER OF ACCURACY: GLOBAL AND LOCAL

In this section, we will consider a simple model problem to illustrate the concepts of order of accuracy of a difference scheme in both a local and a global sense. Consider the following differential equation along with an initial condition

$$U'(x) + AU = 0 , \quad 0 \le x \le 1 \tag{3.1}$$
$$U(0) = 1$$

where A is a constant of order 1. The exact solution is given by

$$\textit{Exact Solution} : \quad U(x) = e^{-Ax} \tag{3.2}$$

Now, if we plug the finite difference operators discussed earlier into the above Equation (3.1), then we can obtain the exact solution of the difference equation as well.

3.1 Scheme I. $\varDelta$–operator

For the $\varDelta$–operator, we have the following:

$$\frac{u(x + h) - u(x)}{h} + A\, u(x) = 0$$
$$\textit{where } h = 1/N , \quad N : \textit{integer} \tag{3.3}$$

Solution :
$$\begin{aligned}
u_n &= (1 - Ah)\, u_{n-1} \\
&= (1 - Ah)\left[(1 - Ah)u_{n-2}\right] \\
&\;\;\vdots \\
&= (1 - Ah)^n\, u_0 \quad \textit{with } u_0 = 1
\end{aligned}$$

Hence
$$u_n = (1 - Ah)^{x_n/h} \tag{3.4}$$

The question is now to estimate the size of the error associated with this operator. We define the error as follows:

$$E(x_n) \;=\; U(x_n) \;-\; u_n \tag{3.5}$$

Thus, for the Δ–operator, the error is

$$E(x_n) \;=\; e^{-Ax_n} \;-\; (1 \;-\; Ah)^{x_n/h} \tag{3.6}$$

The main interest here is not only the values of the error $E(x_n)$ but also the *rate* at which $E(x_n)$ decreases as h decreases. Note that

$$(1 \;-\; Ah)^{x_n/h} \;=\; \exp\left[\frac{x_n}{h}\,\ln\,(1 \;-\; Ah)\right]$$

$$= \exp\left[\frac{x_n}{h}\left\{\;-Ah \;+\; \frac{(Ah)^2}{2} \;+\; \mathcal{O}(h^3)\right\}\right]$$

$$= [\exp(-\,Ax_n)]\,[\exp(A^2hx_n/2)]\,[\exp(\mathcal{O}(h^{2)})] \tag{3.7}$$

Also note that

$$e^x \;=\; 1 \;+\; x \;+\; \frac{x^2}{2!} \;+\; \frac{x^3}{3!} \;+\; \dots$$

Thus Eq. (3.7) becomes

$$(1 \;-\; Ah)^{x_n/h} \;=\; e^{-Ax_n} \;+\; h\left(\frac{A^2x_n}{2}\right)e^{-Ax_n} \;+\; \mathcal{O}(h^2)$$

$$\therefore \qquad E(x_n) \;=\; h\frac{A^2x_n}{2}e^{-Ax_n} \;+\; \mathcal{O}(h^2) \tag{3.8}$$

i.e., the normalized error becomes

$$E_{norm}(x_n) \;=\; \frac{E(x_n)}{U(x_n)} \;=\; \mathcal{O}(h)$$

Here $E_{norm}(x_n)$ tends to zero as h tends to zero; furthermore, its magnitude is of the order of the first power of h. Thus, Scheme I is first–order accurate *globally*. This error is global because we only use the initial condition and want to know $E_{norm}(x_n)$ in terms of its behavior at grid points many h distance away from the boundary.

Now we come back to the issue of local accuracy of Scheme I. Assume $u(x_n){=}U(x_n)$, i.e, $u(x_n)$ is known exactly and we want to know $u(x_n{+}h)$ and compare it with $U(x_n{+}h)$. Using Taylor series expansion, we get

$$U(x_n + h) \;=\; U(x_n) \;+\; U'(x_n)h \;+\; U''(x_n)\frac{h^2}{2} \;+\; \dots \tag{3.9a}$$

Using the finite difference formula (3.3), we have

$$u(x_n + h) \;=\; u(x_n) \;-\; hAu(x_n) \tag{3.9b}$$

Comparing (3.9a) and (3.9b) and using the facts that

$$U(x_n) \;=\; u(x_n) \;,\; \text{ and}$$
$$U'(x_n) \;=\; -\,AU(x_n) \;=\; -\,Au(x_n) \;,$$

the first two terms in (3.9a) and (3.9b) cancel out and we get

$$E(x_n) = U(x_n + h) - u(x_n + h) = O(h^2)$$

i.e., Scheme I is second–order accurate locally.

Since the local error is really the error with x_n only a distance h away from the boundary, i.e, $n=1$, the error obtained for the global behavior can be used here also, i.e,

$$local\ error \quad : \quad E(h)$$
$$global\ error \quad : \quad E(nh) , \quad n \geq O(1)$$

Recall Eq. (3.8). Here, $x_n = nh$. For estimating the local error, we have $n=1$. Thus,

$$E(x_1) = h^2 \frac{a^2}{2} e^{-Ax_n} + \dots$$

$$\therefore \quad E_{norm}(x_1) = O(h^2) : \quad second - order\ accurate$$

In general we have the following relation between the local and global orders of accuracy:

$$Global\ order\ of\ accuracy = Local\ order\ of\ accuracy - 1 , \qquad (3.10)$$

provided that the starting procedure, if needed, does not introduce "substantially" more error into the solution. We will address this point next.

3.2 Scheme II : δ– operator

Using the δ– operator, the finite difference scheme can be written as

$$\frac{u(x + h) - u(x - h)}{2h} + Au(x) = 0 \qquad (3.11)$$

This is a second–order difference equation and hence needs two initial conditions. One of them is available from the problem statement, Eq. (3.1), namely, $u(0)=1$. The second one is yet to be determined and we will come to that issue in a moment. For now, we simply assume that both u_0 and u_1 are already known.

The general solution of Eq. (3.11) is

$$u_n = \alpha \lambda_1^n + \beta \lambda_2^n$$

$$= \left(\frac{\lambda_2 u_0 - u_1}{\lambda_2 - \lambda_1} \right) \lambda_1^n - \left(\frac{\lambda_1 u_0 - u_1}{\lambda_2 - \lambda_1} \right) \lambda_2^n \qquad (3.12)$$

where

$$\lambda_1 = \sqrt{1 + (Ah)^2} - Ah$$

$$= 1 - Ah + \frac{(Ah)^2}{2} + O(h^4)$$

and

$$\lambda_2 = (-1) \left[1 + Ah + \frac{(Ah)^2}{2} \right] + O(h^4)$$

$$\therefore \quad \lambda_1^n = \lambda_1^{x_n/h} = \left[1 - Ah + \frac{(Ah)^2}{2} + O(h^4)\right]^{x_n/h}$$

$$= \exp\left\{\frac{x_n}{h} \ln\left[1 - Ah + \frac{(Ah)^2}{2} + O(h^4)\right]\right\}$$

Note that

$$\ln(1 + z) = z - \frac{z^2}{2} + \frac{z^3}{3} + O(z^4)$$

$$\Rightarrow \quad \ln\left[1 - Ah + \frac{(Ah)^2}{2} + O(h^4)\right] = -Ah + \frac{(Ah)^3}{6} + O(h^4)$$

$$\therefore \quad \lambda_1^n = \exp\left\{\frac{x_n}{h}\left[-Ah + \frac{(Ah)^3}{6} + O(h^4)\right]\right\}$$

$$= \left(e^{-Ax_n}\right)\left[1 + h^2\left(\frac{A^3 x_n}{6}\right)\right] + O(h^3) \tag{3.13}$$

Thus
$$\lambda_1^n = e^{-Ax_n} + O(h^2)$$

Similarly
$$\lambda_2^n = (-1)^n e^{+Ax_n} + O(h^2)$$

Hence, we have

$$u_n = \frac{\lambda_2 u_0 - u_1}{\lambda_2 - \lambda_1}\left[e^{-Ax_n} + O(h^2)\right] - \frac{\lambda_1 u_0 - u_1}{\lambda_2 - \lambda_1}\left[(-1)^n e^{Ax_n} + O(h^2)\right] \tag{3.14}$$

From the above, the following can be noted:

1. Out of the two terms on the right hand side of the finite difference solution (3.14), the first term tends to the analytical solution if

$$\frac{\lambda_2 u_0 - u_1}{\lambda_2 - \lambda_1} \to 1 \quad as \quad h \to 0 . \tag{3.15}$$

2. However, the second term has the factor $(-1)^n$, indicating that it approaches different values depending on whether n is even or odd. Thus, the only acceptable result is that

$$\frac{\lambda_1 u_0 - u_1}{\lambda_2 - \lambda_1} \to 0 \quad as \quad h \to 0 . \tag{3.16}$$

Furthermore, we require that $exp(Ax_n)$ be $O(1)$; otherwise, the second term will grow exponentially fast and become unbounded even if (3.16) holds because the term in (3.16) approaches zero at the rate h^p (where p is the order of accuracy of the scheme) whereas the $exp(Ax_n)$ term grows exponentially with x.

3.3 Boundary Treatment

One unresolved problem is that u_1 is not specified; we need to obtain it in some way. Here, we examine two possibilities:

3.3.1 Possibility (i): Taylor Series Method

$$U_1 = U(h) = U_0 - hU_0' + \mathcal{O}(h^2)$$

$$= U_0 - hAU_0 + \mathcal{O}(h^2)$$

$$\text{Let } u_1 = u_0 - hAu_0 \quad \text{where } u_0 = 1$$

$$\rightarrow \quad \text{error } e_1 = \mathcal{O}(h^2) \; : \quad \text{locally second} - \text{order formula}$$ (3.17)

This actually amounts to using the ∇–operator which is globally a first–order operator. Thus, although we use a first–order operator to obtain u_1, it does not degrade the order of overall global accuracy, as seen from above. The reason lies in the fact that we apply the boundary treatment only at the point x_1; hence, the *local* accuracy of the boundary scheme, which is second–order, controls the performance of the overall scheme. Now, we need to check whether the two conditions, (3.15) and (3.16), are satisfied by this method. We see that

$$\frac{\lambda_2 u_0 - u_1}{\lambda_2 - \lambda_1} = 1 + \mathcal{O}(h^2) \rightarrow 1 \quad as \quad h \rightarrow 0 .$$ (3.18a)

$$\frac{\lambda_1 u_0 - u_1}{\lambda_2 - \lambda_1} = \mathcal{O}(h^2) \rightarrow 0 \quad as \quad h \rightarrow 0 .$$ (3.18b)

Thus, both the requirements given by (3.15) and (3.16) are satisfied at the rate of $\mathcal{O}(h^2)$. Finally, according to Eq. (3.14) we have

$$u_n = \left[1 + \mathcal{O}(h^2)\right]\left[e^{-Ax_n} + \mathcal{O}(h^2)\right] - \left[\mathcal{O}(h^2)\right]\left[(-1)^n e^{Ax_n} + \mathcal{O}(h^2)\right]$$

$$= e^{-Ax_n} + \mathcal{O}(h^2)$$ (3.19)

Thus, it is a second–order accurate solution in the global sense.

Next, we examine the local error of this approach. Substituting n=1 in Eq. (3.13), we get the error associated with each root:

$$E_{norm_{\lambda_1}} = h^2\left(\frac{A^3 x_n}{6}\right) = \mathcal{O}(h^3) \quad since \quad x_n = x_1 = h$$ (3.20a)

Similarly $\quad E_{norm_{\lambda_2}} = \mathcal{O}(h^3) .$ (3.20b)

Thus, the local order of accuracy contributed by λ_1 and λ_2 is 3. However, since u_1, which appears in the coefficients of Eq. (3.12), is now only $\mathcal{O}(h^2)$ accurate, the overall local accuracy is reduced to 2. Again, the global order of accuracy remains as 2.

3.3.2 Possibility (ii): Simple Extrapolation

Another possibility is to assign $u_1 = u_0 = 1$. This is locally a first–order scheme since we have

$$E(x_1) = E(h) = hAU_1 = \mathcal{O}(h) . \tag{3.21}$$

We again check for conditions (3.15) and (3.16):

$$\frac{\lambda_2 u_0 - u_1}{\lambda_2 - \lambda_1} = 1 + \frac{1}{2}Ah = 1 + \mathcal{O}(h) \rightarrow 1 \quad as \quad h \rightarrow 0 .$$

$$\frac{\lambda_1 u_0 - u_1}{\lambda_2 - \lambda_1} = \frac{1}{2}Ah + \mathcal{O}(h^2) = \mathcal{O}(h) \rightarrow 0 \quad as \quad h \rightarrow 0 .$$

Both the requirements are satisfied, but only at the rate of $\mathcal{O}(h)$. Consequently, we have

$$u_n = e^{-Ax_n} + \mathcal{O}(h) \tag{3.22}$$

and hence this scheme is first–order accurate globally.

Conclusion: From the above two possibilities of the boundary treatment, we can see that the boundary treatment can be one order lower in the global sense compared to the interior treatment without degrading the order of overall global accuracy. However, if the boundary condition is two or more orders lower in accuracy, it degrades the overall global accuracy. Some related material on the issue of numerical boundary treatment can be found in a paper by Blottner (1982).

4 STABILITY OF DIFFERENCE SCHEMES

A perfectly reasonable finite difference scheme approximating the differential equation can result in a numerical solution (or *exact* discrete solution) not converging to the correct analytical solution as $h \rightarrow 0$. The reason for the above is that the error can accumulate at a rate faster than tolerable. Thus, $E_{norm}(x_n)$ grows so fast that it becomes $\mathcal{O}(1)$ or higher despite h being small because as the size of h decreases, the total number of grid points, n, increases in order to reach the same value of x_n.

4.1 Illustration of the Instability of a Finite Difference Scheme

Suppose that in order to compute $U'(x) = f(x, U)$, we use a 4–point (3–backward, 1–forward) operator to update the value $u(x_n+1)$ based on $U'(x) = 0$, then we expect to get $\mathcal{O}(h^3)$ accuracy. The operator can be constructed as follows:

$$U_{n+1} = U_n + hU'_n + \frac{h^2}{2}U_n'' + \frac{h^3}{6}U_n''' + \mathcal{O}(h^4)$$

$$U_n = U_n$$

$$U_{n-1} = U_n - hU'_n + \frac{h^2}{2}U_n'' - \frac{h^3}{6}U_n''' + \mathcal{O}(h^4) \tag{4.1}$$

$$U_{n-2} = U_n - (2h)U'_n + \frac{(2h)^2}{2}U_n'' - \frac{(2h)^3}{6}U_n''' + \mathcal{O}(h^4)$$

This gives us

$$a_1 u_{n+1} + a_2 u_n + a_3 u_{n-1} + a_4 u_{n-2} \cong U_x|_{x_n}$$

$$(i) \quad U_n \quad : \quad a_1 + a_2 + a_3 + a_4 = 0$$

$$(ii) \quad U_n' \quad : \quad a_1 \quad\quad - a_3 - 2a_4 = 1/h$$

$$(iii) \quad U_n'' \quad : \quad a_1 \quad\quad + a_3 + 4a_4 = 0 \quad\quad (4.2)$$

$$(iv) \quad U_n''' \quad : \quad a_1 \quad\quad - a_3 - 8a_4 = 0$$

The solution to the above is

$$a_1 = \frac{1}{3h} \quad\quad\quad a_2 = \frac{1}{2h}$$
$$\quad\quad\quad\quad\quad\quad\quad\quad\quad\quad\quad\quad (4.3)$$
$$a_3 = \frac{-1}{h} \quad\quad\quad a_4 = \frac{1}{6h}$$

Hence the resulting scheme is

$$u_{n+1} = -\frac{3}{2} u_n + 3 u_{n-1} - \frac{1}{2} u_{n-2} + 3hf(x_n, u_n) \quad\quad (4.4)$$

The question now is that how "good" is this scheme? To see this, we use a trial case, say $f \equiv 0$ and look at the following two cases:

(A) Assume that u_0, u_1 and u_2 are all known exactly, i.e. $u_0 = u_1 = u_2 = 0$. Thus, the true solution should be $u_n = 0$. Substituting in Eq. (4.4), we see that this is indeed the case. Hence, mathematically, the scheme is quite reasonable.

(b) Now, if we introduce a round–off error into the computation to check the sensitivity of u_n with respect to the round–off error, e.g., $u_0 = 0$, $u_1 = 0$, $u_2 = \varepsilon$ (a small number), then it turns out that

$$u_6 \cong 300\varepsilon , \quad u_{10} \cong 10^4 \varepsilon , \quad u_{15} \cong 10^6 \varepsilon \quad\quad (4.5)$$

Notice that this error grows extraordinarily fast and it has nothing to do with the size of h since h does not appear in the Eq.(4.5) if $f \equiv 0$. In fact, the smaller the size of h, the higher the number of grid points, and consequently the error grows to higher levels. In short, the above scheme would only give unreasonable results.

The obvious question is– how does this happen? To answer this, let us look at the characteristic equation of Eq. (4.4) with $f \equiv 0$, which is

$$\lambda^3 + \frac{3}{2}\lambda^2 - 3\lambda + \frac{1}{2} = 0 \quad\quad (4.6)$$

The roots of the above characteristic equation are

$$\lambda_1 = \frac{-5 - \sqrt{29}}{4} \cong -2.59$$

$$\lambda_1 = \frac{-5 + \sqrt{29}}{4} \cong 0.09 \quad\quad (4.7)$$
$$\lambda_3 = 1$$

Hence the general solution is

$$u_n = a_1 \lambda_1^n + a_2 \lambda_2^n + a_3 \lambda_3^n \tag{4.8}$$

(i) If $u_0=0$, $u_1=0$, $u_2=0$, then $a_1 = 0$, $a_2 = 0$, $a_3 = 0$. Then $u_n = 0$, which is the correct solution for this case.

(ii) If $u_0=0$, $u_1=0$, $u_2=\varepsilon$, then $a_1 \cong \varepsilon$, $a_2 \cong \varepsilon$, $a_3 \cong \varepsilon$. Now the solution becomes

$$u_n \cong \varepsilon(-2.59)^n + \varepsilon(0.09)^n + \varepsilon$$

The error $E_{norm}(x_n)$ grows at the rate of $\varepsilon(-2.59)^{x_n/h}$. If we let $h \to 0$ then the error grows exponentially. Hence, this scheme, while mathematically very reasonable, is computationally very unsuitable because it is very prone to any small disturbance caused by an inexact prescription of the local values of u at any grid point. That means, for example, during the starting procedure as required by the scheme, any small errors will eventually propagate and grow with the grid index to an unacceptably high level. This issue is at the heart of the stability analysis of finite difference schemes.

4.2 Stability

In many physical problems, linear as well as nonlinear, we have the case of exponentially decaying solutions. In such situations, it is desirable to have a super–stable numerical method which yields a numerical solution guaranteed to converge to zero, independent of the mesh size h. Consider the following model problem:

$$U'(x) + AU(x) = 0 , \quad A = constant \text{ and } Re(A) > 0 . \tag{4.9}$$

We seek a solution of this model problem using a numerical scheme by computing an approximate solution u_n at $x_n=nh$. We desire the numerical method to be such that

$$\lim_{n \to \infty} u_n = 0 \tag{4.10}$$

We will look at some of the alternatives to solve the above model problem. Before we do that, we define A–stable methods.

Definition. A–Stable Method: A numerical method, used to solve Eq. (4.9), is said to be A–stable if its region of absolute stability contains the whole of the right–hand half plane, $Re(Ah) > 0$, i.e., the whole region in which the exact solution decays with x.

4.3 Forward Difference (Explicit Euler Method)

In this finite difference method, we use

$$(U_x)_n \cong \frac{u_{n+1} - u_n}{h} \tag{4.11}$$

Thus the model problem can be written as

$$u_{n+1} = (1 - Ah) u_n$$

$$\vdots$$

$$= (1 - Ah)^{n+1} u_0$$

$$\cong e^{-Ax_n} \tag{4.12}$$

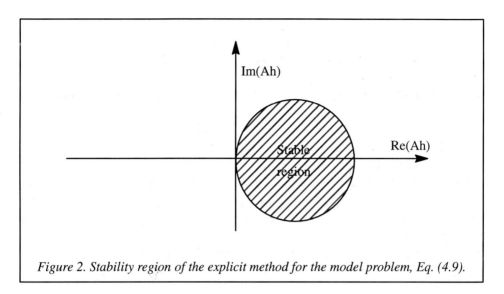

Figure 2. Stability region of the explicit method for the model problem, Eq. (4.9).

Thus, this method is good when $h \to 0$. From the above, it can be seen that the Euler method is A–stable when

$$| 1 - Ah | < 1 . \tag{4.13}$$

The reason for the above is that the solution grows at the rate of $(1–Ah)$ and hence if the above Eq. (4.13) holds, then the errors arising from the grid points other than the one under consideration do not become unbounded as $n \to \infty$. On the other hand if $| 1 - Ah | > 1$, then the errors grow exponentially. Fig. 2 illustrates the above statements.

At this point, it is appropriate to address the issue of computing the stability region for a given differential equation. The procedure to compute the stability region is as follows: write down the finite difference equation corresponding to the ordinary differential equation given by Eq. (4.9) and solve the associated characteristic equation $P(\lambda) = 0$ to obtain the roots $\lambda's$. The finite difference scheme is stable if all the roots satisfy the inequality $| \lambda | < 1$.

For the differential Equation (4.9), any finite difference scheme must satisfy the following two requirements:
1) If $Re(A) < 0$, then the unstable modes, if they contribute significantly to the composite solution, must be accurately represented by the numerical solution, regardless of their intrinsic time scales.
2) The value of Ah must lie within the region of absolute stability when $Re(A) > 0$.

4.4 Backward Difference (Implicit Euler Method)

In this method, we use

$$(U_x)_n \cong \frac{u_n - u_{n-1}}{h} \tag{4.14}$$

Now the model problem can be written as

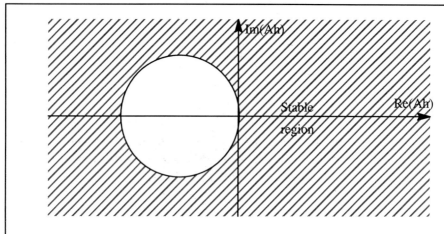

Figure 3. Stability region of the implicit method for the model problem, Eq. (4.9).

$$u_n = (1 + Ah)^{-1} u_{n-1}$$

$$\vdots$$

$$= \left(\frac{1}{1 + Ah}\right)^n u_0 \tag{4.15}$$

Now the stability requirement becomes

$$\left|\frac{1}{1 + Ah}\right| < 1$$

or

$$|1 + Ah| > 1 . \tag{4.16}$$

Fig. 3 illustrates the above condition.

It should be noted at this point that this method is actually too stable in the sense that it will yield a decaying (stable) numerical solution even if $Re(A) < 0$, i.e., even if the exact solution grows exponentially with x. While the present model problem is too trivial to distinguish the computing cost per nodal point between the explicit and the implicit scheme, in general, as will become clear later, the implicit scheme has higher computing cost but does have much broader choice of the mesh size to remain stable.

4.5 Central Difference

For this scheme, we have the following:

$$\frac{u_{n+1} - u_{n-1}}{2h} + Au_n = 0 \tag{4.17}$$

As we have computed earlier, the two roots of the characteristic equation associated with the above finite difference equation with real values of A, are

$$\lambda_1 = 1 - Ah + \frac{(Ah)^2}{2} + \mathcal{O}(h^4) \tag{4.18a}$$

$$\lambda_2 = (-1)\left(1 + Ah + \frac{(Ah)^2}{2} + \mathcal{O}(h^4)\right) \tag{4.18b}$$

For the above two roots, we have

$$|\lambda_1| < 1 , \quad |\lambda_2| > 1 . \tag{4.19}$$

Hence this finite difference scheme will *always* be unstable for any h. Note that we have analyzed the above scheme in a previous section, namely Scheme II in Section 3.2, where we concluded that this scheme is satisfactory in terms of accuracy (second–order) and starting procedure. In that example, we had $0 \le x \le 1$ and $A = \mathcal{O}(1)$, and hence the above scheme behaved reasonably. If these restrictions are removed, then as indicated by Eq. (4.19), the numerical solution grows and becomes unbounded.

4.6 Trapezoid Difference

For this scheme, we have

$$\frac{u_{n+1} - u_n}{h} + Ah\left(\frac{u_{n+1} + u_n}{2}\right) = 0 \tag{4.20a}$$

$$u_{n+1} = \left(\frac{1 - Ah/2}{1 + Ah/2}\right)u_n \tag{4.20b}$$

The stability requirement for the above scheme is

$$|\lambda_1| = \left|\frac{1 - Ah/2}{1 + Ah/2}\right| < 1 \tag{4.21}$$

The above condition will always be met as long as $Ah > 0$, and hence this scheme is A–stable, as shown in Fig. 4.

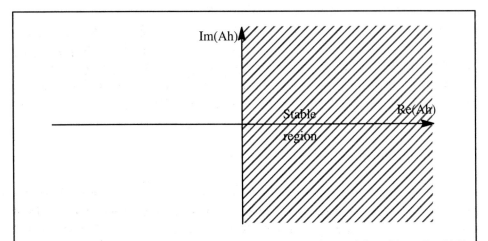

Figure 4. Stability region of the trapezoidal method for the model problem, Eq. (4.9).

Although the trapezoid scheme is stable, it does not guarantee accurate results. For example, if $Ah > 2$, the numerical solution profiles exhibit oscillations, and if $Ah \gg 1$, then we get $u_{n+1} \cong - u_n$, which leads to an odd–even fluctuation in the solution profile.

4.7 Some Relevant Facts

1) No explicit scheme, single step or multistep, can be A–stable.

2) For the general linear implicit multistep method given by

$$\alpha_k u_{n+1} + \ldots\ldots + \alpha_0 u_{n+1-k} + Ah\big(\beta_k u_{n+1} + \ldots\ldots + \beta_0 u_{n+1-k}\big) = 0 \quad (4.22)$$

no A–stable method can be of order greater than two.

3) The second–order A–stable implicit linear multistep with the smallest error is the trapezoidal rule.

For details on the above facts, the interested reader is referred to Dahlquist (1963).

4.8 The Issue of Accuracy Versus Stability

Consider the Euler method to approximate the following differential equation:

$$U' + AU = 0 , \qquad U(0) = 1 \qquad\qquad (4.23)$$

whose exact solution is given by

$$U(x) = e^{-Ax} \qquad\qquad (4.24a)$$
$$u(h) = e^{-Ah} \qquad\qquad (4.24b)$$

The solution for the finite difference equation associated with the forward Euler method is

$$u_1 = 1 - Ah \qquad\qquad (4.24c)$$

Note that

$$e^{-Ah} \rightarrow 1 - Ah \quad only \ when \ Ah \ll 1 .$$

Thus, for the forward Euler method, the accuracy requirement also goes hand in hand with the stability requirement given by Eq. (4.13). On the other hand, when $u(x)$ tends to zero, i.e., as $x \gg 1$, then the accuracy is no longer a problem. However, since the forward Euler method still possesses the same stability restriction, it is very inefficient in such a situation. The issue of stability versus accuracy can be illustrated by the following example.

Example 2 Consider the following equation

$$U' + 100U = 0 , \qquad U(0) = 1$$

The solution over the whole domain x can easily be divided into two regions as shown in Fig. 5. In region (i) the solution is varying rapidly and demands a good accuracy from the numerical scheme being used to simulate it; we do not have to worry about the stability in this region of the solution. On the other hand, in region (ii), the solution varies by a very insignificant amount; however, we still need to satisfy the same

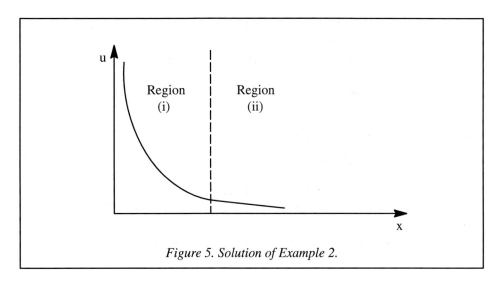

Figure 5. Solution of Example 2.

stability criterion, which for this case is $h < 1/100$, which is very stringent and thus inefficient. This is a typical example of the so–called "stiff" ordinary differential equation, where the calculation in region (i) is limited by accuracy requirement whereas that in region (ii) is limited by stability requirement. In a practical sense, one can view the stiff equations as equations where certain implicit methods can perform (much) better than explicit ones.

4.9 Systems of Equations

Consider a system of ordinary differential equations

$$\vec{U}' = [A]\,\vec{U} \tag{4.25}$$

with N unknowns, i.e.,

$$\vec{U} = \begin{bmatrix} U_1 \\ \vdots \\ U_N \end{bmatrix}$$

If λ denotes the eigenvalues of the matrix $[A]$ and $\vec{\xi}$ its eigenvectors, then we have

$$[A]\,\vec{\xi} = \lambda\,\vec{\xi} \tag{4.26}$$

Usually, for an $N \times N$ matrix $[A]$, there are N linearly independent eigenvectors $\vec{\xi}$ and a matrix composed of these N eigenvectors can be written as

$$[T] = \begin{bmatrix} \vec{\xi}_1 & \vec{\xi}_2 & \cdots & \vec{\xi}_N \end{bmatrix} \tag{4.27}$$

and we have

$$[T]^{-1}[A]\,[T] = [\Lambda] \tag{4.28}$$

where $[\Lambda]$ is the diagonal matrix composed of eigenvalues:

$$[A] = \begin{bmatrix} \lambda_1 & & \\ & \cdot & 0 \\ 0 & \cdot & \\ & & \lambda_N \end{bmatrix} \qquad (4.29)$$

Thus we have

$$[T]^{-1}\vec{U'} = [T]^{-1} [A] \vec{U} = [T]^{-1}[A] [I] \vec{U}$$

$$= [T]^{-1} [A] \left([T] [T]^{-1}\right) \vec{U}$$

$$= \left([T]^{-1} [A] [T]\right) [T]^{-1} \vec{U}$$

$$= [A] [T]^{-1} \vec{U} \qquad (4.30)$$

If we define

$$\vec{\phi} = [T]^{-1} \vec{U} \qquad (4.31)$$

then we have

$$\vec{\phi'} = [A] \vec{\phi} \qquad (4.32)$$

and hence

$$\phi'_i = \lambda_i \phi_i , \qquad i = 1, ..., N \qquad (4.33)$$

which is an uncoupled set of first–order ordinary differential equations and thus all the knowledge about single first–order ordinary differential equations can be directly applied to the above general system of ordinary differential equations. The solution to the above is given by

$$\phi_i = a_i e^{\lambda_i t} \qquad (4.34)$$

where a_i are constants to be determined by the initial condition. Hence, we have

$$\vec{U} = \sum_i a_i \vec{\xi}_i \, e^{\lambda_i t} \qquad (4.35)$$

It should be noted that $1/\lambda$ indicates the characteristic scales that control the solution profile for the system of equations.

Example 3 Consider the following system of equations:

$$y_1' = -1000 \, y_1 + y_2$$
$$y_2' = 994 \, y_1 - 2 \, y_2$$

The eigenvalues for this case are

$$\lambda_1 \cong -1.001$$
$$\lambda_2 \cong -1001.999$$

and it can be seen that they differ by three orders of magnitude. Here, initially, λ_2 is the controlling factor for maintaining accuracy, i.e., we must satisfy the requirement that $h < 1/\lambda_2$. But, then λ_1 becomes dominant and for maintaining accuracy, we must satisfy the condition $h < 1/\lambda_1$. However, with regard to stability, λ_2 is always the controlling factor since it is the largest eigenvalue of the system, and hence we always have $h < 1/\lambda_2$ (which is quite a stringent condition for the present example).

Example 4 Local Analysis of a Nonlinear Problem:

$$y_1' = -0.04y_1 + 10^4 y_2 y_3 = F_1$$

$$y_2' = 0.04y_1 - 10^4 y_2 y_3 - 3\times10^7 y_2^2 = F_2$$

$$y_3' = 3\times10^7 y_2^2 = F_3$$

This system is nonlinear, and hence, in order to find out the stability bound of a numerical scheme, we need to first linearize the system locally and then conduct the necessary analysis (see, e.g., Boyce and DiPrima 1986). The equations can be linearized by finding the Jacobian of the system:

$$\begin{bmatrix} (F_1)_{y_1} & (F_1)_{y_2} & (F_1)_{y_3} \\ (F_2)_{y_1} & (F_2)_{y_2} & (F_2)_{y_3} \\ (F_3)_{y_1} & (F_3)_{y_2} & (F_3)_{y_3} \end{bmatrix} = \begin{bmatrix} -0.04 & 10^4 y_3 & 10^4 y_2 \\ 0.04 & -10^4 y_3 - 6\times10^7 y_2 & -10^4 y_2 \\ 0 & 6\times10^7 y_2 & 0 \end{bmatrix}$$

Now if we are interested in the neighborhood of the point $(y_1, y_2, y_3) = (1, 3.65\times10^{-5}, 0)$, its Jacobian is

$$\begin{bmatrix} -0.04 & 0 & 0.365 \\ 0.04 & -2190 & -0.365 \\ 0 & 2190 & 0 \end{bmatrix}$$

Its eigenvalues are $\lambda_1 = 0$, $\lambda_2 = -0.405$, $\lambda_3 = -2189.6$. So, again, there is a wide disparity among the eigenvalues, causing difficulty in computing the solution numerically.

It can be seen from above that stiffness corresponds to the situation in which eigenvalues of the system differ from each other by orders of magnitude, thus leading to drastically different time or length scales in the problem. Examples of such stiff problems are some situations in chemical kinetics where the rate constants of chemical reaction of, say, two different species differ by six or seven orders of magnitude. For such systems, A–stable numerical methods are very desirable.

5 DISCRETIZATION OF PDEs AND ERROR NORMS

5.1 Discretization of PDEs: an Example

By methodically applying the techniques described above, finite difference schemes can be constructed to approximate partial differential equations. We illustrate this with an example.

Example 5 Consider the following PDE:

$$\frac{\partial U}{\partial t} = \frac{\partial^2 U}{\partial x^2} \tag{5.1}$$

In order to approximate the above PDE with a finite difference equation, we can choose from many different combinations of treatments for x– and t–derivative terms. In this example, we choose forward differencing (Δ–operator) for the time derivative and central differencing (δ–operator) for the spatial derivative. If we define $\Delta t = k$, $\Delta x = h$, $t_n = nk$ and $x_m = mh$, then Eq. (5.1) can be approximated as

$$\frac{u_m^{n+1} - u_m^n}{k} = \frac{u_{m+1}^n - 2u_m^n + u_{m-1}^n}{h^2} \tag{5.2}$$

which can be written as

$$u_m^{n+1} = ru_{m+1}^n + (1 - 2r)u_m^n + ru_{m-1}^n \tag{5.3}$$

where $r = k/h^2$. Now, we wish to assess the local order of accuracy of Eq. (5.3), for which we resort to a Taylor series expansion about the point (x_m, t_n). The local error can be obtained by representing all the dependent variables in a given finite difference equation with the variable and its derivatives at the reference location (x_m, t_n). Thus, we can write

$$U_m^{n+1} = U_m^n + kU_t\big|_m^n + \frac{k^2}{2}U_{tt}\big|_m^n + \ldots \tag{5.4a}$$

$$U_{m+1}^n = U_m^n + hU_x\big|_m^n + \frac{h^2}{2}U_{xx}\big|_m^n + \frac{h^3}{6}U_{xxx}\big|_m^n + \frac{h^4}{4!}U_x^{IV}\big|_m^n + \ldots \tag{5.4b}$$

$$U_{m+1}^n = U_m^n - hU_x\big|_m^n + \frac{h^2}{2}U_{xx}\big|_m^n - \frac{h^3}{6}U_{xxx}\big|_m^n + \frac{h^4}{4!}U_x^{IV}\big|_m^n - \ldots \tag{5.4c}$$

Substituting the above in Eq.(5.3) after replacing u with U, we get

$$U_m^{n+1} - (1 - 2r)U_m^n - r\big(U_{m-1}^n + U_{m+1}^n\big) =$$
$$k\underbrace{(U_t - U_{xx})\big|_m^n} + \frac{k^2}{2}\left(U_{tt} - \frac{1}{6r}U_x^{IV}\right)\bigg|_m^n + \ldots \tag{5.5}$$
$$= 0$$

The right hand size side of Eq. (5.5) divided by k is the local truncation error of Eq. (5.3), which is

$$\frac{k}{2}\left(U_{tt} - \frac{1}{6r}U_x^{IV}\right)\bigg|_m^n \tag{5.6}$$

5.2 Local Error and its Relationship to Global Error

From the above it can be seen that the local error is the error introduced via the difference equation used for approximating the differential equation, by assuming that all the information at $t = nk$ is known in Eq.(5.5). Another interpretation of the local error is that it is the error produced when the exact analytical solution of the differential equation at all the nodes are introduced into the difference equation.

The global error is defined as

$$E_m = U(x_m) - u_m \qquad \text{for all m} \tag{5.7a}$$

and the local error is related to the global error as follows

$$e_m = E_m - E_{m-1} \tag{5.7b}$$

i.e., the local error is the error introduced by the difference equation at the point m without accounting for the errors generated at the points before m – hence the name "local" error.

5.3 Error Norms

Error norms are a means to represent the global error. We wish to associate with the global error one single positive number which will be a definite indicator of the magnitude of the global error for a particular finite difference scheme. If we define an error vector as

$$\vec{E} = \left[E_1, \ ..... \ , \ E_N \right]^T \tag{5.8a}$$

where

$$E_i = U(x_i) - u_i, \qquad 1 \le i \le N \tag{5.8b}$$

then the single positive number mentioned above is called the norm of the error vector $\vec{E}$ and is denoted by $\| \vec{E} \|$. If this number (norm) is small, then the numerical approximation is good; if it is large, then the approximation is bad.

The norm of the vector $\vec{E}$ must satisfy the following axioms:

(i) $\| \vec{E} \| > 0$ if $\vec{E} \ne 0$ and $\| \vec{E} \| = 0$ if $\vec{x} = \vec{0}$. $\tag{5.9a}$

(ii) $\| c\vec{E} \| = |c| \| \vec{E} \|$ for a real or complex scalar quantity c. $\tag{5.9b}$

(iii) $\| \vec{E_1} + \vec{E_2} \| \le \| \vec{E_1} \| + \| \vec{E_2} \|$. $\tag{5.9c}$

Three most commonly used norms are as follows:

(1) 1–norm: it is the sum of the absolute value of all the components of the error vector:

$$\| \vec{E} \|_1 = |E_1| + |E_2| + ... + |E_N| = \sum_{i=1}^{N} |E_i| \tag{5.10a}$$

(2) 2–norm: it is the square root of the sum of the squares of all the components of the error vector:

$$\| \vec{E} \|_2 = \left(|E_1|^2 + |E_2|^2 + ... + |E_N|^2 \right)^{1/2} = \left[\sum_{i=1}^{N} |E_i|^2 \right]^{1/2} \tag{5.10b}$$

(3) Infinity–norm: it is the maximum of the magnitudes of the components of $\vec{E}$:

$$\| \vec{E} \|_\infty = \underset{1 \le i \le N}{Max} \ |E_i| \tag{5.10c}$$

6 FURTHER READING

In the following, we cite some general references which offer much more detailed accounts of the various aspects of the finite difference methods. In the area of fundamentals of numerical analysis, books by Conte and de Boor (1980), Dahlquist and Bjorck (1974), Golub and van Loan (1989), Isaacson and Keller (1966), Schwarz (1989), Stoer and Bulirsch (1980), and Young and Gregory (1988) are useful. Numerical methods for ordinary differential equations have been discussed in Boyce and DiPrima (1986), Gear (1971), Hairer and Wanner (1991), Henrici (1962), and Lambert (1973). As to the solution techniques for partial differential equations, books by Ames (1992), Forsythe and Wasow (1960), Godunov and Ryabenkii (1987), Greenspan and Casulli (1988), Lapidus and Pinder (1982), Mitchell and Griffiths (1980), Richtmyer and Morton (1967), and Smith (1985) are good sources of information. Regarding the methods for solving large systems of linear equations, one should consult Birkhoff and Lynch (1984), Briggs (1987), Hackbusch (1985), Hageman and Young (1981), Varga (1962), Wachspress (1966), and Young (1971).

CHAPTER II
PARABOLIC EQUATIONS

1 BACKGROUND ON PARTIAL DIFFERENTIAL EQUATIONS

Any differential equation containing partial derivatives is called a partial differential equation. The order of a partial differential equation is equal (by analogy with the theory of ODE's) to the order of the highest partial differential coefficient occurring in it. The dependent variable in any partial differential equation must be a function of at least two independent variables, and in general may be a function of n (≥ 2) independent variables.

1.1 Three Operators and Classes of Equations

The following three operators usually serve as the bases for a study of PDE's:

(a) Laplace operator, $\Delta = \dfrac{\partial^2}{\partial X_1^2} + \ldots\ldots + \dfrac{\partial^2}{\partial X_n^2}$

(b) Diffusion operator, $\dfrac{\partial}{\partial t} - \Delta$

(c) D'Alembert operator, $\Box = \dfrac{\partial^2}{\partial t^2} - \Delta$

These three operators typify the three general classes of partial differential operators that one encounters. These are, in the same order as the operators listed above:

(a) elliptic operators which are encountered, for example, in potential flow problems in fluid mechanics.

(b) parabolic operators arising in heat conduction and other diffusion dominated situations in various fields of the physical sciences.

(c) hyperbolic operators of which the most popular physical application is in the phenomenon of wave transmission.

1.2 Classification of Equations : Second–Order PDEs

The general form of a second–order partial differential equation is,

$$A U_{xx} + 2BU_{xy} + CU_{yy} + DU_x + E U_y + FU + G = 0 \tag{1.1}$$

Let $d = -AC + B^2$ be called the discriminant. The classification of the partial differential equations is based on the value assumed by this quantity. Specifically, we have

(i) $d > 0$ for hyperbolic PDE's.

(ii) $d = 0$ for parabolic PDE's.

(iii) $d < 0$ for elliptic PDE's.

The discriminant involves coefficients of the second–order derivatives only. Thus the lower order terms are simply ignored in classifying the differential equations into the categories above.

An elliptic equation of the general form shown above may be transformed by a change of variables from (x, y) to (ξ, η) to yield the Laplacian form, whereupon the resulting equation is called the canonical form of the elliptic equation. The canonical form is written as

$$U_{\xi\xi} + U_{\eta\eta} + (lower\ order\ terms) = 0 \qquad (1.2)$$

Conversion of a parabolic equation to the canonical form results in the expression,

$$U_{\xi\xi} + (lower\ order\ terms) = 0 \qquad (1.3)$$

A hyperbolic equation, when written in the canonical form reads,

$$U_{\xi\xi} - U_{\eta\eta} + (lower\ order\ terms) = 0 \qquad (1.4a)$$

or

$$U_{\xi\eta} + (lower\ order\ terms) = 0 \qquad (1.4b)$$

Thus the Laplacian , heat and wave operators, respectively, are the canonical forms for the three types of PDE's classified earlier.

If an operator has non–constant coefficients, the classification of the nature of the PDE is only local, i.e., it may change as one moves from one part of the domain to another.

Example 1 Consider the equation

$$L\ U = y\ U_{xx} + U_{yy}$$

The difference in type can be quite crucial and the physical as well as numerical aspects relating to the system under investigation may markedly change from one region to another. A striking example is flow over an airfoil at high subsonic velocities. The different regions over the airfoil surface will experience different physical phenomena based on the local Mach number. The classification of the differential equations governing flow in these regions is as follows:

–elliptic region : subsonic flow
–parabolic boundary : sonic barrier
–hyperbolic region : supersonic flow

The distinct regions are depicted in Fig. 1.

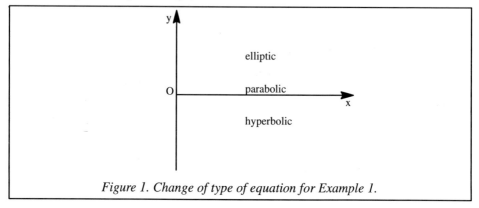

Figure 1. Change of type of equation for Example 1.

Example 2 As another example, the equation

$$y\frac{\partial^2 U}{\partial x^2} + 2x\frac{\partial^2 U}{\partial x \partial y} + y\frac{\partial^2 U}{\partial y^2} = 0$$

is (i) elliptic in the region where $y^2 - x^2 > 0$, (ii) parabolic along lines $y^2 - x^2 = 0$, (iii) hyperbolic in the region where $y^2 - x^2 < 0$.

2 ANALYTICAL BACKGROUND FOR PARABOLIC PDES

The typical equation used as a model to design solution procedures for parabolic equations in general is the the heat equation, which in one–dimensional form is

$$U_t = U_{xx}, \quad U(0, x) = U_0, \; U_0 \to 0 \quad as \quad x \to \pm \infty \quad (2.1)$$

Various properties of such an equation are illustrated in Fig. 2. We now provide some analytical background for methods of solution of parabolic PDEs.

2.1 Fourier Analysis

The standard method of separation of variables yields the following form for the solution of the heat equation:

$$U \sim e^{\tau t}e^{i\omega x}, \quad \omega \in R \quad (2.2)$$

Substituting in the heat equation then gives the relationship, $\tau = (i\omega)^2 = -\omega^2$. Hence, we get

$$U \sim e^{-\omega^2 t}e^{i\omega x} \quad (2.3)$$

This indicates that high frequency components decay fast, due to their shorter time scales which are proportional to $\frac{1}{\omega^2}$. Therefore after some time, only low frequency components are left, resulting in smearing of sharp profiles as shown in Fig. 2. Based on this property, we know that a solution of the parabolic equation can not generate sharp profiles from a smooth initial shape. An observation like this can guide our understanding of the problem and help design suitable numerical methods.

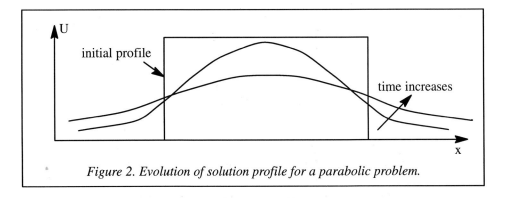

Figure 2. Evolution of solution profile for a parabolic problem.

2.2 The Min–Max Principle

The statement of this principle is as follows:

$$\min U_0 \le U \le \max U_0, \forall t \ge 0, x \in R$$

Thus, any of the maxima or minima of the dependent variable must appear in the initial condition (IC) or the boundary condition (BC). This property is very useful for checking the stability of a numerical scheme, because it addresses the issue of physical realizability.

2.3 Energy Identity

The energy identity states the following:

$$\frac{d}{dt}\int_{-\infty}^{\infty} U^2(t,x)dx = \int_{-\infty}^{\infty} 2UU_t\ dx = \int_{-\infty}^{\infty} 2UU_{xx}\ dx = -2\int_{-\infty}^{\infty} U_x U_x dx \quad (2.4)$$

Integrating (2.4) from 0 to τ with respect to t, we get

$$\int_{-\infty}^{\infty} U^2(\tau,x)dx + 2\int_{0}^{\tau}\int_{-\infty}^{\infty} U_x^2 dxdt = \int_{-\infty}^{\infty} U_0^2(x)dx \quad (2.5)$$

Since the right hand side is a constant and the second term on the left hand side is always positive and increases with increasing τ and we obtain the result, the following holds:

$$\int_{-\infty}^{\infty} U^2(\tau,x)dx \quad decreases\ as\ \tau\ increases.$$

Thus, $U \to 0$ *as* $t \to \infty$, i.e., U will converge to its equilibrium distribution.

2.4 Characteristics of Solutions of Parabolic PDEs

From the previous derivation, it is also clear that for any $t > 0$, the solution, $U(t,x)$, of a parabolic PDE has the following properties:

(i) $U(t,x)$ depends on $U(x)$ for all x.

(ii) The speed of signal propagation is infinite, but the magnitude of signal decays exponentially fast.

3 EXPLICIT NUMERICAL SCHEMES FOR PARABOLIC PDE'S

3.1 An Example

A stability analysis may be performed for the explicit schemes applied to the heat conduction equation for specified initial and boundary conditions. The one–dimensional equation is specified as

$$U_t = U_{xx} \quad (3.1)$$

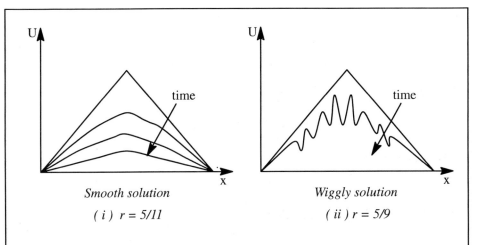

Figure 3. Solution profile for the parabolic equation using the explicit scheme with different r.

The domain under consideration is $0 \leq x \leq 1$, $t > 0$. Let h = 1 / N and

$$x_m = mh, \quad m = 0,1,..... N$$

$$t_n = nk, \quad n=0,1,.......$$

The value $u_m^n \cong U(t_n, x_m) = U(nk, mh)$ is computed, for example by discretizing the equation using forward differencing in time and central differencing in space. The finite difference equation then takes the form

$$\frac{u_m^{n+1} - u_m^n}{k} = \frac{u_{m+1}^n - 2u_m^n + u_{m-1}^n}{h^2} \tag{3.2}$$

We let $r= k/h^2$ and obtain

$$u_m^{n+1} = ru_{m+1}^n + (1 - 2r)u_m^n + ru_{m-1}^n \tag{3.3a}$$

which, in terms of the exact solution of the original differential equation, can be written as

$$U_m^{n+1} = rU_{m+1}^n + (1 - 2r)U_m^n + rU_{m-1}^n + \mathcal{O}(k^2 + kh^2) \tag{3.3b}$$

The choice of the parameter r controls the performance of the finite difference scheme. For example choosing (i) $r = 5/11$ or (ii) $r = 5/9$, the resulting computed profiles show the behaviors depicted in Fig. 3.

Example 3

Let us use the following as the model test case:

$$u_m^1 = (1 - 2r)u_m^0 + ru_{m-1}^0 + ru_{m+1}^0$$

$(u_m)^0$ is an initial condition specified in the form of the zig–zag alternating peaks from $+\varepsilon$ to $-\varepsilon$ as shown

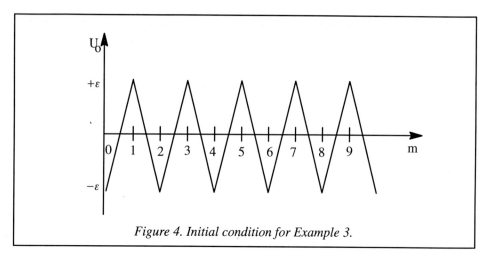

Figure 4. Initial condition for Example 3.

in Fig. 4., i.e., $u_m^0 = (-1)^{m+1}\varepsilon$. Therefore, we have

$$u_m^1 = (1 - 4r)u_m^0$$

and

$$u_m^n = (1 - 4r)^n u_m^0 = (1 - 4r)^n \varepsilon$$

From the above we can draw the following conclusions:

a) If $r > 1/2$, then $|1-4r| > 1$. Therefore, the peaks are magnified without limit as $n \to \infty$

b) If $r < 1/2$, then $|1-4r| < 1$. Therefore, the peaks disappear as $n \to \infty$.

3.2 Methods Employed to Analyze Stability of Difference Schemes

3.2.1 The Min–Max Principle

In order to preserve the stability, all coefficients in Eq.(3.3) should be positive (a sufficient condition). Therefore, $1-2r \geq 0$, i.e., $r \leq 1/2$. Otherwise, $(u_m)^{n+1}$ may not be bounded by $(u_{m-1})^n$, $(u_m)^n$ and $(u_{m+1})^n$. Negative coefficients on the right hand side of Eq. (3.3) means negative numerical diffusivity (or conductivity etc.).

<u>*Theorem:*</u> If we have the following

$$u_m^{n+1} = Au_{m+1}^n + Bu_m^n + Cu_{m-1}^n \tag{3.4}$$

where A, B, C are positive and $A+B+C \leq 1$, then the scheme is stable and the errors die out.

<u>*Proof:*</u> From the difference form, Eq. (3.4) above, taking the absolute value on both sides we get

$$\left|u_m^{n+1}\right| \leq \left|Au_{m+1}^n\right| + |Bu_m^n| + \left|Cu_{m-1}^n\right|$$

$$= A\left|u_{m+1}^n\right| + B|u_m^n| + C\left|u_{m-1}^n\right|$$

$$\leq (A + B + C)Z^n, \quad where \ Z^n = Max \ u_m^n, \quad 1 \leq m \leq N - 1$$

$$\leq Z^n$$

But the above inequality holds for each m, including $\underset{m}{Max} \left| u_m^{n+1} \right| \le Z^n$. Hence,

$$Z^{n+1} \le Z^n$$

Thus, we get

$$\underset{m}{Max} |u_m^{n+1}| \le \underset{m}{Max} |u_m^n|$$

$$\le \underset{m}{Max} |u_m^{n-1}|$$

.

.

.

$$\le \underset{m}{Max} |u_m^0|$$

3.2.2 Application of Concepts from ODE's

Equation (3.2) is equivalent to using the Euler method (explicit) to solve the following ODE, namely,

$$\frac{dU_m}{dt} \cong \frac{U_{m+1} - 2U_m + U_{m-1}}{h^2}$$

where $m = 0,1,......N$. We may rewrite this equation as ,

$$\frac{d\vec{U}}{dt} = \frac{[A]}{h^2} \vec{U} + \frac{\vec{f}}{h^2} \tag{3.5}$$

where

$$\vec{U} = [U_1(t), ..., U_{N-1}(t)]^T$$

$$\vec{f} = [U_0, 0, 0, ..., 0, U_n]^T$$

and

$$[A] = \begin{bmatrix} -2 & 1 & & 0 \\ 1 & -2 & 1 & \\ & & \diagdown & \\ 0 & & & \diagdown \end{bmatrix}$$

Here A is a symmetric, tridiagonal matrix of $(N-1)\times(N-1)$ elements. This represents a Hermitian system.

Properties of a Hermitian System

The important properties of a Hermitian system are as follows:

(a) Eigenvalues $\lambda_1, ..., \lambda_{N-1}$ are all real.

(b) Regardless of multiplicity, i.e., repeated values of λ, there is always a full set of $N-1$ eigenvectors $\left(\vec{\xi}^{(1)}, ..., \vec{\xi}^{(N-1)} \right)$ that are linearly independent and orthogonal, i.e.

$$[A]\vec{\xi}^{(i)} = \lambda_i \vec{\xi}^{(i)} \tag{3.6}$$

$$\left(\vec{\xi}^{(i)}, \vec{\xi}^{(j)}\right) = \begin{cases} 0, & \text{if } i \neq j \\ 1, & \text{if } i = j \end{cases}$$

For a system of first order ODEs

$$\frac{d\vec{U}}{dt} = \frac{[A]}{h^2}\vec{U} \tag{3.7}$$

the solution can be written as

$$\vec{U} = \sum_{i=1}^{N-1} C_i \vec{\xi}^{(i)} \tag{3.8}$$

Substituting (3.8) into (3.7) and using the orthogonality property by taking the product of the result with $\vec{\xi}^{(i)}$ we get,

$$\frac{dC_i}{dt} = \frac{\lambda_i}{h^2} C_i \tag{3.9}$$

which has the solution

$$C_i = (C_i)_0 \exp\left(\frac{\lambda_i t}{h^2}\right) \tag{3.10}$$

The question of stability, i.e., whether errors amplify or die out depends on the values of λ_i. It thus remains to find the magnitudes of the eigenvalues λ_i.

The eigenvalues of the matrix A are given by (Smith, 1985, p. 59 and p. 154–156),

$$\lambda_i = -2 + 2\cos(i\pi)/N \quad , \quad i = 1, \ldots, N-1$$

$$= -2(1 - \cos(i\pi)/N)$$

$$= -4\sin^2(i\pi/2N) \tag{3.11}$$

Therefore, for $i=1$, we obtain $\lambda_1 \cong -4(\pi/2N)^2 = -(h\pi)^2$, and so on with $\lambda_{N-1} \cong -4$ for $i=N-1$. As can be seen, with a small h, λ_1 is quite small in magnitude. Hence $\lambda_{min} = \lambda_{N-1}$, and $\lambda_{max} = \lambda_1$.

The ratio $\lambda_{min}/\lambda_{max} \cong 4/(h\pi)^2 = (2N/\pi)^2$. Thus, for large N, the system is very stiff, i.e., the eigenvalues are very widely spaced. This situation is typical with parabolic PDEs. In implementing a computational scheme one needs to be able to satisfy the stability restriction imposed by λ_{max} if the forward Euler method is used. This property of stiffness is computationally taxing since the stability restriction posed by the maximum eigenvalue leads to wasted effort as far as other low eigenvalue components are concerned. According to Eq. (3.9) and our previous discussion in Chapter I, the stability restriction for the forward Euler time stepping scheme is therefore controlled by the minimum eigenvalue and takes the form given by

$$|1 + k(\lambda_{min}/h^2)| < 1$$

and with $\lambda_{min} = -4$, we obtain

$$k/h^2 = r < 2/|\lambda_{min}| = 1/2 \tag{3.12}$$

All explicit methods for parabolic PDEs must obey a stability bound, of the type

$$k/h^2 = r \leq some\ constant$$

If we use the backward difference in time, then since it is A–stable for the parameters of interest here, we will see no stability restriction for the parabolic PDE either!

3.2.3 *The von Neumann Stability Analysis*

This method may be used to obtain a stability criterion by studying the propagation of a single row of errors, say along the line t=0. To do so, one represents the errors by a finite Fourier series. The issue of stability then reduces to investigating the growth rate of magnitudes of the Fourier components with time.

Consider the initial value problem (IVP) with initial spatial data $\varrho e^{i\beta x}$, $\beta \in R$ at t=0 where ϱ is the Fourier coefficient. Then the solution of the finite difference equation at any given time $t=nk$ should have the magnitude ϱ^n times the original solution at $t=0$ for any given location x_m. The factor ϱ^n can be seen to be appropriate, if we recall that this is the form of the solution to finite difference equations. Separation of variables tells us that $u_m^n \sim \varrho^n e^{i\beta mh}$ while the exact solution to the PDE is of the form $e^{-\beta^2 t} e^{i\beta x}$. Therefore we need to solve for the amplification rate in order to draw conclusions regarding stability. To do this we substitute the Fourier expression into the difference equation to get

$$\frac{u_m^{n+1} - u_m^n}{k} = e^{i\beta mh}\left(\frac{\varrho^{n+1} - \varrho^n}{k}\right) = e^{i\beta mh}\varrho^n\left(\frac{\varrho - 1}{k}\right) \tag{3.13}$$

$$\Rightarrow \frac{u_{m+1}^n - 2u_m^n + u_{m-1}^n}{h^2} = e^{i\beta mh}\varrho^n\left(\frac{e^{i\beta h} - 2 + e^{-i\beta h}}{h^2}\right)$$

$$\therefore \frac{\varrho - 1}{k} = \frac{e^{i\beta h} - 2 + e^{-i\beta h}}{h^2}$$

$$= \frac{2}{h^2}(\cos(\beta h) - 1)$$

Hence $$\varrho = 1 + 2r\left[\cos(\beta h) - 1\right] = 1 - 4r\sin^2(\beta h/2) \tag{3.14}$$

For stability, the criterion $\varrho \leq 1$ must hold.

Since it is required that $-1 \leq \varrho \leq 1$, we have

$$-1 \leq 1 - 4r\sin^2\frac{\beta h}{2} \leq 1 \quad for\ all\ \beta h \tag{3.15}$$

In this inequality the RHS is guaranteed as long as r $\geq$ 0. Satisfaction of the condition on the LHS implies that

$$r \leq \frac{1}{2 \sin^2 \frac{\beta h}{2}} \tag{3.16}$$

Here β runs over the entire domain R. It can be seen that $\beta = 0$ and $\beta = \pi/h$ produce limiting conditions. Thus, the most restrictive condition is $r \leq 1/2$.

Important features of the von Neumann method:
(a) It is applicable only if the coefficients of the linear difference equation are constant and the initial condition is periodic. If the difference equation has variable coefficients the method can still be applied locally and it might be expected that a method will be stable if the von Neumann condition, derived as though the coefficients were constant, is satisfied at every point in the field.
(b) For two level difference schemes with one dependent variable, the von Neumann condition is sufficient and necessary for stability.
(c) Boundary conditions are neglected by the analysis. The von Neumann stability analysis does not provide necessary conditions for stability of constant coefficient problems regardless of the type of boundary conditions.
More information can be found in Richtmyer and Morton (1967).

3.3 Consistency, Stability and Convergence

Consistency

A finite difference scheme is termed consistent if the discretized equation tends to the corresponding differential equation when Δt and Δx tend to zero. In order to check whether a finite difference scheme is consistent one uses the Taylor series. A scheme is consistent if the truncation error tends to zero for Δt and Δx tending to zero. For example, for the PDE

$$\frac{\partial U}{\partial t} + a \frac{\partial U}{\partial x} = 0 \tag{3.17}$$

using forward differencing in time ($\Delta t \equiv k$) and central differencing in space ($\Delta x \equiv h$), the difference form is,

$$\frac{u_m^{n+1} - u_m^n}{k} = -\frac{a}{2h}(u_{m+1}^n - u_{m-1}^n) \tag{3.18}$$

Expanding in Taylor series around U_m^n, we get

$$U_m^{n+1} = U_m^n + k(U_t)_m^n + \frac{k^2}{2}(U_{tt})_m^n + \dots \tag{3.19a}$$

$$U_{m+1}^n = U_m^n + h(U_x)_m^n + \frac{h^2}{2}(U_{xx})_m^n + \frac{h^3}{6}(U_{xxx})_m^n + \dots \tag{3.19b}$$

$$U^n_{m-1} = U^n_m - h(U_x)^n_m + \frac{h^2}{2}(U_{xx})^n_m - \frac{h^3}{6}(U_{xxx})^n_m + \ldots \qquad (3.19c)$$

The difference between the discretized and the differential forms of the equation is given by

$$\frac{U^{n+1}_m - U^n_m}{k} + a\frac{U^n_{m+1} - U^n_{m-1}}{2h} - (U_t + aU_x)^n_m = \frac{k}{2}(U_{tt})^n_m + \frac{h^2}{6}a(U_{xxx})^n_m + \mathcal{O}(k^2, h^4)$$

$$(3.20)$$

Thus, the RHS vanishes as $h \rightarrow 0$ and $k \rightarrow 0$. Therefore the difference equation is consistent. The discretized equation differs from the differential equation by the terms on the RHS.

Stability

A difference scheme is said to be stable if it does not permit errors to grow indefinitely, that is , to be amplified without bound, as the solution is progressed from one step to another in time, i.e., the error between the numerical solution and exact solution should remain finite as the number of time steps $n \rightarrow \infty$.

Convergence

Convergence of a numerical solution procedure is said to be achieved when the numerical solution approaches the exact solution at any point and time when the time step and space mesh size $\rightarrow 0$.

Consistency, stability and convergence are three key elements in analyzing the suitability and performance of any numerical scheme. Furthermore, they are actually related to one another (Richtmyer and Morton 1967), as will be discussed below in the form of Lax's equivalence theorem.

3.4 Lax's Equivalence Theorem

Statement

For a one–step scheme, the combination of consistency and stability is equivalent to convergence. In other words, a consistent scheme is convergent if and only if it is stable.

When the magnitudes of perturbations in the initial conditions are made arbitrarily small as h and $k \rightarrow 0$, the resulting perturbations in the computed solution must vanish rather than grow. In such a case, computed solutions based on a consistent difference approximation converge to the solution of the original parabolic PDE.

To illustrate the point made by this theorem, let the error at any point in space and time be given by

$$Z^n_m = U^n_m - u^n_m$$

where U is the exact differential solution and u represents the approximate discrete solution. The second–order difference scheme is,

$$U_m^{n+1} = (1 - 2r)U_m^n + r(U_{m+1}^n - U_{m-1}^n) + \mathcal{O}(k^2 + kh^2) \tag{3.21}$$

We have demonstrated previously that this scheme yields a consistent approximation. Furthermore, if $0 < r \le 1/2$, then it is also stable. Also, we have

$$\left|Z_m^{n+1}\right| \le (1 - 2r)|Z_m^n| + r\left|Z_{m+1}^n\right| + r\left|Z_{m-1}^n\right| + A(k^2 + kh^2) \tag{3.22}$$

where A depends on the upper bound for $\dfrac{\partial^2 U}{\partial t^2}$ and $\dfrac{\partial^4 U}{\partial x^4}$ because the leading truncation error terms are $\dfrac{1}{2}k^2\left(\dfrac{\partial^2 U}{\partial t^2} - \dfrac{1}{6r}\dfrac{\partial^4 U}{\partial x^4}\right)_m^n$.

Let ψ^n be the maximum value of Z_m^n over the entire range of m, i.e., $1 \le m \le N - 1$. Since Eq. (3.14) is true for all m, we have

$$\psi^{n+1} \le \psi^n + A(k^2 + kh^2) \tag{3.23}$$

So, if $\psi^0 = 0$, i.e., the exact information regarding initial condition is specified for the difference equation, we have

$$\psi^n \le nA(k^2 + kh^2) = nkA(k + h^2) = tA(k + h^2) \to 0 \quad as \quad k, h \to 0 \tag{3.24}$$

for fixed t (time). Hence we have shown that a consistent and stable scheme is also convergent.

3.5 DuFort–Frankel Explicit Scheme

We next consider an explicit scheme which is quite interesting. For the 1–D heat equation, the Dufort–Frankel discretization is given by

$$\frac{u_m^{n+1} - u_m^{n-1}}{2k} = \frac{u_{m+1}^n - u_m^{n+1} - u_m^{n-1} + u_{m-1}^n}{h^2} \tag{3.25}$$

which is obtained by replacing u_m^n by $\dfrac{1}{2}(u_m^{n+1} + u_m^{n-1})$ on the RHS of Eq. (3.2).

Truncation Error

The difference between the approximate and differential equations is obtained on performing the Taylor series expansion around U_m^n and results in

$$\frac{U_m^{n+1} - U_m^{n-1}}{2k} - \frac{U_{m+1}^n - U_m^{n+1} - U_m^{n-1} + U_{m-1}^n}{h^2} - \left(\frac{\partial U}{\partial t} - \frac{\partial^2 U}{\partial x^2}\right)_m^n$$

$$= \left(\frac{k}{h}\right)^2\left(\frac{\partial^2 U}{\partial t^2}\right)_m^n + \mathcal{O}(k^2, h^2) + \mathcal{O}\left(\frac{k^4}{h^2}\right) \tag{3.26}$$

Consistency

In accordance with the definition, consistency requires that $k/h \to 0$ as $k \to 0$ in order to let the truncation errors approach zero. Therefore, Dufort–Frankel scheme is consistent with the differential equation if and only if k goes to zero faster than h. If k/h is kept fixed, say equal to a constant β, then the scheme is consistent not with the original PDE but with the hyperbolic equation,

$$\frac{\partial U}{\partial t} - \frac{\partial^2 U}{\partial x^2} + \beta^2 \frac{\partial^2 U}{\partial t^2} = 0 \tag{3.27}$$

However, the Dufort–Frankel scheme has the advantage that it is unconditionally stable. The von Neumann stability analysis of the Dufort–Frankel three level scheme is as follows. The difference equation may be written as

$$(2r + 1)u_m^{n+1} - 2r(u_{m-1}^n + u_{m+1}^n) + (2r - 1)u_m^{n-1} = 0 \tag{3.28}$$

Consider the Fourier component, $u_m^n \sim \varrho^n e^{i\beta mh}$, $m = 1, ..., N - 1$. Substituting in the difference form yields,

$$(2r + 1)\varrho^2 - 4r\varrho \cos(\beta h) + (2r - 1) = 0$$

$$\therefore \ \varrho = \frac{2r \cos(\beta h) \pm \sqrt{(1 - 4r^2 \sin^2(\beta h))}}{(1 + 2r)} \tag{3.29}$$

There are two possibilities that need to be considered here, namely,
(a) $4r^2 \sin^2(\beta h) \geq 1$. Then, we have

$$|\varrho^2| = \frac{4r^2 \cos^2(\beta h) + (4r^2 \sin^2(\beta h) - 1)}{(1 + 2r)^2} = \frac{2r - 1}{2r + 1} < 1 \tag{3.30}$$

(b) $4r^2 \sin^2(\beta h) < 1$. Then, we have

$$|\varrho| \leq \frac{|2r \cos(\beta h)| + \sqrt{(1 - 4r^2 \sin^2(\beta h))}}{1 + 2r} \leq \frac{|2r \cos(\beta h)| + 1}{1 + 2r} \leq 1 \tag{3.31}$$

Hence $|\varrho|$ is always no larger than 1. Therefore the Dufort–Frankel method is unconditionally stable.

The following points may be noted:
(a) Since the scheme is a three–level scheme, it needs two initial conditions; one is given by the problem description and the other needs to be supplied via computational treatment.
(b) The von Neumann analysis does not account for the effect of initial and boundary conditions, hence this aspect can not be analyzed here using the von Neumann analysis.
(c) One can devise an alternative approach by converting the original three–level scheme to a two–level scheme by writing,

$$(2r + 1)u_m^{n+1} - 2r\left(u_{m-1}^n + u_{m+1}^n\right) + (2r - 1)v_m^n = 0 ,$$

$$v_m^{n+1} = u_m^n \tag{3.32}$$

In this case, instead of having a scalar amplification factor ϱ as for the two–level scheme, we now have a 2×2 amplification matrix $[G]$. The numerical stability requires that the two eigenvalues of the matrix $[G]$ should not be greater than 1.

4 GENERAL TWO–STEP SCHEMES FOR PARABOLIC PDE'S

For the 1–D heat equation, the general form of the difference equation, using central differencing in space is

$$u_m^{n+1} - u_m^n = r\left\{\xi\left[u_{m-1}^{n+1} - 2u_m^{n+1} + u_{m+1}^{n+1}\right] + (1 - \xi)\left[u_{m-1}^n - 2u_m^n + u_{m+1}^n\right]\right\} \tag{4.1}$$

where $0 \le \xi \le 1$. Three cases may be identified with regard to timewise discretization, namely,

(a) $\xi=0 \Rightarrow$ fully explicit , i.e., forward Euler in time.
(b) $\xi= 1/2 \Rightarrow$ Crank–Nicolson, i.e. trapezoidal rule in time.
(c) $\xi=1 \Rightarrow$ fully implicit , i.e., backward Euler in time.

4.1 Stability

4.1.1 Results from ODEs

According to the results obtained from ODE solution methods, we can draw the following conclusions:

(a) The fully explicit scheme ($\xi=0$) is stable for $r \le 1/2$
(b) Crank–Nicolson scheme ($\xi= 1/2$) is stable for all r.
(c) The fully implicit scheme ($\xi=1$) is stable for all r.

Using the ODE concept, as given in Section 3.2.2, the solution to the heat equation can be obtained by solving

$$\frac{dC_m}{dt} = \frac{\lambda_m}{h^2} C_m, \quad m = 1, ..., N - 1$$

The *Crank–Nicolson* discretization yields:

$$\frac{C_m^{n+1} - C_m^n}{k} = \frac{\lambda_m}{2h^2}(C_m^{n+1} + C_m^n) \tag{4.2}$$

which is stable for all λ_m and $r = k/h^2$, i.e., it is A–stable. The *fully implicit method,* on the other hand, yields

$$\frac{C_m^{n+1} - C_m^n}{k} = \frac{\lambda_m}{h^2} C_m^{n+1} \tag{4.3}$$

and is also A–stable.

4.1.2 von Neumann analysis

The general formula of the von Neumann stability bound can be obtained as

$$\varrho^{n+1} - \varrho^n = 2r\varrho^{n+1}\xi[\cos(\beta h) - 1] + 2r\varrho^n(1 - \xi)[\cos(\beta h) - 1] \tag{4.4}$$

Thus, we get

$$\varrho = \frac{1 - 4r(1 - \xi)\sin^2\frac{\beta h}{2}}{1 + 4r\xi\sin^2\frac{\beta h}{2}} \tag{4.5}$$

We get the following results for the different schemes:

(a) Crank–Nicolson scheme

$$\varrho = \frac{1 - 2r\sin^2\frac{\beta h}{2}}{1 + 2r\sin^2\frac{\beta h}{2}} \tag{4.6}$$

Hence $|\varrho| < 1$ for all choices of $r > 0$, and it is stable as well.

(b) Fully implicit scheme

$$\varrho = \frac{1}{1 + 4r\sin^2\frac{\beta h}{2}} \tag{4.7}$$

Clearly, this scheme is always stable as well. In fact, $|\varrho| \le 1$ for all βh if $r(1 - 2\xi) \le 1/2.$, i.e., if $\xi \ge 1/2$ then the scheme is unconditionally stable.

4.1.3 The Min–Max principle

Using the Min–Max principle, we obtain the following:

(a) Crank –Nicolson method. Equation (4.1) can be written as

$$(1 + r)u_m^{n+1} = (1 - r)u_m^n + \frac{r}{2}\left[u_{m-1}^{n+1} + u_{m+1}^{n+1} + u_{m-1}^n + u_{m+1}^n\right] \tag{4.8}$$

If $r > 1$, then the Min–Max principle will be violated and wiggles will appear in the numerical results. However, since the Min–Max principle is a sufficient condition, $r > 1$ does not imply instability. As in the case of trapezoidal scheme for the first order ODE studied earlier, the scheme is unconditionally stable, but can generate numerical oscillations in the solution profile. Quantities such as U_x (heat flux) can be completely wrong in a qualitative sense even though the scheme is stable.

(b) Fully implicit method. The discretized form is,

$$(1 + 2r)u_m^{n+1} = r\left[u_{m-1}^{n+1} + u_{m+1}^{n+1}\right] + u_m^n \tag{4.9}$$

which can be rewritten as

$$u_m^{n+1} = \frac{r}{1 + 2r}\left[u_{m-1}^{n+1} + u_{m+1}^{n+1}\right] + \frac{1}{1 + 2r}u_m^n \tag{4.10}$$

As already discussed, two conditions are crucial in this regard: (i) all u^{n+1} coefficients must be positive, and (ii) coefficient of u_m^{n+1} must be no smaller than the sum of the other coefficients. Hence, the fully implicit scheme can always satisfy the Min–Max principle.

4.2 Accuracy Improvement Via Manipulation of the Degree of Implicitness

For the equation, $U_t = U_{xx}$, with the general difference equation given by

$$\nabla_t u_m^{n+1} = r\left[\xi\delta_x^2 u_m^{n+1} + (1 - \xi)\delta_x^2 u_m^n\right] \tag{4.11}$$

where ∇_t implies backward differencing in time and δ_x implies central differencing in space. To obtain the truncation error, we rewrite the difference equation in terms of u as follows,

$$(1 - r\xi\delta_x^2)u_m^{n+1} = \left[1 + (1 - \xi)r\delta_x^2\right]u_m^n \tag{4.12}$$

In terms of U the solution to the discretized form is given by

$$\left(1 - r\xi\delta_x^2\right)U_m^{n+1} = \left[1 + (1 - \xi)r\delta_x^2\right]U_m^n + k\tau_m^n \tag{4.13}$$

where τ is the local truncation error. Now,

$$\frac{1}{k}\nabla_t U = U_t + \frac{k}{2}U_{tt} + \dots \tag{4.14a}$$

and

$$\frac{1}{h^2}\delta_x^2 U = U_{xx} + \frac{h^2}{12}U_{xxxx} = \dots \tag{4.14b}$$

Therefore, we get

$$\tag{4.15}$$
$$\tau_m^n = \frac{1}{k}\nabla_t U_m^{n+1} - \frac{1}{h^2}\left[\xi\delta_x^2 U_m^{n+1} + (1 - \xi)\delta_x^2 U_m^n\right]$$

$$= \left[U_t + \frac{k}{2}U_{tt} - (1 - \xi)\left(U_{xx} + \frac{h^2}{12}U_{xxxx} + \dots\right)\right]_m^n - \left[\xi\left(U_{xx} + \frac{h^2}{12}U_{xxxx}\right)\right]_m^{n+1}$$

Note that $U_{tt} = (U_t)_t = (U_t)_{xx} = U_{xxxx}$. Using Taylor series, we get

$$U_{xx}^{n+1} = U_{xx}^n + kU_{xxt}^n + \dots = U_{xx}^n + kU_{xxxx}^n + \dots \tag{4.16a}$$

$$U_{xxxx}^{n+1} = U_{xxxx}^n + kU_{xxxxt}^n + \dots \tag{4.16b}$$

$$\therefore \quad \tau_m^n = U_t - U_{xx} + U_{xxxx}\left[\frac{k}{2} + (\xi - 1)\frac{h^2}{12} - \xi\frac{h^2}{12} - \xi k\right] \tag{4.17}$$

But, $U_t - U_{xx} = 0$. Therefore, we get

$$\tau_m^n = U_{xxxx}\left[(\tfrac{1}{2} - \xi)k - \frac{h^2}{12}\right] + \dots$$

$$= \mathcal{O}(k, h^2) \ \text{in general} \ \text{and} \ \mathcal{O}(k^2, h^2) \ \text{if} \ \xi = 1/2 \tag{4.18}$$

i.e., for Crank–Nicolson scheme, we achieve second–order accuracy in both space and time. However, if we choose $\left(\frac{1}{2} - \xi\right)k - \frac{h^2}{12} = 0$, i.e.,

$$\xi = \frac{1}{2} - \frac{1}{12r}, \qquad r = k/h^2 \tag{4.19}$$

then the leading term, which is $\mathcal{O}(k, h^2)$, of the expression for τ_m^n disappears. τ_m^n now becomes $\mathcal{O}(k^2, h^4)$. This means that such a choice of ξ can yield a higher order of local accuracy. Since ξ and r are related, it can be verified that this scheme is only conditionally stable, i.e., there is a restriction on r.

As a further example, if $r = 1/\sqrt{20}$ then the next term in τ_m^n also disappears, resulting in a scheme of even higher order. τ_m^n now is $\mathcal{O}(k^3, h^6)$! It is a very high order scheme, but since $r = 1/\sqrt{20}$, the size of the time step k is also very restrictive.

We must realize that a high order scheme is only good for smooth solutions, as shown in Fig. 5. Hence, if , for example, the initial condition contains non–smooth data then the usefulness of a higher order scheme disappears. However, since a parabolic equation is diffusive in nature, it usually benefits from high order schemes compared to the hyperbolic equations.

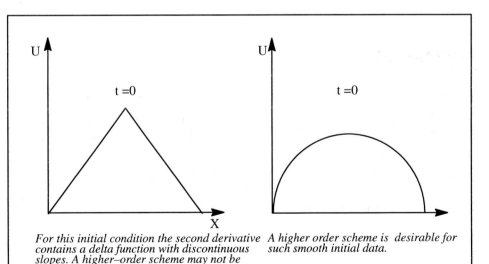

For this initial condition the second derivative contains a delta function with discontinuous slopes. A higher–order scheme may not be suitable.

A higher order scheme is desirable for such smooth initial data.

Figure 5. Two examples of initial conditions.

5 KELLER'S BOX METHOD

This method (Keller 1971) is an alternative to the Crank–Nicolson method. Here, one converts the second–order PDEs to a system of first order PDEs. Considering the equation, $U_t = U_{xx}$, we define, $V = \dfrac{\partial U}{\partial x}$. The equation is then of the form,

$$U_t = V_x \qquad\qquad\qquad (5.1a)$$

$$V = U_x \qquad\qquad\qquad (5.1b)$$

The time and space levels are shown in Fig. 6. This system of equations is now integrated over the 'box' shown in Fig. 6, using the trapezoidal rule throughout for integration in space. Thus,

$$\int_{x_{m-1}}^{x_m}\int_{t_n}^{t_{n+1}} \frac{\partial U}{\partial t}\,dt\,dx = \int_{x_{m-1}}^{x_m} [U(x,t_{n+1}) - U(x,t_n)]dx \cong \frac{h}{2}[(u_m^{n+1} + u_{m-1}^{n+1}) - (u_m^n + u_{m-1}^n)]$$

$$(5.2a)$$

$$\int_{t_n}^{t_{n+1}}\int_{x_{m-1}}^{x_m} \frac{\partial V}{\partial x}\,dx\,dt = \int_{t_n}^{t_{n+1}} [V(x_m,t) - V(x_{m-1},t)]dt \cong \frac{k}{2}[(v_m^{n+1} + v_m^n) - (v_{m-1}^{n+1} + v_{m-1}^n)]$$

$$(5.2b)$$

The equation (5.1) after integration therefore becomes,

$$(u_m^{n+1} + u_{m-1}^{n+1}) - (u_m^n + u_{m-1}^n) = \lambda[(v_m^{n+1} + v_m^n) - (v_{m-1}^{n+1} + v_{m-1}^n)] \qquad (5.3)$$

where $\lambda = k/h$. Equation (5.1b) after integration with respect to x from node $m–1$ to m and at time step $n+1$ yields,

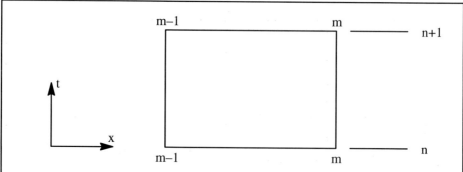

Figure 6. The spatial and temporal nodes employed by Keller's box method.

$$\frac{h}{2}(v_m^{n+1} + v_{m-1}^{n+1}) = (u_m^{n+1} - u_{m-1}^{n+1}) \tag{5.4}$$

Therefore, with regard to this scheme the following points are noted:

(a) We now have two variables at each node but each finite difference equation only needs two nodes in space $(m, m-1)$.

(b) To compare with the Crank–Nicolson method, we sum Eq. (5.3) over two boxes, $(m-1,m)$ and $(m,m+1)$. The spatial and temporal nodes considered are shown in Fig. 7 and may be contrasted with Keller's method. We get,

$$(u_m^{n+1} + u_{m-1}^{n+1}) - (u_m^n + u_{m-1}^n) + (u_{m+1}^{n+1} + u_m^{n+1}) - (u_{m+1}^n + u_m^n)$$

$$= \lambda\left[(v_m^{n+1} - v_m^n) - (v_{m-1}^{n+1} + v_{m-1}^n) + (v_{m+1}^{n+1} + v_{m+1}^n) - (v_m^{n+1} + v_m^n)\right]$$

We rearrange this equation to obtain,

$$(u_{m+1}^{n+1} + 2u_m^{n+1} + u_{m-1}^{n+1}) - \lambda(v_{m+1}^{n+1} - v_{m-1}^{n+1})$$

$$= (u_{m+1}^n + 2u_m^n + u_{m-1}^n) \quad + \lambda(v_{m+1}^n - v_{m-1}^n) \tag{5.5}$$

Subtracting Eq. (5.4) over box $(m-1, m)$ from that over box $(m, m+1)$, we get,

$$\frac{h}{2}(v_{m+1}^{n+1} - v_{m-1}^{n+1}) = (u_{m+1}^{n+1} - 2u_m^{n+1} + u_{m-1}^{n+1}) \tag{5.6}$$

Substituting Eq. (5.6) , with indices n and $n+1$ into Eq. (5.5) yields

$$[u_{m+1}^{n+1} + 2u_m^{n+1} + u_{m-1}^{n+1}] - 2r(u_{m+1}^{n+1} - 2u_m^{n+1} + u_{m-1}^{n+1})$$

$$= [u_{m+1}^n + 2u_m^n + u_{m-1}^n] + 2r(u_{m+1}^n - 2u_m^n + u_{m-1}^n) \tag{5.7}$$

If the term in square brackets on the RHS and LHS (with indices n and $n+1$) are replaced with $4u_m$, then the above equation is identical to the Crank–Nicolson scheme. The Box

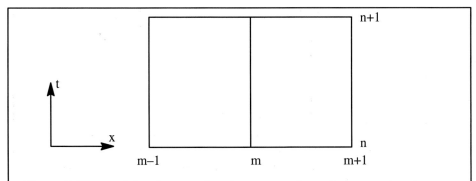

Figure 7. The spatial and temporal nodes over two boxes for comparison between Keller's box method and the Crank–Nicolson scheme.

method requires slightly more computation, but its compact form can handle nonuniform grid more accurately.

6 LEAP– FROG SCHEME

This is a three–level scheme and thus requires a starting procedure. We describe the application of this scheme for three different equations. As can be clearly observed, very different results can be obtained from the same basic concept.

6.1 Heat Conduction Equation

For the heat conduction equation, $U_t = U_{xx}$, the scheme yields the discretized form,

$$\frac{u_m^{n+1} - u_m^{n-1}}{2k} = \frac{1}{h^2}[u_{m+1}^n - 2u_m^n + u_{m-1}^n] \tag{6.1}$$

Compared to the explicit schemes the accuracy is improved from $O(k)$ to $O(k^2)$, to the original partial differential equation, locally in time. However, improved local time accuracy does not guarantee global accuracy improvement, since one needs to address the issue of stability, as follows.

Fourier Analysis: Substituing $u_m^n \sim \varrho^n \exp(i\beta mh)$ into Eq. (6.1) we obtain

$$\varrho^2 - 1 = 2r\varrho\left[e^{i\beta h} - 2 + e^{-i\beta h}\right] = 4r\varrho(\cos\beta h - 1) = -8r\varrho\sin^2\frac{\beta h}{2} \tag{6.2}$$

Therefore, we have

$$\varrho = -4r\sin^2\frac{\beta h}{2} \pm \sqrt{\left(16r^2\sin^4\frac{\beta h}{2} + 1\right)}$$

$$= -4r\sin^2\frac{\beta h}{2} \pm \left[1 + 8r^2\sin^4\frac{\beta h}{2} + O(r^4)\right] \tag{6.3}$$

Considering the negative sign, we have,

$$\varrho = -1 - 4r\sin^2\frac{\beta h}{2}\left(1 + 2r\sin^2\frac{\beta h}{2}\right) - O(r^4) \tag{6.4}$$

Thus, $|\varrho| > 1$ for all $r > 0$. This implies that the scheme is unconditionally unstable.

6.2 First–Order ODE

$$U' + AU = 0 \quad , A = constant > 0 \tag{6.5}$$

This problem has already been studied in Chapter I. We include it here for completeness. With the central difference approximation, we get

$$U' \cong \frac{u_{n+1} - u_{n-1}}{2h} \tag{6.6}$$

Hence Eq. (6.5) can be approximated as

$$u_{n+1} + 2hAu_n - u_{n-1} = 0 \tag{6.7}$$

The characteristic equation takes the form, $q^2 + 2hAq - 1 = 0$. Therefore, we have

$$q = -hA \pm \sqrt{((hA)^2 + 1)} \tag{6.8a}$$

$$q_1 = 1 - hA + \frac{1}{2}(hA)^2 + \dots \tag{6.8b}$$

and,

$$q_2 = -1 - hA - \frac{1}{2}(hA)^2 + \dots \tag{6.8c}$$

Evidently, $|q_2| > 1$ for any $h > 0$, which implies unconditional instability.

6.3 The Wave Equation

$$U_t + U_x = 0 \tag{6.9}$$

The finite difference form is,

$$\frac{u_m^{n+1} - u_m^{n-1}}{2k} + \frac{u_{m+1}^n - u_{m-1}^n}{2h} = 0 \tag{6.10}$$

The von Neumann stability analysis yields

$$\varrho^2 - 1 + \lambda \varrho(2i \sin \beta h) = 0 \tag{6.11}$$

which implies that

$$\varrho = -i(\lambda \sin \beta h) \pm \sqrt{\left(1 - (\lambda \sin \beta h)^2\right)} \tag{6.12a}$$

i.e.,

$$\varrho = -i(\lambda \sin \beta h) \pm (1 - \frac{1}{2}(\lambda \sin \beta h)^2 - O(\lambda^4)) \tag{6.12b}$$

Hence, $|\varrho| \leq 1$ if $\lambda \leq 1$. Thus the scheme is conditionally stable.

Apparently, the same idea leads to quite different results, depending on the exact form of the differential equation under consideration. Clearly, when analyzing the performance of a given discretization scheme, the whole equation, instead of the individual terms, needs to be taken into account.

7 MULTI–DIMENSIONAL PROBLEMS

The parabolic equation in two dimensions is of the form,

$$U_t = U_{xx} + U_{yy} \tag{7.1}$$

Discretizing by the explicit method :

$$\frac{u_{i,j}^{n+1} - u_{i,j}^n}{k} = \frac{u_{i+1,j}^n - 2u_{i,j}^n + u_{i-1,j}^n}{h_1^2} + \frac{u_{i,j}^n - 2u_{i,j}^n + u_{i,j-1}^n}{h_2^2} \tag{7.2}$$

where $k = \Delta t$, $h_1 = \Delta x$, $h_2 = \Delta y$. Introducing discrete Fourier decomposition such that

$$u_{i,j}^n \sim \varrho^n e^{\sqrt{(-1)}\beta i h_1} e^{\sqrt{(-1)}\beta j h_2} \tag{7.3}$$

where i and j denote the spatial grid indices; substituting (7.3) into (7.2), and letting $r = k/(h_1)^2$, we get

$$\varrho - 1 = r\left[\left(e^{\sqrt{(-1)}\beta h_1} + e^{-\sqrt{(-1)}\beta h_1} - 2\right) + \left(\frac{h_1}{h_2}\right)^2\left(e^{\sqrt{(-1)}\beta h_2} + e^{-\sqrt{(-1)}\beta h_2} - 2\right)\right]$$

$$= -4r\left\{\sin^2\frac{\beta h_1}{2} + \left(\frac{h_1}{h_2}\right)^2\sin^2\frac{\beta h_2}{2}\right\} \tag{7.4}$$

Therefore, stability of the difference scheme requires that ,

$$\left|1 - 4r\left\{\sin^2\frac{\beta h_1}{2} + \left(\frac{h_1}{h_2}\right)^2\sin^2\frac{\beta h_2}{2}\right\}\right| \le 1$$

i.e.,

$$-1 \le 1 - 4r\left\{\sin^2\frac{\beta h_1}{2} + \left(\frac{h_1}{h_2}\right)^2\sin^2\frac{\beta h_2}{2}\right\} \le 1 \tag{7.5}$$

The right hand side is always satisfied. For the left hand side to be valid, we must satisfy the condition that

$$r \le \frac{1}{2\left\{\sin^2\frac{\beta h_1}{2} + \left(\frac{h_1}{h_2}\right)^2\sin^2\frac{\beta h_2}{2}\right\}} \tag{7.6}$$

At worst, therefore, we have

$$r \leq \frac{1}{2\left\{1 + \left(\frac{h_1}{h_2}\right)^2\right\}} \tag{7.7}$$

Therefore the condition for stability in the two–dimensional case becomes

$$k\left[\frac{1}{h_1^2} + \frac{1}{h_2^2}\right] \leq \frac{1}{2} \tag{7.8}$$

This shows that the restriction on the time step for two–dimensional problems is more severe than that for the one–dimensional case. For example, if $h_1 = h_2 = h$, then $r = \frac{k}{h^2} \leq \frac{1}{4}$ is required for 2–D, while for 1–D the requirement was previously found to be $r \leq \frac{1}{2}$. In fact for 3–D problems the requirement becomes $r = \frac{k}{h^2} \leq \frac{1}{6}$.

8 CHOICE OF METHOD : EXPLICIT OR IMPLICIT ?

For the heat conduction equation, $U_t = U_{xx}$, the characteristics of implicit and explicit temporal discretization may be contrasted as follows. For the explicit discretized form given by

$$u_m^{n+1} = r u_{m+1}^n + (1 - 2r)u_m^n + r u_{m-1}^n$$

the solution is obtained by a time marching procedure. There is no coupling among points at the two time levels, namely, n and $n+1$. For the implicit scheme, Eq. (4.1) can be rewritten as:

$$u_m^{n+1} = (1 - \xi)\left[r u_{m+1}^n + (1 - 2r)u_m^n + r u_{m-1}^n\right]$$
$$+ \xi\left[r u_{m+1}^{n+1} + (1 - 2r)u_m^{n+1} + r u_{m-1}^{n+1}\right]$$

In the implicit procedure, one needs to solve a set of simultaneous linearized equations. The computational effort expended per time step is therefore greater than the explicit schemes. This shortcoming can be mitigated by using a tridiagonal solver or a point iterative method. If the system of equations to be solved can be conveniently cast into a tridiagonal matrix, the former method is indeed highly attractive. The computational requirement is also offset by the fact that the implicit scheme being unconditionally stable, larger time steps are permitted, and thus the solution up to a given time may be advanced with fewer time steps than in the purely explicit methods.

With the implicit method, therefore, the issue of computational effort per time step versus allowable value of k is of concern in determining its viability. For instance, for 1–D problems, if one uses a tridiagonal solver, then for each time step the implicit method is about 5 times as time consuming as an explicit method. The explicit method is of course restricted by the condition $r = \frac{k}{h^2} \leq \frac{1}{2}$. Therefore, if $h = 0.01$, then k should not exceed

5×10^{-5}. In the 2–D case, with $h_1 = h_2 = h$, the explicit scheme faces the restriction, $r = \dfrac{k}{h^2} \leq 1/4$, while in 3–D, $r = \dfrac{k}{h^2} \leq 1/6$. Thus, for the multidimensional case, as the explicit method is restricted to progressively smaller time step sizes, the computational expense grows. Thus, in the interest of computational economy, the implicit method is generally favoured for problems where the allowable values of k are large. However, employing large k values may not be entirely preferable in view of the resulting loss of accuracy. The local error estimates for the fully implicit scheme being $\mathcal{O}(k, h^2)$, employing large k values implies that $\mathcal{O}(k) > \mathcal{O}(h^2)$ engendering poor accuracy. In this respect the Crank–Nicolson scheme is better since the local error is $\mathcal{O}(k^2, h^2)$. But for 2–D and 3–D cases, the coefficient matrix of the implicit scheme is no longer tridiagonal and thus more computational effort is involved in obtaining solutions.

An alternative is to use the DuFort–Frankel method. This is an explicit, unconditionally stable scheme and thus the solution may be obtained by time marching without the stability restrictions commonly applying to explicit schemes. However, this scheme is not necessarily consistent as indicated by the Taylor–series error estimates. Nevertheless, the scheme does work quite well in many situations.

Example 4 Consider a system of 2 PDE's as follows,
$$U_t = k_l U_{xx}$$

and
$$V_t = k_s V_{xx}$$

Such a situation is encountered in a conjugate heat transfer problem where the ratio k_{solid}/k_{liquid} is generally very high. Then if an explicit scheme is used, both equations require that $k_i/h^2 \leq 0.5$. So, if for the solid phase $k_s = 10^3 k_l$ and $h_s = 0.1 h_l$, then extremely small time steps (i.e., k values) must be used for both equations. Such a requirement is obviously wasteful in the liquid phase where the temperature field relaxes much slower than in the solid. Thus, the choice of a difference scheme is also governed by the physics of the problem.

9 SOLUTION METHODS FOR IMPLICIT SCHEME : THE ADI METHOD

Consider the heat equation in 2–D, Eq. (7.1). The Alternating Direction Implicit (ADI) methods are two–step methods involving the solution of tridiagonal sets of equations along lines parallel to the x– and y–axes at the first and second steps respectively. The procedure reduces the original multi–dimensional problems to multiple one–dimensional problems and repetitively uses tridiagonal matrix solver each time.

The Implicit scheme for 2–D case results in the discretized form,
$$u_{i,j}^{n+1} - u_{i,j}^{n} = r\left(\delta_x^2 u_{i,j}^{n+1} + \delta_y^2 u_{i,j}^{n+1}\right) \tag{9.1}$$

where

$$r = k/h^2, h = \Delta x = \Delta y,$$
and
$$\delta_x^2 u_{i,j} = u_{i+1,j} - 2u_{i,j} + u_{i-1,j}$$

The resulting system of equations produces a coefficient matrix with penta–diagonal structure in 2–D, and a tri–diagonal matrix in 1–D. Thus, in 2–D, the procedure is to replace the fully implicit difference equation (9.1) by two semi–implicit equations with half the time increment, i.e., $k/2$, in each equation. Such, a procedure, put forward by Peaceman and Rachford (1955) yields

$$u_{i,j}^{n+\frac{1}{2}} - u_{i,j}^{n} = \frac{r}{2}\left(\delta_x^2 \, u_{i,j}^{n+\frac{1}{2}} + \delta_y^2 \, u_{i,j}^{n}\right) \tag{9.2a}$$

as the first step, which is implicit in x and explicit in y and the next step is

$$u_{i,j}^{n+1} - u_{i,j}^{n+\frac{1}{2}} = \frac{r}{2}\left(\delta_x^2 \, u_{i,j}^{n+\frac{1}{2}} + \delta_y^2 \, u_{i,j}^{n+1}\right) \tag{9.2b}$$

which is implicit in y and explicit in x. The above expressions can be rewritten as

$$\left(1 - \frac{r}{2}\delta_x^2\right)u_{i,j}^{n+\frac{1}{2}} = \left(1 + \frac{r}{2}\delta_y^2\right)u_{i,j}^{n} \tag{9.3a}$$

$$\left(1 - \frac{r}{2}\delta_y^2\right)u_{i,j}^{n+1} = \left(1 + \frac{r}{2}\delta_x^2\right)u_{i,j}^{n+\frac{1}{2}} \tag{9.3b}$$

The single composite expression equivalent to the above two difference expressions may be written as,

$$\left(1 - \frac{r}{2}\delta_x^2\right)\left(1 - \frac{r}{2}\delta_y^2\right)u_{i,j}^{n+1} = \left(1 + \frac{r}{2}\delta_x^2\right)\left(1 + \frac{r}{2}\delta_y^2\right)u_{i,j}^{n} \tag{9.4}$$

The method is unconditionally stable in 2–D but not in 3–D. The accuracy of the method is second order in both k and h with respect to the original PDE.

Comparing this method with the Crank–Nicolson scheme which reads:

$$u_{i,j}^{n+1} - u_{i,j}^{n} = \frac{r}{2}\left(\delta_x^2 \, u_{i,j}^{n+1} + \delta_y^2 \, u_{i,j}^{n+1}\right) + \frac{r}{2}\left(\delta_x^2 \, u_{i,j}^{n} + \delta_y^2 \, u_{i,j}^{n}\right) \tag{9.5}$$

and may also be written as,

$$\left(1 - \frac{r}{2}\delta_x^2 - \frac{r}{2}\delta_y^2\right)u_{i,j}^{n+1} = \left(1 + \frac{r}{2}\delta_x^2 + \frac{r}{2}\delta_y^2\right)u_{i,j}^{n} \tag{9.6}$$

Thus the Peaceman–Rachford method is identical to Crank–Nicolson method except for an extra higher order term, $\frac{r^2}{4}\delta_x^2\delta_y^2\left(u_{i,j}^{n+1} - u_{i,j}^{n}\right)$. Therefore both schemes are second–order accurate in both k and h.

10 NONUNIFORM MESHES

The one–dimensional heat equation for a non–uniform mesh assumes the discretized form,

$$\left.\frac{dU}{dx}\right|_i \cong \frac{1}{\Delta x + \nabla x}\left[\frac{\Delta x}{\nabla x}\nabla u + \frac{\nabla x}{\Delta x}\Delta u\right] \tag{10.1}$$

where

$$\Delta x = x_{i+1} - x_i, \quad \nabla x = x_i - x_{i-1}$$

The local truncation error involved in differencing the derivative as above is obtained as follows

$$\nabla U = U_i - U_{i-1} = (\nabla x)U'|_i - \frac{(\nabla x)^2}{2}U''|_i + \frac{(\nabla x)^3}{6}U'''|_i + \ldots\ldots \qquad (10.2a)$$

and

$$\Delta U = U_{i+1} - U_i = (\Delta x)U'|_i + \frac{(\Delta x)^2}{2}U''|_i + \frac{(\Delta x)^3}{6}U'''|_i + \ldots \qquad (10.2b)$$

Substituting the above in Eq. (10.1), we have,

$$\begin{aligned}
\frac{dU}{dx}\Big|_i &\sim \frac{1}{\Delta x + \nabla x}\Bigg[\Delta x U'|_i - \frac{(\Delta x)(\nabla x)}{2}U''|_i + \frac{(\Delta x)^2(\nabla x)}{6}U'''|_i + \ldots\ldots \\
&\qquad + (\nabla x)U'|_i + \frac{(\Delta x)(\nabla x)}{2}U''|_i + \frac{(\Delta x)(\nabla x)^2}{6}U'''|_i + \ldots\Bigg] \\
&= U'|_i + \frac{1}{6}\frac{(\Delta x)(\nabla x)}{(\Delta x) + (\nabla x)}\Big[(\nabla x)U'''|_i + (\Delta x)U'''|_i + \ldots.\Big] \\
&= U'|_i + \frac{1}{6}(\nabla x)(\Delta x)U'''|_i + \ldots.. \qquad\qquad (10.3)
\end{aligned}$$

Therefore the local error is of second order, i.e., $\mathcal{O}\{(\Delta x)(\nabla x)\} = \mathcal{O}(h^2)$.

$$\frac{d^2U}{dx^2} \cong \frac{2}{(\Delta x) + (\nabla x)}\Bigg[\frac{\Delta U}{\Delta x} - \frac{\nabla U}{\nabla x}\Bigg] = U''|_i + \frac{1}{3}[(\Delta x) - (\nabla x)]U'''|_i + \ldots \qquad (10.4)$$

For this approximation, the local error is first– or second–order; for example, the local error is second–order, i.e.,

$$\mathcal{O}((\Delta x) - (\nabla x)) = \mathcal{O}(h^2) \quad \text{if} \quad \Delta x = \mathcal{O}(\nabla x) = \mathcal{O}(h) \qquad (10.5a)$$

and first order, i.e.,

$$\mathcal{O}((\Delta x) - (\nabla x)) = \mathcal{O}(\Delta x) \quad or \quad \mathcal{O}(\nabla x) \quad \text{if} \quad \Delta x \neq \mathcal{O}(\nabla x) \qquad (10.5b)$$

Hence, the rate of nonuniformity can have a noticeable impact on the local order of accuracy.

11 TREATMENT OF BOUNDARY CONDITIONS

Consider the 1–D heat equation along with initial and boundary conditions:

$$U_t = U_{xx} \quad , \quad 0 \leq x \leq 1, t \geq 0 \qquad (11.1)$$

I.C. $$\qquad U(x, 0) = f(x), 0 \leq x \leq 1$$

$$p_1 \frac{\partial U}{\partial x} - q_1 U = g_1(t), \quad x = 0, \quad t \geq 0,$$

B. C. s

$$p_2 \frac{\partial U}{\partial x} + q_2 U = g_2(t), \quad x = 1, \quad t \geq 0$$

Here $g_1(t)$ and $g_2(t)$ are continuous and bounded as $t \rightarrow \infty$.

(a) Specification of Dirichlet type boundary conditions would imply that $p_1, p_2 = 0$, $q_1, q_2 \neq 0$, i.e., the value of U is prescribed on the boundaries x=0 and x=1 for all t.

(b) The Neumann type boundary conditions will involve specifying $p_1, p_2 \neq 0$, $q_1, q_2 = 0$, i.e., the gradient $\frac{\partial U}{\partial x}$ is specified on the boundaries.

(c) The mixed form of boundary condition on the boundaries will involve $p_1, p_2, q_1, q_2 \neq 0$.

In general, while imposing the difference form of the boundary conditions, it is desirable to match the accuracy of the boundary condition treatment with that of the governing equation. For example,

$$\frac{\partial U}{\partial x} \cong \frac{u_{m+1} - u_{m-1}}{2h} + \mathcal{O}(h^2) \tag{11.2a}$$

$$\cong \frac{u_{m+1} - u_m}{h} + \mathcal{O}(h) \tag{11.2b}$$

If the heat equation $U_t = U_{xx}$ is approximated by central difference method for the U_{xx} term, then the differencing at the boundary employing Eq. (11.2b) will cause the treatment there to be one order lower than in the interior of the domain. Thus, the overall order of accuracy may turn out to be $\mathcal{O}(k, h)$ instead of $\mathcal{O}(k, h^2)$. This may however not be the case always as discussed in relation to the boundary treatment in Chapter I. If the boundary condition is imposed in accordance with (a) above, the overall order of accuracy of the governing equation differencing, namely $\mathcal{O}(k, h^2)$ will be maintained.

11.1 Central Difference Treatment of the Boundary Condition

The discretized form of the boundary condition of the general form above, using central differencing is as follows:

$$p_1 \frac{u_1^n - u_{-1}^n}{2h} - q_1 u_o^n = g_1^n \tag{11.3}$$

The governing equation at $m=0$ is discretized as ,

$$u_o^{n+1} = r u_1^n + (1 - 2r) u_o^n + r u_{-1}^n \tag{11.4}$$

where u_{-1} is located outside the domain, and needs to be defined. From Eq. (11.3), we get u_{-1} in terms of u_0 and u_1,

$$u_{-1}^n = u_1^n - \frac{2hq_1u_o^n - 2hg_1^n}{P_1} \tag{11.5}$$

We combine Eqs. (11.4) and (11.5) to eliminate u_{-1} to obtain, at $m=0$,

$$u_o^{n+1} = ru_1^n + (1 - 2r)u_o^n + r\left[u_1^n - \frac{2hq_1u_o^n - 2hg_1^n}{P_1}\right]$$

$$= 2ru_1^n + \left[(1 - 2r) - \frac{2rhq_1}{P_1}\right]u_o^n - \frac{2rhg_1^n}{P_1} \tag{11.6}$$

In the interior, i.e., for $m=1, ..., N-1$, we have the difference form,

$$u_m^{n+1} = ru_{m+1}^n + (1 - 2r)u_m^n + ru_{m-1}^n \tag{11.7}$$

Similarly at the other boundary, namely at $m=N$, we have,

$$u_N^{n+1} = 2ru_{N-1}^n + \left[(1 - 2r) + \frac{2rhq_2}{P_2}\right]u_N^n + \frac{2rhg_2^n}{P_2} \tag{11.8}$$

11.2 One–Sided Treatment of the Boundary Condition

$$P_1\frac{u_1^n - u_o^n}{h} - q_1u_o^n = g_1^n \qquad for\ m=0. \tag{11.9a}$$

$$u_m^{n+1} = ru_{m+1}^n + (1 - 2r)u_m^n + ru_{m-1}^n, m = 1, ..., N - 1 \tag{11.9b}$$

The procedure is as follows. Compute U_1^{n+1} based on Eq. (11.9b) . Then using Eq. (11.9a) compute U_o^{n+1} based on U_1^{n+1}. Do the same for U_N^{n+1}.

12 EQUATIONS CONTAINING VARIABLE COEFFICIENTS

The following are examples of a parabolic PDE with variable coefficients:

12.1 Equations of Variable Diffusivity

$$U_t = (DU_x)_x \tag{12.1}$$

Here D is a function of U, namely $D(U)$. The explicit form of differencing is as follows

$$\frac{u_m^{n+1} - u_m^n}{k} = \frac{1}{h^2}\left[D\left(u_{m+\frac{1}{2}}^n\right)\{u_{m+1}^n - u_m^n\} - D\left(u_{m-\frac{1}{2}}^n\right)\{u_m^n - u_{m-1}^n\}\right] \tag{12.2}$$

If D is a constant then the truncation error for the right hand side approximation is $O(h^2)$; otherwise, if D is a function of U, then we can use

$$\frac{\partial D}{\partial x}\bigg|_m^n = \frac{D\left(U^n_{m+\frac{1}{2}}\right) - D\left(U^n_{m-\frac{1}{2}}\right)}{h} \tag{12.3}$$

Here, we have

$$D\left(U^n_{m+\frac{1}{2}}\right) = D(U^n_m) + \frac{\partial D}{\partial U}\frac{\partial U}{\partial x}\left(\frac{h}{2}\right)\bigg|_m^n + \frac{1}{2}\frac{\partial^2 D}{\partial U^2}\frac{\partial^2 U}{\partial x^2}\left(\frac{h}{2}\right)^2\bigg|_m^n + O(h^3) \tag{12.4a}$$

and

$$D\left(U^n_{m-\frac{1}{2}}\right) = D(U^n_m) - \frac{\partial D}{\partial U}\frac{\partial U}{\partial x}\left(\frac{h}{2}\right)\bigg|_m^n + \frac{1}{2}\frac{\partial^2 D}{\partial U^2}\frac{\partial^2 U}{\partial x^2}\left(\frac{h}{2}\right)^2\bigg|_m^n + O(h^3) \tag{12.4b}$$

The local truncation errors are obtained on the two sides of the equation. On the LHS, the error is estimated from,

$$\frac{\partial U}{\partial t}\bigg|_m^n + \frac{k}{2}\frac{\partial^2 U}{\partial t^2}\bigg|_m^n + O(k^2)$$

The right hand side of Eq. (12.2) equals the following:

$$D\left(U_{m+\frac{1}{2}}\right)\left[\frac{1}{h}\frac{\partial U}{\partial x}\bigg|_m^n + \frac{1}{2}\frac{\partial^2 U}{\partial x^2}\bigg|_m^n + \frac{h}{6}\frac{\partial^3 U}{\partial x^3}\bigg|_m^n\right]$$

$$- D\left(U^n_{m-\frac{1}{2}}\right)\left[\frac{1}{h}\frac{\partial U}{\partial x}\bigg|_m^n - \frac{1}{2}\frac{\partial^2 U}{\partial x^2}\bigg|_m^n + \frac{h}{6}\frac{\partial^3 U}{\partial x^3}\bigg|_m^n\right] + O(h^2)$$

The overall truncation error is therefore $O(k,h^2)$.

As a note, the nonlinear equation $\frac{\partial U}{\partial t} = \frac{\partial^2 U^M}{\partial x^2}$ (where M is an exponent, not an index) can be converted into $\frac{\partial U}{\partial x} = \frac{\partial}{\partial x}\left[MU^{M-1}\frac{\partial U}{\partial x}\right]$, and MU^{M-1} can be treated as D.

12.2 The 1–D Unsteady Conduction Equation in Different Coordinates

$$U_t = U_{\xi\xi} + \frac{a}{\xi}U_\xi \tag{12.5}$$

The value of a determines the coordinate system employed, viz.,

        (i) a=0 for Cartesian coordinates

        (ii) a=1 for cylindrical coordinates

        (iii) a=2 for spherical coordinates

The forward in time, central in space differencing procedure yields the relation,

$$\frac{u^{n+1}_m - u^n_m}{k} = \frac{u^n_{m+1} - 2u^n_m + u^n_{m-1}}{h^2} + \frac{a}{\xi_m}\frac{u^n_{m+1} - u^n_{m-1}}{2h} \tag{12.6}$$

In solving the above difference equation, one needs to keep track of the factor ξ_m as the calculation proceeds. The von Neumann stability analysis indicates the stability bound, $0 < r \leq 1/2$ for $a=0,1,2$. Thus the stability restriction is the same for all cases. The local truncation error for such differencing is $\mathcal{O}(k, h^2)$.

For cylindrical or spherical coordinates, usually, the symmetry condition is satisfied at $\xi = 0$, i.e., $U_\xi = 0$ at $\xi = 0$. However, the product $\frac{a}{\xi} U_\xi$ at $\xi = 0$ becomes undetermined. In order to circumvent this problem, the L'Hospital rule can be applied to obtain a limit value. Thus, we have the following

$$\lim_{\xi \to 0} \left(\frac{a}{\xi} U_\xi \right) = \lim_{\xi \to 0} aU_{\xi\xi} \tag{12.7}$$

Substituting this limit in Eq. (12.5), we obtain at $\xi = 0$, the following:

$$U_t = (a + 1)U_{\xi\xi} \quad \text{at } \xi = 0 \tag{12.8}$$

Hence the finite difference form becomes,

$$\frac{u_m^{n+1} - u_m^n}{k} = (a + 1)\frac{u_1^n - 2u_0^n + u_{-1}^n}{h^2} \tag{12.9}$$

For $m=0$, i.e., $\xi = 0$, we get

$$\frac{u_0^{n+1} - u_0^n}{k} = (a + 1)\frac{u_1^n - 2u_0^n + u_{-1}^n}{h^2} \tag{12.10}$$

Application of the symmetry condition implies that

$$\frac{\partial U}{\partial \xi} = 0, \quad i.e., \quad \frac{u_1 - u_{-1}}{2h} = 0 \tag{12.11}$$

i.e., $u_{-1} = u_1$. Therefore, we get

$$\frac{u_0^{n+1} - u_0^n}{k} = 2(a + 1)\frac{u_1^n - u_0^n}{h^2} \tag{12.12}$$

In obtaining solutions, therefore, one uses Eq. (12.6) for $m=1,2,......N-1$ and Eq. (12.12) for $m=0$. The stability restriction of Eq. (12.12) is

$$0 < r \leq \frac{1}{2(1 + a)} \tag{12.13}$$

For $a=0$, the boundary treatment therefore has no effect on stability. For $a=1$, stability imposes $r \leq \frac{1}{4}$, and for $a=2$, one requires $r \leq \frac{1}{6}$. This severe restriction on stability is imposed upon the entire field of calculation by a single boundary point and thus the overall computational costs are likely to be affected substantially. To avoid this burden due to boundary treatment, one can use an implicit treatment for $m=0$, in which case the difference form becomes

$$\frac{U_0^{n+1} - U_0^n}{k} = 2(a + 1)\frac{U_1^{n+1} - U_0^{n+1}}{h^2} \qquad (12.14)$$

Now the overall stability limit is again $r \leq \frac{1}{2}$ for $a = 0, 1$ and 2.

13 SUMMARY OF SCHEMES FOR THE HEAT EQUATION $U_t = U_{xx}$

A summary of the performance characteristics of the various schemes studied in this chapter is given below.

Explicit Scheme :
        (i) Forward difference in time

        (ii) Stability limit: $r = k/h^2 \leq \frac{1}{2}$

        (iii) Order of accuracy: $\mathcal{O}(k, h^2)$

Crank–Nicolson Scheme and Box Scheme :
        (i) Trapezoidal differencing in time
        (ii) Unconditionally stable
        (iii) Order of accuracy: $\mathcal{O}(k^2, h^2)$

Implicit Scheme :
        (i) Backward difference in time
        (ii) Unconditionally stable
        (iii) Order of accuracy: $\mathcal{O}(k, h^2)$

Dufort– Frankel Scheme :
        (i) Three–level scheme, explicit in time, mixed in U_{xx} .
        (ii) Unconditionally stable
        (iii) Order of accuracy: $\mathcal{O}(k^2, h^2, (k/h)^2)$

Leap – Frog Scheme :
        (i) Three–level scheme, explicit in time.
        (ii) Always unstable.
        (iii) Order of accuracy: $\mathcal{O}(k^2, h^2)$

The stencils for the above schemes (i.e., the spatial and temporal nodes employed by the above schemes) are summarized in Fig. 8.

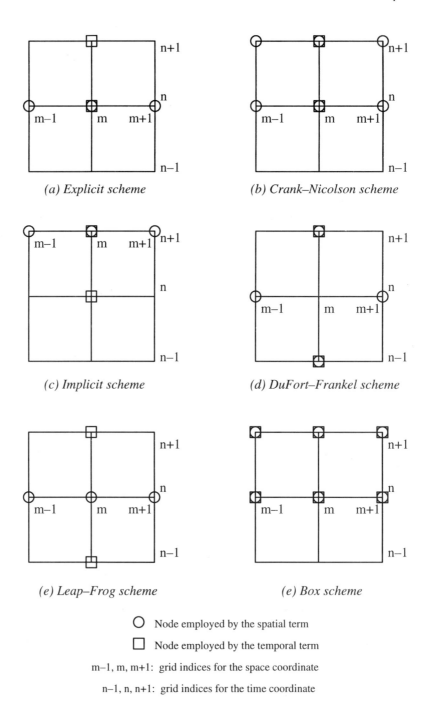

Figure 8. A schematic of the stencils used by the various schemes presented in Chapter II.

CHAPTER III
ELLIPTIC EQUATIONS

1 INTRODUCTION

The prototype equation to be used to study elliptic PDE's is the Laplace equation, shown below. The domain on which the equation is to be solved is shown in Fig. 1.

$$\Delta U = U_{xx} + U_{yy} = 0 , \quad (x, y) \in \Omega$$

$$U(x, y) = f(x, y) \quad on \quad \partial\Omega$$

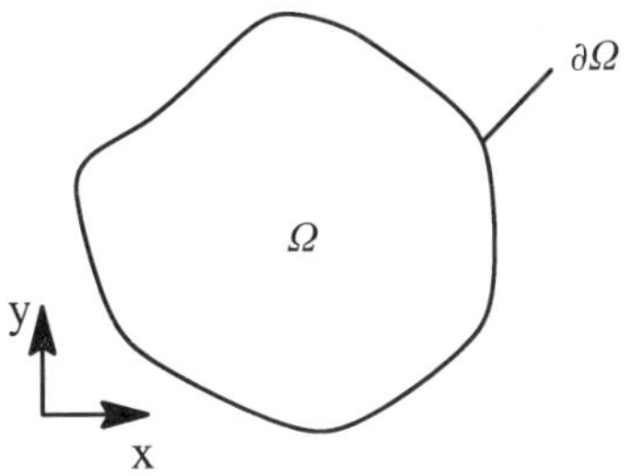

Figure 1. Domain of the PDE.

First, we review some fundamental features related to the solution of the Laplace equation.

1. Any particular solution of the Laplace equation is called a harmonic function.

2. Min–Max property: If Ω is a bounded, simply–connected region whose boundary is $\partial\Omega$, and if U is harmonic on Ω and continuous on $\Omega \cup \partial\Omega$, then U takes on its maximum and minimum values on $\partial\Omega$.

3. A second–order elliptic equation requires one condition at every point of the closed boundary which can be specified by either boundary values (Dirichlet type), boundary gradients (Neumann type), or a combination of the two (Robin type).

4. A closed boundary is required for a well–posed elliptic problem. To be a well–posed problem a solution must exist, be unique, and continuously dependent on the boundary conditions.

5. A solution of the Laplace equation subject to a Dirichlet type boundary condition, i.e. U values are specified on $\partial\Omega$, always exists and is unique.

2 BASIC ANALYSIS AND ANALOGY BETWEEN DIFFERENCE AND DIFFERENTIAL EQUATIONS

Consider the one–dimensional, linear, elliptic equation shown below with prescribed Dirichlet boundary conditions and coefficient constraints:

$$L(U) = U_{xx} + q(x)U = f(x) , \quad 0 \leq x \leq 1 \tag{2.1}$$

$$q(x) \leq 0 \qquad\qquad U(0) = U(1) = 0$$

where the operator L is defined as $L = \dfrac{d^2}{dx^2} + q$

Two important qualitative properties of the solution of the equation above can be proven, the first being the positivity of the solution over the domain, and the second the bounded nature of the solution. Since it is desirable that any finite difference solution display the same qualitative properties as the exact solution, we will also construct a particular finite difference representation of equation (2.1) and prove that the same qualitative characteristics do indeed hold.

2.1 Positivity Property

Positivity Property for the Exact Solution

Theorem: If $f(x) \leq 0$ then $U(x) \geq 0$.
Proof (by contradiction): Suppose that $f(x) \leq 0$ and there exists an interval (α,β), $0 \leq \alpha < \beta \leq 1$, such that $U(x_1) < 0$, $\alpha < x_1 < \beta$ and $U(\alpha) = U(\beta) = 0$.
Then, within (α,β), $U_{xx} = f - qU \leq 0$ (*since* $f \leq 0, q \leq 0$ *and* $U < 0$).
The above inequality means that either (i) $U \equiv 0$ *in* (α,β), or (ii) $U > 0$ *in* (α,β).
But, it has already been assumed that $U(x_1) < 0$ which contradicts the above conclusion.
QED.
Note: Based on the same procedure, one can show that if $f(x) \geq 0$ then $U(x) \leq 0$.

Positivity Property for the Difference Solution

We define a difference operator L_h as, $L_h = \dfrac{1}{h^2}\delta^2 + q_m$. Hence, the corresponding difference equation to Eq. (2.1) is $L_h u\,_m = f_m$ subject to the boundary conditions $u_0 = u_N = 0$, where the integer subscripts 0 and N represent the starting and ending grid points, respectively of the finite difference domain.

Theorem: If $f_m \leq 0$ then $u_m \geq 0$, and if $f_m \geq 0$ then $u_m \leq 0$.
Proof: Suppose this is not true, i.e. $f_m \leq 0$ for all m, but there exists some m_1 such that $u_{m_1} < 0$. Then there are integers $0 \leq \alpha < m_1 < \beta \leq N$ such that $u_{m_1} < 0$ in (α , β) and $u_\alpha , u_\beta \geq 0$. Now, according to Eq. (3.1)

$$\frac{1}{h^2}\left(u_{m_1-1} - 2u_{m_1} + u_{m_1+1}\right) + q_{m_1}u_{m_1} = f_{m_1} \tag{2.2}$$

Solving for the value of u at grid point m yields the following:

$$u_{m_1} = \frac{u_{m_1+1} + u_{m_1-1}}{2} + h^2\left(q_{m_1}u_{m_1} - f_{m_1}\right) \tag{2.3}$$

Since $u_{m_1} < 0$, it is clear that with negative q_m and negative f_m, $h^2\left(q_{m_1}u_{m_1} - f_{m_1}\right) \geq 0$, and hence,

$$u_{m_1} \geq \frac{u_{m_1+1} + u_{m_1-1}}{2} \tag{2.4}$$

Combined with $u_\alpha = u_\beta = 0$ the only possibilities are (i) $u_{m_1} = 0$, $\alpha \leq m_1 \leq \beta$, or (ii) $u_{m_1} > 0$ $\alpha \leq m_1 \leq \beta$. Both of these possibilities contradict the assumption that $u_{m_1} < 0$. QED.

These two theorems show that the finite difference solutions of the elliptic equation will always behave *qualitatively* the same as the differential solutions, i.e., unlike that of the parabolic equations, spurious oscillations will *not* appear. The key remaining issues are hence (i) accuracy, and (ii) computational efficiency. To examine these issues, we next show the boundedness property for both the exact solution and the finite difference solution.

2.2 Boundedness Property

2.2.1 Boundedness Property for the Exact Solution

<u>Theorem:</u> $|U(x)| \leq \underset{[0,1]}{Max} |f| \cdot \left[\frac{-x(x-1)}{2} \right]$, $0 \leq x \leq 1$

<u>Proof:</u> Let $K = \underset{[0,1]}{Max} |f(x)|$ and $W = \frac{x(x-1)}{2}$. Since $W \leq 0$ within the interval of $0 \leq x \leq 1$, $KW \leq 0$. Furthermore, $KW_{xx} = K$, because $W_{xx} = 1$. Noting the form of Eq. (2.1), and using the fact that $q \leq 0$, we obtain the following:

$$L(KW) = K + qKW \geq K \geq f = L(U) \tag{2.5}$$

In other words,

$$L(U - KW) \leq 0 \tag{2.6}$$

Here we note that L is a linear operator, thus allowing us to interchange the operation order, i.e., $L(U) - L(KW) = L(U - KW)$.

Based on the positivity principle previously shown, we deduce that $U - KW \geq 0$ and hence that $U \geq KW$. We note here that KW has a negative value.

Similarly, one can show that $L(U + KW) \geq 0$ which leads to $U \leq -KW$, and thus overall, $|U| \leq KW$. QED.

2.2.2 Boundedness Property for the Difference Solution

<u>Theorem:</u> $|u_m| \leq Max |f_m| \cdot \left[\frac{-x_m(x_m - 1)}{2} \right]$, $0 \leq m \leq N$

<u>Proof:</u> Let $K = Max |f_m|$ and $W = \frac{x_m(x_m - 1)}{2}$. Hence, $KW \leq 0$ for all x_m. Realizing that $f \leq K$ and $qKW > 0$ we obtain the following:

$$L_h(u_m - KW) = f_m - K - (qKW)_m \leq 0 \qquad (2.7)$$

Again according to the maximum principle, we deduce that $u_m - KW \geq 0$ and thus that $u_m \geq KW$.

Similarly, one can show that $L(u_m + KW) = f_m + K \geq 0$ and thus that $u_m \leq -KW$. Now, overall we obtain $|u_m| \leq KW$. QED.

Now that the bounded nature of both the exact and finite difference solutions has been proven, what do we do with these results? We next show that the above results can be used to investigate both the stability and convergence characteristics of the finite difference representation of the exact equation.

2.3 Stability and Convergence

2.3.1 Stability analysis

A question to ask at this point is that if f_m produces solution u_m, does $f_m + \delta$ produce $u_m + \varepsilon_m$? From the finite difference equation, we see that

$$L_h(u_m + \varepsilon_m) = f_m + \delta \quad \text{and hence} \quad L_h(\varepsilon_m) = \delta \qquad (2.8)$$

Using the boundedness property, we know that ε_m is bounded by $\delta\left[\dfrac{-x(x-1)}{2}\right]$ and hence ε_m does not grow without bound. Based on this observation, one concludes that the scheme is always stable.

2.3.2 Convergence

Recall the difference equation

$$L_h(u_m) = \frac{1}{h^2}\delta^2 u_m + q_m u_m = f_m \qquad (2.9)$$

and the differential equation

$$L(U)|_{x_m} = U_{xx}|_{x_m} + (qU)|_{x_m} = f_{x_m} \qquad (2.10)$$

Since the right–hand sides of both equations are the same, one knows that

$$|L_h(U_m) - L(U(x_m))| = |\frac{1}{h^2}\delta^2 U_m - U_{xx}|_{x_m}| \leq \frac{h^2}{12} Max|U_{xxxx}| \qquad (2.11)$$

In other words,

$$L_h(U_m) = f_m + \tau_m \quad \text{where} \quad |\tau_m| \leq \frac{h^2}{12} Max|U_{xxxx}| \qquad (2.12)$$

So, $L_h(U_m - u_m) = \tau_m$, and thus by the boundedness property, we have the following:

$$|U_m - u_m| \leq \frac{h^2}{12} Max|U_{xxxx}| \cdot \left|\frac{-x_m(x_m - 1)}{2}\right| \qquad (2.13)$$

which means that

$$|U_M - u_m| \leq \frac{h^2}{96} Max|U_{xxxx}| \tag{2.14}$$

hence, the difference solution u_m approaches the differential solution U_m as $h \to 0$, and thus the convergence requirement is always satisfied.

In summary, *when solving a linear elliptic equation, the standard central difference scheme guarantees numerical solutions to converge and to be free from any spurious oscillations. Hence, the main issue that remains to be resolved is how to obtain the numerical solutions in economical ways.* In the following, we shall concentrate on the various iterative techniques suitable for solving a large system of linear equations.

3 LAPLACE EQUATION IN A SQUARE

In the following, we shall use the two–dimensional Laplace equation, shown below, as a prototype to facilitate our study of elliptic PDE's.

$$U_{xx} + U_{yy} = 0 \quad , \quad 0 \leq x, y \leq 1 \tag{3.1}$$

To obtain a numerical approximation to the exact equation, we must first construct a grid over the square domain. In this case, we choose a uniformly spaced grid with $M+1$ nodes in each of the coordinate directions, as shown in Figure 2. The uniform grid spacing is denoted as h, and the number of internal grid nodes is $(M–1) \times (M–1)$, i.e., $Mh = 1$.

A standard practice of approximating the second–order derivatives is to use the second–order central difference operator, which results in the following difference equation:

$$\frac{u_{l+1,m} - 2u_{l,m} + u_{l-1,m}}{h^2} + \frac{u_{l,m+1} - 2u_{l,m} + u_{l,m-1}}{h^2} = 0 \tag{3.2}$$

A local truncation error analysis yields

$$\frac{u_{l+1,m} - 2u_{l,m} + u_{l-1,m}}{h^2} + \frac{u_{l,m+1} - 2u_{l,m} + u_{l,m-1}}{h^2}$$

$$= (U_{xx} + U_{yy})|_{l,m} + \frac{h^2}{12}(U_{xxxx} + U_{yyyy})|_{l,m} + \cdots \tag{3.3}$$

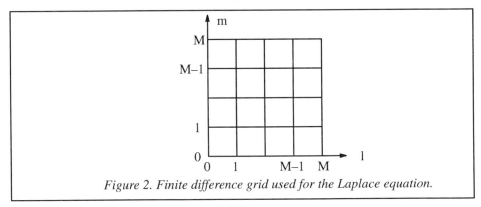

Figure 2. Finite difference grid used for the Laplace equation.

i.e., the local truncation error (LTE) to the differential equation is $\frac{h^2}{12}(U_{xxxx} + U_{yyyy})|_{l,m} + ...$, indicating, again, that the numerical scheme is second–order accurate.

Next we discuss the implementation of the boundary conditions. As stated previously, two fundamental types of boundary conditions are used for elliptic equations, Dirichlet conditions, and Neumann conditions. The implementation of each of these conditions will be discussed individually. Implementation of the mixed (Robin) boundary condition follows directly from a knowledge of the Dirichlet and Neumann conditions.

3.1 Dirichlet BC

The Dirichlet boundary condition can be accommodated via a straightforward assignment of boundary values at the appropriate grid locations. Using the notation given in Fig. 2, the unknown vector $\vec{u}$ is defined as follows:

$$\vec{u} = [\underbrace{u_{1,1}, u_{2,1}, ..., u_{M-1,1}}_{\text{1st row}}, \underbrace{u_{1,2}, ..., u_{M-1,2}}_{\text{2nd row}}, ..., \underbrace{u_{1,M-1}, ..., u_{M-1,M-1}}_{\text{(M–1)th row}}]^T \quad (3.4)$$

It is clear that with the Dirichlet boundary condition, overall there are $(M–1)\times(M–1)$ unknowns. There are many ways to order the sequence of the unknowns $u_{l,m}$; Eq. (3.4) is only one of them. The expression given in Eq. (3.4) is called the *natural ordering* i.e., the unknowns are arranged column by column or row by row. The resulting finite difference equations are of the form ,

$$[A]\vec{u} = \vec{b} \quad (3.5)$$

where [A] is a coefficient matrix of $[(M–1)\times(M–1)]^2$ elements:

$$[A] = \begin{bmatrix} B & -I & & & \\ -I & B & -I & & \mathbf{0} \\ & \ddots & \ddots & \ddots & \\ \mathbf{0} & & -I & B & -I \\ & & & -I & B \end{bmatrix} \quad (3.6)$$

which is a pentadiagonal matrix, i.e., in general there are five non–zero elements along a column or row. The matrix I is an identity matrix of $(M–1)\times(M–1)$ elements. The matrix B is a tridiagonal matrix of $(M–1)\times(M–1)$ elements, and is shown below:

$$B \equiv \begin{bmatrix} 4 & -1 & & & \\ -1 & 4 & -1 & & \mathbf{0} \\ & \ddots & \ddots & \ddots & \\ \mathbf{0} & & -1 & 4 & -1 \\ & & & -1 & 4 \end{bmatrix} \quad (3.7)$$

Finally, the source, or forcing term on the right hand side appears as follows:

$$\vec{b} = [(u_{0,1} + u_{1,0}), u_{2,0}, ..., u_{M-2,0}, (u_{M-1,0} + u_{M,1}), u_{0,2}, 0, ..., 0, u_{M,2}, ...]^T \quad (3.8)$$

<u>B.C. of first 1st row of u</u>   <u>2nd row</u>

As we can clearly observe, $\vec{b}$ results from the specification of the boundary values.

3.2 Neumann BC

The Neumann boundary condition can be generally written as follows:

$$\frac{\partial U}{\partial n} = g(x, y) \quad \text{on the boundary } \partial\Omega \tag{3.9}$$

where $\frac{\partial}{\partial n}$ denotes differentiation along the normal to the boundary directed away from the interior of the square. The basic procedure of assembling the difference equation here is the same as in the previous case, except that the number of unknowns needed to be determined is now $(M+1) \times (M+1)$ instead of $(M-1) \times (M-1)$. For internal nodes, Eq. (3.2) remains unchanged, however, for boundary nodes we use a finite difference representation of the Neumann condition to prescribe the unknown values in the original equation for each node in terms of the interior unknowns. For example, at the node $(m, 0)$ along the lower boundary line, a central difference operator can be used to approximate the first partial derivative Neumann condition, resulting in

$$\frac{u_{m,1} - u_{m,-1}}{2h} = g_{m,0} \tag{3.10}$$

Substituting Eq. (3.10) into Eq. (3.2) to eliminate $u_{m,-1}$ provides the appropriate equation for the boundary node. The entire system of equations for both interior and boundary nodes appears as follows:

$$[A]\vec{u} = 2h\vec{b} \tag{3.11}$$

The unknown vector $\vec{u}$ is defined as follows:

$$\vec{u} = [u_{0,0}, u_{1,0}, ..., u_{M,0}, u_{0,1}, ..., u_{M,1}, ..., u_{0,M}, ..., u_{M,M}]^T \tag{3.12}$$

<u>0th row</u>   <u>1st row</u>   <u>M+1 th row</u>

The matrix $[A]$, shown below, is again, a pentadiagonal matrix, however, it now contains $[(M + 1) \times (M + 1)]^2$ elements. The matrix I is an identity matrix of $(M+1) \times (M+1)$ elements. The matrix B is a tridiagonal matrix of $(M+1) \times (M+1)$ elements, and is also shown below.

$$[A] \equiv \begin{bmatrix} B & -2I & & & \\ -I & B & -I & & \mathbf{0} \\ & \ddots & \ddots & \ddots & \\ & & -I & B & -I \\ \mathbf{0} & & & -2I & B \end{bmatrix} \tag{3.13}$$

$$B \equiv \begin{bmatrix} 4 & -2 & & & \\ -1 & 4 & -1 & & \mathbf{0} \\ & \ddots & \ddots & \ddots & \\ & & -1 & 4 & -1 \\ \mathbf{0} & & & -2 & 4 \end{bmatrix} \tag{3.14}$$

The source vector $\vec{b}$ now takes the form

$$\vec{b} = [2g_{0,0} \, , \, g_{1,0} \, , \, \cdots \, , \, g_{M-1,0} \, , \, 2g_{M,0} \, , \, g_{0,1} \, , \, 0 \, , \, \cdots \, , \, 0 \, , \, g_{M,1} \, , \, \ldots \ldots \, ,$$

$$g_{0,M-1} \, , \, 0 \, , \, \cdots \, , \, 0 \, , \, g_{M,M-1} \, , \, 2g_{0,M} \, , \, g_{1,M} \, , \, \cdots \, , \, g_{M-1,M} \, , \, 2g_{M,M}]^T \tag{3.15}$$

An important point to note is that the matrix A for the Dirichlet problem is nonsingular, while for the Neumann problem, it is singular. This is consistent with the solution to the differential equation which has an arbitrary constant for the Neumann problem.

4 SOLUTION OF LINEAR EQUATIONS: CLASSICAL ITERATIVE METHODS

4.1 Basics

In this section we discuss techniques for solving linear systems of equations of the form

$$[A]\vec{u} = \vec{b} \tag{4.1}$$

using iterative solution methods. An iterative solution method is a technique whereby an initial guess is specified for the unknowns and the algebraic equations are used repetitively to obtain improved estimates of the unknowns. As the number of iterations increases, the accuracy of the estimates should approach the exact solution of the system of equations.

An important question to ask is whether the iterative solution technique is guaranteed to converge to the appropriate exact solution. For some iterative methods (e.g. Gauss–Seidel), a sufficient condition for convergence can be proven, namely that,

$$|a_{ii}| \geq \sum_{j \neq i} |a_{ij}| \tag{4.2}$$

where a_{ii} represents the main diagonal term, and the summation is taken over the other terms in the row. It should be noted here, that these conditions are naturally guaranteed by using the standard central difference scheme for approximating the equation

$$U_{xx} + U_{yy} = f(x, y) \tag{4.3}$$

Now, suppose that the matrix $[A]$ is properly scaled so that all entries along the main diagonal are of unity value. Therefore, $[A]$ can now be expressed as follows

$$[A] = [I] - [R] - [S] \tag{4.4}$$

where the matrix $[I]$ is an identity matrix, $[R]$ is a lower triangular matrix where only entries below the main diagonal are nonzero, and $[S]$ is an upper triangular matrix. Using this expression for matrix $[A]$, the Eq. (4.1) can now be written as

$$[A]\vec{u} = ([I] - [R] - [S])\vec{u} = \vec{b} \tag{4.5}$$

From Eq. (4.5) several different iterative methods can be constructed. The first, and simplest is the Jacobi iteration method. With the Jacobi method, the unknown solution values at some iteration level n are used to obtain new estimates of the solution at iteration $n+1$. In matrix notation, this method is constructed by moving the lower and upper diagonal terms in Eq. (4.5) to the right–hand side, resulting in

$$[I]\vec{u} = \vec{u} = \vec{b} + ([R] + [S])\,\vec{u} \tag{4.6}$$

Thus, the Jacobi iteration method is

Jacobi Method: $\qquad \vec{u}^{(n+1)} = \vec{b} + ([R] + [S])\,\vec{u}^{(n)} \tag{4.7}$

It is easily seen that the matrix representation (4.6) corresponds to that fact that only values at the nth iteration level are used in calculating the solution at the $(n+1)^{th}$ iteration level.

The next method we will discuss is the Gauss–Seidel iterative method. In any pointwise iterative method the unknowns are usually computed in a line by line order, either by column or by row. Thus, one proceeds to solve for the unknowns point by point across one line, and when that line is done, moves on to the next. With the Jacobi point iterative method, the unknowns at each point on the grid at iteration level $n+1$ are computed using only the unknown values at the nth iteration level. However, it is clear that since the equations for the unknowns on one line involve the values of the unknowns on previously calculated lines, we should be able to use these newly calculated values immediately. This is the idea behind the Gauss–Seidel iterative method. In the matrix notation of Eq. (4.4) and using the concept of row by row natural ordering, we can observe that the lower triangular matrix $[R]$, represents the contribution from terms that are computed at the $(n+1)^{th}$ iteration level, and the upper triangular matrix $[S]$, represents the contribution from terms that are computed at the nth iteration level. Thus, the

Gauss–Seidel iteration method, in matrix notation becomes

Gauss–Seidel Method: $\quad ([I] - [R])\, \vec{u}^{(n+1)} = \vec{b} + [S]\vec{u}^{(n)}$ $\qquad$ (4.8)

The next iterative method we present is a modification to the Gauss–Seidel method and is called the successive overrelaxation (SOR) method. With the SOR method, instead of retaining the solution $\vec{u}^{(n+1)}$ computed from Eq. (4.8) as the solution at the $(n+1)^{th}$ iteration level, the solution is taken to be a linear weighting between the new solution obtained, $\vec{u}^{(n+1)}$, and the solution at the previous iteration level, $\vec{u}^{(n)}$. Thus, we have

$$\vec{u}^{(n+1)} = (1 - \omega)\vec{u}^{(n)} + \omega\vec{u}^{(n+1)} \qquad (4.9)$$

In Eq. (4.9), ω is called the relaxation factor and can take values from 0 to 2. Using values in the range $0 \le \omega < 1$ is termed underrelaxation while using values in the range $1 < \omega \le 2$ is termed overrelaxation. The original Gauss–Seidel method is recovered for $\omega = 1$. With the above concept in mind, the SOR method can be written in matrix notation as follows:

SOR Method: $\quad \vec{u}^{(n+1)} = \omega\left([R]\vec{u}^{(n+1)} + [S]\vec{u}^{(n)} - [I]\vec{u}^{(n)} + \vec{b}\right) + \vec{u}^{(n)}$ $\qquad$ (4.10)

4.2 Convergence of Iterative Methods

In this section we examine the convergence characteristics of the various iterative methods which have been detailed above. In general, all of the iterative methods described above can be expressed in the following form:

$$\vec{u}^{(n+1)} = \vec{b} + [H]\vec{u}^{(n)} \qquad (4.11)$$

where the coefficient matrix $[H]$, which operates on the vector $\vec{u}^{(n)}$, and the source term $\vec{b}$ depend on the method being used. We will now prove that the matrix $[H]$ for any iterative method must meet a certain condition in order for convergence to be guaranteed. We begin by defining the spectral radius $\varrho([H])$ of the coefficient matrix $[H]$ to be the absolute value of the largest magnitude eigenvalue:

$$\varrho([H]) = Max\{ |\lambda| : \lambda \text{ is an eigenvalue of } [H] \} \qquad (4.12)$$

With the definition above, we now state, and prove the following theorem:

<u>Theorem</u>: The iterative method $\vec{u}^{(n+1)} = \vec{b} + [H]\vec{u}^{(n)}$ converges if and only if $\varrho([H]) \le 1$.

<u>Proof</u>: Let us define an error vector $\vec{e}^{(n)}$ as $\vec{e}^{(n)} = \vec{u}^{(n)} - \vec{u}$, where $\vec{u}$ is the exact solution of the matrix equation $[A]\vec{u} = \vec{b}$. Therefore, we can write the following:

$$\vec{e}^{(n+1)} = [H]\, \vec{e}^{(n)}$$

$$= [H]\Big([H]\, \vec{e}^{(n-1)}\Big)$$

$$\vdots$$

$$= [H]^{n+1}\, \vec{e}^{(0)} \tag{4.13}$$

Now, the eigenvalues and eigenvectors of $[H]$ can be obtained by solving the eigenvalue problem,

$$[H]\vec{\xi} = \lambda\vec{\xi} \tag{4.14}$$

We now assume that the matrix $[H]$, of order M, has M linearly independent eigenvectors, $\vec{\xi}$, for $i=1,...,M$. These eigenvectors can be used as a basis for our M–dimensional vector space, and the arbitrary error vector $\vec{e}^{(0)}$, with its M components, can be expressed uniquely as a linear combination of them, namely,

$$\vec{e}^{(0)} = \sum_{i=1}^{M} c_i\vec{\xi}_i \tag{4.15}$$

where the c_i are the scalar weighting values associated with each eigenvector. Using Eq. (4.15), we can express the error at iteration level 1, denoted $\vec{e}^{(1)}$, as

$$\vec{e}^{(1)} = [H]\, \vec{e}^{(0)} = [H]\left[\sum_{i=1}^{M} c_i\vec{\xi}_i\right]$$

$$= \sum_{i=1}^{M} c_i\Big([H]\vec{\xi}_i\Big) \tag{4.16}$$

$$= \sum_{i=1}^{M} c_i\lambda_i\vec{\xi}_i$$

Similarly, we can express the error at the n+1 the iteration level, $\vec{e}^{(n+1)}$, as

$$\vec{e}^{(n+1)} = [H]^{n+1}\, \vec{e}^{(0)} = [H]^{n+1}\left[\sum_{i=1}^{M} c_i\vec{\xi}_i\right]$$

$$= \sum_{i=1}^{M} c_i[H]^{n}\Big([H]\vec{\xi}_i\Big)$$

$$= \sum_{i=1}^{M} c_i[H]^{n}\Big(\lambda_i\vec{\xi}_i\Big)$$

$$= \sum_{i=1}^{M} c_i\lambda_i[H]^{n-1}\Big([H]\vec{\xi}_i\Big)$$

$$\vdots$$

$$= \sum_{i=1}^{M} c_i \lambda_i^{n+1} \vec{\xi}_i \tag{4.17}$$

Hence, if $|\lambda_i| < 1$ for all i, then

$$\lim_{n \to \infty} \vec{e}^{(n+1)} = \lim_{n \to \infty} \left[\sum_{i=1}^{M} c_i \lambda_i^{n+1} \vec{\xi}_i \right] = 0 \tag{4.18}$$

i.e., for *arbitrary* $\vec{e}^{(0)}$, the iterative method converges. Also, it should be observed that $\vec{b}$ does *not* affect the convergence.

4.3 Asymptotic Rate of Convergence

Let us order the M eigenvalues associated with matrix H, so that

$$|\lambda_1| > |\lambda_2| \geq |\lambda_3| \geq \cdots \geq |\lambda_M| \tag{4.19}$$

then the error, $\vec{e}^{(n+1)}$, can be expressed as

$$\vec{e}^{(n+1)} = \lambda_1^{n+1} \left\{ c_1 \vec{\xi}_1 + c_2 \left(\frac{\lambda_2}{\lambda_1} \right)^{n+1} \vec{\xi}_2 + \ldots + c_M \left(\frac{\lambda_M}{\lambda_1} \right)^{n+1} \vec{\xi}_M \right\} \tag{4.20}$$

$$\cong \lambda_1^{n+1} c_1 \vec{\xi}_1 \qquad \text{as n becomes large}$$

From this result, we can also see that

$$\vec{e}^{(n)} \cong \lambda_1^{n} c_1 \vec{\xi}_1 \tag{4.21}$$

and therefore

$$\vec{e}^{(n+1)} \cong \lambda_1 \vec{e}^{(n)} \tag{4.22}$$

In terms of the vector norm, we can write

$$\| \vec{e}^{(n+1)} \| \cong \lambda_1 \| \vec{e}^{(n)} \| = \varrho(H) \| \vec{e}^{(n)} \| \tag{4.23}$$

From Eq. (4.23), we can observe that the value of $\varrho(H)$ indicates how fast the iterative scheme will converge. Using Eq. (4.23), we can obtain an estimate of the number of decimal digits by which the error is decreased by each convergent iteration. We obtain,

$$\begin{pmatrix} \textit{number of decimal digits of the error} \\ \textit{to be reduced per iteration} \end{pmatrix} = \log_{10}\left(\frac{1}{\varrho}\right) = -\log_{10}\varrho \tag{4.24}$$

To illustrate this point, suppose we want to reduce the size of the error by 10^{-q}. Then the number of iterations needed to do this will be the value of p for which

$$|u_1^p| = \varrho^p \leq 10^{-q} \tag{4.25a}$$

which leads to

$$p \geq \left(\frac{q}{-\log_{10}\varrho} \right) \tag{4.25b}$$

From Eq. (4.25) we can see that p decreases as $(-\log_{10}\varrho)$ increases, i.e. as ϱ decreases. With a sufficiently large value of n, $(-\log_{10}\varrho)$ provides a measure for comparing the rates of convergence of the various iterative methods. The value of $(-\log_{10}\varrho)$ corresponding to the infinity–norm of the matrix H is called the *asymptotic rate of convergence*.

4.4 A Sufficient Condition for Convergence: Diagonal Dominance

In this section we give a more complete discussion of the concept of diagonal dominance which was briefly touched upon in the introduction. We will show here, that if

$$\Theta_{max} = \underset{1 \leq i \leq M}{Max} \sum_{\substack{j=1 \\ j \neq i}}^{M} \frac{|a_{ij}|}{a_{ii}} < 1 \tag{4.26}$$

where a_{ij} and a_{ii} are the elements of the matrix $[A]$, then both the Jacobi iterative method and the Gauss–Seidel iterative method converge. The proof will be given for the Gauss–Seidel method, but can be similarly shown for the Jacobi method. Using the Gauss–Seidel iterative method, the errors at iteration level n can be expressed as

$$e_i^{(n)} = -\sum_{j=1}^{i-1} \frac{a_{ij}}{a_{ii}} e_j^{(n)} - \sum_{j=i+1}^{M} \frac{a_{ij}}{a_{ii}} e_j^{(n-1)} \tag{4.27}$$

where $e_i^{(n)} = u_i - u_i^{(n)}$. Now, let $\varrho_i = \sum_{\substack{j=1 \\ j \neq i}}^{M} \frac{|a_{ij}|}{|a_{ii}|}$ and $\varrho = \underset{1 \leq i \leq M}{Max} \varrho_i$. Using these

definitions, and Eq. (4.27), we can write the following for the error component e_1

$$|e_1^{(n)}| \leq \sum_{j=2}^{M} \left| \frac{a_{1j}}{a_{11}} \right| |e_j^{(n-1)}|$$

$$\leq \| \vec{e}^{(n-1)} \|_\infty \sum_{j=2}^{M} \left| \frac{a_{1j}}{a_{11}} \right| \tag{4.28}$$

$$\leq \varrho \| \vec{e}^{(n-1)} \|_\infty$$

where

$$\| \vec{e}^{(n-1)} \|_{\infty} = \underset{1 \le j \le M}{Max} \, e_j^{(n-1)} \tag{4.29}$$

Now, suppose that

$$|e_l^{(n)}| \le \varrho \, \| \vec{e}^{(n-1)} \|_{\infty} \, , \quad l = 1,...,i-1 \tag{4.30}$$

then

$$|e_i^{(n)}| \le \sum_{j=1}^{i-1} \left| \frac{a_{ij}}{a_{ii}} \right| \cdot |e_j^{(n)}| + \sum_{j=i+1}^{M} \left| \frac{a_{ij}}{a_{ii}} \right| \cdot |e_j^{(n-1)}|$$

$$\le \varrho \, \| \vec{e}^{(n-1)} \|_{\infty} \sum_{j=1}^{i-1} \left| \frac{a_{ij}}{a_{ii}} \right| + \| \vec{e}^{(n-1)} \|_{\infty} \sum_{j=i+1}^{M} \left| \frac{a_{ij}}{a_{ii}} \right| \tag{4.31}$$

$$\le \| \vec{e}^{(n-1)} \|_{\infty} \left\{ \sum_{\substack{j=1 \\ j \ne i}}^{M} \left| \frac{a_{ij}}{a_{ii}} \right| \right\} \quad \text{since } |\varrho| < 1$$

$$\le \varrho \, \| \vec{e}^{(n-1)} \|_{\infty} \qquad QED$$

4.5 Analogy Between Iterative Methods for Elliptic Equations and Time Marching Methods for Parabolic Equations

In this section we develop an analogy between the iterative methods which have been presented, and the time–marching methods developed for parabolic equations in Chapter II. We will show that the numerical equations obtained using iterative techniques for solving steady (time independent) elliptic problems, are mathematically identical to those obtained using the techniques previously presented for parabolic problems, and thus, iterative techniques for elliptic problems can be viewed in a similar light as time marching techniques. Recall from Chapter II, that the basic numerical form of the parabolic problem is

$$\Delta U = f(x, y) \tag{4.32}$$

where a forward difference has been used to represent the time derivative term. Now, consider the Jacobi iterative method, written for a typical grid point l,m as follows:

$$u_{l,m}^{n+1} = \frac{1}{4}(u_{l+1,m}^n + u_{l-1,m}^n + u_{l,m+1}^n + u_{l,m-1}^n) - \frac{h^2}{4} f_{l,m} \tag{4.33}$$

Eq. (4.33) can be also be written in the following form:

$$u_{l,m}^{n+1} - u_{l,m}^n = \frac{h^2}{4} \Delta_h u_{l,m}^n - \frac{h^2}{4} f_{l,m} \tag{4.34}$$

where $\Delta_h = \frac{1}{h^2}(\delta_x^2 + \delta_y^2)$. Eq. (4.34) can also be put in the following form:

$$\frac{u_{l,m}^{n+1} - u_{l,m}^{n}}{k} = \Delta_h u_{l,m}^n - f_{l,m} \tag{4.35}$$

where $k = \frac{h^2}{4}$. Therefore, with the Jacobi iterative method, it looks as though we are solving the equation

$$U_t^* = \Delta U^* - f(x,y) \text{ in } \Omega$$
$$U^* = g(x,y) \quad \text{on } \partial\Omega \tag{4.36}$$

which is a parabolic equation. It should be noted here, that with the value of $k = \frac{h^2}{4}$, $r = \frac{k}{h^2} = \frac{1}{4}$, indicating that the scheme presented in Eq. (4.35) is stable for the 2–D problem.

Now, let $\Phi = U^* - U$, the difference of the solutions to Eqs. (4.36) and (4.32). The variable Φ can be interpreted as the error between the solution at the iteration $n+1$, $\vec{u}^{n+1}$, and the final solution $\vec{u}$. Subtracting Eq. (4.32) from Eq. (4.36) yields the following:

$$\Phi_t = \Delta\Phi \text{ in } \Omega$$
$$\Phi(x,y) = 0 \text{ on } \partial\Omega \tag{4.37}$$

For Eq. (4.37), separation of variables says that Φ decays as $e^{-\lambda t}$, where $t = nk = nh^2/4$, and therefore, $U^* - U \sim e^{-\lambda nh^2/4}$. From this relation for the decay rate, we can see that as the grid spacing h goes down, so does the decay rate. This result is consistent with that obtained using a rigorous analysis. In addition, we can also observe, that to obtain the same rate of decay for different grid spacings h, the product nh^2 must be held constant, indicating that the number of iterations required for convergence, n, is proportional to h^{-2}, which means that if one doubles the number of spatial grid points, the required number of iteration for convergence roughly quadruples.

4.6 Convergence Rate of Classical Iterative Methods

Consider again, Eq. (4.13) relating the error at iteration level n to that at level $n+1$.

$$\vec{e}^{(n+1)} = [H] \vec{e}^{(n)}$$

For the 2–D Laplace equation in the rectangle $0 \le x \le a, 0 \le y \le b$, with constant grid spacing $h = \frac{a}{L} = \frac{b}{M}$, the coefficient matrix $[H]$ contains $[(L-1) \times (M-1)]^2$ entries. We will now examine in more detail the convergence characteristics of the three iterative methods previously presented, the Jacobi, Gauss–Seidel, and SOR methods.

4.6.1 Jacobi Method

Since the convergence rate of any iterative method is directly tied to the eigenvalues of its corresponding coefficient matrix [H], this is the point from which we should start. From the eigenvalue problem

$$[H]\vec{\xi} = \lambda\vec{\xi} \tag{4.38}$$

the eigenvalues $\lambda_{l,m}$, where $l = 1,...,L-1$, $m = 1,...,M-1$, are found to be

$$\lambda_{l,m} = \frac{1}{2}\left(\cos\frac{l\pi h}{a} + \cos\frac{m\pi h}{b}\right) \tag{4.39}$$

Recalling that the spectral radius is defined by $\varrho(H_J) = Max|\lambda_{l,m}|$, then for Eq. (4.39), we see that the largest magnitude eigenvalue corresponds to $l=m=1$, and thus the spectral radius is given by

$$\varrho(H_J) = \frac{1}{2}\left[\cos\frac{\pi h}{a} + \cos\frac{\pi h}{b}\right]$$

$$\tag{4.40}$$

$$\cong 1 - \frac{1}{4}\left[\frac{\pi^2}{a^2} + \frac{\pi^2}{b^2}\right]h^2 \quad \text{for small h}$$

From Eq. (4.40), the asymptotic rate of convergence for the Jacobi method is given by

$$-\log_{10}[\varrho(H_J)] \cong -\log_{10}[1 - ah^2], \quad \text{where } a = \frac{1}{4}\left[\frac{\pi^2}{a^2} + \frac{\pi^2}{b^2}\right]$$

$$= ah^2 + O(h^4) \tag{4.41}$$

and thus the asymptotic rate of convergence of the Jacobi iterative method is proportional to h^2.

4.6.2 Gauss–Seidel Method

To determine the convergence characteristics for the Gauss–Seidel iterative method, we follow the same line of thought as for the Jacobi method just presented. For the Gauss–Seidel method, the coefficient matrix [H] is given as follows:

$$[H]_{G-S} = ([I] - [R])^{-1}[S] \tag{4.42}$$

For the 2–D Laplace equation with Dirichlet boundary conditions, the eigenvalues $\lambda_{l,m}$ are found to be

$$\lambda_{l,m} = \frac{1}{4}\left(\cos\frac{l\pi h}{a} + \cos\frac{m\pi h}{b}\right)^2 \tag{4.43}$$

Again, the maximum magnitude eigenvalue corresponds to the values $l=m=1$, resulting in

$$\lambda_{max} = \lambda_{1,1} = \frac{1}{4}\left(\cos\frac{\pi h}{a} + \cos\frac{\pi h}{b}\right)^2$$

$$\cong 1 - \frac{1}{2}\left(\frac{\pi^2}{a^2} + \frac{\pi^2}{b^2}\right)h^2 \tag{4.44}$$

$$= 1 - 2\alpha h^2$$

Thus, the asymptotic rate of convergence for the Gauss–Seidel method is

$$- \log_{10}[\varrho(H_{G-S})] \cong - \log_{10}[1 - 2\alpha h^2]$$

$$= 2\alpha h^2 + O(h^4) \tag{4.45}$$

i.e. $$\left\{- \log_{10}[\varrho(H_{G-S})]\right\} \cong 2\left\{- \log_{10}[\varrho(H_J)]\right\}$$

From the last part of Eq. (4.45) we see that the asymptotic rate of convergence for the Gauss–Seidel iterative method is about twice that of the Jacobi method. The fact that the convergence rate is increased is consistent with the fact that with the Gauss–Seidel method, the most up–to–date information is being used when iterating at each point, which is not the case for the Jacobi iterative method. We will next examine the convergence characteristics of the SOR method to see if a further increase in the convergence rate can be obtained.

4.6.3 *Successive Overrelaxation (SOR) Method*

For this method, we have

$$[H]_{SOR} = \{[I] - \omega[R]\}^{-1}\{(1 - \omega)[I] + \omega[S]\} \tag{4.46}$$

$$where \ \ 1 < \omega < 2 \tag{4.47}$$

Young's theorem for Laplace equation states the following:
(a) The optimum value of ω, ω_0, is

$$\omega_0 = \frac{2}{1 + \left[1 - \varrho([H]_{G-S})\right]^{1/2}} \tag{4.48}$$

(b)The spectral radius for the SOR method is given by

$$\varrho\left([H(\omega_0)]_{SOR}\right) = \omega_0 - 1 \tag{4.49}$$

From the above, we get

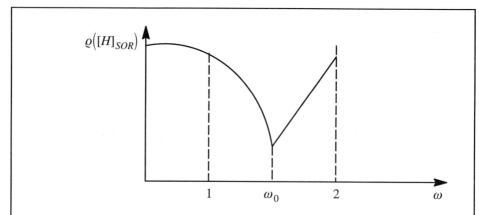

Figure 3. Variation of spectral radius with relaxation factor for the SOR method.

$$\varrho\left(\left[H(\omega_0)\right]_{SOR}\right) = \frac{2}{1 + \left[1 - \varrho\left([H]_{G-S}\right)\right]^{1/2}} - 1 \qquad (4.50)$$

For $a = b = 1$, we have

$$\varrho\left([H]_{G-S}\right) = \cos^2 \pi h \qquad (4.51)$$

and thus Eq. (4.50) yields

$$\varrho\left(\left[H(\omega_0)\right]_{SOR}\right) = \frac{1 - \sin \pi h}{1 + \sin \pi h}$$

$$\cong 1 - 2\pi h \qquad (4.52)$$

Finally, we get

$$- \log_{10}\left\{\varrho\left(\left[H(\omega_0)\right]_{SOR}\right)\right\} \cong 0.4342(2\pi h) \qquad (4.53)$$

which is proportional to h.

The asymptotic rates of convergence for the three iterative methods for $a = b = 1$ can be summarized as follows

$$Jacobi \quad \cong \quad \frac{\pi^2 h^2}{2} \qquad (4.54a)$$

$$Gauss - Seidel \quad \cong \quad 2\left(\frac{\pi^2 h^2}{2}\right) \qquad (4.54b)$$

$$Optimum\ SOR \quad \cong \quad \pi h \qquad (4.54c)$$

Thus, the Gauss–Seidel method increases the rate of convergence by a factor of 2 whereas the optimum SOR method increases it by an order of magnitude compared to the Jacobi method.

4.7 A Simple Acceleration Technique

Consider an iterative method given by Eq. (4.11), i.e.,

$$\vec{u}^{(n+1)} = \vec{b} + [H]\,\vec{u}^{(n)}$$

Define an error vector

$$\vec{e}^{\,n+1} = \vec{u}^{\,n+1} - \vec{u} \qquad\qquad (4.55)$$

where $\vec{u}$ is the exact solution of the finite difference equation. Thus, we can write

$$\vec{e}^{\,n+1} = [H]\,\vec{e}^{\,n}$$
$$\cong \lambda_1 \vec{e}^{\,n} \qquad\qquad (4.56)$$

where λ_1 is the largest eigenvalue.

Now we define another vector using values of $\vec{u}$ at two successive iterations as

$$\vec{\varepsilon}^{\,n+1} = \vec{u}^{\,n+1} - \vec{u}^{\,n} \qquad\qquad (4.57)$$

This vector can be estimated, using Eq. (4.11), as

$$\vec{u}^{\,n+1} - \vec{u}^{\,n} = [H]\left(\vec{u}^{\,n} - \vec{u}^{\,n-1}\right)$$
$$= [H]\left\{\left(\vec{u}^{\,n} - \vec{u}\right) - \left(\vec{u}^{\,n-1} - \vec{u}\right)\right\}$$
$$= [H]\left(\vec{e}^{\,n} - \vec{e}^{\,n-1}\right)$$
$$\cong [H]\left(\vec{e}^{\,n} - \frac{1}{\lambda_1}\vec{e}^{\,n}\right) \qquad\qquad (4.58)$$

The above can be rewritten as

$$\lambda_1 \vec{\varepsilon}^{\,n+1} \cong [H](\lambda_1 - 1)\,\vec{e}^{\,n}$$

or

$$\frac{\lambda_1 \vec{\varepsilon}^{\,n+1}}{(\lambda_1 - 1)} \cong [H]\,\vec{e}^{\,n} \qquad\qquad (4.59)$$

However, we know that

$$\vec{e}^{\,n+1} = [H]\,\vec{e}^{\,n}$$

Thus, we get

$$\vec{e}^{\,n+1} \cong \frac{\lambda_1 \vec{\varepsilon}^{\,n+1}}{\lambda_1 - 1} \qquad\qquad (4.60)$$

From the above we can draw two conclusions:
(a) small differences between successive iterations do not necessarily imply a close approximation to the final equation, and
(b) If λ_1 is close to 1, then the convergence rate is slow, and consequently, $|\vec{\varepsilon}|$ can be quite small.
However, realizing that

$$|\vec{e}^{\,n+1}| \cong \left| \frac{\lambda_1 \vec{\varepsilon}^{\,n+1}}{1 - \lambda_1} \right| \tag{4.61}$$

we can attempt to overshoot the iterative correction, i.e., instead of advancing $\vec{\varepsilon}^{\,n+1}$ at the step $n+1$, we advance $\dfrac{\lambda_1 \vec{\varepsilon}^{\,n+1}}{\lambda_1 - 1}$ (again, if $\lambda_1 \cong 1$, then this is a very large correction).

Since $\vec{e}^{\,n+1} \cong \lambda_1 \vec{e}^{\,n}$ can be quite approximate a formula, it is better that one use Eq. (4.61) once every sufficient number of iterations, i.e., do not overshoot the correction until the influences from other eigenvalues have sufficiently diminished.

4.8 Line or Block Iterative Methods

In these methods, we solve the whole line or block simultaneously and iterate among the lines. Consider the equation

$$U_{xx} + U_{yy} = 0 \tag{4.62}$$
$$0 \leq x \leq a , \quad 0 \leq y \leq b , \quad \textit{Dirichlet B.C.}$$

The iterations can be performed in different manner, as follows:

4.8.1 Jacobi Row Iteration on a Uniform Mesh

For this method, we have the following

$$u_{l,m}^{n+1} = \frac{1}{4}\left\{ u_{l+1,m}^{n+1} + u_{l-1,m}^{n+1} + u_{l,m+1}^{n} + u_{l,m-1}^{n} \right\} \tag{4.63}$$

$$0 \leq l \leq L , \quad 0 \leq m \leq M$$

i.e.,

$$-\frac{1}{4}u_{l+1,m}^{n+1} + u_{l,m}^{n+1} - \frac{1}{4}u_{l-1,m}^{n+1} = \frac{1}{4}\left\{ u_{l,m+1}^{n} + u_{l,m-1}^{n} \right\}$$

The whole set of simultaneous equations is solved from $l=1$ to $l=L-1$, $l=0$ and L being the boundary points. The iterations are performed along the m–direction.

4.8.2 Gauss–Seidel Iteration

For the Gauss–Seidel iteration, we have

$$u_{l,m}^{n+1} = \frac{1}{4}\left\{ u_{l+1,m}^{n+1} + u_{l-1,m}^{n+1} + u_{l,m+1}^{n} + u_{l,m-1}^{n+1} \right\} \tag{4.64}$$

As a natural extension, we can also have a line SOR procedure given by the following:

$$u_{l,m}^{n+1} = u_{l,m}^{n} + \omega\left[\left(u_{l,m}\right)_{G-S}^{n+1} - u_{l,m}^{n} \right] \tag{4.65}$$

where $\left(u_{l,m}\right)_{G-S}^{n+1}$ is the value of $u_{l,m}^{n+1}$ obtained using the Gauss–Seidel row iteration method.

4.8.3 Asymptotic Rate of Convergence

For $a = b = 1$, the asymptotic rates of convergence for the various methods are as follows:

(a) line Jacobi $\quad\quad = 2 \times$ (point Jacobi)
(b) line Gauss–Seidel $= 2 \times$ (point Gauss–Seidel)
(c) Line SOR $\quad\quad = \sqrt{2} \times$ (point SOR)

However, the rate of convergence can vary if $a \neq b$, i.e., if the aspect ratio of the mesh is not 1, then the convergence rate of the line iteration method depends on the direction of the iterative sweep.

4.9 Effect of Grid Aspect Ratio on Convergence Rate

The grid aspect ratio is defined as

$$\gamma = \frac{(\Delta y)^2}{(\Delta x)^2} \tag{4.66}$$

If we use line Gauss–Seidel relaxation method, then sweeping from one row to the next yields

$$\gamma\left(u_{l-1,m}^{n+1} + 2u_{l,m}^{n+1} + u_{l+1,m}^{n+1}\right) + \left(u_{l,m-1}^{n+1} - 2u_{l,m}^{n+1} + u_{l,m+1}^{n}\right) = 0 \tag{4.67}$$

Performing von Neumann stability analysis, i.e., if we let

$$u_{l,m}^{n} \sim \varrho^{n} \, e^{\,i(\beta_1 l h_1 + \beta_2 m h_2)}$$

where $h_1 = \Delta x$ and $h_2 = \Delta y$, then we have

$$\gamma\left(\varrho\, e^{\,-i\beta_1 h_1} - 2\varrho + \varrho\, e^{\,i\beta_1 h_1}\right) + \left(\varrho\, e^{\,-i\beta_2 h_2} - 2\varrho + e^{\,i\beta_2 h_2}\right) = 0$$

which yields

$$\varrho = \frac{e^{\,i\beta_2 h_2}}{2\gamma\left(1 - \cos\beta_1 h_1\right) + 2 - e^{\,-i\beta_2 h_2}} \tag{4.68}$$

Depending on the value of γ, we have the following two cases:

(a) If $\gamma \gg 1$, then ϱ is generally smaller than 1, resulting in a good convergence rate.
(b) If $\gamma \ll 1$, then under the condition that $\cos\beta_1 h_1 = 0$ and $\cos\beta_2 h_2 = 1$, we have

$$|\varrho| = \left|\frac{1}{2\gamma + 1}\right| \cong 1 - O(\gamma) \tag{4.69}$$

which is not good at all.

Thus, the line relaxation method is desirable if we solve along the lines with large coefficients.

4.10 First–Order Derivatives

For elliptic equations with strong first–derivative terms, a modification of the relaxation factor is often needed. Consider the following simple linear equation solved by a point iterative method:

$$RU_x = U_{xx} , \qquad R : positive \ constant \tag{4.70}$$

$$U(0) = 0 , \qquad U(1) = 1$$

If we define $R_h = Rh$, then using central differencing for both U_x and U_{xx}, we get

$$\frac{1}{2}\left(1 + \frac{1}{2}R_h\right)u_{i-1} - u_i + \frac{1}{2}\left(1 - \frac{1}{2}R_h\right)u_{i+1} = 0$$

If one uses a SOR type of relaxational iterative method, it turns out that the optimum value of the relaxation factor ω_{opt} for large R_h is no longer between 1 and 2 as in the case of the Laplace equation. This fact can be partially expected if one realizes that when $R_h > 2$, the diagonal dominance can no longer be maintained. Although this is only a sufficient condition, the trend, that the coefficients of u_{i+1} and u_{i-1} are of comparable magnitudes but of opposite signs, indicates some difficulties associated with a standard iterative procedure. In fact, the optimum ω is given by

$$\omega_{opt} \cong \omega_{max} \cong \frac{2}{1 + \frac{R_h}{2}}$$

This formula is different from that of the Laplace equation in more than one way. Not only is ω_0 smaller than 1 for large R_h, but the sensitivity of the spectral radius of the iterative procedure is also much higher on the right side of ω_0, instead of the left side as is the case for the Laplace equation. For more details, the interested reader is referred to Botta & Veldman (1982) and Shyy (1984).

5 STONE'S STRONGLY IMPLICIT PROCEDURE (SIP)

This procedure is a factorization procedure to convert a two–dimensional problem into two one–dimensional problems. It is similar in spirit to ADI for the parabolic equation. But SIP does not, of course, affect the convergent solution.

Consider the system of equations given by

$$[A] \ \vec{u} = \vec{b} \tag{5.1}$$

The SIP replaces the matrix $[A]$ by a modified form $[A+P]$ which can be decomposed into $[L]$ and $[U]$, i.e., $[A+P] = [L] [U]$:

$$[A + P] \ \vec{u}^{n+1} = \vec{b} + [P] \ \vec{u}^n \tag{5.2a}$$

$$[LU] \ \vec{u}^{n+1} = \vec{b} + [P] \ \vec{u}^n \tag{5.2b}$$

The above can be expressed in two steps as follows:

$$[L] \ \vec{v}^{n+1} = \vec{b} + [P] \ \vec{u}^n \qquad forward \ substitution \tag{5.3a}$$

$$[U] \ \vec{u}^{n+1} = \vec{v}^{n+1} \qquad backward \ substitution \tag{5.3b}$$

The following points can be noted regarding the above:

(a) The matrix $[P]$ is selected so that $[L]$ and $[U]$ have only three non–zero diagonals with the principal diagonal of $[U]$ being the unity diagonal.

(b) The elements of $[L]$ and $[U]$ are determined so that the coefficients in the $[A+P]$ matrix in the locations of the non–zero entries of matrix $[A]$ are identical to those in $[A]$.

(c) Overall, two additional non–zero diagonals appear in $[A+P]$. The matrix $[P]$ can be determined from forming the LU product.

6. MULTIGRID METHOD

6.1 Heuristic Analysis

The Jacobi method for solving the equation $U_{xx} = 0$ (an elliptic equation) is equivalent to solving $U_t = U_{xx}$ (a parabolic equation) with $k=h^2/4$. Now, for the equation $U_t = U_{xx}$, the separation of variables method indicates that

$$U \sim e^{\tau t}\, e^{i\omega x} \qquad (6.1)$$

with $\tau = (i\omega)^2 = -\omega^2$. This means that the high wave number components decay faster because the time scale is not a constant but is instead proportional to $1/\omega^2$. This is demonstrated in Fig. 4.

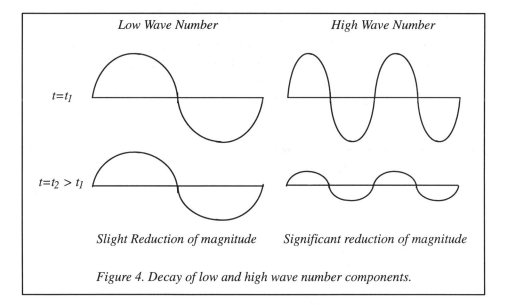

Figure 4. Decay of low and high wave number components.

If we use the Gauss–Seidel method for $U_{xx} = 0$, then we get

$$u_m^{n+1} = \frac{1}{2}\left(u_{m-1}^{n+1} + u_{m+1}^n\right) \tag{6.2}$$

$$u_m^{n+1} - u_m^n = \frac{1}{2}\left[\left(u_{m-1}^{n+1} - u_{m-1}^n\right) + \left(u_{m-1}^n + u_{m+1}^n - 2u_m^n\right)\right]$$

$$U_t|_m = \frac{1}{2}U_t\Big|_{m-1} + U_{xx}|_m , \qquad k = \frac{h^2}{2}$$

$$= \frac{1}{2}\left(U_t|_m - hU_{tx}|_m + ...\right) + U_{xx}|_m \tag{6.3}$$

Therefore, we have

$$\frac{1}{2}(U_t + hU_{tx}) = U_{xx} \tag{6.4}$$

which is a wave equation with U_{tx} term.

6.2 Spectral Analysis

We now present some spectral analysis in the context of the multigrid method. For this purpose, suppose we solve the equation $\nabla^2 U = f(x,y)$ in a square domain $0 \leq x, y \leq 1$. Using the five–point central difference approximation for the Laplacian operator, we get

$$\left(\delta_x^2 + \delta_y^2\right)u_{lm} = h^2 f\left(x_l, y_m\right) \tag{6.5}$$

The above equation is solved iteratively. Let us define the error vector

$$\vec{e}^{\,(n)} = \vec{u} - \vec{u}^{\,(n)} \tag{6.6}$$

If we use the Gauss–Seidel method for iteratively solving the equation, then we have

$$e_{lm}^{n+1} = \frac{1}{4}\left(e_{l-1,m}^{n+1} + e_{l,m-1}^{n+1} + e_{l+1,m}^n + e_{l,m+1}^n\right) \tag{6.7}$$

Consider a Fourier component of the error vector:

$$e_{lm}^n \sim \varrho^n \, e^{\,i(\beta_1 lh + \beta_2 mh)}$$

Thus, for the Gauss–Seidel method we have

$$\varrho\left(4 - e^{-i\beta_1 h} - e^{-i\beta_2 h}\right) = e^{i\beta_1 h} + e^{i\beta_2 h} \tag{6.8}$$

Thus, we have

$$|\varrho| = \left|\frac{e^{i\beta_1 h} + e^{i\beta_2 h}}{4 - e^{-i\beta_1 h} - e^{-i\beta_2 h}}\right|$$

$$= \frac{\sqrt{\left(\cos\beta_1 h + \cos\beta_2 h\right)^2 + \left(\sin\beta_1 h + \sin\beta_2 h\right)^2}}{\sqrt{\left(4 - \cos\beta_1 h - \cos\beta_2 h\right)^2 + \left(\sin\beta_1 h + \sin\beta_2 h\right)^2}} \tag{6.9}$$

We will consider the following two possibilities:

(i) Small Wave Number
The largest wavelength that can be present is 1, i.e., the length of the whole domain. Thus,

$$\beta_1 = \beta_{\min} = \frac{2\pi}{1} = 2\pi$$

which gives us

$$\beta_1 h = 2\pi h \ll 1 \;, \quad since \; h \ll 1 \qquad (6.10a)$$

Likewise $\qquad\qquad \beta_2 h = 2\pi h \ll 1 \qquad\qquad\qquad\qquad (6.10b)$

This yields the following:

$$\cos\beta_1 h = 1 - \mathcal{O}(h^2)$$

$$\cos\beta_2 h = 1 - \mathcal{O}(h^2)$$

$$\sin\beta_1 h = \mathcal{O}(h)$$

$$\sin\beta_2 h = \mathcal{O}(h)$$

Hence

$$|\varrho| = \frac{\sqrt{\left[2 - \mathcal{O}(h^2)\right]^2 + [\mathcal{O}(h)]^2}}{\sqrt{\left[2 + \mathcal{O}(h^2)\right]^2 + [\mathcal{O}(h)]^2}}$$

$$= 1 - \mathcal{O}(h^2) \qquad\qquad\qquad (6.11)$$

which means that the Fourier components with small wave numbers decay very slowly.

(ii) Large Wave Number
The smallest wavelength resolvable in the numerical scheme is $2h$ where h is the mesh spacing. Thus, we have

$$\beta_1 = \beta_2 = \beta_{\max} = \frac{2\pi}{2h} = \frac{\pi}{h} \qquad\qquad (6.12)$$

and so

$$\cos\beta_1 h = \cos\beta_2 h = -1$$

$$\sin\beta_1 h = \sin\beta_2 h = 0$$

which yields

$$|\varrho| = \frac{1}{3} \qquad\qquad\qquad (6.13)$$

The above indicates that the Fourier components with large wave numbers decay fast.

The next Fourier component, beginning from the largest wave number, has the wavelength $4h$ and so for this component

$$\beta_1 = \beta_2 = \frac{2\pi}{2(2h)} = \frac{\pi}{2h} \qquad (6.14)$$

and

$$\cos\beta_1 h = \cos\beta_2 h = 0$$

$$\sin\beta_1 h = \sin\beta_2 h = 1$$

which yields

$$|\varrho| = \frac{\sqrt{2^2}}{\sqrt{4^2 + 2^2}} = \frac{1}{\sqrt{5}} \qquad (6.15)$$

Jacobi Method

Using the Jacobi method, we get

$$e_{l,m}^{n+1} = \frac{1}{4}\left(e_{l-1,m}^{n} + e_{l+1,m}^{n} + e_{l,m-1}^{n} + e_{l,m+1}^{n}\right) \qquad (6.16)$$

Again considering Fourier components given by

$$e_{l,m}^{n} \sim \varrho^{n} e^{i(\beta_1 lh + \beta_2 mh)}$$

we obtain

$$|\varrho| = \left|\frac{2\cos\beta_1 h + 2\cos\beta_2 h}{4}\right| \qquad (6.17)$$

For small wave numbers, $\beta = 2\pi$ which yields $|\varrho| = 1 - O(h^2)$ indicating the slow decay of the small wave number components.

For the large wave number, we have $\beta_{max} = 2\pi/2h$ which yields $|\varrho| = 1$ which indicates no decay for this component. But, for, say, $l=1$ and $m=1$, we have $e_{l-1,m} = e_{l,m-1} = 0$ (Dirichlet boundary condition). Thus, there can be no error associated with this wave number since $e_{l+1,m} = e_{l-1,m} = 0$ and $e_{l,m+1} = e_{l,m-1} = 0$ for $\beta = 2\pi/2h$.

For the next wave number, we have $\beta = 2\pi/2(2h) = \pi/2h$ which yields $|\varrho| = 0$. For the next one, we have $\beta = 2\pi/2(3h) = \pi/3h$ which yields $|\varrho| = 1/2$ which indicates the fast convergence for high wave numbers.

Some Observations

The following observations can be made:

(a) The errors of high wave number components decay fast and those of low wave numbers decay slowly.

(b) Depending on the grid resolution, a low wave number component in a fine mesh system looks like a high wave number component in a coarse mesh system. Note that the smallest wavelength (i.e., the highest wave number) resolvable is 2h; it depends on the value of h.

    (c) We can design a hierarchy of grid systems by solving the equation on the finest mesh system to eliminate high wave number error components and then use the coarser mesh system to eliminate low wave number error. In this manner, we can improve the computational efficiency substantially.

6.3 Restriction, Prolongation and Scheduling

    The key point in the multigrid method is that a low frequency component on a fine grid becomes a high frequency component on a coarse grid. Consider a sequence of grids denoted by $p=1,...,P$, where the ratio of successive grid sizes, h_p / h_{p-1} is 0.5. The finest grid, i.e., the grid with index P is associated with the system of equations given by

$$[A]_P \vec{u}_P = \vec{b}_P \tag{6.18}$$

The final converged solution is obtained on this (finest) grid but we do not always have to conduct all the computations on the finest grid.

    Let $\vec{u}*$ be an estimated solution of $\vec{u}$ and $\varDelta\vec{u}$ be a correction, i.e.,

$$\vec{u} = \vec{u}* + \varDelta\vec{u} \tag{6.19}$$

Then, Eq. (5.19) can be written as

$$[A]_P \vec{u}_P* + [A]_P \varDelta\vec{u}_P = \vec{b}_P \tag{6.20}$$

The key components of the multigrid method are the so–called
         (i) Restriction step
         (ii) Prolongation step, and
         (iii) Scheduling of Grids

6.3.1 Restriction Step

    The first step is to represent the above correction equation on the next coarser grid system (i.e., the grid indexed $P–1$) as

$$[A]_{P-1} \varDelta\vec{u}_{P-1} + I_{P-1}^P \left([A]_P \vec{u}_P* - \vec{b}_P\right) = 0 \tag{6.21}$$

where I_{P-1}^P is the operator performing the transfer between grids; here the transfer takes place from the fine (P) to the coarse $(P–1)$ grid system. I_{P-1}^P is called the *restriction operator*.

    There are several ways by which the above mentioned inter–grid transfer can take place; some of these methods are described below:

(a) *Injection*: This is the easiest way to achieve the inter–grid transfer; the coarse grid vector simply takes its value directly from the corresponding fine grid point, i.e.,

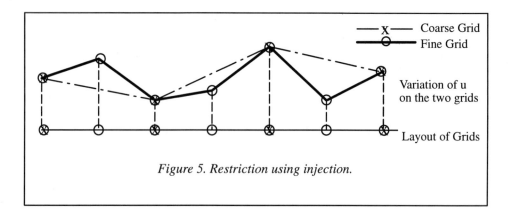

Figure 5. Restriction using injection.

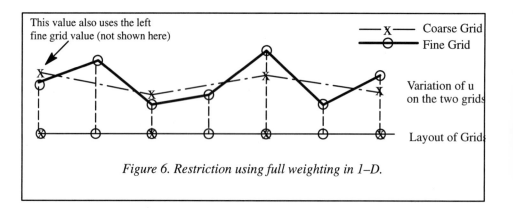

Figure 6. Restriction using full weighting in 1–D.

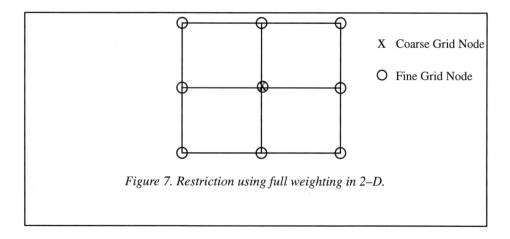

Figure 7. Restriction using full weighting in 2–D.

$$\left(u_j\right)_{P-1} = \left(u_{2j}\right)_P \qquad (6.22)$$

(b) *Full Weighting*: This method interpolates the values of fine grid to obtain the values on the coarse grid. For the one–dimensional case, we get

$$\left(u_j\right)_{P-1} = \frac{1}{4}\left\{\left(u_{2j-1}\right)_P + 2\left(u_{2j}\right)_P + \left(u_{2j+1}\right)_P\right\} \qquad (6.23)$$

For the two–dimensional case, we get

$$\left(u_{ij}\right)_{P-1} = \frac{1}{16}\left\{\left(u_{2i-1,2j-1}\right)_P + \left(u_{2i-1,2j+1}\right)_P + \left(u_{2i+1,2j-1}\right)_P + \left(u_{2i+1,2j+1}\right)_P + \right.$$
$$\left. 2\left[\left(u_{2i,2j-1}\right)_P + \left(u_{2i,2j+1}\right)_P + \left(u_{2i-1,2j}\right)_P + \left(u_{2i+1,2j}\right)_P\right] + 4\left(u_{2i,2j}\right)_P\right\}$$

$$(6.24)$$

6.3.2 Prolongation Step

After the restriction operator has been applied, the next step is to interpolate the correction back to the fine grid

$$\vec{u}_P = \vec{u}_P^* + I_P^{P-1}\left(\Delta\vec{u}_{P-1}\right) \qquad (6.25)$$

where I_P^{P-1} is called the *prolongation* or the *interpolation operator*. This operator transfers the correction $(\Delta u)_{P-1}$ calculated on the $(P-1)^{\text{th}}$ –level (coarse) grid to the Pth–level (fine) grid. Again, there are different ways of doing this, but a simple and effective choice is linear interpolation, i.e.,

$$\left(u_{2j}\right)_P = \left(u_j\right)_{P-1} \qquad Same\ Location \qquad (6.26a)$$

$$\left(u_{2j+1}\right)_P = \frac{1}{2}\left(u_j + u_{j+1}\right)_{P-1} \qquad Mid-Point\ Location \quad (6.26b)$$

6.3.3 Scheduling of Grids

By scheduling of grids we mean the sequence of the grid levels on which the restriction and prolongation operators are applied in each cycle of operations. Of the several sequences that are possible, the two most common are the so–called V–cycle and W–cycle scheduling. These are illustrated in Figs. 8 and 9 and are self–explanatory.

Using the multigrid method, one hopes to achieve a small spectral radius as well as the convergence rate which is independent of the grid size. More detailed account in the context of fluid flow equations is given in Chapter V, Sections 7 and 8.

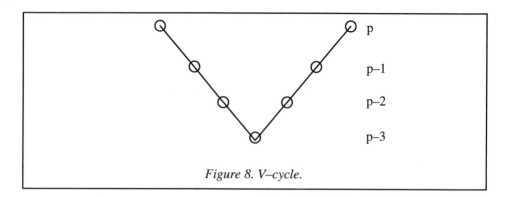

p

p–1

p–2

p–3

Figure 8. V–cycle.

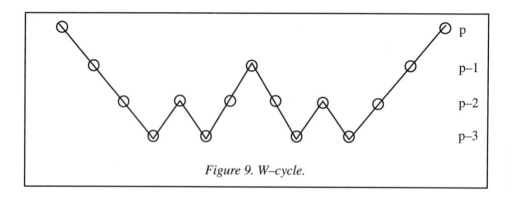

p

p–1

p–2

p–3

Figure 9. W–cycle.

CHAPTER IV
HYPERBOLIC EQUATIONS

1 INTRODUCTION AND ANALYTICAL BACKGROUND

Hyperbolic equations are a special class of equations with a very unique characteristic, namely, they allow wave propagation at finite speed and the occurrence of discontinuities in the solution profiles. Let us consider two test problems involving first–and second–order hyperbolic equations.

1.1 First–Order Hyperbolic Equation

$$U_t + aU_x = 0 , \qquad a = constant > 0 \qquad (1.1a)$$

$$U(0,x) = U_0(x) , \qquad -\infty < x < \infty \qquad (1.1b)$$

The solution to the above equation is given by

$$U(x,t) = U_0(x - at) \qquad (1.2)$$

i.e., U is constant along the characteristic lines which are defined by

$$x - at = C \ (constant) \qquad (1.3)$$

i.e., the value of U at a point P, shown in Fig. 1, is solely dependent on the initial value at the point Q.

1.2 Second–Order Hyperbolic Equation

$$U_{tt} - a^2 U_{xx} = 0 , \qquad a = constant > 0 \qquad (1.4)$$

$$U(0,x) = f(x)$$

$$U_t(0,x) = g(x) , \qquad -\infty < x < \infty$$

The exact solution to this problem is given by

$$U(x,t) = \frac{1}{2}[f(x + at) + f(x - at)] + \frac{1}{2a} \int_{x-at}^{x+at} g(s)ds \qquad (1.5)$$

The domains of influence and dependence for this problem are shown in Fig. 2. The characteristic lines, shown in Fig. 2, can be defined as the curves along which singularities in the solution propagate. These lines describe the edges of the domains of influence and dependence.

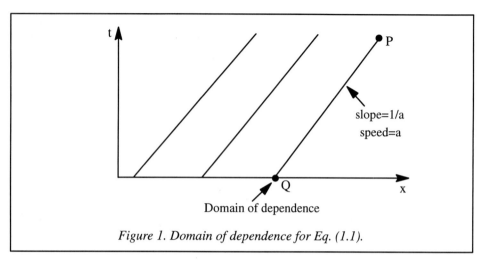

Figure 1. Domain of dependence for Eq. (1.1).

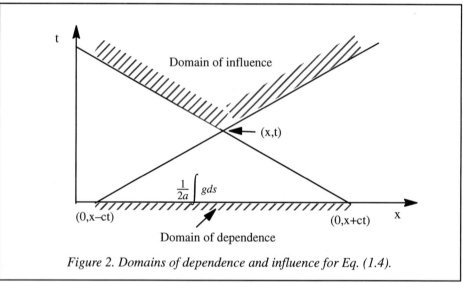

Figure 2. Domains of dependence and influence for Eq. (1.4).

1.3 Conservation Laws and Weak Form Solutions

A conservation law asserts that the rate of change of the total amount of a substance contained in a fixed domain Ω is equal to the "flux" of that substance across the boundary of Ω. Denoting the "density" of that substance by U, and the flux by $\vec{F}$, the conservation law can be written as

$$\frac{d}{dt}\int_{\Omega} U \, dV = -\int_{\partial\Omega} \vec{F} \cdot \hat{n} \, dS \qquad (1.6)$$

where $\hat{n}$ is the outward unit normal vector to the boundary of Ω, denoted as $\partial\Omega$, and dS is the surface element . The integral on the right hand side of Eq. (1.6) measures outflow, and hence it assumes the minus sign. Eq. (1.6) is the integral form of the conservation law. Applying the divergence theorem, we obtain

$$\int_{\Omega} \left(U_t + \nabla \cdot \vec{F} \right) dV = 0 \tag{1.7}$$

which holds for any arbitrary volume element dV. Hence the differential conservation law can be written as

$$\left(U_t + \nabla \cdot \vec{F} \right) = 0 \tag{1.8}$$

Note that in deriving Eq. (1.7) from Eq. (1.6), an implicit assumption is that the the flux F possesses continuous first–derivatives in the domain Ω. Thus, Equations (1.6) and (1.8) are generally equivalent except for the cases in which the solutions contain discontinuous first–derivatives, and hence Eqs. (1.7) and (1.8) do not apply anymore. Such cases are commonly encountered in mathematical description of some "inviscid" phenomena, with shocks and breaking waves being notable examples. For these cases, we need to modify the meaning of the "solution" to the conservation law, since it is no longer differentiable at some locations in the domain. In order to accommodate this situation, we use the integral, instead of the differential, form of the conservation law where less smoothness is required in order for a solution to exist. It is noted that for the first–order hyperbolic equation, only the initial condition is required and in general the spatial domain is unbounded. However, one can employ a finite space to perform the integration in Eq. (1.6) since the conservation law, whether in a weak form or not, must apply in any region. This fact can also be mathematically deduced via some extra manipulation using the notion of weighted integration with the requirement that the weighting function has "compact support". More information can be found in Lax (1973), Whitham (1974), Smoller (1983) and LeVeque (1990).

Now, in devising a numerical scheme to approximate a wave propagation problem, one can nominally use the identical concept to discretize the various terms without considering whether the solutions are "weak" or not. However, this lack of precise distinction can also lead to "qualitatively" erroneous results. For example, consider the simple equation

$$\frac{\partial U}{\partial t} + U \frac{\partial U}{\partial x} = 0 \tag{1.9}$$

The conservation law form of the above equation is

$$\frac{\partial U}{\partial t} + \frac{\partial}{\partial x} \left(\frac{U^2}{2} \right) = 0 \tag{1.10}$$

If we assign the initial conditions

$$U(x,0) = \begin{cases} a & \text{if } -\infty < x < 0 \\ b & \text{if } \quad 0 \le x < \infty \end{cases} \tag{1.11}$$

then the solution is given by

$$U(x,t) = \begin{cases} a & \text{if } x - Ut < 0 \\ b & \text{if } x - Ut > 0 \end{cases} \tag{1.12}$$

with the speed of the discontinuity, V, to be determined. Suppose the discontinuity is contained in the domain $-L \leq x \leq L$, then

$$\frac{\partial}{\partial t} \int_{-L}^{L} U \, dx = -\frac{U^2}{2} \Big|_{x=-L}^{x=L} = -\frac{b^2 - a^2}{2} \tag{1.13}$$

However, it is obvious that

$$\frac{\partial}{\partial t} \int_{-L}^{L} U \, dx = V(a - b) \tag{1.14}$$

since inside the domain the fluid state remains unchanged until the shock runs through it. Hence

$$V = \frac{(b^2 - a^2)/2}{(b - a)} = \frac{[U^2/2]}{[U]} \tag{1.15}$$

where $[\bullet]$ denotes the jump of a quantity across the discontinuity.

If we solve the equation in conservation law form, in the domain defined by the nodes $-(M+1)$ to $(M+1)$ with Δt and Δx as the time and space increments, respectively, then we have the following forward–in–time and central–in–space finite difference equation:

$$\frac{u_m^{n+1} - u_m^n}{\Delta t} + \frac{1}{2\Delta x} \left\{ \frac{(u_{m+1}^n)^2}{2} - \frac{(u_{m-1}^n)^2}{2} \right\} = 0 \tag{1.16}$$

$$-M \leq m \leq M, \quad n \geq 0$$

While the above equation is not appropriate for this problem, as will become clear later, it does satisfy the basic conservation law, as follows:

$$\frac{\partial}{\partial t} \int_{-L}^{L} U \, dx \cong \frac{1}{\Delta t} \left[\sum_{m=-M}^{M} u_m^{n+1} - \sum_{m=-M}^{M} u_m^n \right] \Delta x \tag{1.17a}$$

$$= -\left\{ \frac{(u_{M+1}^n)^2}{4} + \frac{(u_M^n)^2}{4} - \frac{(u_{-M}^n)^2}{4} - \frac{(u_{-M-1}^n)^2}{4} \right\} \tag{1.17b}$$

$$= \frac{1}{2}(a^2 - b^2) \tag{1.17c}$$

The linkage between Eqs. (1.17a) and (1.17b) is via Eq. (1.16), summing over all the nodes. We only have the four terms shown above in Eq. (1.17b) because the differences all "telescope" out except for the nodes around the boundaries. Hence, the integrals and the shock speed can be correctly reproduced by the numerical scheme.

However, if we were to use Eq. (1.9) directly to devise the difference scheme, we would obtain the following:

$$\frac{\partial}{\partial t} \int_{-L}^{L} U \, dx \cong \frac{1}{\Delta t} \left\{ \sum_{m=-M}^{M} u_m^n \left(u_{m-1}^n - u_{m+1}^n \right) \right\} \tag{1.18}$$

The right hand side of Eq. (1.18) obviously does not telescope out to the quantity $(a^2-b^2)/2$, so there is now an error of $\mathcal{O}(1)$ in the shock speed.

2 NAIVE SCHEMES

As usual, we have a wide variety of finite difference schemes to choose from. However, for the case of hyperbolic equations, the selection or design of finite difference schemes is more critical than in the case of elliptic or parabolic problems due to the important reason that hyperbolic equations allow the existence of discontinuities in the solution profile. In this section we demonstrate that certain seemingly satisfactory finite difference schemes turn out to be unsuitable for application to hyperbolic problems. We show this using three finite difference schemes.

Consider the hyperbolic equation given by

$$U_t + aU_x = 0 , \qquad a = constant > 0 \tag{2.1}$$
$$U(nk, mh) = u_m^n$$

The three schemes that we choose to solve the above equation in this section are as follows:
Scheme I: Forward in time and space

$$\left(\frac{1}{k}\Delta_t + \frac{a}{h}\Delta_x \right) u_m^n = 0 \qquad \Rightarrow \ \mathcal{O}(k, h) \tag{2.2}$$

Scheme II: Forward in time and central in space

$$\left(\frac{1}{k}\Delta_t + \frac{a}{2h}\delta_x \right) u_m^n = 0 \qquad \Rightarrow \ \mathcal{O}(k, h^2) \tag{2.3}$$

Scheme III: Forward in time and backward in space

$$\left(\frac{1}{k}\Delta_t + \frac{a}{h}\nabla_x \right) u_m^n = 0 \qquad \Rightarrow \ \mathcal{O}(k, h) \tag{2.4}$$

For all the above three schemes consistency is satisfied in the sense that as $h \to 0$ and $k \to 0$, then $k/h \to$ constant, which means that all of the above schemes are consistent with the physical domain of dependence. At this point it should be noted that for the explicit approximation of parabolic equations, $r=k/h^2$ must be smaller than some critical value in order for the scheme to remain stable, i.e., $k/h \sim h \to 0$ as $k,h \to 0$. This means that for parabolic equations, the domain of dependence is infinite, i.e., all the information from the previous time step affects the solution at the present time step. This is equivalent to saying that for parabolic problems, the speed of signal propagation is infinite.

In order to analyze the stability aspects of the above schemes applied to the hyperbolic equation given by Eq. (2.1), we briefly examine some tools that will assist us in the stability analysis of these schemes, namely

(i) **Min–Max principle**, which is a sufficient condition for stability and is discussed in Chapter II,

(ii) **Von Neumann stability analysis** which is a necessary condition, and for a two–step method without the influence of the boundary conditions (as discussed earlier), it is a sufficient and necessary condition, and

(iii) **CFL (Courant–Friedrichs–Lewy) condition** which states that the (numerical) region of dependence of the finite difference equation, for the line $x=0$ (in the context of the present simple model problem), must contain the (physical) region of dependence of the differential equation. If this condition is not satisfied, then signals from some portion, however small, of the physical domain of dependence, will not influence the numerical solution, making it unstable. This condition is also a necessary condition for stability.

2.1 Scheme I: Forward in Time and Space

This scheme, defined by Eq. (2.2), can be rewritten as

$$u_m^{n+1} = (1 + \lambda) u_m^n - \lambda u_{m+1}^n \tag{2.5}$$

where $\lambda = ak/h$ is defined as the CFL number or Courant number. Depending on the sign of a, we have two possibilities, as discussed below.

2.1.1 Case I: $a > 0$

If $a > 0$, then the above scheme is called the downwind scheme and for this case, a stability analysis yields the following:

(i) Min–Max principle: This principle pronounces this scheme to be always unstable.

(ii) von Neumann analysis yields the following:

$$u_m^n \sim \varrho^n e^{i\beta mh} \tag{2.6a}$$

$$\therefore \quad \varrho = (1 + \lambda) - \lambda e^{i\beta h}$$

$$|\varrho| = \sqrt{[1 + \lambda(1 - \cos \beta h)]^2 + (\lambda \sin \beta h)^2} \tag{2.6b}$$

$$> 1 \quad \text{for } all \ \lambda \neq 0 .$$

which again indicates that the scheme is always unstable. The situation is illustrated in Fig. 3.

(iii) CFL condition: The situation for the CFL condition is illustrated in Fig. 4. It can be seen that this scheme will never satisfy the CFL condition and hence the scheme is always unstable.

2.1.2 Case II: $a < 0$

If $a < 0$, the scheme is called the upwind scheme and stability analysis yields the following:

(i) Min–Max principle: According to this principle, the scheme is stable if $|\lambda| < 1$.

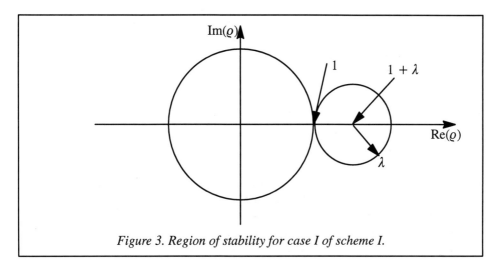

Figure 3. Region of stability for case I of scheme I.

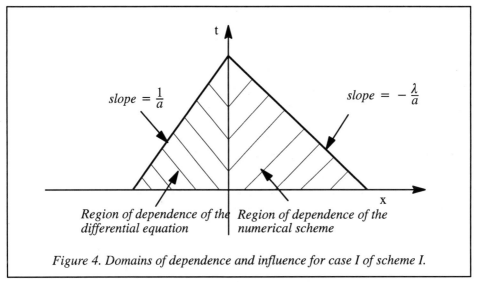

Figure 4. Domains of dependence and influence for case I of scheme I.

(ii) von Neumann analysis: It yields the following:

$$\varrho = (1 + \lambda) - \lambda\, e^{i\beta h} \tag{2.7}$$

which indicates that the region of stability now is defined by a circle whose center is located at $(1+\lambda)$ and the radius is equal to λ, as shown in Fig. 5. The above gives us the following:

$$|\varrho| = \sqrt{[1 + \lambda(1 - \cos\beta h)]^2 + (\lambda \sin\beta h)^2} \tag{2.8}$$

from which it can be again deduced that this scheme is stable if $|\lambda| \le 1$.

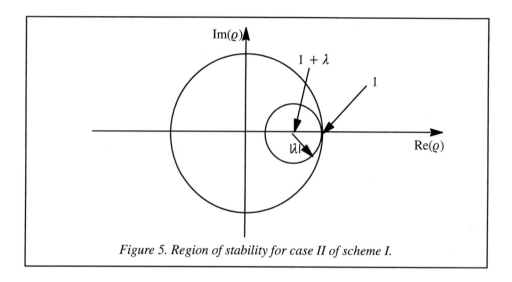

Figure 5. Region of stability for case II of scheme I.

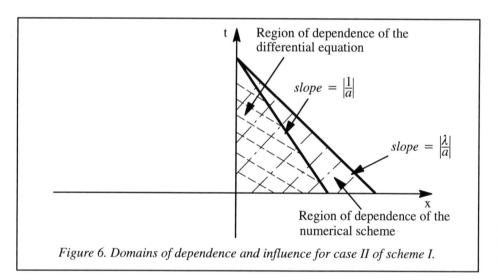

Figure 6. Domains of dependence and influence for case II of scheme I.

(iii) CFL condition: The situation according to the CFL condition is illustrated in Fig. 6. If $|\lambda| = 1$, then the numerical characteristic line is identical to the differential characteristic line, i.e.

$$u_m^{n+1} = u_{m+1}^n \tag{2.9}$$

With the present scheme under consideration, problem arises when "a" is not a constant or when the equation is not linear. In such a situation, no constant value of k and h can fit the exact characteristic lines.

2.2 Scheme II: Forward in Time and Central in Space

This scheme, given by Eq. (2.3), can be rewritten as follows:

$$u_m^{n+1} = u_m^n + \frac{\lambda}{2}u_{m-1}^n - \frac{\lambda}{2}u_{m+1}^n \qquad (2.10)$$

Stability Analysis

As for Scheme I, we can do the following analyses:

(i) Min–Max principle: according to this, the scheme is unstable for either $\lambda > 0$ or $\lambda < 0$.

(ii) von Neumann analysis: this yields the following:

$$\varrho = 1 + \frac{\lambda}{2}\left(e^{-i\beta h} - e^{i\beta h}\right)$$

$$= 1 - \lambda \, i \, \sin\beta h \qquad (2.11)$$

which indicates that $|\varrho| > 1$ for all $\lambda \neq 0$ and hence the scheme is always unstable.

(iii) CFL condition: According to the CFL condition, the scheme is stable as long as $|\lambda| < 1$. Thus, the CFL condition does not guarantee stability for this scheme.

2.3 Scheme III. Forward in Time and Backward in Space

This scheme is given by Eq. (2.4). Doing analyses similar to above, the stability of this scheme can be summarized as follows:

(a) stable if $a > 0$ and $\lambda \leq 1$

(b) unstable if $a < 0$ or $\lambda > 1$.

3. MORE COMPLICATED SCHEMES

3.1 Courant–Isaacson–Rees (CIR) Scheme

If we combine Schemes I and III described above, the resulting scheme is called the Courant–Isaacson–Rees (CIR) scheme. If we approximate the differential equation

$$U_t + aU_x = 0 \qquad (3.1)$$

by the CIR scheme, then we get the following:

$$\left(\frac{1}{k}\Delta_t + \frac{a^+}{h}\nabla_x + \frac{a^-}{h}\Delta_x\right)u_m^n = 0 \qquad (3.2)$$

where

$$a^+(t,x) = \begin{cases} a & \text{if } a \geq 0 \\ 0 & \text{if } a < 0 \end{cases} \qquad (3.3a)$$

$$a^-(t,x) = \begin{cases} 0 & \text{if } a \geq 0 \\ a & \text{if } a < 0 \end{cases} \qquad (3.3b)$$

Thus, in the CIR scheme, the $\dfrac{a^+}{h}\nabla_x$ operator simulates the wave moving towards the right whereas the $\dfrac{a^-}{h}\Delta_x$ operator simulates the wave moving towards the left. Thus, the CIR scheme essentially amounts to always employing the upwind operator.

In this section, we discuss some schemes which are more complicated in their original concept than the schemes discussed in the previous section.

3.2 Lax–Friedrichs Scheme

In this scheme, u_m^n is replaced by $\dfrac{u_{m+1}^n + u_{m-1}^n}{2}$ and the U_t term can be expressed as:

$$U_t = \frac{u_m^{n+1} - \frac{1}{2}\left(u_{m+1}^n + u_{m-1}^n\right)}{k} + \mathcal{O}(k , h^2) \tag{3.4}$$

while the U_x term is approximated by the central difference operator and the resulting scheme can be expressed as follows:

$$\frac{u_m^{n+1} - \frac{1}{2}\left(u_{m+1}^n + u_{m-1}^n\right)}{k} + \frac{a}{2h}\left(u_{m+1}^n - u_{m-1}^n\right) = 0 \tag{3.5}$$

which can be further written as

$$u_m^{n+1} = \frac{1 - \lambda}{2}u_{m+1}^n + \frac{1 + \lambda}{2}u_{m-1}^n \tag{3.6}$$

Stability Analysis

As for the schemes in the previous section, we can do a stability analysis for the present scheme by three methods:

(i) Min–Max Principle: The maximum principle for this scheme yields the condition that the scheme is stable if $|\lambda| \le 1$.

(ii) von Neumann Analysis: As usual, if we substitute

$$u_m^n \sim \varrho^n e^{i\beta mh} \tag{3.7}$$

into the above scheme, then we get

$$\begin{aligned} \varrho &= \frac{1 - \lambda}{2}e^{i\beta h} + \frac{1 + \lambda}{2}e^{-i\beta h} \\ &= \cos\beta h - i\lambda \sin\beta h \end{aligned} \tag{3.8}$$

which yields

$$\begin{aligned} |\varrho| &= \sqrt{(\cos\beta h)^2 + (\lambda \sin\beta h)^2} \\ &\le 1 \quad \text{if } \lambda \le 1 . \end{aligned} \tag{3.9}$$

(iii) CFL Condition: According to the CFL condition the scheme is stable if $|\lambda| \le 1$.

The truncation error of the finite difference equation for this scheme representing the original PDE is $\mathcal{O}(k, h)$, i.e., the scheme is first–order accurate in both the independent variables and is thus not very accurate.

3.3 Lax–Wendroff Scheme

This scheme has been designed in order to achieve second–order accuracy in both h and k. The design philosophy of this scheme rests on the Taylor series expansion of U_m^{n+1} as follows:

$$U_m^{n+1} = U_m^n + kU_t \big|_m^n + \frac{k^2}{2} U_{tt} \big|_m^n + \mathcal{O}(k^3) \tag{3.10}$$

Now, from the original differential equation, we get the relation

$$U_t = -aU_x \tag{3.11}$$

and extending it to higher–order, we get

$$
\begin{aligned}
U_{tt} &= -a(U_x)_t \\
&= -a(U_t)_x \\
&= a^2 U_{xx}
\end{aligned}
\tag{3.12}
$$

Thus, the crux of the Lax–Wendroff scheme is to replace U_t and U_{tt} by spatial derivatives by using the original PDE (as above) and then to approximate these spatial derivatives by second–order accurate finite difference operators. Thus, U_m^{n+1} can now be written as

$$U_m^{n+1} = U_m^n - (ak)U_x \big|_m^n + \frac{(ak)^2}{2} U_{xx} \big|_m^n + \mathcal{O}(k^3) \tag{3.13}$$

and U_x is replaced by

$$\frac{u_{m+1} - u_{m-1}}{2h} + \mathcal{O}(h^2)$$

and U_{xx} by

$$\frac{u_{m+1} - 2u_m + u_{m-1}}{h^2} + \mathcal{O}(h^2)$$

Thus, the final scheme can be expressed as follows:

$$u_m^{n+1} = \tfrac{1}{2}\lambda(\lambda + 1)u_{m-1}^n + (1 - \lambda^2)u_m^n + \tfrac{1}{2}\lambda(\lambda - 1)u_{m+1}^n \tag{3.14}$$

which has a truncation error of $\mathcal{O}(k^2, h^2)$.

Stability Analysis

As earlier we use the following three methods for stability analysis:

(i) *Min–Max Principle:* According to this, the scheme is always unstable.

(ii) *von Neumann Analysis:* This yields the following:

$$|\varrho| = \left| 1 - \lambda^2 + \tfrac{1}{2}(\lambda^2 - \lambda)e^{i\beta h} + \tfrac{1}{2}(\lambda^2 + \lambda)e^{-i\beta h} \right|$$

$$\leq 1 \quad \text{if } \lambda \leq 1. \tag{3.15}$$

(iii) *CFL Condition:* This condition also yields the same condition as obtained by the von Neumann analysis.

3.4 Unified Expression: q–Schemes

The above schemes, used to approximate the original PDE (3.1) can be written as a unified expression as follows:

$$u_m^{n+1} = u_m^n - \frac{\lambda}{2}\left(u_{m+1}^n - u_{m-1}^n\right) + \frac{q}{2}\left(u_{m+1}^n - 2u_m^n + u_{m-1}^n\right) \qquad (3.16)$$

where the expression for q varies with each scheme:

1) *Lax − Friedrichs scheme* : $q = 1$ $\Rightarrow$ *first − order*   (3.17a)
2) *CIR scheme* : $q = |\lambda|$ $\Rightarrow$ *first − order*   (3.17b)
3) *Lax − Wendroff scheme* : $q = |\lambda|^2$ $\Rightarrow$ *second − order*   (3.17c)

Because of the above form of these schemes, they are called "*q*–schemes". The above schemes can be interpreted as solving the following modified equation:

$$U_t + aU_x = \left(\frac{qh^2}{2k}\right)U_{xx} \qquad (3.18)$$

which is a parabolic equation (recall that the original PDE is hyperbolic). The above modified equation is constructed by using the central difference equation as the basis for representing U_x, and thus the q term on the right hand side of the above equation is the effective artificial viscosity introduced by the respective schemes, over and above the central difference scheme for the spatial derivative. This dissipation involving U_{xx} is introduced in these schemes to help stabilize the numerical solution. On account of this artificial dissipation, these schemes are called dissipative schemes and they do not provoke nonlinear instabilities even in the presence of rapid transitions and/or discontinuities in the solution.

3.5 Leapfrog Scheme

This scheme uses central difference operators for both the t– and x–derivatives. Thus, the numerical approximation to the PDE (3.1) is now given by

$$\frac{u_m^{n+1} - u_m^{n-1}}{2k} + a\frac{u_{m+1}^n - u_{m-1}^n}{2h} = 0 \qquad (3.19)$$

It is evident from the above expression that this scheme is a three–level scheme and thus requires a starting procedure. The truncation error of the above scheme with respect to the original PDE is $\mathcal{O}(k^2, h^2)$. The above expression can be further written as

$$u_m^{n+1} = u_m^{n-1} + \lambda u_{m-1}^n - \lambda u_{m+1}^n \qquad (3.20)$$

Stability Analysis

(i) Min–Max principle is more complicated to use due to the three–level nature of this scheme.

(ii) von Neumann analysis yields the following:

$$\varrho^2 + 2i\lambda\varrho \sin\beta h - 1 = 0 \tag{3.21}$$

From this it can be deduced that since the sum of the two roots, ϱ_1 and ϱ_2, of the above equation equals $2i\lambda \sin\beta h$, whose modulus is more than 2 if $|\lambda \sin\beta h| > 1$, then the magnitude of one root must be greater than 1. Thus, the scheme is unstable if $|\lambda| > 1$.

On the other hand, if $|\lambda \sin\beta h| \leq 1$, then we have

$$\varrho = i\lambda \sin\beta h \pm \sqrt{-(\lambda \sin\beta h)^2 + 1} \tag{3.22a}$$

$$|\varrho| = \sqrt{(\lambda \sin\beta h)^2 + \left(\sqrt{-(\lambda \sin\beta h)^2 + 1}\right)^2} \tag{3.22b}$$

Thus, if $|\lambda| \leq 1$, then $|\varrho| = 1$. This means that the initial error in the numerical solution will not decay. Thus, the leapfrog scheme is expected to behave satisfactorily for linear problems with smooth initial conditions but for nonlinear problems it can give meaningless results.

3.6 Crank–Nicolson Scheme for $U_t + aU_x = 0$

If we use the Crank–Nicolson scheme for the time–derivative, then the numerical approximation to the original PDE can be written as follows:

$$u_m^{n+1} = u_m^n - \frac{1}{2}\left\{\frac{\lambda}{2}\left(u_{m+1}^n - u_{m-1}^n\right) + \frac{\lambda}{2}\left(u_{m+1}^{n+1} - u_{m-1}^{n+1}\right)\right\} \tag{3.23}$$

3.6.1 von Neumann analysis

For the above numerical scheme, we get

$$\varrho = \frac{1 - i\frac{\lambda}{2}\sin\beta h}{1 + i\frac{\lambda}{2}\sin\beta h} \tag{3.24a}$$

$$|\varrho| = 1 \quad \text{for } all\ \lambda . \tag{3.24b}$$

Thus, the scheme is unconditionally stable but the initial error will not decay; in other words, the scheme is neutrally stable.

3.6.2 Round–off Analysis

For an explicit scheme, the round–off error will simply accumulate from the level nk to $(n+1)k$. Here, since the scheme is implicit, we are concerned about the round–off error created at the same time level, due to the solution of coupled equations. Consider the possibility that the solution u_m^{n+1} computed at the time level $(n+1)$ consists of the exact solution of the finite difference equation plus some round–off error, denoted by $\varepsilon_m^{n+1} \equiv \varepsilon_m$. Here ε_m must satisfy the homogeneous form of Eq. (3.23), i.e.,

$$\lambda\varepsilon_{m+1} + 4\varepsilon_m - \lambda\varepsilon_{m-1} = 0 \tag{3.25}$$

The characteristic equation corresponding to the above is

$$-\lambda q^2 + 4q + \lambda = 0 \tag{3.26}$$

whose roots q_1 and q_2 satisfy the following relations

$$q_1 q_2 = 1 \quad \text{and} \quad q_1 + q_2 = \frac{4}{\lambda} \tag{3.27}$$

according to which one of the two roots must be less than 1 and the other greater than 1. Thus, regardless of whether λ is positive or negative, there is always one root which is greater than 1. Thus, the round–off error ε_m which is given by

$$\varepsilon_m = \alpha q_1^m + \beta q_2^m \tag{3.28}$$

will continue to grow along the spatial direction (at the same time level) for both the right– and left–moving waves.

3.7 Box Scheme for $U_t + aU_x = 0$

If we employ the box scheme, which is described in Chapter II, to the differential equation given by Eq. (3.1), we get the following:

$$\frac{1}{2}\left\{ \frac{u_{m+1}^{n+1} - u_{m+1}^{n}}{k} + \frac{u_m^{n+1} - u_m^{n}}{k} \right\} + \frac{1}{2}a\left\{ \frac{u_{m+1}^{n+1} - u_m^{n+1}}{h} + \frac{u_{m+1}^{n} - u_m^{n}}{h} \right\} = 0 \tag{3.29}$$

which can be further written as

$$(1 + \lambda)\, u_{m+1}^{n+1} + (1 - \lambda)\, u_m^{n+1} = (1 - \lambda)\, u_{m+1}^{n} + (1 + \lambda)\, u_m^{n} \tag{3.30}$$

3.7.1 von Neumann Analysis

Performing the von Neumann stability analysis for the above scheme, we obtain

$$(1 + \lambda)\, \varrho e^{i\beta h} + (1 - \lambda)\, \varrho = (1 - \lambda)\, e^{i\beta h} + (1 + \lambda) \tag{3.31}$$

which gives

$$\varrho = \frac{(1 + \lambda) + (1 - \lambda)\, e^{i\beta h}}{(1 - \lambda) + (1 + \lambda)\, e^{i\beta h}} \tag{3.32a}$$

$$\therefore \quad |\varrho| = 1 \quad \text{for all } \lambda . \tag{3.32b}$$

Thus, this scheme is neutrally stable.

3.7.2 Round–off Analysis

Performing the analysis similar to the Crank–Nicolson scheme, we get the following:

$$(1 + \lambda)\, \varepsilon_{m+1} + (1 - \lambda)\, \varepsilon_m = 0 \tag{3.33}$$

which gives

$$\varepsilon_m = \alpha q^m , \quad q = \frac{\lambda - 1}{\lambda + 1} \tag{3.34}$$

According to the expression above, we have the following two cases:

(i) If $\lambda > 0$, then $|q| < 1$. Thus ε decays as m increases, i.e., the round–off error decreases from left to right which is desirable but ε grows from right to left. This is not a problem, in theory, for an infinite domain. But since computationally we deal with a finite domain, we need to be careful as far as the boundary condition treatment is concerned.

(ii) If $\lambda < 0$, then $|q| > 1$. This case, too, does not cause any problem because ε decays from right to left, which is the direction of wave propagation. Hence, the box method in general performs better than the Crank–Nicolson method.

4 NUMERICAL DISSIPATION AND DISPERSION

4.1 Background

It may be recalled that non–smooth initial and boundary conditions can be imposed on the elliptic and parabolic partial differential equations, but they are smoothed out due to the inherent nature of these systems of equations. Thus, non–smooth initial and boundary conditions are less of an issue for these types of equations, although it is still a relevant issue; recall, for example, that a higher–order scheme for a parabolic equation does not perform well if the initial condition has discontinuities. For hyperbolic partial differential equations, on the other hand, a discontinuity in the initial condition propagates directly into the domain of interest. Thus, the issue for hyperbolic PDEs is how to maintain a sharp solution profile as required by the governing equation and not produce non–physical oscillations. Since the high wave–number modes are not resolvable by the finite spacing of the numerical mesh, one needs to add extra numerical dissipation to eliminate the high wave–number oscillations.

4.2 Monotone Schemes

A finite difference scheme is defined to be monotone if the solution at all time levels is monotonically increasing or decreasing whenever the initial data is monotonically increasing or decreasing in space. For a q–scheme, the general finite difference formula can be expressed as follows:

$$u_m^{n+1} = \sum_{q=-l}^{l} C_q\, u_{m+q}^n \tag{4.1}$$

The requirement for the q–scheme to be monotone is that all C_q's are non–negative, i.e, the scheme should satisfy the min–max principle.

Theorem: A scheme is monotone if and only if $C_q \geq 0$ for all q.

Proof: Consider the difference between the finite difference solutions at two neighboring grid points:

$$u_{m+1}^{n+1} - u_m^{n+1} = \sum C_q u_{m+1+q}^n - \sum C_q u_{m+q}^n$$

$$= \sum C_q \left(u_{m+1+q}^n - u_{m+q}^n \right) \tag{4.2}$$

Now we consider the following two cases:

(a) if $C_q \geq 0$ and $\left(u_{m+q+1}^n - u_{m+q}^n \right)$ has the same sign for all m (i.e., the scheme is monotone), then $\left(u_{m+1}^{n+1} - u_m^{n+1} \right)$ has that sign as well.

(b) If $C_q < 0$ for, say, $q=k$, then if we consider the following initial conditions:

$$u_m^n = 0 , \qquad m \leq k \qquad\qquad (4.3a)$$
$$u_m^n = 1 , \qquad m > k \qquad\qquad (4.3b)$$

then $\left(u_{m+1}^n - u_m^n \right) = 0$ for all m, except $m=k$, where we have $\left(u_{k+1}^n - u_k^n \right) = 1$. Thus, we have the following:

$$u_{k+1}^{n+1} - u_k^n = C_k \left(u_{k+1}^n - u_k^n \right) = C_k < 0 . \qquad\qquad (4.4)$$

which shows that the scheme is no longer monotonic. This completes the proof.

As an example of monotone schemes, the first–order upwind (CIR) scheme described in the previous section is monotone in its region of stability. Lax–Wendroff scheme, on the other hand, is not monotone.

4.3 Godunov's Theorem

Statement: A monotone scheme can not be higher than first–order accurate.
Proof: Consider a q–scheme, for which we have:

$$u_m^{n+1} = \sum_{q=-l}^{l} C_q\, u_{m+q}^n$$

Using a Taylor series expansion, we get

$$U_m^{n+1} = U_m^n + k(U_t)_m^n + \frac{k^2}{2}(U_{tt})_m^n + \dots$$
$$= U_m^n + k(- aU_x)_m^n + \frac{k^2}{2}\left(a^2 U_{xx} \right)_m^n + \dots \qquad (4.5a)$$
$$U_{q+m}^n = U_m^n + qh(U_x)_m^n + \frac{(qh)^2}{2}(U_{xx})_m^n + \dots \qquad (4.5b)$$

Thus, the scheme will produce second–order accuracy if the following hold:

$$\text{(i)} \qquad \sum_{q=-l}^{l} C_q \doteq 1 \qquad\qquad \text{zeroth–order} \qquad (4.6a)$$

$$\text{(ii)} \qquad \sum_{q=-l}^{l} qC_q = -\lambda \qquad\qquad \text{first–order} \qquad (4.6b)$$

$$\text{(iii)} \qquad \sum_{q=-l}^{l} q^2 C_q = \lambda^2 \qquad\qquad \text{second–order} \qquad (4.6c)$$

For a monotone scheme, $C_q \geq 0$, so we can define the following:

$$\alpha_q = \sqrt{C_q} \tag{4.7a}$$
$$\beta_q = q\sqrt{C_q} \tag{4.7b}$$

and the previous expressions can be rewritten as follows:

from (i)
$$\sum_{q=-l}^{l} (\alpha_q)^2 = 1 \tag{4.8a}$$

from (iii)
$$\sum_{q=-l}^{l} (\beta_q)^2 = \lambda^2 \tag{4.8b}$$

from (ii)
$$\left[\sum_{q=-l}^{l} \alpha_q \beta_q\right]^2 = \left(\sum qC_q\right)^2 = (-\lambda)^2 \tag{4.8c}$$

i.e.,
$$\sum(\alpha_q)^2 \sum(\beta_q)^2 = \left(\sum \alpha_q \beta_q\right)^2 \tag{4.9}$$

By Schwartz inequality, we have

$$\sum(\alpha_q)^2 \sum(\beta_q)^2 \geq \left(\sum \alpha_q \beta_q\right)^2 \tag{4.10}$$

In the above, the equality holds only if the following holds:

$$\beta_q = C\,\alpha_q \tag{4.11}$$

where C is a constant. According to the original definition of α_q and β_q, given in Eqs. (4.7a) and (4.7b), this is possible if only one C_q is non–zero since q varies with the grid location. Hence, the scheme will be at most first–order accurate. This completes the proof of Godunov's theorem.

Equation (4.9) can not hold except for $q=-\lambda$, i.e.,

$$u_m^{n+1} = C_{-\lambda}\, u_{m-\lambda}^n \tag{4.12}$$

But $\sum C_q = 1$ which gives us $C_{-\lambda} = 1$, and hence we have

$$u_m^{n+1} = u_{m-\lambda}^n \tag{4.13}$$

4.4 High Resolution Schemes

As shown above, monotone schemes can not be better than first–order accurate. Hence, the CIR scheme is the best scheme possible among all the monotone schemes. Thus, the suppression of spurious oscillations near discontinuities in the solution profile by employing a monotone numerical scheme is accompanied by a loss of accuracy in the overall solution profile and manifests itself in the form of an excessive amount of smearing of profiles near sharp gradients. However, one cannot use a higher–order scheme (which obviously will not possess the monotonicity property) in general, since it will very likely lead to spurious oscillations near discontinuities. The situation is resolved by noting that the condition that a scheme be monotone is a stronger condition than necessary for the scheme to guarantee a suppression of spurious oscillations; a weaker condition is that the so–called *total variation* (Harten, 1983) of a numerical scheme be bounded. An important class of schemes based on this idea is the so–called *TVD (total variation diminishing) schemes* which require that the total variation of the numerical scheme should not increase with time. The crux of this concept is that one does not have to enforce the monotonicity condition on the numerical scheme and thus schemes with higher formal order of accuracy can be designed in order to achieve better accuracy than that obtained by monotone schemes; however, the TVD schemes reduce locally to first–order accuracy near local extrema in the solution profile, in order to maintain the TVD property which is required to suppress spurious oscillations near sharp gradients. This is achieved in general by constructing the TVD scheme as a nonlinear combination of a higher–order scheme (Lax–Wendroff scheme, for example) and a first–order monotone scheme. Thus, the numerical dispersion of the higher–order component of the TVD scheme is suppressed by numerical dissipation of the first–order component in regions where necessary, i.e., in the regions surrounding rapid variations in the solution profile. More information on the theory and the design philosophy of the TVD schemes can be found in Harten (1983, 1984), Hirsch (1990), LeVeque (1990) and Yee (1987) among others.

Of course, TVD is only one of several formal concepts proposed in the literature to treat the convective problems. Besides the CIR scheme, which is the least accurate, the method of Godunov, the Flux Correct Transport (FCT) scheme (Boris and Book 1973, Zalesak 1979), Harten's scheme (1983, 1984), the Monotonic Upwind Schemes of Conservation Laws (MUSCL) scheme by van Leer (1979), and approximate Riemann solvers proposed by Roe (1981) and by Osher (1984) have all made noticeable impact. These approaches, while based on different ideas originally, can often be interpreted on common ground now. Notably, all these achemes share the following properties. First, they do not generate spurious oscillations. They are TVD in the nonlinear scalar case and the constant coefficients system case. Furthermore, they are consistent with the conservation law form and satisfy the entropy inequality (Lax 1973), which guarantees that the numerical weak solutions of the hyperbolic conservation law are physically realizable.

4.5 Relation of Cell Peclet Number with Numerical Dispersion and Dissipation

Consider the one–dimensional , steady–state linear Burgers' equation:

$$aU_x = bU_{xx} , \qquad a, b = constants > 0 \qquad\qquad (4.14)$$

Numerical schemes (to any finite degree of accuracy) introduce numerical diffusion and dispersion in roughly the same manner as physical diffusion and dispersion (Shyy 1985). This may be examined by expanding the finite difference equation corresponding to the

Table I. Coefficients of modified equation for different schemes.

Method	a	q	r	d	e	f
First–order upwind	a	$\dfrac{a\varDelta x^2}{6}$	$\dfrac{a\varDelta x^4}{120}$	$b+\dfrac{a\varDelta x}{2}$	$\dfrac{b\varDelta x^2}{12}\left(\dfrac{P_{e\varDelta x}}{2}+1\right)$	$\dfrac{b\varDelta x^4}{360}\left(\dfrac{P_{e\varDelta x}}{2}+1\right)$
Second–order upwind	a	$-\dfrac{a\varDelta x^2}{3}$	$-\dfrac{7a\varDelta x^4}{60}$	b	$\dfrac{b\varDelta x^2}{12}(-3p_{e\varDelta x}+1)$	$\dfrac{b\varDelta x^4}{360}(-15p_{e\varDelta x}+1)$
Second–order central	a	$\dfrac{a\varDelta x^2}{6}$	$\dfrac{a\varDelta x^4}{120}$	b	$\dfrac{b\varDelta x^2}{12}$	$\dfrac{b\varDelta x^4}{360}$

above partial differential equation in Taylor series to get the original differential equation plus higher–order terms which represent the truncation errors introduced in the course of numerical approximation. The resulting equation is called the modified equation. Table I shows the coefficients of the first six derivatives in the modified equation given by

$$aU_x + qU_{xxx} + rU^V = dU_{xx} + eU^{IV} + fU^{VI} + HOT \tag{4.15}$$

where *HOT* represents the higher–order terms in Taylor series, derived by the combination of one of the finite difference approximations to the convection and the diffusion terms of the linear Burgers' equation, to be described next. The quantity $p_{e\varDelta x}$ is the so–called local cell Peclet number and is defined as follows:

$$P_{e\varDelta x} = \frac{a}{b}\varDelta x \tag{4.16}$$

It represents the ratio of the effects of local convection and diffusion in the numerical scheme.

For the diffusion term in the governing equation (4.14), the standard central differencing is employed. As to the finite difference approximation to the convection terms, the following schemes can be considered:

(a) first–order upwind

$$\left.\frac{\partial(aU)}{\partial x}\right|_i = \frac{a_iU_i - a_{i-1}U_{i-1}}{\varDelta x} + T_c, \quad \text{for } u > 0 \tag{4.17}$$

and likewise for $u < 0$. Here T_c is the truncation error inherent in replacing the convection term with the finite difference approximation.

(b) second–order central differencing

$$\left.\frac{\partial(aU)}{\partial x}\right|_i = \frac{a_{i+1}U_{i+1} - a_{i-1}U_{i-1}}{2\varDelta x} + T_c \tag{4.18}$$

(c) second–order upwind

$$\frac{\partial(aU)}{\partial x}\bigg|_i = \frac{3a_iU_i - 4a_{i-1}U_{i-1} + a_{i-2}U_{i-2}}{2\Delta x} + T_c, \qquad \text{for } u > 0 \quad (4.19)$$

If we expand the dependent variable U in Eq. (4.15) as a Fourier series in the x–direction, we obtain

$$U = \sum_{j=-\infty}^{\infty} a_j\, e^{ik_j x} \qquad (4.20)$$

The Fourier component of the shortest wavelength resolved by a finite difference mesh is of wavelength $l = 2\,\Delta x$. The corresponding wavenumber is $k_{max} = \pi/\Delta x$. The longest wavelength is $l_{max} = L$, which is the total length spanned by the meshes. The corresponding minimum wavenumber is $k_{min} = 2\pi/L$. If we substitute a Fourier component $U_j = a_j e^{ik_j x}$ into Eq. (4.15), then one can easily find that all the schemes in Table I except the first–order upwind give good approximations to Eq. (4.14) for $k = k_{min}$ with the errors proportional to $1/N^2$ $(L = N\,\Delta x)$. On the other hand, for $k = k_{max}$, no scheme can yield reasonable accurate approximations. This may be seen from Table II, which shows the values of α and β for the Fourier components with $k = k_{max}$, for the various schemes, where

$$\alpha = \frac{qU_{xxx} + rU^V}{aU_x} \qquad (4.21a)$$

$$\beta = \frac{(d - b)U_{xx} + eU^{IV} + fU^{VI}}{bU_{xx}} \qquad (4.21b)$$

All the values of α and β are at least of the order of unity; in the region where those high–wavenumber Fourier components are important for the solution, the numerical diffusion and dispersion are by no means negligible. Furthermore, this difficulty cannot be resolved by using formally a higher–order scheme. It is precisely because of this that the TVD schemes, mentioned earlier, are designed such that they reduce to first–order accuracy near local extrema. More discussion in the context of fluid flow problems will be given in Chapter V, Section 6.

Table II. Values of Fourier components for different schemes.

Method	α	β
First–order upwind	−0.83	$0.23\ P_{e\Delta x} - 0.55$
Second–order upwind	−8.07	$-1.59\ P_{e\Delta x} - 0.55$
Second–order central	−.83	−0.55

5 SYSTEM OF EQUATIONS

5.1 Introduction

Until now, we have discussed various finite difference schemes applied to a scalar hyperbolic partial differential equation. In this section, we illustrate the technique by which any finite difference scheme developed for a scalar PDE can be extended to a system of PDEs. Consider the following system of equations:

$$\vec{U}_t + [A]\vec{U}_x = 0 \tag{5.1}$$

where $[A]=(a_{ij})$ is a constant coefficient matrix. By definition of a hyperbolic system, all the eigenvalues of the matrix $[A]$ are real and the set of right eigenvectors is complete. Thus, the hyperbolic nature of the above system guarantees the existence of a similarity transformation given by

$$[T]^{-1}[A][T] = [\Lambda] \tag{5.2}$$

where $[\Lambda]$ is a diagonal matrix consisting of the eigenvalues of the matrix $[A]$ as its diagonal elements. If we define

$$\vec{W} = [T]^{-1}\vec{U} \tag{5.3}$$

then the original equation (5.1) can be written as

$$[T]^{-1}\vec{U}_t + [T]^{-1}[A]\vec{U}_x = 0$$

$$\Rightarrow \quad [T]^{-1}\vec{U}_t + [T]^{-1}[A][T][T]^{-1}\vec{U}_x = 0$$

$$\Rightarrow \quad \vec{W}_t + [\Lambda]\vec{W}_x = 0 \tag{5.4}$$

Eq. (5.4) represents a set of uncoupled PDEs.

5.2 Application of the CIR Scheme to the System of Equations

Now, we will illustrate how any scheme, originally designed for a scalar equation, such as the CIR scheme, can be applied to a system of equations. If we define

$$|\Lambda| = diag\left[|a_{11}|, \ldots, |a_{mm}|\right] \tag{5.5}$$

then we can apply the CIR scheme to each of the decoupled equations with w_i as the independent variable, i.e., $\vec{w} = (w_i)$. Thus, at the mth point, we can write

$$\vec{w}_m^{m+1} = \vec{w}_m^n - \frac{k}{2h}[\Lambda]\left(\vec{w}_{m+1}^n - \vec{w}_{m-1}^n\right) + \frac{k}{2h}|\Lambda|\left(\vec{w}_{m+1}^n - 2\vec{w}_{m+1}^n + \vec{w}_{m-1}^n\right) \tag{5.6}$$

Using the original variable $\vec{u}$, we can rewrite the above as

$$\vec{u}_m^{m+1} = \vec{u}_m^n - \frac{k}{2h}[A]\left(\vec{u}_{m+1}^n - \vec{u}_{m-1}^n\right) + \frac{k}{2h}|A|\left(\vec{u}_{m+1}^n - 2\vec{u}_{m+1}^n + \vec{u}_{m-1}^n\right) \tag{5.7}$$

where

$$|A| = [T] |A| [T]^{-1} \tag{5.8}$$

or, equivalently, we can write

$$\vec{u}_m^{m+1} = \vec{u}_m^n - \frac{k}{h}[A]^+ \left(\vec{u}_m^n - \vec{u}_{m-1}^n\right) - \frac{k}{h}[A]^- \left(\vec{u}_{m+1}^n - \vec{u}_m^n\right) \tag{5.9}$$

where

$$[A]^+ = [T]\,\Lambda^+\,[T]^{-1} = \frac{1}{2}([A] + |A|) \tag{5.10a}$$

$$[A]^- = [T]\,\Lambda^-\,[T]^{-1} = \frac{1}{2}([A] - |A|) \tag{5.10b}$$

$$[A] = [A]^+ + [A]^- \tag{5.10c}$$

$[A]^+$: diagonal matrix with positive elements $\qquad$ (5.10d)

$[A]^-$: diagonal matrix with negative elements $\qquad$ (5.10e)

Following the stability analysis conducted for the scalar equation, the solution is stable if each individual component of the CIR scheme satisfies the CFL condition, i.e,

$$\underset{i}{Max}\, |a_i|\left(\frac{k}{h}\right) \leq 1 \tag{5.11}$$

As an example of application of the above, for the system of inviscid gas dynamics equations, the eigenvalues of the matrix $[A]$ and the similarity transformation matrix $[T]$ are easily obtained.

PART II

PRESSURE–BASED ALGORITHMS AND THEIR APPLICATIONS

This part deals with issues related to computational modeling for fluid flow and transport phenomena at the macroscopic scale. The pressure–based method is emphasized because, first, this method has not been covered in depth in most texts currently available, and, second, it has been well developed and can adequately handle many issues encountered both for the applications demonstrated in this part and for the physical problems considered in Part III. Recent advancement made to improve the performance of the pressure–based algorithm, both qualitatively and quantitatively, is emphasized. The approach taken is to use S.V. Patankar's well known and successful book (Patankar, 1980) as the basis, offering a more updated account of the various elements of the pressure–based algorithm without duplicating the material already discussed in that book. The topics addressed include (i) formulation of the pressure–based algorithm and its generalization to all flow speeds, (ii) choice of the coordinate system and primary velocity variables, (iii) illustrations of adaptive grid solutions, (iv) open boundary treatment and the role of global mass conservation, (v) convection treatment, (vi) convergence and (vii)composite grid. Part II also discusses practical applications; examples discussed include (i) gas–turbine combustor flow, (ii) heat transfer and convection in high pressure discharge lamps, (iii) thermal management under microgravity, and (iv) flow through hydraulic turbines. For those who are mainly interested in the pressure–based algorithms, it is recommended that they first get familiar with the material covered in Part I and Patankar's book before reading Part II.

CHAPTER V
PRESSURE–BASED ALGORITHMS

1 INTRODUCTION

With recent advances in computer hardware, numerical solution algorithms and grid generation schemes, many thermofluid flow problems with both physical as well as geometrical complexities can now be modelled and solved. Categorically, two different approaches are needed to advance the current state–of–the–art of scientific computing in transport processes. One approach tends to simplify the geometry to, say, a square cavity (Ghia *et al.* 1982, DeVahl Davis and Jones 1983, Shyy 1988), or a simple straight channel (Kim *et al.* 1987), so as to concentrate on some basic features of the numerical algorithm and/or the physical mechanism. The goal of these studies is to develop new predicting techniques and fundamental understanding based on idealized conditions. The hope is that by conducting a step–by–step, logically planned research, complexity can be gradually added, once the understanding of simpler cases matures, eventually leading to a satisfactory and complete knowledge of the whole subject of the momentum, heat and mass transport of fluid flows. This approach, while appealing, cannot satisfy some other needs. Firstly, since much of the research in the area of transport processes is motivated by practical engineering need, the gradual, and at times, meandering pace of the progress of the aforementioned approach cannot meet the desire of obtaining some approximate but useful information before the complete knowledge can be acquired. Secondly, at a fundamental level, it is not clear that this type of approach can always be helpful for elucidating some key mechanisms contained by complex flowfields. Recognizing that most transport processes of relevance are characterized by the simultaneous presence of multiple dependent variables, and that their characteristics are often represented by more than one length scale and dictated by the strong nonlinearity of the system, the various physical and chemical mechanisms cannot be treated simply as "modules" that can be added or deleted without affecting qualitatively the characteristics of the others. Often, the mechanisms or processes identified from the study of simpler problems cannot be individually incorporated into those of more complicated ones, because they interact with other mechanisms. Hence, there is a clear need to acquire the capability of computing complex flowfields, containing multiple physicochemical features under conditions of practical importance. It is this second approach of obtaining numerical solutions in practically motivated configurations, often with the aid of some approximate physical models, that is the focus here.

One unique role computational fluid dynamics can play is that, by devising effective methods to graphically display the numerical solutions, many aspects of the flow structure can be carefully studied. This availability of detailed information of the whole flowfield is especially useful for situations in which direct physical measurements are difficult to make. Despite the many potential pitfalls of relying on numerical solutions to study flowfields and transport mechanisms, the philosophy advocated here is that by carefully exercising a well defined comparison between prediction and measurement, the level of the numerical accuracy of a given calculation can in general be established. Hence, based on this assessment, one can further study the various aspects of the flowfields that are not easily amenable to direct experimental measurements. It is this unique capability of the computational tool that can most impact engineering practice. This article reviews the various elements important for conducting successful computations for complex flow problems involving momentum, heat and mass transfer. They often contain the characteristics of dominant convection, recirculating flow and wide distribution of length scales. We are particularly interested in those issues applicable to complicated geometries.

At the outset, a critical issue that must be addressed is the choice of algorithms for solving the system of fluid flow equations involving several dependent variables, which requires the extra consideration of coordinating the coupling and signal propagation among the equations. The existing algorithms to solve the Navier–Stokes equations can be generally classified as density–based methods and pressure–based methods. For these methods, the velocity field is normally specified using the momentum equations. The density–based methods, usually employed for compressible flows, use the continuity equation to specify the density and extract pressure information using the equation of state. The system of equations is usually solved simultaneously. These methods can be extended with modification to low Mach number regimes where the flows are incompressible and hence density has no role to play in determining the pressure field (Merkle and Choi 1987, Hirsch 1990, Withington *et al.* 1991). The pressure based methods, initially developed for incompressible flow regimes, obtain the pressure field via a pressure or a pressure correction equation which is formulated by manipulating the continuity and momentum equations (Patankar 1980). The solution procedure is conventionally sequential in nature, and hence can more easily accommodate a varying number of equations depending on the physics of the problem involved, without the necessity of reformulating the entire algorithm. The pressure–based methods can be extended to compressible flows by taking the dependence of density on pressure, via the equation of state, into account (Rhie 1986, Shyy and Braaten 1988, Shyy *et al.* 1992d). The above delineation of algorithms is a very general one. Many formulations different in specific aspects have been proposed to deal with the Navier–Stokes flows, as reviewed by Gresho (1991). It is useful to discuss the role played by the continuity equation in determining the pressure field. In essence, as is well known, much of the difficulty of obtaining pressure distribution is caused by the fact that while the continuity equation regulates the pressure distribution, it does not explicitly contain pressure variable. However, by manipulating the momentum and continuity equations a second–order equation containing pressure, or pressure correction, as primary dependent variable can be derived (Roache 1972, Patankar 1980). This practice is also at the heart of the pressure–based algorithms. There are varying degrees of difficulty in solving this type pressure equations, and one primary source of difficulty comes from the appropriate prescription of boundary condition. Among the fluid flow equations, the continuity equation, in differential form, does not have an explicit boundary condition. However, in order to facilitate the field computations, appropriate boundary treatments need to be identified. Many algorithms simply adopt some sort of extrapolation schemes for pressure on the boundary of flow domain; such practices lack necessary rigor and may adversely impact the solution accuracy. More satisfactorily, one can follow the same idea used in deriving the pressure (or pressure correction) equation. In other words, one can formally apply, say, the normal component of the momentum equation as the boundary condition for pressure. This will apply on both solid wall and open boundary. However, in a way fundamentally more consistent with the mathematical structure of the original differential equations, as discussed by Patankar(1980) and Shyy(1987b), with the aid of staggered grid, one can set up appropriate procedures to utilize the mass continuity constraint itself to uniquely determine the pressure distribution without resorting to other artificially assigned conditions. This point will be discussed further in Section 5 (Open Boundary Treatment). The pressure–based algorithm pioneered by Patankar and Spalding (Patankar and Spalding 1972, Patankar 1980) and by Gosman and Ideriah (1976) will be emphasized here. Some recent advancements made to improve the performance of the pressure–based algorithm, both qualitatively and quantitatively, will be discussed. We will first give an overview of the computational framework standardized by the work of Patankar and Spalding, and Gosman and coworkers, then discuss some recent progress. The topics that will be addressed include (1) formulation of the pressure–based algorithm and its generalization to all flow speeds, (2) choice of the coordinate system and primary velocity variables, (3) issues of grid layout, (4) open boundary treatment and the role of global mass conservation, (5) convection treatment and (6) convergence.

Reviews of the various aspects of pressure–based computational methods for fluid flow and heat/mass transport problems are available (Peyret and Taylor 1983, Anderson *et al.* 1984, Minkowycz *et al.* 1988, Patankar 1980, 1988, Fletcher 1988, Hirsch 1990, Merkle *et al.* 1992). The emphasis of the present chapter is on some recent developments in handling complex geometry, treating wide flow speed variations, yielding accurate solutions, and producing results efficiently. Examples will be given, based on the research of the author's group, to illustrate the points made in each area. Practical applications will be demonstrated in Chapter VI.

2 PRESSURE–BASED FORMULATION FOR ALL FLOW SPEEDS

2.1 Overview of a Popular Pressure–Based Method

In order to help appreciate the progress made recently, it is useful to give a brief overview of the SIMPLE family of algorithms originated at the Imperial College in the form of a computer code, TEACH (Gosman and Ideriah 1976), and a book (Patankar 1980). Although some concepts were already published (e.g., Harlow and Welch 1965), the philosophy and actual implementation embodied by the work of Patankar and Spalding (1972), Gosman and Ideriah (1976) and Patankar (1980) are the most widely used for the computation of complex transport problems.

The generic algorithm utilized by Gosman and Ideriah and by Patankar and Spalding for solving the partial differential equations of subsonic fluid flow were originally developed for Cartesian or cylindrical coordinates. They contain the basic steps listed below, some of which cause numerical error and slow convergence and for which replacements have been developed. The recent progress made in those areas will be reviewed next.

Staggered grid: The domain of interest is covered with a mesh or grid; however, the variables are not defined at the same mesh points. The staggered grid concept (Harlow and Welch 1965) originally developed at Los Alamos and employed in TEACH and SIMPLE, locates scalar variables such as pressure, density, turbulence kinetic energy and so on at common mesh points. Offset half a mesh spacing to the left and below are respectively, the (axial) u and the (lateral) v velocity points. In three dimensions, w is similarly staggered along the third coordinate.

Control volume formulation: The partial differential equations are integrated over control volumes centered on each grid point. The advantages of the staggered grid are then realized, since the velocity and scalar variables are located exactly where they are needed – at the surfaces of the control volumes, resulting in a more compact difference operator. This formulation also ensures global continuity without the requirement of artificial boundary conditions, which makes it consistent with the mathematical structure of the Navier–Stokes equations.

Discretization schemes: The control volume formulation of the governing equations requires evaluation of the fluxes of the various quantities across the control surfaces. In the TEACH code, the fluxes are evaluated using the "hybrid" scheme, which is based on the solution of the one–dimensional convection–diffusion equation (Spalding 1972). The use of the hybrid scheme causes the loss of one order of accuracy in regions of high (> 2) cell Reynolds or Peclet numbers, often leading to overly diffusive solutions. Furthermore, because the interpolation used in evaluating fluxes is between sets of nodes along Cartesian grid lines, significant errors arise when the flow is skewed with respect to the grid lines (Raithby 1976, DeVahl Davies and Mallinson 1976).

Pressure–Velocity coupling: In the set of fluid flow equations, pressure is the only variable that does not have a governing equation. Furthermore, at low Mach numbers, the pressure is essentially constant and so the continuity equation cannot be solved directly.

The so–called "Semi–Implicit Method for Pressure Linked Equations" (SIMPLE) algorithm (Patankar 1980) is intended to overcome this problem. The essential feature is the replacement of the continuity equation (which does not contain the pressure) with a pressure correction equation, and subsequent sequential manipulations of the velocity field.

Matrix solver: The algebraic equations that are the result of discretization are cast into a tridiagonal form and solved by successive line under–relaxation (SLUR) (Varga 1962) using the tridiagonal matrix algorithm (TDMA).

This outline of the TEACH code introduces several major numerical disadvantages which prevent it from being useful for simulating flows in complicated geometries. These are the use of rectilinear coordinates and inaccurate discretization. The grid distribution is often as important. The boundaries of the flow devices of practical engineering interest may be curved in a complex manner and fitting them with a rectilinear grid is an unnecessarily complicated problem in grid generation. Another very serious disadvantage of Cartesian coordinates is that the grid cannot be clustered locally without affecting the rest of the domain. Thus a solution requiring a non–uniform grid (e.g. concentrated in regions of steep gradient) would lead to a dense grid in regions where it is not needed. Although this aspect can be partially improved with the use of methods such as the composite grid, the desire of having more flexibility in the coordinate system is clearly justifiable.

Obviously, the non–orthogonal curvilinear coordinates can far more easily accomodate the irregular shapes of the flow domain. Furthermore, besides the obvious flexibility of grid distribution, there are additional advantages in curvilinear coordinates. With the algorithm constructed in the same spirit as in Cartesian coordinates, the convergence rate for calculations in curvilinear coordinates is less sensitive to the choice of the underrelaxation factors than for calculations in Cartesian coordinates (Shyy 1985b, Braaten and Shyy 1986b). Furthermore, the number of iterations needed to reach a fixed level of convergence is usually proportional to $\sqrt{N}$ where N is the total number of grid points used. Using the SIMPLE algorithm in Cartesian coordinates, (Patankar 1980), the number of iterations needed is usually proportional to N (Braaten 1985, Vanka 1985). The superior performance of the algorithm in curvilinear coordinates apparently occurs because the coordinates follow the streamline direction of signal propagation more closely. As will be discussed later, without an appropriate grid distribution, there are cases where the numerical algorithm may not be able to converge at all.

In the following, the governing equations are first given in both Cartesian and general non–orthogonal curvilinear coordinates to illustrate the implication of coordinate transformation. Then a finite volume based pressure correction procedure capable of computing flows at all speeds (Shyy *et al.* 1992) is presented.

2.2 Governing Equations

2.2.1 Continuity Equation

The continuity equation in Cartesian coordinates:

$$\frac{\partial \varrho}{\partial t} + \frac{\partial}{\partial x}(\varrho u) + \frac{\partial}{\partial y}(\varrho v) + \frac{\partial}{\partial z}(\varrho w) = 0 \tag{2.1}$$

can be rewritten in (ξ, η, γ) curvilinear coordinates as follows:

$$J \frac{\partial \varrho}{\partial t} + \frac{\partial}{\partial \xi}(\varrho U) + \frac{\partial}{\partial \eta}(\varrho V) + \frac{\partial}{\partial \gamma}(\varrho W) = 0 \tag{2.2}$$

where

$$\begin{aligned}
U &= u\,(y_\eta z_\gamma - y_\gamma z_\eta) + v\,(z_\eta x_\gamma - z_\gamma x_\eta) + w\,(x_\eta y_\gamma - x_\gamma y_\eta) \\
V &= u\,(y_\gamma z_\xi - y_\xi z_\gamma) + v\,(z_\gamma x_\xi - z_\xi x_\gamma) + w\,(x_\gamma y_\xi - x_\xi y_\gamma) \\
W &= u\,(y_\xi z_\eta - y_\eta z_\xi) + v\,(z_\xi x_\eta - z_\eta x_\xi) + w\,(x_\xi y_\eta - x_\eta y_\xi) \\
J &= x_\xi y_\eta z_\gamma + x_\gamma y_\xi z_\eta + x_\eta y_\gamma z_\xi - x_\xi y_\gamma z_\eta - x_\gamma y_\eta z_\xi - x_\eta y_\xi z_\gamma
\end{aligned} \tag{2.3}$$

2.2.2 Momentum Equation in x–Direction

The momentum equation in *x*–direction in Cartesian coordinates:

$$\frac{\partial(\varrho u)}{\partial t} + \frac{\partial}{\partial x}(\varrho uu) + \frac{\partial}{\partial y}(\varrho uv) + \frac{\partial}{\partial z}(\varrho uw) = \frac{\partial}{\partial x}(\mu \frac{\partial u}{\partial x}) + \frac{\partial}{\partial y}(\mu \frac{\partial u}{\partial y}) + \frac{\partial}{\partial z}(\mu \frac{\partial u}{\partial z})$$
$$- \frac{\partial p}{\partial x} + G_1(x, y, z) \tag{2.4}$$

where $G_1(x, y, z)$ is the body force component in the *x*–direction per unit volume. In curvilinear coordinates this becomes:

$$\begin{aligned}
J \frac{\partial(\varrho u)}{\partial t} + \frac{\partial}{\partial \xi}(\varrho Uu) + \frac{\partial}{\partial \eta}(\varrho Vu) + \frac{\partial}{\partial \gamma}(\varrho Wu) &= \frac{\partial}{\partial \xi}[\frac{\mu}{J}(q_{11}u_\xi + q_{12}u_\eta + q_{13}u_\gamma)] \\
&+ \frac{\partial}{\partial \eta}[\frac{\mu}{J}(q_{21}u_\xi + q_{22}u_\eta + q_{23}u_\gamma)] \\
&+ \frac{\partial}{\partial \gamma}[\frac{\mu}{J}(q_{31}u_\xi + q_{32}u_\eta + q_{33}u_\gamma)] \\
&- [\frac{\partial}{\partial \xi}(f_1 p) + \frac{\partial}{\partial \eta}(f_4 p) + \frac{\partial}{\partial \gamma}(f_7 p)] \\
&+ G_1(\xi, \eta, \gamma).J
\end{aligned} \tag{2.5}$$

where *U*, *V*, *W* and *J* are defined in (2.3), and

$$\begin{aligned}
f_1 &= (y_\eta z_\gamma - y_\gamma z_\eta) \\
f_2 &= (z_\eta x_\gamma - z_\gamma x_\eta) \\
f_3 &= (x_\eta y_\gamma - x_\gamma y_\eta) \\
f_4 &= (y_\gamma z_\xi - y_\xi z_\gamma) \\
f_5 &= (z_\gamma x_\xi - z_\xi x_\gamma)
\end{aligned} \tag{2.6}$$

$$f_6 = (x_\gamma y_\xi - x_\xi y_\gamma)$$
$$f_7 = (y_\xi z_\eta - y_\eta z_\xi)$$
$$f_8 = (z_\xi x_\eta - z_\eta x_\xi)$$
$$f_9 = (x_\xi y_\eta - x_\eta y_\xi)$$

(2.6 contd.)

$$q_{11} = (y_\eta z_\gamma - y_\gamma z_\eta)^2 + (z_\eta x_\gamma - z_\gamma x_\eta)^2 + (x_\gamma y_\eta - x_\eta y_\gamma)^2$$

$$q_{22} = (y_\gamma z_\xi - y_\xi z_\gamma)^2 + (z_\gamma x_\xi - z_\xi x_\gamma)^2 + (x_\gamma y_\xi - x_\xi y_\gamma)^2$$

$$q_{33} = (y_\xi z_\eta - y_\eta z_\xi)^2 + (z_\xi x_\eta - z_\eta x_\xi)^2 + (x_\xi y_\eta - x_\eta y_\xi)^2$$

$$q_{12} = (y_\gamma z_\xi - y_\xi z_\gamma)(y_\eta z_\gamma - y_\gamma z_\eta)$$
$$+ (z_\gamma x_\xi - z_\xi x_\gamma)(z_\eta x_\gamma - z_\gamma x_\eta)$$
$$+ (x_\gamma z_\xi - x_\xi z_\gamma)(x_\eta y_\gamma - x_\gamma y_\eta)$$

$$q_{13} = (y_\xi z_\eta - y_\eta z_\xi)(y_\eta z_\gamma - y_\gamma z_\eta)$$
$$+ (z_\xi x_\eta - z_\eta x_\xi)(z_\eta x_\gamma - z_\gamma x_\eta)$$
$$+ (x_\xi y_\eta - x_\eta y_\xi)(x_\eta y_\gamma - x_\gamma y_\eta)$$

(2.7)

$$q_{23} = (y_\xi z_\eta - y_\eta z_\xi)(y_\gamma z_\xi - y_\xi z_\gamma)$$
$$+ (z_\xi x_\eta - z_\eta x_\xi)(z_\gamma x_\xi - z_\xi x_\gamma)$$
$$+ (x_\xi y_\eta - x_\eta y_\xi)(x_\gamma y_\xi - x_\xi y_\gamma)$$

$$q_{21} = q_{12}$$
$$q_{31} = q_{13}$$
$$q_{32} = q_{23}$$

2.2.3 Momentum Equation in y–Direction

The momentum equation in y–direction in Cartesian coordinates:

$$\frac{\partial}{\partial t}(\varrho v) + \frac{\partial}{\partial x}(\varrho u v) + \frac{\partial}{\partial y}(\varrho v v) + \frac{\partial}{\partial z}(\varrho v w) = \frac{\partial}{\partial x}\left(\mu \frac{\partial v}{\partial x}\right) + \frac{\partial}{\partial y}\left(\mu \frac{\partial v}{\partial y}\right) + \frac{\partial}{\partial z}\left(\mu \frac{\partial v}{\partial z}\right)$$
$$- \frac{\partial p}{\partial y} + G_2(x, y, z)$$

(2.8)

where $G_2(x, y, z)$ is the body force component in the y–direction per unit volume. In curvilinear coordinates this becomes:

$$J\frac{\partial}{\partial t}(\varrho v) + \frac{\partial}{\partial \xi}(\varrho U v) + \frac{\partial}{\partial \eta}(\varrho V v) + \frac{\partial}{\partial \gamma}(\varrho W v) = \frac{\partial}{\partial \xi}[\frac{\mu}{J}(q_{11}v_\xi + q_{12}v_\eta + q_{13}v_\gamma)]$$
$$+ \frac{\partial}{\partial \eta}[\frac{\mu}{J}(q_{21}v_\xi + q_{22}v_\eta + q_{23}v_\gamma)]$$
$$+ \frac{\partial}{\partial \gamma}[\frac{\mu}{J}(q_{31}v_\xi + q_{32}v_\eta + q_{33}v_\gamma)]$$
$$- [\frac{\partial}{\partial \xi}(f_2 p) + \frac{\partial}{\partial \eta}(f_5 p) + \frac{\partial}{\partial \gamma}(f_8 p)]$$
$$+ G_2(\xi, \eta, \gamma).J$$

(2.9)

where U, V, W and J are defined in (2.3), f_i are defined in (2.6), and q_{ij} are defined in (2.7).

2.2.4 Momentum Equation in z–Direction

The momentum equation in z–direction in Cartesian coordinates:

$$\frac{\partial}{\partial t}(\varrho w) + \frac{\partial}{\partial x}(\varrho uw) + \frac{\partial}{\partial y}(\varrho vw) + \frac{\partial}{\partial z}(\varrho ww) = \frac{\partial}{\partial x}(\mu \frac{\partial w}{\partial x}) + \frac{\partial}{\partial y}(\mu \frac{\partial w}{\partial y}) + \frac{\partial}{\partial z}(\mu \frac{\partial w}{\partial z})$$

$$- \frac{\partial p}{\partial z} + G_3(x,y,z) \qquad (2.10)$$

where $G_3(x,y,z)$ is the body force component in z–direction per unit volume. In curvilinear coordinates this becomes:

$$J\frac{\partial}{\partial t}(\varrho w) + \frac{\partial}{\partial \xi}(\varrho Uw) + \frac{\partial}{\partial \eta}(\varrho Vw) + \frac{\partial}{\partial \gamma}(\varrho Ww) = \frac{\partial}{\partial \xi}[\frac{\mu}{J}(q_{11}w_\xi + q_{12}w_\eta + q_{13}w_\gamma)]$$

$$+ \frac{\partial}{\partial \eta}[\frac{\mu}{J}(q_{21}w_\xi + q_{22}w_\eta + q_{23}w_\gamma)]$$

$$+ \frac{\partial}{\partial \gamma}[\frac{\mu}{J}(q_{31}w_\xi + q_{32}w_\eta + q_{33}w_\gamma)]$$

$$- [\frac{\partial}{\partial \xi}(f_3 p) + \frac{\partial}{\partial \eta}(f_6 p) + \frac{\partial}{\partial \gamma}(f_9 p)]$$

$$+ G_3(\xi,\eta,\gamma).J \qquad (2.11)$$

where U, V, W and J are defined in (2.3), f_i are defined in (2.6), and q_{ij} are defined in (2.7).

Energy, species and other transport equations can all be written in a similar form; for brevity, they will not be presented here.

With the introduction of curvilinear coordinates, new issues appear and need to be addressed. For example, the formulation of the pressure and the pressure–correction equations must be developed. Furthermore, the choice of the velocity components as the primary dependent variables is no longer straightforward. The extra terms appearing due to the chain rule type of coordinate transformation also needs to be treated and can impact the design of the matrix solver of the linearized equations. There have been an increasing number of studies devoted to these topics (Rhie and Chow 1983, Shyy et al. 1985, Peric 1985, Raithby et al. 1986, Demirdzic et al. 1987, Karki and Patankar 1988, Yang and Yang 1988, Leschziner and Dimitriades 1989, Rodi et al. 1989, Shyy and Vu 1991, Kelkar and Choudhary 1992, Shuen et al. 1992, Yang et al. 1992, Jeng and Chen 1992a). Some of these issues will be addressed in the following.

2.3 Formulation of the Pressure Correction Equation

The steady state equations will be used as the basis for the present discussion. The addition of the unsteady terms does not change the basic framework described here. The general equation for the pressure correction for a flow expressed in a non–orthogonal curvilinear coordinate system can be derived in the following manner. For a guessed pressure field p^*, velocities u^*, v^*, and w^* satisfy equations of the form:

$$u_p^* = \sum \frac{A_{nb}^u}{A_p^u} u_{nb}^* + D^u + B^u p_\xi^* + C^u p_\eta^* + E^u p_\gamma^* \qquad (2.12)$$

Here, the notation used is consistent with that used by Shyy *et al.* (1985), Braaten and Shyy (1987) and Shyy and Vu (1991). The D^u are the coefficients resulting from the coordinate transformation and the source terms, and B^u, C^u, and E^u represent the projected areas acted upon by the pressure gradients in the ξ, η, and γ directions respectively. The corrected velocity u_p is given by the following equation:

$$u_p = \sum \frac{A_{nb}^u}{A_p^u} u_{nb} + D^u + B^u p_\xi + C^u p_\eta + E^u p_\gamma \qquad (2.13)$$

Subtracting equation (2.12) from (2.13) leads to the following complete correction formula for u_p' :

$$u_p' = \sum \frac{A_{nb}^u}{A_p^u} u'_{nb} + B^u p_\xi' + C^u p_\eta' + E^u p_\gamma'$$

$$= \bar{u}_p' + B^u p_\xi' + C^u p_\eta' + E^u p_\gamma' \qquad (2.14)$$

where
$$\bar{u}_p' = \sum \frac{A_{nb}^u}{A_p^u} u'_{nb}$$

The correction formulas for v_p' and w_p' follow similarly and need not be shown here.

The discretized continuity equation can be written in terms of the contravariant velocities U, V, and W in the following finite volume form:

$$(\varrho U)_e - (\varrho U)_w + (\varrho V)_n - (\varrho V)_s + (\varrho W)_f - (\varrho W)_b = 0 \qquad (2.15)$$

Since U, V and W appear naturally in the continuity equation, it is convenient to express the correction formulas for u_p', v_p' and w_p' in terms of the corresponding correction formulas for U_p', V_p' and W_p' . As an example, the correction formula for U' can be written as:

$$U' = A_{1x} u' + A_{1y} v' + A_{1z} w' \qquad (2.16)$$

The correction formulas for V' and W' follow similarly.

The final form of the correction formula for U' becomes:

$$U' = (A_{1x} B^u + A_{1y} B^v + A_{1z} B^w)\, p'_\xi + (A_{1x} C^u + A_{1y} C^v + A_{1z} C^w)\, p'_\eta$$
$$+ (A_{1x} E^u + A_{1y} E^v + A_{1z} E^w)\, p'_\gamma + A_{1x} \bar{u}' + A_{1y} \bar{v}' + A_{1z} \bar{w}' \qquad (2.17)$$

This correction formula and the corresponding formulas for V' and W' are then substituted into the continuity equation, leading to the complete pressure correction equation, which can be written in the following form:

$$a_p p_p' = a_E p_E' + a_W p_W' + a_N p_N' + a_S p_S' + a_F p_F' + a_B p_B'$$

$$+ \sum a_{nb} p_{nb}'$$ (2.18)

$$+ [(\varrho U^*)_w - (\varrho U^*)_e + (\varrho V^*)_s - (\varrho V^*)_n + (\varrho W^*)_b - (\varrho W^*)_f]$$

$$+ [(\varrho \overline{U}')_w - (\varrho \overline{U}')_e + (\varrho \overline{V}')_s - (\varrho \overline{V}')_n + (\varrho \overline{W}')_b - (\varrho \overline{W}')_f]$$

where $$\overline{U}' = A_{1x}\overline{u}' + A_{1y}\overline{v}' + A_{1z}\overline{w}'$$

Here p_{nb}' terms include all of the additional neighboring points introduced as a result of the use of the curvilinear coordinate system, U^*, V^*, W^* represent mass imbalance source terms resulting from the starred velocity field, and U', V', W' terms represent mass imbalance source terms arising from the velocity corrections at the neighboring points. The complete pressure correction equation is virtually intractable in the present form. The equation directly contains links between 27 neighboring points. In addition, since the terms with tildes contain velocity corrections at the neighboring points, which in turn depend on the neighboring pressure corrections, the system of equations represents a totally full matrix. Any attempt to directly solve such a system of equations will require too large a computing resource to be practical.

Several approximations can be made to lead to a suitable form of the pressure correction equation. The approach developed by Shyy *et al.* (1985) and Braaten and Shyy (1986a) follows the spirit of the SIMPLE algorithm as applied in Cartesian coordinates. The velocity corrections, such as U', at the neighboring points are neglected (Patankar 1980). In Shyy *et al.* (1985) it is found that dropping the P_{nb}' terms also leads to a more efficient and stable algorithm because the resulting pressure correction equation (5–point in 2D and 7–point in 3D) is easier to solve. By directly correcting the contravariant velocity components via the D'yakanov iteration procedure (Concus and Golub 1973), an efficient and consistent algorithm is obtained, as will be discussed next.

The introduction of curvilinear coordinates requires extra consideration of pressure and velocity correction procedure. To convert between the contravariant (U, V) and cartesian (u, v) velocity components, care must be taken in ensuring that the mass continuity is satisfied.

To facilitate discussion, the two–dimensional, steady–state continuity equation can be written in a discretized form over each control volume, according to the grid definition given in Fig. 1, as follows:

$$\left(y_\eta \varrho u - x_\eta \varrho v\right)_e - \left(y_\eta \varrho u - x_\eta \varrho v\right)_w + \left(- y_\xi \varrho u + x_\xi \varrho v\right)_n - \left(- y_\xi \varrho u + x_\xi \varrho v\right)_s = 0$$

(2.19)

It is assumed that u^* and v^* are the velocity components that satisfy the momentum equations with a given distribution of p^*. Since u^* and v^* in general will not satisfy the continuity equation, the pressure p^* must be corrected. In the SIMPLE procedure, the corrected pressure p is obtained from

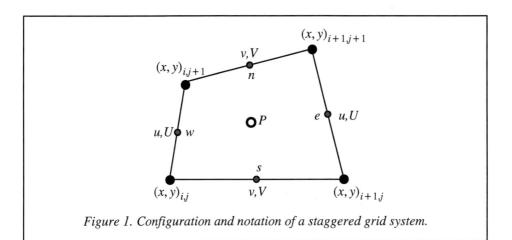

Figure 1. Configuration and notation of a staggered grid system.

$$p = p^* + p' \tag{2.20}$$

where p' is called the pressure correction. The corresponding velocity corrections u' and v' can be introduced in a similar way:

$$u = u^* + u' \qquad\qquad v = v^* + v' \tag{2.21}$$

To derive the pressure correction equation, it is assumed that u^* and v^* are obtained from the momentum equations as follows:

$$u_P^* = \sum_{i=E,W,N,S} \frac{A_i^u}{A_P^u} u_i^* + D^u + \left(B^u p_\xi^* + C^u p_\eta^*\right) \tag{2.22a}$$

$$v_P^* = \sum_{i=E,W,N,S} \frac{A_i^v}{A_P^v} v_i^* + D^v + \left(B^v p_\xi^* + C^v p_\eta^*\right) \tag{2.22b}$$

where A, B, C, and D are coefficients resulting from the discrete operators chosen for the various terms in the momentum equations. The velocity components can be assumed to be corrected by the following formulas:

$$u = u^* + \left(B^u p_\xi' + C^u p_\eta'\right) \tag{2.23a}$$

$$v = v^* + \left(B^v p_\xi' + C^v p_\eta'\right) \tag{2.23b}$$

Subsequently, the corresponding correction forms for U and V are obtained by substituting Eqs. (2.23) into the following equations:

$$U = uy_\eta - vx_\eta \tag{2.24a}$$

$$V = vx_\xi - uy_\xi \tag{2.24b}$$

to yield

$$U = U^* + \left(B^u y_\eta - B^v x_\eta\right)p_\xi' + \left(C^u y_\eta - C^v x_\eta\right)p_\eta' \tag{2.25a}$$

$$V = V^* + \left(B^v x_\xi - B^u y_\xi\right)p_\xi' + \left(C^v x_\xi - C^u y_\xi\right)p_\eta' \tag{2.25b}$$

where U^* and V^* are calculated based on u^* and v^*.

It is noted that the continuity equation, Eq. (2.19), can be rewritten in the following finite–difference form:

$$(\varrho U)_e - (\varrho U)_w + (\varrho V)_n - (\varrho V)_s = 0 \tag{2.26}$$

To retain a five–point approximation in the pressure correction equation, the p_η' term in Eq. (2.25a) and the p_ξ' term in Eq. (2.25b) are dropped, leading to the following simplified correction equations for U and V:

$$U = U^* + \left(B^u y_\eta - B^v x_\eta\right)p_\xi' \tag{2.27a}$$

$$V = V^* + \left(C^v x_\xi - C^u y_\xi\right)p_\eta' \tag{2.27b}$$

These equations are then substituted into Eq. (2.26) to obtain the pressure correction equation.

The pressure corrections should be used to update U and V by using the simplified correction equations, Eqs. (2.27). Provided the pressure correction equation is solved to convergence at each iteration, the resulting U and V fields satisfy continuity exactly. The u and v velocity fields corresponding to U and V are then found in the manner described below.

U and V, the contravariant velocity components, are defined in Eqs. (2.24). Analytically, these relations may be inverted to obtain

$$u = U\frac{x_\xi}{J} + V\frac{x_\eta}{J} \tag{2.28a}$$

$$v = V\frac{y_\eta}{J} + U\frac{y_\xi}{J} \tag{2.28b}$$

Because of the staggered nature of the grid (Fig. 1), it is necessary to interpolate for the value of v when calculating U and likewise to interpolate for the value of u when calculating V. If Eqs. (2.28) were used to calculate u and v from U and V, it causes an inconsistency in the procedure, in that if known u and v fields were used to calculate U and V from Eqs. (2.24), and then Eqs. (2.28) were used to recalculate u and v, the original fields would not be recovered. To avoid this inconsistency, it is necessary to invert the finite–difference forms of Eqs. (2.24), instead of adopting Eq. (2.28) to obtain the values of u and v from U and V.

Equation (2.24) can be rewritten as a system of matrix equations:

$$[M]\vec{q} = \vec{Q} \tag{2.29a}$$

where

$$\vec{q} = \begin{bmatrix} u \\ v \end{bmatrix} \quad \text{and} \quad \vec{Q} = \begin{bmatrix} U \\ V \end{bmatrix} \quad (2.29b)$$

Obtaining the u and v velocity fields from the corrected U and V fields requires finding

$$\vec{q} = [M]^{-1} \vec{Q} \quad (2.30)$$

Since the matrix $[M]$ is large and sparse, and since the u and v fields must be extracted from the U and V fields at the end of each iteration, the use of matrix inversion or a direct solution procedure such as Gaussian elimination is clearly impractical.

The use of conventional point– or line–iterative methods is also unattractive since the $[M]$ matrix is not guaranteed to be diagonally dominant. Here an efficient iterative scheme based on D'yakonov iteration is used to solve for u and v, as described in the following.

The problem is first reformulated in the form

$$[H]\left(\vec{q} - \vec{q}^{*}\right) = [M]\vec{q}^{*} \quad (2.31)$$

Here $\vec{q}^{*}$ represents a guessed field for $\vec{q}$. Iterative methods of this type can be shown to converge quickly when the matrix $[H]$ is similar to the matrix $[M]$ in the sense discussed by Concus and Golub (1973).

A suitable matrix $[H]$ can be found from the finite–difference form of the inverted relations, Eqs. (2.28). These equations can be expressed as

$$\vec{\omega} = [G]\vec{Q} \quad (2.32)$$

where the variable $\vec{\omega}$ approximates the $\vec{q}$ field which is not consistent with Eq. (2.29). In the limit as the mesh size is reduced to zero, the errors caused by the interpolations vanish and $\vec{\omega} \to \vec{q}$. Hence, in the limit of zero mesh size

$$[G]^{-1}\vec{\omega} \to [G]^{-1}\vec{q} = \vec{Q} = [M]\vec{q} \quad (2.33)$$

so that

$$[G]^{-1} \to [M] \quad (2.34)$$

For finite mesh size $[G]^{-1}$ serves as a close approximation to $[M]$ and hence is a suitable choice for $[H]$.

Equation (2.31) becomes

$$[G]^{-1}\left(\vec{q} - \vec{q}^{*}\right) = \vec{Q} - [M]\vec{q}^{*} \quad (2.35)$$

Multiplying Eq. (2.35) through by $[G]$ leads to

$$\Delta\vec{q} = [G]\vec{Q} - [G]\left([M]\vec{q}^{*}\right) \quad (2.36)$$

The first term $[G]\vec{Q}$ represents the inconsistent field $\vec{\omega}$ computed from Eq. (2.32). The term $[M]\vec{q}^{*}$ represents the U, V field computed from the guessed $\vec{q}^{*}$ field by using Eq. (2.29). Finally, $[G]\left([M]\vec{q}^{*}\right)$ represents the u, v field computed from $[M]\vec{q}^{*}$. Through the use of this iterative procedure, the "inversion" of Eq. (2.29) is accomplished by nothing more than a series of function evaluations using Eqs. (2.29) and (2.32). In practice, this procedure converges very rapidly (typically less than five cycles) and adds very little additional effort to the overall flow calculation. For examples and demonstration of the D'yakonov iteration procedure, the reader is referred to Braaten and Shyy (1986a).

2.4 Compressibility Effect

The above discussion is restricted to the flows of low Mach number effects. For the highly compressible flow, the formulation of the pressure correction equation needs to be revised (Rhie 1986, Van Doormaal *et al.* 1987, Shyy and Braaten 1988, Shyy *et al.* 1992d). In the following, the salient features of this formulation are summarized.

The essence of the pressure correction algorithm for solving the incompressible flows can be briefly described as follows. Discretization of the momentum and continuity equations yields the following difference equations:

$$([D] - [E])\{\vec{q}\} + [B]\{\vec{p}\} = \{\vec{F}\} \tag{2.37a}$$

$$[C]\{\vec{q}\} = \{\vec{G}\} \tag{2.37b}$$

where

$\{\vec{q}\}, \{\vec{p}\}$: vector with nodal value of (u,v,w) and p, respectively, as its components

$[D]$: diagonal matrix with positive elements

$[E]$: matrix with zero entries on its diagonal; difference operator $[D]–[E]$ accounts for both the convection and viscous effects

$[B]$: difference operator for gradient

$[C]$: difference operator for divergence

$\{\vec{F}\}$ and $\{\vec{G}\}$: explicit forcing function terms from source and boundary conditions.

Next, one can formulate a predictor/corrector procedure to iteratively update both the velocity and static pressure fields by splitting $\{\vec{q}\}$ and $\{\vec{p}\}$ into two parts:

$$\{\vec{q}\} = \{\vec{q}^{*}\} + \{\vec{q}'\} \tag{2.38a}$$

$$\{\vec{p}\} = \{\vec{p}^{*}\} + \{\vec{p}'\} \tag{2.38b}$$

By reformulating the momentum equation (2.37a) to the form of

$$([D] - [E])\{\vec{q}^{*}\} + [B]\{\vec{p}^{*}\} = \{\vec{F}\} \tag{2.39}$$

One obtains $\{\vec{q}\}$ based on a given $\{\vec{p}\}$. Furthermore, the relationship between the pressure correction and the velocity correction can also be derived:

$$([D] - [E])\{\vec{q}'\} + [B]\{\vec{p}'\} = 0 \tag{2.40}$$

Here, the SIMPLE algorithm takes a simplified form of equation (2.40) to link $\{\vec{q}'\}$ and $\{\vec{p}'\}$, namely,

$$[D]\{\vec{q}'\} + [B]\{\vec{p}'\} = 0 \tag{2.41}$$

Similarly, the continuity equation can be written as:

$$[C]\{\vec{q}'\} = \{\vec{G}\} - [C]\{\vec{q}*\} \tag{2.42}$$

Hence, a pressure correction equation for incompressible flow can be derived by combining equations (2.41) and (2.42)

$$[C][D]^{-1}[B]\{\vec{p}'\} = [C]\{\vec{q}*\} - \{\vec{G}\} \tag{2.43}$$

For highly compressible flows, the density is a strong function of pressure. Therefore, in the formulation of the pressure correction equation, it is necessary to correct both the density and velocity fields simultaneously to satisfy the continuity equation. This practice holds a key to the successful solution of the compressible flow problems. Each flux term in the continuity equation can be decomposed into four parts, e.g.,

$$\varrho U = (\varrho^* + \varrho')\,(U^* + U') = \varrho^* U^* + \varrho^* U' + \varrho' U^* + \varrho' U' \tag{2.44}$$
$$\quad (i)\qquad (ii)\qquad (iii)\qquad (iv)$$

where U is the contravariant velocity along the ξ–direction. These four terms represent: *(i)* the mass flux calculated based on the given density and velocity fields, *(ii)* the contribution from the velocity corrections, *(iii)* the linear contribution from the density corrections stemming from the compressibility effect, and *(iv)* the nonlinear contribution from the compressibility effect, respectively.

It is useful to examine the relative importance of the various terms as a function of the Mach number. The combination of terms *(i)* and *(ii)* leads to the incompressible form of the pressure correction equation given earlier. With the aid of the equation of state and definition of Mach number, the velocity correction term $\varrho^* U'$ is found to be inversely proportional to the local Mach number, Ma, i.e.,

$$\varrho^* U' \sim \frac{1}{Ma\sqrt{T}}\,p'_\xi \tag{2.45}$$

The equation of state for an ideal gas can be written as

$$\varrho = C^\varrho P \tag{2.46}$$

where

$$C^\varrho = \frac{1}{RT}$$

Hence, the density correction ϱ' can be expressed as

$$\varrho' = C^\varrho p' \tag{2.47}$$

While $\varrho^* U'$ contributes a diffusion term to the pressure correction equation, $\varrho' U^*$ contributes a convection–like term. This occurs because the velocity correction is proportional to the gradient of the pressure correction, while the density correction is proportional to the pressure correction directly. Since ϱ is inversely related to T, the contribution of $\varrho' U^*$ to the pressure correction equation is found to be proportional to the local Mach number i.e.,

$$\varrho' U^* \sim \frac{Ma}{\sqrt{T}} \, p'$$

(2.48)

Hence it is clear that the ratio of *(iii)* to *(i)* in equation (2.44) is proportional to the square of local Mach number; this characteristic explains why by omitting *(iii)* and *(iv)* in equation (2.44), the original SIMPLE algorithm works well for incompressible flows but not for high speed flows.

For the nonlinear correction term $\varrho' U'$, since neither ϱ' nor U' is necessarily small during the early iterations, it is useful to include these terms explicitly in the source term of the pressure correction equation to help stabilize the computational procedure.

The major remaining issue is how to interpolate the transport coefficients as well as the p' values at the control volume faces for the pressure correction equation. Due to the convection–diffusion nature of the pressure correction equation for compressible flow, a two–point average does not guarantee the positiveness of the coefficients, and hence the criterion for the numerical stability of an iterative solution procedure may be violated. It is important to note that it is the accuracy of the divergence of the mass flux calculation, namely the terms such as $\varrho^* U^*$, that determines the accuracy of the pressure field. The formulation of the other pressure correction terms does not affect the final solution accuracy; however, they can critically affect the numerical stability. Based on this realization, one can devise any suitable approximation for the pressure correction terms to expedite the convergence rate, provided the source term $\{G\}$ in equation (2.37b) is handled accurately. For example, based on the local Mach number, the values at the control volume faces are taken as either two–point central average (for low *Ma*) or one–sided upwind value (for high *Ma*). The switch point of the interpolation formulas is found to be necessarily subsonic; however the convergence path is not sensitive to the change of the switch point as long as a converged solution can be obtained. Furthermore, for the supersonic cases, the convection part of the transport equation for the pressure correction is linearly weighted according to the local Mach number. These procedures have been found effective in yielding stable convergence. As to the nonlinear contribution from the compressibility effect, part *(iv)* in equation (2.44), the present algorithm treats it as an explicit source term with the same switch criterion of interpolation formula adopted, as discussed above. This treatment of the nonlinear term is useful in supersonic flow cases, preventing divergence of the algorithm that may occur otherwise.

3 CHOICE OF THE VELOCITY VARIABLES

3.1 Basic Notions

Next, we discuss the choices of the velocity components suitable for the primary dependent variables. In generic terms, one has the option of using the Cartesian, contravariant or covariant components for this purpose. While mathematically these options must yield the same results, in practical calculations with finite degree of numerical accuracy, these choices may not yield the same, or even comparable solutions.

First, using the two–dimensional case as an illustration, the relationship between the Cartesian (u, v) and the contravariant (U, V) velocity components is defined as follows:

$$U = y_\eta\, u - x_\eta\, v$$
$$V = -\, y_\xi\, u + x_\xi\, v \tag{3.1}$$

Then the continuity equation in curvilinear coordinates can be written in a form similar to that in Cartesian coordinates as:

Cartesian:
$$\frac{\partial(\varrho u)}{\partial x} + \frac{\partial(\varrho v)}{\partial y} = 0$$

$$\tag{3.2}$$

contravariant:
$$\frac{\partial(\varrho U)}{\partial \xi} + \frac{\partial(\varrho V)}{\partial \eta} = 0$$

With regard to the covariant velocity components, defined as:

$$\tilde{U} = x_\xi u + y_\xi v$$
$$\tilde{V} = x_\eta u + y_\eta v \tag{3.3}$$

the continuity equation written in covariant velocity components is:

$$\frac{\partial}{\partial \xi}\,(\varrho\alpha_1\tilde{U} + \varrho\beta_1\tilde{V}) + \frac{\partial}{\partial \eta}\,(\varrho\alpha_2\tilde{U} + \varrho\beta_2\tilde{V}) = 0 \tag{3.4}$$

where

$$a_1 = q_{11}\, q_{22}^2\, /\, J$$
$$\beta_1 = a_1\,(\vec{e}_\xi \cdot \vec{e}_\eta)$$
$$a_2 = q_{11}^2\, q_{22}\, /\, J$$
$$\beta_2 = a_2\,(\vec{e}_\xi \cdot \vec{e}_\eta)$$
$$q_{11} = (x_\xi^2 + y_\xi^2)^{1/2} \tag{3.5}$$
$$q_{22} = (x_\eta^2 + y_\eta^2)^{1/2}$$
$$J = x_\xi\, y_\eta + x_\eta\, y_\xi$$

and $\vec{e}_\xi$ and $\vec{e}_\eta$ are unit vectors along the ξ and η directions respectively,

$$\vec{e}_\xi = \frac{x_\xi\vec{e}_x + y_\xi\vec{e}_y}{q_{11}}$$

$$\tag{3.6}$$

$$\vec{e}_\eta = \frac{x_\eta\vec{e}_x + y_\eta\vec{e}_y}{q_{22}}$$

3.2 Physical and Geometric Conservation Laws

When considering the various possible choices of velocity variables, one of the primary criteria is that in the framework of the finite volume formulation, a full conservation law form of the governing equations is desirable as it can satisfy the physical laws more easily and accurately. This consideration has a particularly important implication on the convection terms of the momentum equations since they are nonlinear and are usually a major source of numerical difficulty. With the Cartesian coordinates, the convection terms in the momentum equations are of the form $(\varrho uu)_x + (\varrho vu)_y$ which is fully conservative. In a curvilinear coordinate system, these terms can be transformed in a straightforward manner with the use of the Cartesian velocity components as the primary dependent variables to the form $(\varrho Uu)_\xi + (\varrho Vu)_\eta$ which is also fully conservative.

However, when either the contravariant or the covariant velocity components are used as the primary dependent variables, the fully conservative form can no longer be guaranteed since linear momentum is conserved along a straight line, not a curved line. Thus the differential equations for both the contravariant and the covariant velocity components involve source terms arising from the curvature of the coordinate lines. Furthermore, in the numerical implementation, the contravariant components ϱU and ϱV on each boundary of the mesh are defined as the mass flux between the two end points of the boundary (Shyy et al. 1985) and their values can artificially change with different grid systems. Hence, for the same flowfield, the values of those contravariant and covariant velocity components can be greatly affected by the ways that the grid systems are generated. These aspects can cause difficulties in preserving a high degree of numerical accuracy in satisfying the conservation laws.

To demonstrate this point, consider the purely convective equation

$$(\varrho uu)_x + (\varrho vu)_y = 0 \tag{3.7}$$

One of the most basic tests of the numerical accuracy of any computational algorithm for this equation can be made by generating a grid system with arbitrary skewness and nonuniformity and then to use this grid system to check the numerical accuracy by solving a uniform flow field of say, $\varrho=1$, $u=1$ and $v=1$. With this condition, equation (3.7) is trivially satisfied in the differential sense. Hence it serves as a good case to test whether an algorithm can honor the geometric aspect of the conservation laws in a discrete form. Here we call this requirement the geometric conservation law (Thomas and Lombard 1979, Shyy and Vu 1991) since the governing equations retain the conservation law form but contain only the geometric quantities. The transformed form of equation (3.7) with the Cartesian velocity components as dependent variables in curvilinear coordinates becomes:

$$(\varrho Uu)_\xi + (\varrho Vu)_\eta = 0 \tag{3.8}$$

which with the uniform flowfield is reduced to

$$(y_\eta - x_\eta)_\xi + (-y_\xi + x_\xi)_\eta = 0 \tag{3.9}$$

Referring to Fig. 1 (Section 2), Equation (3.9) is discretized as:

$$(y_\eta - x_\eta)_e - (y_\eta - x_\eta)_w + (-y_\xi + x_\xi)_n - (-y_\xi + x_\xi)_s = 0 \tag{3.10}$$

where e, w, n and s refer to the east, west, north and south faces of the control volume respectively. If a consistent finite volume formulation is adopted by approximating the derivative of the metric terms in (3.10), with the difference between two end points of the mesh line, then equation (3.10) becomes:

$$[(y_{i+1,j+1} - y_{i+1,j}) - (x_{i+1,j+1} - x_{i+1,j})] - [(y_{i,j+1} - y_{i,j}) - (x_{i,j+1} - x_{i,j})] +$$
$$[- (y_{i+1,j+1} - y_{i,j+1}) + (x_{i+1,j+1} - x_{i,j+1})] - [- (y_{i+1,j} - y_{i,j}) + (x_{i+1,j} - x_{i,j})]$$
$$= 0 \tag{3.11}$$

which is satisfied *exactly* regardless of how skew or nonuniform the mesh may be. It is also noted that one of the merits of this test problem is that since the flowfield is uniform, the whole focal point is directed toward the satisfaction of the geometric requirements; other issues such as the appropriate approximation of the convection effects do not arise here.

Since our primary interest is for the Navier–Stokes flow computation, it is useful to point out that the above geometric conservation law is applicable to the pressure gradient terms as well. However, the same requirements cannot be rigorously satisfied by the viscous terms (for flowfields of constant velocity gradients) due to the appearance of the nonlinear metric products associated with the coordinate transformation of the second derivative terms. Overall, one can summarize the situation by stating that with the use of Cartesian velocity components, the Navier–Stokes equations can be written in the strong conservation law form in the curvilinear coordinate system. The first derivative terms, including the convection and the pressure gradient terms can always numerically satisfy the geometric conservation law. The degree of the viscous terms, however, is dependent on the actual grid distribution.

Considering the use of the curvilinear components, say, the contravariant terms, the transformed form of the convective terms can be obtained by performing a chain rule type of coordinate transformation:

$$\left[\frac{\varrho U^2}{q_{11}}\right]_\xi + \left\{\begin{matrix}1\\1\ 1\end{matrix}\right\}\frac{\varrho U^2}{q_{11}} + \left\{\begin{matrix}1\\1\ 2\end{matrix}\right\}\frac{\varrho UV}{q_{11}} + \frac{q_{11}}{q_{22}}\left[\frac{\varrho UV}{q_{11}}\right]_\eta + \left\{\begin{matrix}1\\1\ 2\end{matrix}\right\}\frac{\varrho UV}{q_{22}} +$$
$$\left\{\begin{matrix}1\\2\ 2\end{matrix}\right\}\frac{\varrho V^2 q_{11}}{q_{22}^2} - \left[\frac{\varrho U}{q_{11}}\right]U_\xi - \left[\frac{\varrho U}{q_{22}}\right]V_\eta = 0 \tag{3.12}$$

where the Christoffel symbols of the second kind are defined as:

$$\left\{\begin{matrix}1\\1\ 1\end{matrix}\right\} = \frac{y_\eta x_{\xi\xi} - x_\eta y_{\xi\xi}}{J}$$

$$\left\{\begin{matrix}1\\1\ 2\end{matrix}\right\} = \frac{y_\eta x_{\xi\eta} - x_\eta y_{\xi\eta}}{J} \tag{3.13}$$

$$\left\{\begin{matrix}1\\2\ 2\end{matrix}\right\} = \frac{y_\eta x_{\eta\eta} - x_\eta y_{\eta\eta}}{J}$$

It is now obvious that Equation (3.12) not only possesses more terms than equation (3.8), but more critically it contains source terms resulting from the curvature of the coordinate lines. Hence, it is no longer of the fully conservative form which can cause difficulties with the finite–volume formulation, especially if the grid is of substantial nonuniformity and skewness. The fact that q_{11} and q_{22} are nonlinear with respect to the metric terms resulting from the coordinate transformation further compounds the difficulty of exactly satisfying the conservation law in the discrete form. A similar case can be made with the equation cast in terms of the covariant velocity components.

The other observation related to the satisfaction of the geometric conservation law can be made by studying the continuity equation written in terms of the covariant velocity components. Equation (3.4) demonstrates that the conservation law can be preserved in differential form for the covariant velocity components. However, because the terms α and β involve nonlinear combinations of metric terms, the geometric conservation law cannot always be honored in a skewed mesh system. It is clear that since the physical conservation laws are the ones that we ultimately strive to satisfy, the numerical algorithms not only preferably should be written to satisfy the strong conservation law in discrete form, but should also satisfy the geometric conservation law in discrete form. The latter requirement cannot be satisfied as long as the equations contain nonlinear metric terms regardless of whether the full conservation law form is adopted in the differential equations or not.

In summary, it has been demonstrated here that with either the contravariant or covariant velocity vectors, the momentum equations with curvilinear velocity vectors are no longer of full conservation law form. The curvature of the grid lines introduces extra source terms into the governing equations, which cause the degree of satisfaction of the physical conservation law to be sensitive to the uniformity and skewness of the mesh distribution. For the continuity equation, on the other hand, the contravariant velocity can maintain both the fully conservative form and the compactness of the equation. Table I summarizes the points discussed above. It appears that a combined use of the contravariant

Table I. Choice of primary velocity variable and implications in conservation laws

		Momentum Equations		**Continuity Equation**
	Convection	Pressure	Viscous	
Cartesian	satisfies both physical and geometric laws	satisfies both physical and geometric laws	satisfies physical but not geometric laws	satisfies both physical and geometric laws
Contravariant	does not satisfy either law	satisfies both laws	does not satisfy either law	satisfies both laws
Covariant	does not satisfy either law	satisfies both laws	does not satisfy either law	satisfies physical but not geometric law

velocity components for the continuity equation and the Cartesian velocity components for the momentum equations is a good balance. This practice has particular merits in a pressure–correction algorithm where the velocity corrections are derived based on the information of the pressure correction. As discussed in Braaten and Shyy (1986 a), the contravariant components should be used in the velocity correction procedure to ensure the satisfaction of the conservation laws.

4 GRID

4.1 Staggered Grid Layout

The choice of a grid system appropriate for the solution of the Navier–Stokes equations has been a subject of research for a long time (Patankar 1980, Peyret and Taylor 1983, Shyy and Vu 1991). Nominally, the non–staggered grid is easier to implement because all the dependent variables are placed at the same locations. On the other hand, on the staggered grid, there is no need for artificial pressure boundary conditions in the continuity equation (Patankar 1980); this feature is fundamentally satisfying since it is consistent with the mathematical property of Navier–Stokes flows. One recalls that, in the Navier–Stokes equations, the pressure field is determined based on the constraint provided by the continuity equation where no explicit boundary condition is applied. Besides this characteristic, the other merit of the staggered grid, that is less mentioned, is that it reduces the effective grid spacing since, for example, the pressure gradients can be computed "naturally" based on the two adjacent pressure nodes surrounding each velocity node. Similarly, in deriving the pressure–correction equation, the mass flux terms across all control surfaces can be directly estimated according to the velocity nodes located on those faces without resorting to interpolation. Of course, some convection terms such as $(\varrho u^2)_x$ in the u–momentum equation will still need to be interpolated to yield the mass flux estimation on the east and west faces. However, the problems of convection treatment are more complicated than a compact differencing scheme can resolve, as discussed in a separate section. Suffice it to say that the staggered grid results in a generally more compact differencing approximations for some key terms in the equations.

Now, with the introduction of the curvilinear coordinate system, the situation is not as clear–cut as in Cartesian coordinates. Using the grid notation adopted by Shyy *et al.* (1985), as shown in Fig. 1 (Section 2), the Cartesian and contravariant velocity components are defined at the middle of the east–west and north–south faces. That is, in 2–D curvilinear coordinates designated as ξ–lines and η–lines, u and U are defined at the middle of the η–lines of the control volumes and v and V components are defined at the middle of the ξ–lines of the control volume. All the scalar variables, including the pressure, temperature and density are defined at the geometric center of the four vertices defining the control volume. Several papers (Karki and Patankar 1988, Yang and Yang 1988, Rodi *et al.* 1989) suggest that with the combined use of the Cartesian velocity components and the staggered grid arrangement, difficulties arise when the grid lines turn ninety degrees from the original orientation and the benefits of grid staggering are lost.

A detailed discussion has been given in this regard by Shyy and Vu (1991). It was demonstrated that, if the metric terms between the (x, y) and the (ξ, η) coordinates are non–constant, then the spurious pressure oscillations do not appear in the staggered grid, even with the Cartesian velocity components used as the primary dependent variables in the momentum equations. For the staggered grids, moreover, the problem of spurious pressure oscillations can be prevented even with the constant metric terms. One can simply define the curvilinear coordinates to be non–parallel to the Cartesian coordinates.

As an illustrative example of the performance of the staggered grid, an isothermal turbulent flow through a circular tube with a 90° bend is computed. Fig. 2(a) shows the grid layouts in both cross–sectional and sideview directions where $51 \times 11 \times 11$ nodes are utilized. The Reynolds number based on the inlet condition is 10^6 and the original k–ε two–equation model, along with the wall function treatment (Launder and Spalding 1974) is adopted in the turbulence closure. The pressure contours are shown in Fig. 2(b). As can be clearly seen, the solutions are of expected appearance, with no spurious pressure oscillations and exhibit normal convergence characteristics. More information and examples about the issue of the staggered grid can be found in Shyy and Vu (1991).

Besides the algorithms utilizing the staggered grid arrangement, methods based on the non–staggered arrangement have also been proposed for both the pressure and the density based algorithms, e.g., Rhie and Chow (1983). These methods require special procedures to couple the velocity and pressure and prevent the appearance of checkerboard oscillations. For example, in Rhie and Chow (1983), an explicit fourth order pressure dissipation term is added to the pressure correction equation to suppress the spurious oscillations. However, with the use of finite mesh sizes, the artificially added fourth–order gradient term may not be smaller than the original lower–order derivative terms, especially in the presence of large gradients in the flowfield. This is demonstrated by a Fourier type of analysis by Shyy (1985a). Thus, the actual degree of numerical accuracy may be affected by the numerical smoothing procedure. Furthermore, it is also well known (Peyret and Taylor 1983) that artificially generated boundary conditions are necessary for the pressure in a non–staggered grid system. With the use of the staggered grid system, there is no need to devise artificial boundary conditions for the pressure–correction equation (Shyy *et al.* 1985) regardless of the orientation of the coordinate system. In the momentum equations, since both P_ξ and P_η terms appear in both the momentum equations, some extrapolation procedures will still be needed for both types of grid arrangement. Table II summarizes the need for prescribing pressure boundary conditions in staggered and non–staggered grid systems. Besides the issue of boundary conditions, there is also a concern about the sensitivity of the solutions with respect to the relaxation factors chosen for the computations (Majumdar 1988, Kobayashi and Pereira 1991). Recently, the non–staggered grid has gained popularity due mainly to convenience of implementation. However, more effort is needed to clarify the issues mentioned above in the context of non–staggered grids.

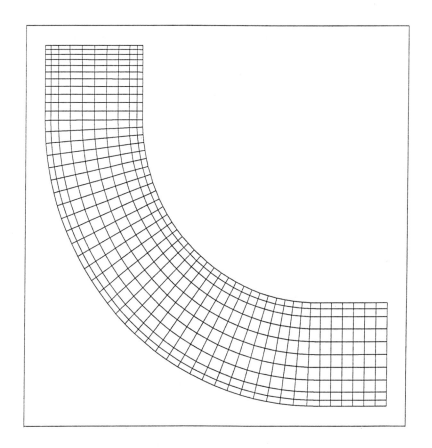

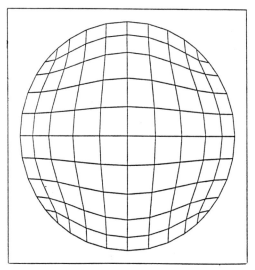

(a) Grid distribution in side–view and cross–sectional planes.
Figure 2. Grid and solution of a circular tube of 90° turn.

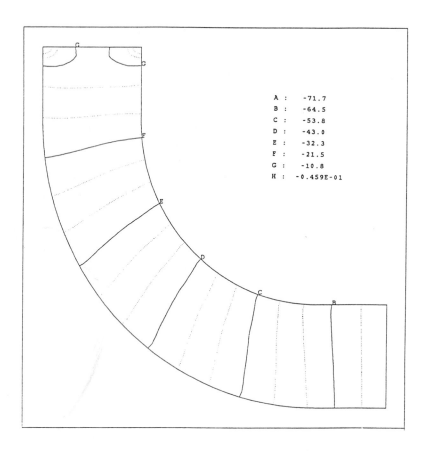

A :	-71.7
B :	-64.5
C :	-53.8
D :	-43.0
E :	-32.3
F :	-21.5
G :	-10.8
H :	-0.459E-01

(b) Pressure contours for flow with Re=10⁶. The solution is obtained using a staggered grid.
Figure 2. continued.

Table II. Grid systems versus need of pressure boundary conditions

	u–momentum equation	**v–momentum equation**	**continuity (or pressure equation)**
Staggered grid	needs artificial condition for P_η only	needs artificial condition for P_ξ only	no artificial condition
Non–staggered grid	needs artificial conditions for both	needs artificial conditions for both	needs artificial condition

4.2 Adaptive Grid

Grid generation is a major topic in complex fluid flow computations (Thompson *et al.* 1985, Shih *et al.* 1990). We will only discuss the use of the adaptive grid method to illustrate the main advantage of having a flexible grid system. Comprehensive reviews of this subject can be found in Thompson (1985), Eiseman (1987) and Hawken *et al.* (1991). The most appealing aspect of the adaptive moving grid is that the grid distribution can be adjusted in an intelligent way without requiring *a priori* knowledge and/or the intuition of the user, and hence one can reduce the size of the grid system that is needed to yield an accurate solution (Dwyer *et al.* 1980, Gnoffo 1983). For multi–dimensional Navier–Stokes flow, the effective application of the adaptive grid method cannot be made unless issues such as the different characteristics of the dependent variables in a coupled system of equations, the non–linear behavior of the flow, and the extra complexities introduced by the irregular flow configurations are appropriately resolved. An adaptive grid method based on the concept of equidistribution has been developed which accounts for the important fact that for the Navier–Stokes equations, the various dependent variables can have different characteristics depending on the flow configurations (Shyy 1986, 1987a, 1988, Shyy *et al.* 1992d). The numerical procedure involves starting the grid adaptation process with uniform grids and the numerical solutions obtained on them. The uniform grid solution is used to estimate the weighting function w and the new grid positions are determined from the following equation:

$$\int_0^{s_i} w \ ds \ = \ \frac{\xi_i}{\xi_{\max}} \int_0^{s_{\max}} w \ ds \tag{4.1}$$

with ξ incremented uniformly from one grid to the next. By solving the above equation, the grid positions, s_i's can be determined one by one. The new grid positions are then in turn used to recalculate the numerical solution of the Navier–Stokes and associated transport equations. This adaptive grid method can be coupled with time accurate calculations, where issues such as preservation of conservation laws between adaptive grid systems at different time instants need to be investigated.

It has been demonstrated that the advantage of this *a posteriori* multiple one–dimensional adaptive grid method is its flexibility in adding grid points along coordinates if desired. It was also shown that in the multi–stage adaptive grid procedure, the resulting grid system is already close to optimum after a couple of adaptations. Furthermore, as the adaptive readjustment of the grid distribution proceeds, not only is the overall error reduced but the error distribution becomes more uniform (Shyy 1987a).

One important issue relevant to the present type of adaptive moving grid computation is the possible appearance of highly skewed meshes in the flow domain. In this regard, several treatments can be employed. In Shyy (1988), a combined use of a variational adaptive grid formulation as a post–processing treatment in conjunction with the aforementioned multiple one–dimensional procedure has been found effective for the natural convection flow within an enclosure. Furthermore, the Laplace equation can be used to improve the mesh skewness resulting from grid movement. The Laplace equation is solved by a point–SOR procedure. It is well established that the point–SOR can quickly eliminate the high wave number components of error during the course of the iterations; it is however, slow in the overall convergence rate. Both features are highly desirable for the purpose of reducing grid skewness while retaining the effectiveness of grid adaptation. Hence, by utilizing the point–SOR for, say, the first five iterations, the resulting grid system can exhibit improved control of grid skewness while largely retaining the same clustering characteristics. Alternatively, an algorithm can be devised where the maximum allowable skewness in each mesh can be assigned *a priori*. Based on this constraint, the smoothing factor embedded within the weighting function can be determined in an iterative manner to ensure that the allowable amount of mesh skewness is not exceeded. This type of procedure is very attractive since highly distorted flow patterns usually appear in complex systems involving high Rayleigh, Reynolds, Stefan and Marangoni numbers.

The current state of the adaptive grid method is quite advanced and three–dimensional applications have been already made, e.g. Chang and Shyy (1991), Davies and Venkatapathy (1992). It is noted that effort has been made to develop

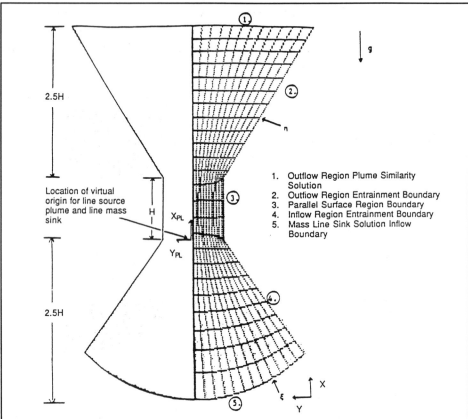

Figure 3.   Schematic of geometry and boundary conditions of the natural convection flow in a vertical slot.

quantitative measures to guide the grid distribution for problems with strong convection effects, and some interesting results have been obtained (Lee and Tsuei 1992, Jeng and Chen 1992b). As an illustrative example of the adaptive grid procedure, the buoyancy induced flow in a vertical slot is presented. A schematic of the geometry and the boundary conditions is given in Fig. 3. Details of the formulation can be found in Shyy *et al.* (1992a). The adaptive grid distributions for the thermal Grashof numbers $(Gr)_T$ of 10^4 and 10^5 (for definition, see Part III) are depicted in Fig. 4(a). The Grashof number is defined based on plate length and the temperature difference between plate wall and far upstream. The fluid flow equations are simplified by the Boussinesq approximation. It may be noted that the clustering of the grid points in the wall region and in the central plume region downstream by the adaptive grid procedure is virtually impossible to create by manually altering the original grid. Also, as the Grashof number is increased, there is a corresponding change in the grid distribution, which reflects the change of the flow structure.

The adaptive grid solutions (isotherms and velocity vectors) for $(Gr)_T = 10^4$ and 10^5 are shown in Fig. 4(b). There are recirculating eddies present inside the slot for the higher Grashof number case. For both Grashof numbers, from the corners of the slot inlet, the temperature convects and diffuses down into the inflow region. In the outflow region, despite the tendency of the thermal field to expand in the cross–stream direction, the side–entrainment causes the temperature contours to be concentrated in the central portion of the flow field. As $(Gr)_T$ increases, the effective velocity and thermal boundary layer thicknesses along the solid surface decrease and the thermal plume downstream of the slot

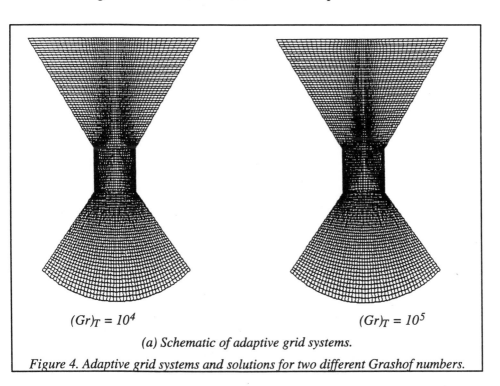

$(Gr)_T = 10^4$ $(Gr)_T = 10^5$

(a) Schematic of adaptive grid systems.

Figure 4. Adaptive grid systems and solutions for two different Grashof numbers.

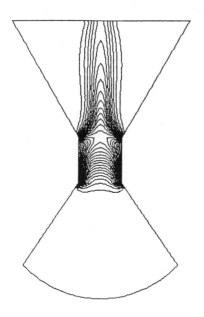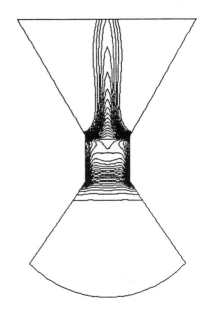

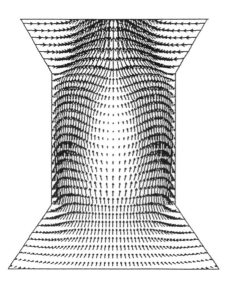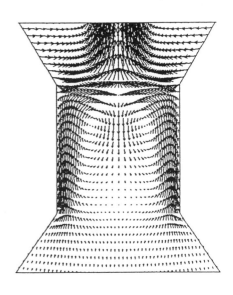

$(Gr)_T = 10^4$ $(Gr)_T = 10^5$

(b) Temperature and velocity contours.
Figure 4. continued.

becomes more confined in width. The use of adaptive grid is found to substantially help exhibit those characteristics.

In the following, we give an illustrative presentation of the adaptive grid refinement for two–dimensional channel flow. As is well–known, the no–slip constraint for the velocity on the solid surface combined with the relatively large Reynolds number can produce a highly nonuniform variation of the velocity profile along a specific direction. On the other hand, since the Poisson equation governs the pressure distribution and since a totally different kind of restraint is imposed on the pressure field by the solid surface, the pressure profile not only has a smoother variation but usually changes along different directions compared to the velocity profile. Due to these fundamental differences between the structure of the pressure field and the velocity field, an adaptive grid strategy that employs multiple one–dimensional grid adjustment procedures using different weighting functions for different directions is useful.

A two–dimensional nonsymmetric divergent planar channel flow is chosen for flow calculations. The inlet velocity profile is taken as a plug flow. The Reynolds number based on the inlet height is 500. The flow is considered as steady–state, incompressible and laminar. A body–fitted curvilinear coordinate system is constructed to specify the flow domain, as shown in Fig. 5(a). The adaptive grid procedure is implemented as a series of pseudo–one–dimensional problems along each coordinate line. In the present flow configuration, the adaptive procedure along the η–lines is conducted first by assigning the following weighting function:

$$w = 1 + |u_\eta| + |v_\eta| \qquad (4.2a)$$

where u and v are the velocity components along the x and y directions respectively. The subscript designates the derivative of the variable along that family of coordinate lines. In the context of the equidistribution constraint, this weighting function is similar to one that produces the constant arc length across each mesh spacing. The adaptive procedure along the ξ–lines is conducted next by assigning the following weighting function:

$$w = 1 + |p_\xi| \qquad (4.2b)$$

where p is the static pressure. Thus, based on the different characteristics of the velocity and the pressure fields, a multiple one–dimensional procedure can be devised to yield a multi–dimensional adaptive grid system.

The grid system shown in Fig. 5(a) has 41x26 nodal points. The grid distribution there is essentially uniform. Two flow calculations, one based on the grid shown in Fig. 5(a), the other based on a system with uniform mesh spacings and with 56x41 nodal points were performed to serve as the basis solutions for comparison. The u–velocity contours obtained on these two grid systems are shown in Fig. 5(b) and 5(c), respectively. In the adaptive readjustment procedure on the 41x26 grid, the number of nodes along the η–lines is increased from 26 to 32, as shown in Fig. 5(d). One of the noticeable differences between the numerical results obtained on the original smooth 41x26 grid, Fig. 5(b), and the adaptive 41x32 grid, Fig. 5(e), is that the adaptive grid solution predicts a smaller recirculation zone which is absent from the solution obtained on the uniform grid. This

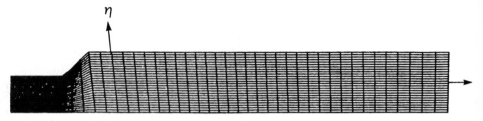

(a) Asymmetric channel with 41×26 smooth grid.

(b) Constant *u*–velocity contours on 41×26 smooth grid.

(c) Constant *u*–velocity contours on 56×41 smooth grid.

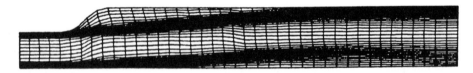

(d) Adaptively refined 41×32 grid based on the 41×26 smooth grid.

(e) Constant *u*–velocity contours on the adaptive grid.

*Figure 5.   Original and adaptively refined grid and comparison of solution
          characteristics.*

flow characteristic is consistent with that on the smooth 56x41 node grid as shown in Fig. 5(c).

Linkage between adaptive grid computations and the theory of modern nonlinear dynamics has been investigated by Shyy (1991a,b). The term strange attractors has been used to describe bounded, chaotic, nonperiodic solutions of deterministic, nonlinear differential equations in contrast to more predictable solutions, such as those near equilibrium points and limit cycles (e.g., Guckenheimer & Holmes 1983, Berge *et al.* 1984, Seydel 1988). The model chosen is a simple convection–diffusion equation; the approach is to conduct an adaptive grid computation based on the aforementioned equidistribution concept. In Shyy (1991a,b) the objective is *not* to show how useful this adaptive computational method is, but instead, to illustrate that a very reasonable computational method can result in *quite poor solution accuracy.* It has been demonstrated that in the course of the adaptive computational procedure, a rich variety of patterns of the structure of the grid distribution can be observed, depending on the various parameters such as Reynolds number and grid control parameters. For example, the intermittency scenario has been identified as a route to chaos in this adaptive procedure. The grid distribution may not reach equilibrium state, and fluctuates around some strange attractors. Furthermore, increasing the total number of grid points does not necessarily stabilize the adaptive procedure; the probability density function of the grid distribution can actually change from single–modal to multiple–modal shapes, indicating an increasingly complex pattern.

5 OPEN BOUNDARY TREATMENT

In computations involving the Navier–Stokes equations, one of the main difficulties is the numerical treatment of boundary conditions. Generally speaking, for a given set of governing equations, there are appropriate boundary conditions that result in a well posed problem with continuous solutions and provide relations that allow some dependent variables to be determined at the boundaries. These boundary conditions are usually determined from the physics of the problem. The remaining relations required to determine the dependent variables at the boundaries must be determined from boundary approximations such as extrapolation, characteristic compatibility relations, or different relations obtained from the governing equations.

One important issue related to the boundary treatment is that in many flows where downstream portions of the flow domain are open, there is no information available to give a precise prescription for boundary conditions there. In this regard, Blottner (1982) investigated three model problems and found that first–order accurate boundary approximations directly derived from the governing equations give second–order accuracy, while first–order accurate boundary approximations resulting from extrapolation give first–order global accuracy. For equations of the hyperbolic type, the concept of characteristics can be introduced to help both stabilize the numerical calculations and accelerate convergence. The so called absorbing boundary conditions were introduced by Engquist and Majda (1977). The idea is that when outgoing waves hit an artificial boundary, there should be very little reflection. Based on this observation, it can be generally stated that for all types of fluid flow problems, since the suitable form for the boundary conditions along open boundaries is not obvious, any artificial condition formulated for the open boundaries should determine the interior flow as though the boundaries were not there at all. Thus Oliger and Sundstrom (1978) pointed out that an open boundary should not produce large boundary–layer type gradients in the flow field.

For incompressible flows, Shyy (1985c) investigated the appropriate open boundary conditions for the multiple branched flow problem. Shyy (1985b) also studied the accuracy and efficiency aspects of the various possible space extrapolation formulas applied to open boundaries for both single and multiple branched flows. It was found that by comparing the conditions of $\phi_\xi = 0$ and $\phi_{\xi\xi} = 0$, where ϕ is a dependent variable (velocity or scalar variables, but not including the static pressure which does not need boundary conditions due to the use of a staggered grid), and ξ is the coordinate along the streamwise direction, the condition of $\phi_{\xi\xi} = 0$ does not always yield unique solutions, whereas $\phi_\xi = 0$ is found to be satisfactory.

A related issue concerns the appropriate conditions along open boundaries for regions where backward flow appears. It is often accepted (Patankar 1980, p 104) that for those parts of the boundary where the fluid flows into the domain, sufficient prescription of the dependent variables, such as velocity components, should be given. The reasoning is that the region of the open boundary where the backward inflow occurs is now upstream of the flow field; hence the validity of a straight forward extrapolation along a computational coordinate line is doubtful.

A series of incompressible Navier–Stokes flow computations has been conducted, by Shyy (1987b), to investigate the effects of a downstream open boundary on the calculated flow characteristics, especially for flows where an inflow appears on the open boundary. The results show that the conventional viewpoint does not always hold. The numerical experiments indicated that even with inflow across some portions of the boundary, the numerical procedure can be well posed and stable and convergent solutions can be obtained. Once the convergent steady state solutions are obtained, the overall flow characteristics are not sensitive to the location of the open boundary. The numerical stability is dependent on the strength of the non–linearity, i.e. convection strength, especially for flows with multiple recirculating eddies across the open boundary. According to the present numerical experiments, a straightforward first–order extrapolation, $\phi_\xi = 0$, for the velocity variables is acceptable on the open boundary. It should be emphasized that this extrapolation procedure only serves the need of the momentum equations. In the course of computing the pressure field, measure needs to be taken to ensure that the mass conservation is satisfied globally around the whole flow domain. In achieving this requirement, the velocity distribution along the open boundary, initially assigned via the extrapolation scheme in the momentum equation, needs to be iteratively corrected; the amount of correction is determined according to the imbalance of the mass fluxes across the whole flow boundaries. Since the velocity profiles on the inlet and wall are known, the correction will be applied to the open boundaries. While different schemes can be used in this correction procedure, they should only affect the efficiency of the computation, not the final converged solutions. This point has been empirically verified in some recirculating flow computations. Although the exact mathematical nature of the appropriate open boundary treatments still requires further clarification, it seems clear that for internal fluid flows with no property variations, the mass continuity constraint is able to yield an unique solution even for the case involving recirculating flows across open boundaries.

As an illustrative example, a backward facing step flow has been computed in a time accurate manner, as shown in Fig. 6. Figures 7 and 8 present the time dependent streamfunction contours. Central differencing has been used for the convection terms and first–order backward differencing for the unsteady terms. A parabolic inflow velocity profile is specified, while outflow boundary velocities are obtained by first–order extrapolation. The Reynolds number, based on the average inflow velocity and the channel height is 800. The expansion ratio H/h is 2 as indicated in the schematic shown in Fig. 6. Time–accurate simulations were performed for two channel configurations, one with length $L = 8$ (41×41 mesh) and the other with $L = 16$ (161×41 mesh). The time–step was twice the viscous time scale, i.e. $\Delta T = 2\Delta y^2/\nu$, thus a fluid particle entering the domain at the average velocity $u = 1$ travels $2\Delta y$ downstream during a time–step. More details can be found in the work of Blosch *et al.* (1993).

Figures 7 and 8 show the formation of alternate bottom/top wall recirculation regions, for both $L = 8$ and $L = 16$ cases during startup, which gradually become thinner and elongated downstream. For the $L = 16$ simulation (Fig. 8), the transient flowfield has as many as four separation bubbles at $T = 32$, the latter three of which are eventually

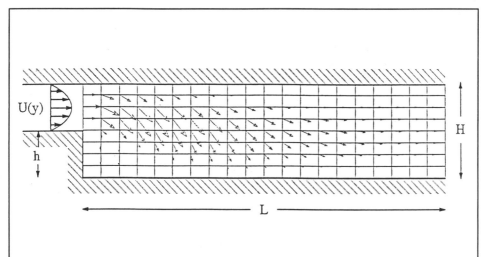

Figure 6. Backward–facing step flow. The expansion ratio H/h is 2, and a parabolic inflow profile is specified.

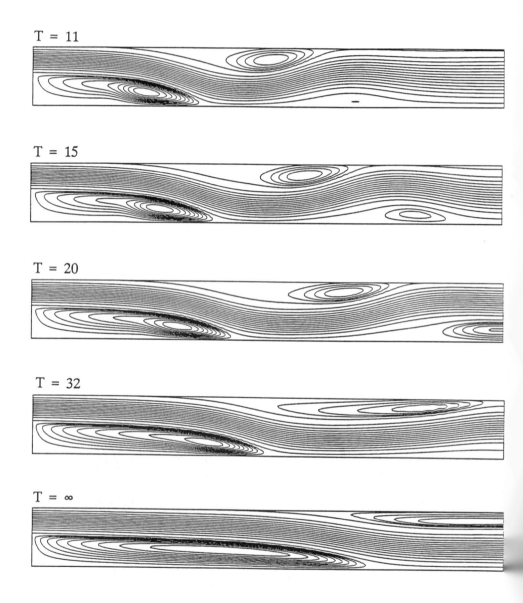

T = 11

T = 15

T = 20

T = 32

T = ∞

*Figure 7.   Time–dependent flowfield for impulsively started backward–facing step flow,
Re=800. The domain has length L=8. Streamfunction contours are plotted at
several instants up to the steady state, which is shown in the last plot.*

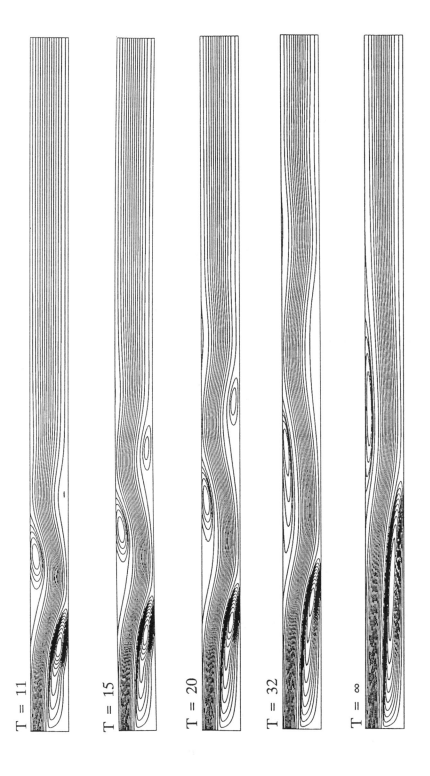

T = 11

T = 15

T = 20

T = 32

T = ∞

Figure 8. Time–dependent flowfield for impulsively started backward–facing step flow, Re=800. The domain has length L=16. Streamfunction contours are plotted at several instants during the evolution to steady state (last plot).

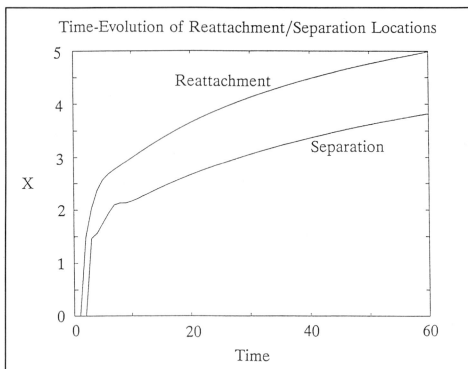

Figure 9. Time–dependent location of bottom wall reattachment point and top wall separation point for Re=800 impulsively started backward–facing step flow. The curves for L=8 and L=16 cases are shown; they overlap identically.

convected out of the domain. In the $L = 8$ simulation (Fig. 7), the stream function plots are at time–instants corresponding to those shown in Fig. 8. It may be noted that between $T=11$ and $T=32$ a secondary bottom wall recirculation zone forms and drifts downstream, exiting without reflection through the downstream boundary. The time evolution of the flow field for the $L = 8$ and $L = 16$ simulations is almost identical.

As can be observed, the fact that a shorter channel length was used and that a recirculating cell may go through the open boundary do not affect the solutions. Figure 9 compares the computed time histories of the bottom wall reattachment and top wall separation points between the two computations. The $L = 8$ and $L = 16$ curves are perfectly overlapped. The steady–state solutions for both the $L = 8$ and $L = 16$ channel configurations are also shown in Figs. 7 and 8 respectively. Although the outflow boundary cuts the top wall separation bubble approximately in half, there is no apparent difference between the computed stream function contours for $0 < x < 8$. Furthermore, the convergence rate is not affected by the choice of outflow boundary location.

Figure 10 compares the steady state u and v velocity profiles at $x = 7$ between the two computations; their accuracy is assessed by comparison with a finite element numerical simulation reported by Gartling (1990). Figure 10 establishes quantitatively

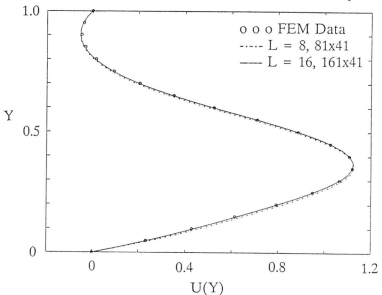

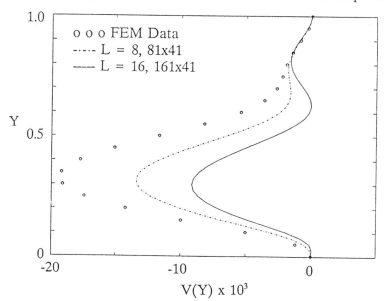

Figure 10. Comparison of u– and v–velocity profiles at x=7.0 for the L=16 and L=8 backward–facing step simulations at Re=800, with central–differencing. (o) indicates the solution obtained by Gartling (1990).

that the two simulations differ negligibly over $0 < x < 8$, the difference in the v–velocity profiles being less than 10^{-3}. Neither v–profile agrees perfectly with the solution obtained by Gartling (1990), which may be attributed to the need for conducting further grid refinement studies in the work of Blosch *et al.* (1993) and/or Gartling (1990).

Evidently the location of the open boundary is not critical to obtaining a converged solution. This means that the downstream information is completely accounted for by the continuity equation. The correct pressure field can develop because the system of algebraic equations for the pressure correction equation requires only the boundary mass flux specification; if the global continuity constraint is satisfied, the continuity equation for the implicit specification of the pressure is well posed regardless of whether there is inflow or outflow at the boundary where extrapolation is applied. This is reflected by the close agreement between the time–accurate $L = 8$ and $L = 16$ simulations as well as the agreement of the steady state solutions with the benchmark calculations of Gartling (1990).

A word of caution is also needed here. As demonstrated by Shyy (1987b), with multiple recirculating cells present at the open boundary, there may be inadequate numerical damping to yield a converged solution. Furthermore, and more fundamentally, as discussed in Shyy *et al.* (1992a), for a flow field completely defined by the open boundaries, the issue of the specification of the net mass flux across the boundaries may not be easily resolvable and needs to be investigated carefully. Nevertheless, it appears to be clear that as far as the outlet boundary condition is concerned, mass conservation is a primary factor that needs to be appropriately accounted for. In a recent work, Jin and Braza (1993) have used a nonreflecting boundary treatment for the outlet. They do not give explicit discussion of the role played by the mass continuity constraint. It is likely that the lack of satisfactory performance of some of the treatment discussed by them may be caused by this reason.

6 CONVECTION TREATMENT

6.1 Background

The accurate representation of sharp gradients caused by strong convection has long been a challenge for the computational fluid dynamicist (Roache 1972, Hirsch 1990). It is well known that conventional numerical techniques often create spurious oscillations in these regions rendering the solution unsatisfactory. It is well known that the second–order central difference operator for the first derivative convection terms, though quite accurate for low cell Peclet numbers, gives rise to wiggles when the local cell Peclet number is greater than some critical value (Gresho and Lee 1981, Shyy 1985a) for the one–dimensional linear Burgers equation, the critical value is 2. To overcome this problem, the first–order upwind scheme has been widely used but the excessive numerical dissipation inherent to this scheme has necessitated the search for more accurate schemes. The so–called hybrid scheme uses the central difference operator for regions in the flowfield with cell Peclet numbers less than 2 and the first–order upwind operator otherwise. But, for convection–dominated flows, this again leads to the use of the first–order upwind scheme for a majority of the flowfield, leading to excessive smearing of the solution profiles, especially the sharp gradients in the flowfield. To take advantage of the concept of upwinding, while achieving higher accuracy, higher–order upwind schemes, such as second–order upwind (Shyy 1985, Warming and Beam 1976, Vanka 1987, Jiang *et al.* 1990, Shyy *et al.* 1992b) and QUICK (Quadratic Upstream Interpolation for Convective Kinematics) (Leonard 1979) schemes have been proposed. The second–order upwind scheme has received mixed reviews in the literature– some researchers have criticized its accuracy and convergence characteristics (Vanka 1987) while others have found it satisfactory (Jiang *et al.* 1990, Shyy *et al.* 1992b). As pointed out by Shyy *et al.* (1992b), the key to this discrepancy lies in the fact that within a control volume formulation, a conservative and consistent formulation of second–order upwinding at the control volume interfaces is required to produce an accurate and well behaved scheme. The QUICK scheme too has received mixed reviews (Han *et al.* 1981, Huang *et al.* 1985, Hayase *et al.* 1992), but more in terms of convergence rates. It seems that the differences in the actual implementation of the same basic conservative form of the scheme may be responsible for these inconsistent findings, regarding in particular, the second–order upwind scheme. Besides the above treatments, various remedies have been proposed following other lines of reasoning. For example, it has been argued (Gresho and Lee 1981) that the appearance of undesirable oscillations can serve, at the most basic level, to indicate the regions that are in need of finer grid spacing to improve solution quality. Furthermore, in the region containing the oscillations, one can improve the solution using various adaptive grid techniques, including grid redistribution (Dwyer *et al.* 1980, Eiseman 1987, Shyy 1988), as well as local refinement (Berger and Jameson 1985, Babuska *et al.* 1986, Holmes and Conner 1989). Spurious oscillations can also be controlled either by using schemes that are intrinsically dissipative (Shyy 1985a), or by explicitly adding artificial viscosity to the governing equations to damp out numerical oscillations (Roache 1972, Hirsch 1990). However, both methods share some commonality if interpreted appropriately (Shyy 1984b). Furthermore, such methods generally tend to smear gradients and cannot adequately represent the complicated flow field unless fine grid spacing is used to resolve the characteristics of the flow in these regions. In order to more accurately control the amount of numerical dissipation, various modern schemes have been developed to capture sharp gradients without oscillations (Van Leer 1974, Osher and Chakravarthy 1983, Roe 1986, Woodward and Collela 1984a,b, Hirsch 1990, Liou et al. 1990, Liou and Steffen 1993).

From a different perspective, one may also choose to extract the "useful", i.e., physically realizable information from oscillatory solutions obtained using unsatisfactory numerical schemes that are excessively dispersive. The idea is to eliminate undesirable portions of the solution while retaining only the desired, i.e., physically realizable ones. To this end, Engquist *et al.* (1989) and Shyy *et al.* (1992c) have devised a nonlinear filtering algorithm designed to work in conjunction with standard numerical schemes. This subject is too complicated to be adequately discussed here. Suffice it to say that in the results to be presented later, convection terms are handled by the second–order schemes which do not generate spurious oscillations under the given conditions.

In order to gain some insight into the dissipative and dispersive characteristics of a given numerical scheme, the von Neumann stability analysis can be utilized to analyze several conventional schemes for the linear Burgers equation :

$$\phi_t + u \phi_x = v \phi_{xx} \quad , \quad u , v : \text{constants} \qquad (6.1a)$$

The elementary solution of equation (6.1a) is:

$$\phi(t,x) = \exp[-(vk_m^2 + ik_mu)t + ik_mx] \qquad (6.1b)$$

where $i = \sqrt{-1}$, $k_m = 2\pi/L_m$ is the wave number of the mth component, and L_m is the wavelength of the mth component. Based on the elementary solution, the amplification factor of the exact solution can be defined as

$$G_e = \frac{\phi(t + \Delta t)}{\phi(t)} = \exp[-\frac{C}{P}\beta^2 - i\beta C] \qquad (6.2)$$

where $|G_e| = exp(-C\beta^2/P)$ is the amplification magnitude, $\theta = -\beta C$ is the phase angle, $\beta = k_m \Delta x$, $2\pi\Delta x/\beta$ is the wavelength, $P = uh/v$ is the cell Peclet number and $C = u\Delta t/h$ is the Courant number.

It is clear that G_e depends on the cell Peclet number, Courant number, and the wave number. The amplification factors obtained by the von Neumann analysis for the backward Euler time stepping scheme along with three convection schemes, namely, first–order upwind, second–order central difference and second–order upwind schemes and second–order central differencing for the viscous term, are shown in Fig. 11 with $P=10^3$, $C=0.25$ and 0.5, and a periodic boundary condition. It is noted that with a high Peclet number, $P=10^3$, and a periodic boundary condition, the exact solution essentially maintains its initial profile just like a pure wave equation. Consequently, the amplification magnitude is very close to unity at all wave numbers. With the use of the central difference scheme, although the amplification magnitudes are close to the exact values, the short waves travel with much smaller phase speeds than the exact solutions. It is these errors in the phase speed for the short waves that cause the central difference scheme to perform unsatisfactorily. This observation explains why the central difference scheme usually yields solutions with highly noticeable $2\Delta x$ oscillations. Furthermore, it is clear that both first– and second–order upwind schemes also exhibit substantial errors in the phase angle for short waves; however, since these schemes are also quite dissipative in the short wave range, their phase angle errors are more effectively suppressed.

The other interesting aspect revealed by this analysis is that the second–order upwind scheme can actually be more dissipative than the first–order upwind scheme at short wavelengths; it is less dissipative than the first–order upwind scheme only in the long wavelength regime. This feature may be useful if a solution profile contains sharp gradients because the phase speed errors associated with short waves will be suppressed. Overall, this study illustrates that it is often the dispersion rather than the dissipation

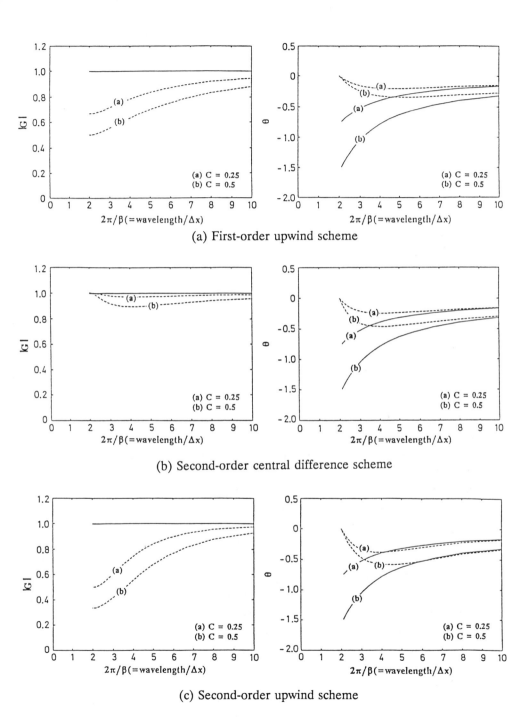

(a) First-order upwind scheme

(b) Second-order central difference scheme

(c) Second-order upwind scheme

Figure 11. Amplification factor and phase angle of the three schemes for the linear Burgers' equation. Solid lines: exact; dotted lines: numerical.

caused by a numerical scheme that makes an unsatisfactory performance more visible. Accordingly, the standard upwind schemes suppress, not eliminate, these problems via enhanced dissipation in the troublesome wavenumber range. On the other hand, from a different perspective, if a selective improvement of phase speed can be made, then the accuracy of an originally wiggle–prone scheme may be qualitatively improved.

6.2 Nonlinear Filtering Method

With this background, a filtering technique has been investigated to study its potential of improving the numerical accuracy, in particular for dispersive schemes. As pointed out in Engquist *et al.* (1989) and Shyy *et al.* (1992c), for a filter to be effective it should have minimal effect on an already smooth solution, and should enforce some criterion to guarantee no spurious oscillations near discontinuities. Let ϕ_j be the variable obtained by solving the conservation equations after n time steps and at grid index *j*. The filtering algorithm proceeds by first scanning the value ϕ_j to correct for any local maxima or minima. When a correction is added at a point, the algorithm ensures that the same correction is subtracted from a neighboring point to maintain conservation. The corrected neighbor is taken as the one with the greater difference from ϕ_j. Furthermore, correction is made so that no value may pass its neighbors. Thereby, overcompensation and creation of new extrema is avoided.

Specifically, let the symbols Δ and ∇ denote the forward and backward differences respectively, i.e., $\Delta\phi_j = (\phi_{j+1} - \phi_j)$ and $\nabla\phi_j = (\phi_j - \phi_{j-1})$. The filter algorithm, in essence, works according to the following procedure :

(i) If $(\Delta\phi_j)(\nabla\phi_j) < 0$, then ϕ_j is a local extremum, and it will be adjusted.

(ii) ϕ_{j+1} or ϕ_{j-1} must be adjusted by the same amount as ϕ_j is corrected; the one of larger difference from ϕ_j is chosen to be adjusted. The extent of correction applied to ϕ_j is limited to the smaller of ε_+ and ε_-, where ε_+ is equal to 0.5 times the larger of $\Delta\phi_j$ and $\nabla\phi_j$, and ε_- is equal to half of the smaller of the two differences.

It is noted that different correction schemes can be devised in step (ii) to adjust the solution profiles. A relaxation procedure has been developed (Shyy *et al.* 1992c) to modify the original scheme given above which can further improve the effectiveness of the filtering algorithm adopted here. The variables fed into the filter at each time or iterative step are all the dependent variables to be solved.

As a demonstration of the filter, a one–dimensional gas dynamics problem with a shock moving into a quiescent gas is solved numerically with the following initial conditions: $P_2/P_1 = 1.5$, $\varrho_2/\varrho_1 = 1.33$, $P_2 = 10^5$ N/m^2, $T_1 = 290$ K, where the subscripts 1 and 2 correspond to conditions downstream and upstream of the shock, respectively. These conditions correspond to an upstream Mach number of 0.28. The values of the Prandtl number, *Pr*, and the ratio of the specific heats, γ, are 0.72 and 1.4, respectively. All calculations are made with $\Delta t/\Delta x = 0.03$, which is within the stability limit of the numerical scheme. The explicit Euler method for time stepping is adopted. The equations are marched in time sequentially, from continuity, to momentum, and then to energy equation. A finite–volume approach is adopted to discretize the equations cast in strong

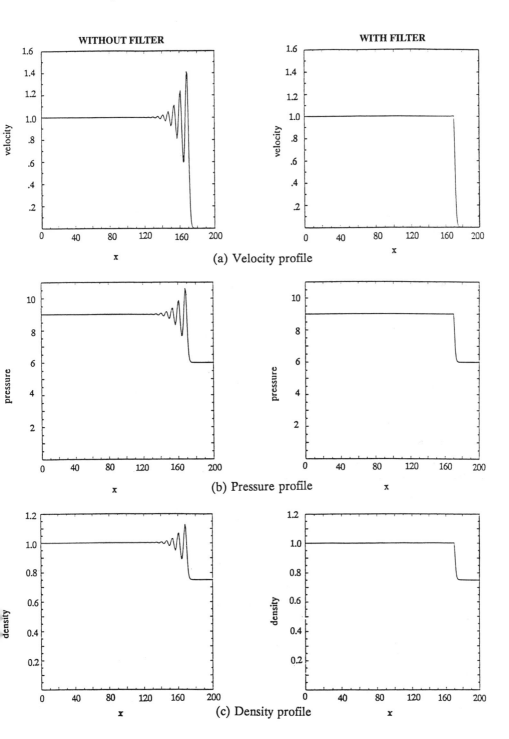

Figure 12. Effect of filtering on a 1–D compressible flow solution; Re = 10^4, timestep = 600.

conservation form. Two hundred uniformly distributed grid points are used. At the upstream boundary, values of density, velocity and pressure are specified, whereas at the downstream boundary, pressure is specified and density and velocity are extrapolated from the interior points. In the above numerical scheme, the dependent variables contained in the convection terms are treated by the first–order upwind scheme. The mass flux and the pressure terms are discretized by the standard second–order central difference schemes. Since there is no consideration for the propagation of different types of waves via flux splitting, spurious oscillations can result.

The Reynolds number for the flow is 10^4. As shown in Fig. 12, solutions without filtering exhibit large oscillations propagating well into the upstream domain. When the filter is applied, sharp shock profiles are recovered without any evidence of oscillations, demonstrating the effectiveness of the filter for this problem. The resulting solution is not only oscillation free but also yields a sharper shock than commonly yielded by a first–order scheme. For problems featuring sharp gradients and without large curvatures, the present filtering technique can work very satisfactorily.

6.3 Some Currently Popular Schemes for Recirculating Flows

In the following, we discuss the characteristics of several convection schemes that have been often applied to convection dominated complex flow problems. Here we summarize the work of Thakur and Shyy (1993b) by first casting these schemes in a conservative and consistent form, based on a unified framework, and then assessing their performance.

Typical control volumes employed for the u–velocity component in the 2–D staggered grid arrangement in the Cartesian coordinates are shown in Fig. 13. Here for simplicity, we shall use the uniform Cartesian grid to illustrate the salient points; the formulas on curvilinear nonuniform grids can be derived based on identical concepts. For illustration, 2–D steady state incompressible flow is considered. The u–momentum and continuity equations are respectively,

$$\frac{\partial}{\partial x}(\varrho u^2 - \mu \frac{\partial u}{\partial x}) + \frac{\partial}{\partial y}(\varrho uv - \mu \frac{\partial u}{\partial y}) = -\frac{\partial P}{\partial x}$$

$$\frac{\partial}{\partial x}(\varrho u) + \frac{\partial}{\partial y}(\varrho v) = 0 \tag{6.3}$$

Integrating these equations over the control volume shown in Fig. 13, we get for the momentum equation:

$$F_e \phi_e - F_w \phi_w + F_n \phi_n - F_s \phi_s = S + D_e(\phi_E - \phi_P) - D_w(\phi_P - \phi_W) \\ + D_n(\phi_N - \phi_P) - D_s(\phi_P - \phi_S) \tag{6.4}$$

where S is the source term, and for the continuity equation:

$$F_e - F_w + F_n - F_s = 0 \tag{6.5}$$

where

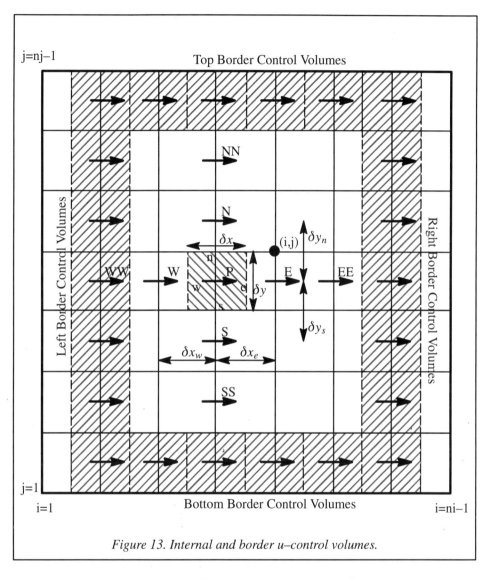

Figure 13. Internal and border u–control volumes.

$$F_e = (\varrho u)_e \delta y \qquad F_w = (\varrho u)_w \delta y$$

$$F_n = (\varrho v)_n \delta x \qquad F_s = (\varrho v)_s \delta x$$

$$D_e = \frac{\mu_e \delta y}{\delta x_e} \qquad D_w = \frac{\mu_w \delta y}{\delta x_w} \qquad (6.6)$$

$$D_n = \frac{\mu_n \delta x}{\delta y_n} \qquad D_s = \frac{\mu_s \delta x}{\delta y_s}$$

and $\delta x_e, \delta y_n$ etc. are the distances shown in Fig. 13. The F's are the convection fluxes and D's the diffusion fluxes at the control volume interfaces and are estimated by a linear interpolation between the two neighbors to either side of the control volume interfaces.

For example, the convection flux on the east interface, is estimated as :

$$F_e = (\varrho u)_e \delta y = \frac{(\varrho_P + \varrho_E)}{2} \frac{(u_P + u_E)}{2} \delta y \qquad (6.7)$$

The variable ϕ at a control volume interface can be determined using an interpolation involving the values of ϕ at one or more of the grid point neighbors of the interface.

In devising and interpreting the property of a convection scheme, one can resort to either an algebraic approach, utilizing say, the standard central difference scheme and an added amount of numerical diffusion (MacCormack and Baldwin 1975, Jameson *et al.* 1981, Harten 1983, Shyy 1984b) or a geometric approach utilizing different interpolation profiles to construct the flux distribution (Boris and Book 1973, Van Leer 1974, Hirsch 1990).

In the framework of algebraic approach, for example, for constant u,

$$L_{u1}(\phi_{ij}) \quad = \quad L_{c2}(\phi_{ij}) \quad - \quad \frac{u\Delta x}{2} D_{c2}(\phi_{ij}) \qquad (6.8)$$

where L_{u1}, L_{c2}, D_{c2} are respectively, first–order upwinding for $(u\phi)_x$, second–order central differencing for $(u\phi)_x$ and second–order central differencing for ϕ_{xx}. Corresponding interpretations can also be made for the QUICK and the second–order upwind schemes (Shyy 1984b). If the fourth order central difference operator approximating $(u\phi)_x$ and the second–order central difference operator approximating ϕ_{xxxx} are defined as $L_{c4}(\phi_{ij})$ and $D_{c4}(\phi_{ij})$, respectively, and the QUICK approximation to $(u\phi)_x$ as L_{QU}, then

$$L_{QU}(\phi_{ij}) \quad = \quad \tfrac{1}{4}L_{c2}(\phi_{ij}) \quad + \quad \tfrac{3}{4}L_{c4}(\phi_{ij}) \quad + \quad \frac{u(\Delta x)^3}{16} D_{c4}(\phi_{ij}) \qquad (6.9)$$

showing that the QUICK scheme is equivalent to using 25% second–order accurate central differencing, 75% fourth order accurate central differencing and some fourth derivative damping. On the other hand, if the second–order upwind approximation to $(u\phi)_x$ is defined as L_{u2} then,

$$L_{u2}(\phi_{ij}) \quad = \quad - \; 2L_{c2}(\phi_{ij}) \quad + \quad 3L_{c4}(\phi_{ij}) \quad + \quad \frac{u(\Delta x)^3}{4} D_{c4}(\phi_{ij}) \qquad (6.10)$$

or, the second–order upwind scheme is equivalent to using 300% fourth order accurate central differencing, then subtracting 200% second–order central differencing and adding some fourth derivative damping.

In the following, the geometric interpolation procedure will be adopted to facilitate the derivation of the various schemes. Different convection schemes use different interpolations, as described below.

(a) First–Order Upwind Scheme

For the first–order upwind scheme, the value of ϕ at any control volume face is estimated using the value at the upwind neighbor. Thus, as is well known, the value ϕ_e is assigned the value ϕ_P if the flux F_e is positive, and the value ϕ_E if F_e is negative. This choice of extrapolation scheme amounts to a piecewise constant profile, with no variation allowed between the values of ϕ at two adjacent nodes along the one–dimensional convection field as illustrated in Fig. 14(a). This can be conveniently summarized as

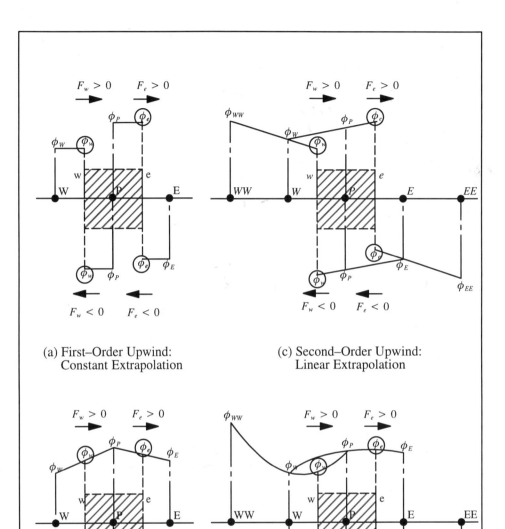

$F_w > 0$ $F_e > 0$

$F_w < 0$ $F_e < 0$

(a) First–Order Upwind:
Constant Extrapolation

(c) Second–Order Upwind:
Linear Extrapolation

(b) Central Difference:
Linear Interpolation

(d) QUICK:
Quadratic Interpolation

Figure 14. Schematic of geometric interpolation / extrapolation employed by the various schemes.

$$F_e \phi_e = \phi_P [\![F_e, 0]\!] - \phi_E [\![-F_e, 0]\!] \tag{6.11}$$

where the operator $[\![a, b]\!]$ yields the larger of a and b.

Using the above equation (6.11), equation (6.4) can be written in the form

$$a_P \phi_P = a_E \phi_E + a_W \phi_W + a_N \phi_N + a_S \phi_S + b \tag{6.12}$$

and for the first–order upwind scheme, after subtracting the continuity equation (6.5) multiplied by ϕ_P from the momentum equation equation (6.4), we get

$$
\begin{aligned}
a_E &= D_E + [\![-F_e, 0]\!] \\
a_W &= D_W + [\![F_w, 0]\!] \\
a_N &= D_N + [\![-F_n, 0]\!] \\
a_S &= D_S + [\![F_s, 0]\!] \\
a_P &= a_E + a_W + a_N + a_S \\
\text{and} \quad b &= S
\end{aligned}
\tag{6.13}
$$

A Taylor series expansion about the point P gives a locally first–order accurate scheme.

(b) Second–Order Central Difference Scheme

Here, as depicted in Fig. 14(b), ϕ at any interface is estimated by a linear interpolation between the two grid point neighbors on either side of the interface, regardless of the direction of local convection; for example

$$F_e \phi_e = F_e \frac{\phi_P + \phi_E}{2} \tag{6.14}$$

and now the coefficients in equation (6.4), after subtracting the continuity equation (6.5) multiplied by ϕ_P from Equation (6.4) become:

$$
\begin{aligned}
a_E &= D_e - \frac{F_e}{2} \\
a_W &= D_w + \frac{F_w}{2} \\
a_N &= D_n - \frac{F_n}{2} \\
a_S &= D_s - \frac{F_s}{2} \\
a_P &= a_E + a_W + a_N + a_S \\
\text{and} \quad b &= S
\end{aligned}
\tag{6.15}
$$

For the central difference scheme, doing a Taylor–series expansion about the point P gives a second–order accurate scheme.

(c) Second–Order Upwind Scheme

In this case, we estimate ϕ at any interface by a linear extrapolation of the ϕ values at two upwind neighbors, as shown in Fig. 14(c). Thus, for example, depending on the sign of F_e, we estimate ϕ_e as

$$\phi_e = \frac{3}{2}\phi_P - \frac{1}{2}\phi_W \;, \qquad \text{if } F_e > 0$$

$$\phi_e = \frac{3}{2}\phi_E - \frac{1}{2}\phi_{EE} \;, \qquad \text{if } F_e < 0 \tag{6.16}$$

which can be summarized as follows,

$$F_e\phi_e = \left(\frac{3}{2}\phi_P - \frac{1}{2}\phi_W\right) [\![F_e, 0]\!] - \left(\frac{3}{2}\phi_E - \frac{1}{2}\phi_{EE}\right) [\![-F_e, 0]\!] \tag{6.17}$$

Similar expressions can be written for the other interfaces.

Many alternatives can be devised to implement the second–order upwind scheme into the standard notation typified by equations (6.13) and (6.15). Some information can be found in Shyy *et al.* (1992b) in this regard. For instance, in one form of the second–order upwind devised in Shyy *et al.* (1992b), only the terms not involving the four immediate neighbors of the grid point in question are placed in the source term; all the other terms constitute the coefficients without regard to diagonal dominance. In the present work, it is found to be advantageous to use the first–order upwind scheme as the "basis" and assign the difference between the two schemes to the source term b, which will be treated explicitly. This arrangement offers a simple and straightforward way to implement more complicated schemes in general, including not only the second–order upwind scheme but also QUICK and other schemes. A merit of such a treatment is that it maintains diagonal dominance in the course of the iterative updating. With the present formulation of the second–order upwind scheme, a slightly faster convergence rate has been observed. For a scheme such as QUICK, the type of formulation of the scheme is very critical for yielding a converged solution for the large grid sizes employed for numerical computations.

As mentioned above, we write the total flux for the second–order upwind scheme as the flux for the first–order upwind scheme with the additional terms going into the source term *b*:

$$F_e u_e = \left[u_p + \frac{1}{2}(u_p - u_w) \right] [\![F_e, 0]\!] - \left[u_E + \frac{1}{2}(u_E - u_{EE}) \right] [\![-F_e, 0]\!] \tag{6.18}$$

Similar to the first–order upwind and second–order central difference schemes, we subtract the continuity equation (6.5) multiplied by ϕ_P and now all the coefficients are the same as that for the first–order upwind scheme, i.e., those given by equation (6.13). However, we now have the following form for the source term b:

$$\begin{aligned}
b = \frac{1}{2} \Big\{ &(\phi_W - \phi_P) [\![F_e, 0]\!] + (\phi_E - \phi_{EE}) [\![-F_e, 0]\!] \\
+ &(\phi_W - \phi_{WW}) [\![F_w, 0]\!] + (\phi_E - \phi_P) [\![-F_w, 0]\!] \\
+ &(\phi_S - \phi_P) [\![F_n, 0]\!] + (\phi_N - \phi_{NN}) [\![-F_n, 0]\!] \\
+ &(\phi_S - \phi_{SS}) [\![F_s, 0]\!] + (\phi_N - \phi_P) [\![-F_s, 0]\!] \Big\} + S
\end{aligned} \tag{6.19}$$

It must be noted that any form of the same basic scheme will be consistent and conservative as long as the following property holds:

$$(F_e u_e)_i = (F_w u_w)_{i+1} \tag{6.20}$$

and likewise for the other faces. Here, in order to evaluate the flux at any interface, we must use an expression of the form given by equation (6.18), which includes the coefficients a_P, a_E, etc. multiplied with the respective ϕ's and the source term b. Thus, any form of the scheme that satisfies the above requirement should converge to the same solution. If formulated appropriately, it is only the convergence and numerical stability properties of the different implementations which are expected to be different. As regards the formal order of accuracy of this scheme, a Taylor series expansion about the point P yields a second–order accurate scheme.

Compared to the first–order upwind scheme, the shape function utilized for the extrapolation procedure for the second–order upwind scheme is of a higher order, allowing some variation in the solution profile between the adjacent nodes. There are some implications in this regard. Firstly, it may be recalled that the use of the first–order upwind scheme is basically justified by the observation that, for the high Reynolds number case of the one–dimensional linear Burgers equation, the solution profile is indeed piecewise constant ("flat") throughout the whole domain except in the thin layer next to the boundary. With the use of a linear extrapolation, instead of a piecewise constant extrapolation, the second–order upwind scheme appears capable of accommodating flow fields that span a wide range of the cell Reynolds number. In this aspect, one should note that for a nominally high Reynolds number flow, the actual dominance of the convection effect relative to the viscous effect can vary considerably from one region to the other, as discussed and demonstrated in Shyy *et al.* (1992b).

(d) QUICK Scheme

The QUICK scheme (Quadratic Upstream Interpolation for Convective Kinematics) has been proposed by Leonard (1979). It uses a quadratic interpolation between two upstream neighbors and one downstream neighbor in order to estimate the value at any control volume interface, as shown in Fig. 14(d). Depending on the sign of the flux at any interface, we estimate by a quadratic interpolation; for example, at the east face, we have

$$\begin{aligned}
\phi_e &= \frac{3}{8}\phi_E + \frac{3}{4}\phi_P - \frac{1}{8}\phi_W, \quad \text{if } F_e > 0 \\
\phi_e &= \frac{3}{4}\phi_E + \frac{3}{8}\phi_P - \frac{1}{8}\phi_{EE}, \quad \text{if } F_e < 0
\end{aligned} \tag{6.21}$$

which can be summarized as

$$F_e\phi_e = \left(\frac{3}{8}\phi_E + \frac{3}{4}\phi_P - \frac{1}{8}\phi_W\right)[\![F_e,0]\!] - \left(\frac{3}{4}\phi_E + \frac{3}{8}\phi_P - \frac{1}{8}\phi_{EE}\right)[\![-F_e,0]\!] \tag{6.22}$$

As for the local order of accuracy, if we do a Taylor series expansion about the velocity at the point P, we get a second–order accurate scheme. Compared to the first– and second–order upwind schemes, the QUICK scheme assumes a parabolic shape profile for conducting the interpolation procedure. From this viewpoint, it seems that although QUICK is still a second–order scheme in terms of the formal order of accuracy, it can yield a higher degree of actual accuracy for many practical computations. However, on the other hand, it is well known that the performance of a convection scheme is not correlated with either the degree of formal order or the complexity of the scheme. It has been proved in Shyy (1985a) for instance, based on the methodology discussed in Chapter I, that for the linear Burgers equation, the critical cell Peclet number for the QUICK scheme is 8/3, which is only slightly higher than that for the second–order central difference scheme (namely, 2). Hence, the actual performance of the scheme depends on the balance between the higher order upwind extrapolation profile and the weaker ability of resisting spurious oscillations. It is not appropriate to make a universal assertion regarding the relative performance among these higher–order upwind schemes since their performance can vary for different physical problems.

6.4 Controlled Variation Scheme

Following the discussion in Section 4 of Chapter IV, we address some implementational aspects of Harten's TVD scheme in a pressure–based algorithm. TVD schemes have been widely implemented in density–based simultaneous solvers, yielding solution profiles with very little smearing (unlike the conventional first–order schemes) and no spurious oscillations (unlike the conventional higher–order schemes). Studies by Thakur and Shyy (1992, 1993a) have demonstrated that similar accuracy in the solution profiles can be achieved with the TVD schemes implemented in a sequential solver, provided the speed of signal propagation in the solution is coordinated by assigning the local convection speed as the characteristic speed for the whole system and the source terms are appropriately handled. Recently, Liou and Steffen (Liou 1992, Liou and Steffen 1993) have proposed a new class of flux splitting which also treats the convective and pressure fluxes seperately. For both approaches, special treatment of terms involving pressure improves the accuracy of the solution. For simultaneous solvers, the eigenvalues of the Jacobian matrix of the system of equations are assigned as the characteristic speeds of the system.

In a sequential solution approach, the gas dynamics equations are considered as a collection of individual scalar equations and each equation is treated as a scalar conservation law. However, it is necessary to prescribe a characteristic speed for the entire system of sequential equations. The obvious choice is the convection speed and the Euler equations can now be written as

$$\frac{\partial \varrho}{\partial t} + \frac{\partial (\varrho u)}{\partial x} = 0$$
$$\frac{\partial m}{\partial t} + \frac{\partial (mu)}{\partial x} = -\frac{\partial p}{\partial x}$$
$$\frac{\partial E}{\partial t} + \frac{\partial (Eu)}{\partial x} = -\frac{\partial (pu)}{\partial x}$$
(6.23)

In Equation (6.23), the terms on the right hand side are treated as source terms. Regardless of the solution technique, if ϱ, m and E are treated as primary variables, the same situation arises. The local characteristic speed, e.g., $a_{j+1/2}$ defined on the right interface of the control volume for all the three equations, is now defined as

$$a_{j+1/2} = \frac{1}{2}\left(u_j + u_{j+1}\right) \tag{6.24}$$

Thus each of the scalar conservation laws has the same characteristic speed, namely u, and the terms involving pressure are now source terms; this is consistent with the conventional sequential pressure–based methods.

The algebraic TVD scheme of Harten is now applied to this system. Explicit Euler time discretization is used, with the source terms appearing explicitly. A conservation law for the dependent variable ϕ, with a source term $\psi(\phi)$ can be written as

$$\phi_t + f(\phi)_x = \psi(\phi) \tag{6.25}$$

To treat the source terms, several approaches have been used in literature (Yee 1987, Strang 1968):

(a) an extension of MacCormack's predictor–corrector method by applying the flux limiter in the corrector step and including the source terms in either an explicit or an implicit manner, and

(b) a time–splitting procedure, proposed by Strang (1968), in which one alternates between solving a system of conservation laws with no source terms and a system of ordinary differential equations modelling the source term. In Thakur and Shyy (1993a), it is found that both of the above source term treatments yield very comparable results. The approach adopted here is Strang's time–splitting, which maintains second–order accuracy.

As a test case, the standard shock tube problem is used. An ideal gas is present in the tube on either side of a diaphragm located at the center of the tube with initial states to the left and the right of the diaphragm as shown in Fig. 15. The total length of the shock tube is 14 meters and is discretized using 141 equally spaced grid points. The value of λ, which is the ratio of $\Delta t/\Delta x$, is chosen to be 0.1 and thus Δt is 0.01. At time $t=0$ the diaphragm is broken and the solution is traced at subsequent time steps. The solution profile consists of a shock wave and a contact discontinuity moving to the right of the initial location of the diaphragm and an expansion fan moving to the left.

The density profiles of the shock tube problem after elapsed time $t=2.0$ for the system of simultaneous Euler equations obtained by using the first–order upwind and the conventional second–order Lax–Wendroff schemes are shown in Fig. 16(a). The first–order scheme smears out the profiles near the discontinuities whereas the second–order Lax–Wendroff scheme yields spurious oscillations. The density profile at $t=2.0$ obtained using Harten's TVD scheme for the system of simultaneous Euler equations with $\delta=0$ is plotted in Fig. 16(b). It can be seen that the TVD scheme gives high–order accuracy in the smooth regions of the flow and a much better resolution of the discontinuities, compared to the conventional schemes and exhibits no spurious oscillations.

The algebraic TVD scheme of Harten is now implemented in the sequential solver. Explicit Euler time discretization is used, with the source terms appearing explicitly. As in the simultaneous case, the value of λ used in these calculations is 0.1 with the timestep size $\Delta t=0.01$. For $\delta=0$, i.e., for the least amount of allowable dissipation in the TVD scheme,

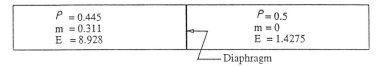

Figure 15. Schematic of the shock tube problem.

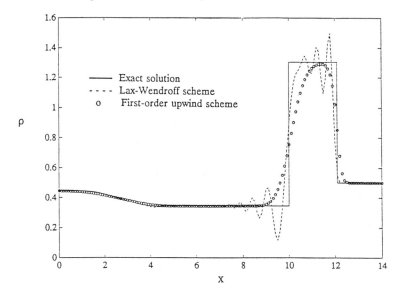

a) First-order upwind and second-order Lax-Wendroff schemes.

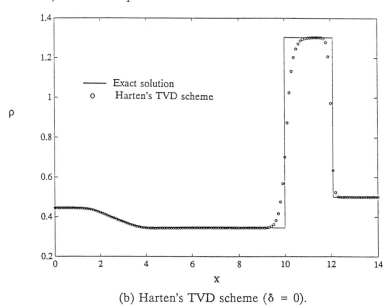

(b) Harten's TVD scheme (δ = 0).

Figure 16. Density profiles using the simultaneous solution approach with different schemes.

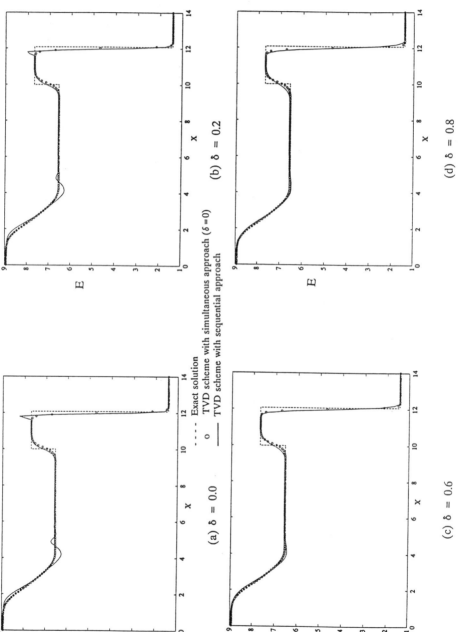

Figure 17. Energy profiles using the sequential approach along with special source term treatment, for various δ.

the sequential approach yields severe oscillations near the discontinuities, as seen in Fig. 17(a). This can be attributed to two factors – one is the lack of dissipation in the scheme to coordinate the different equations and the other is the presence of source terms on the right hand side of equation (6.23). To observe the effect of the amount of dissipation on the accuracy of the scheme, the value of δ is successively increased; the plots for various δ are shown in Fig. 17. The profiles obtained using the simultaneous solution approach (with δ=0) are also plotted in Fig. 17 to compare the relative accuracy of the sequential approach for varying values of δ, both in terms of the presence of oscillations and the amount of smearing of the profiles. It is seen that for lower δ the energy profile compares very well with that obtained using the simultaneous approach, in terms of smearing of the profile in the region of the contact discontinuity and the shock but some oscillations are present. The oscillations are completely suppressed for $\delta \geq 0.8$, as seen in Fig. 17(d). This area is still in a developing stage, awaiting more effort to be devoted. However, the concept presented herein can be extended to the pressure–based algorithm with little conceptual difficulty. It offers a way to control the amount of numerical diffusion in a spatially varying manner. A controlled variation scheme (CVS), based on the concept just presented, has been under development in the context of pressure–based algorithms (Shyy and Thakur 1993), as will be presented in the following.

6.4.1 Harten's Implicit Second–Order TVD Scheme

A scalar conservation law can be written as follows

$$\frac{\partial w}{\partial t} + \frac{\partial f(w)}{\partial x} = 0 \tag{6.26}$$

where w is the dependent variable and f is the flux of w. For the second–order implicit Harten's TVD scheme, the numerical approximation for w_i^{n+1}, where the subscript i represents the spatial node and the superscript $n+1$ and n represent the time levels, is given by

$$w_i^{n+1} + \lambda\theta\left(\tilde{f}_{i+1/2}^{n+1} - \tilde{f}_{i-1/2}^{n+1}\right) = w_i^n - \lambda(1 - \theta)\left(\tilde{f}_{i+1/2}^n - \tilde{f}_{i-1/2}^n\right) \tag{6.27}$$

where $\lambda = \Delta t/\Delta x$ and θ is a measure of implicitness of the scheme. We obtain explicit, fully implicit and Crank–Nicolson schemes for $\theta = 0$, 1 and 1/2, respectively. The numerical fluxes in the above are given by

$$\tilde{f}_{i+1/2} = \frac{1}{2}\left[f_i + f_{i+1} + g_i + g_{i+1} - Q(a_{i+1/2} + \gamma_{i+1/2})\Delta_{i+1/2}w\right] \tag{6.28a}$$

$$\tilde{f}_{i-1/2} = \frac{1}{2}\left[f_i + f_{i-1} + g_i + g_{i-1} - Q(a_{i-1/2} + \gamma_{i-1/2})\Delta_{i-1/2}w\right] \tag{6.28b}$$

For later use, we express the above numerical fluxes as follows:

$$\tilde{f}_{i+1/2} = \overline{F}_{i+1/2} - \left(\overline{Q}_{i+1/2} - \overline{G}_{i+1/2}\right) \tag{6.29a}$$

$$\tilde{f}_{i-1/2} = \overline{F}_{i-1/2} - \left(\overline{Q}_{i-1/2} - \overline{G}_{i-1/2}\right) \tag{6.29b}$$

and also write

$$\overline{F} = \overline{F}_{i+1/2} - \overline{F}_{i-1/2}, \quad \overline{Q} = \overline{Q}_{i+1/2} - \overline{Q}_{i-1/2}, \quad \overline{G} = \overline{G}_{i+1/2} - \overline{G}_{i-1/2} \qquad (6.30)$$

The $\overline{F}$, $\overline{Q}$ and $\overline{G}$ terms in Eqs. (6.29) and (6.30) consist of the f, g and Q terms of Eq. (6.28), respectively. For later reference, we present here the complete forms of $\overline{Q}_{i+1/2}$ and $\overline{G}_{i+1/2}$:

$$\overline{Q}_{i+1/2} = \frac{1}{2}Q\left(a_{i+1/2} + \gamma_{i+1/2}\right)\Delta_{i+1/2}w \qquad (6.31a)$$

$$\overline{G}_{i+1/2} = \frac{1}{2}\left(g_i + g_{i+1}\right) \qquad (6.31b)$$

The various quantities used in the above definitions are given as follows:

$$g_i = s_{i+1/2} \cdot \max\left[0, \min\left(\sigma(a_{i+1/2}) \cdot |\Delta_{i+1/2}w|, s_{i+1/2} \cdot \sigma(a_{i-1/2})\Delta_{i-1/2}w\right)\right] \qquad (6.32)$$

$$= \begin{cases} s_{i+1/2} \cdot \min\left(\sigma(a_{i+1/2}) \cdot |\Delta_{i+1/2}w|, \sigma(a_{i-1/2}) \cdot |\Delta_{i-1/2}w|\right), & \text{if } \Delta_{i+1/2}w\Delta_{i-1/2}w > 0 \\ 0 & \text{if } \Delta_{i+1/2}w\Delta_{i-1/2}w \le 0 \end{cases}$$

with

$$\sigma(a_{i+1/2}) = \begin{cases} \frac{1}{2}Q(a_{i+1/2}) + \left(\theta - \frac{1}{2}\right)\lambda(a_{i+1/2})^2 & \text{for } \textit{unsteady cases} \\ \frac{1}{2}Q(a_{i+1/2}) & \text{for } \textit{steady cases} \end{cases} \qquad (6.33a)$$

$$s_{i+1/2} = sign\left(\Delta_{i+1/2}w\right) \qquad (6.33b)$$

$$a_{i+1/2} = \begin{cases} \dfrac{f_{i+1} - f_i}{\Delta_{i+1/2}w} & \text{if } \Delta_{i+1/2}w \ne 0 \\ \dfrac{\partial f}{\partial w} & \text{if } \Delta_{i+1/2}w = 0 \end{cases} \qquad (6.33c)$$

$$\gamma_{i+1/2} = \begin{cases} \dfrac{g_{i+1} - g_i}{\Delta_{i+1/2}w} & \text{if } \Delta_{i+1/2}w \ne 0 \\ 0 & \text{if } \Delta_{i+1/2}w = 0 \end{cases} \qquad (6.33d)$$

and $\quad \Delta_{i+1/2}w = w_{i+1} - w_i$. $\qquad (6.33e)$

It should be noted that $a_{i+1/2}$ is the physical characteristic speed whereas $(a_{i+1/2} + \gamma_{i+1/2})$ is the modified numerical characteristic speed. Q, called the dissipation function, is a non–negative quantity, given by

$$Q(x) = |x| \qquad (6.34)$$

It should be pointed out here that a modification to the above expression of $Q(x)$ is often made for the solution of Euler equations in simultaneous solvers by selectively increasing the magnitude of $Q(x)$ for vanishingly small characteristic speeds (Harten 1983, 1984, Yee *et al.* 1985). However, there it is necessary to introduce this additional numerical dissipation to prevent the violation of the so–called entropy condition. For computing incompressible, recirculating flows using the multi–dimensional pressure–based sequential solver employed in the present study, no such modification of $Q(x)$ is required.

We employ the fully implicit TVD scheme (i.e., $\theta = 1$) as the basis for development of the controlled variation scheme (CVS) for computing steady state flows. The implicit and highly nonlinear equations resulting from this would require iterations at every timestep if a time–stepping approach to steady state is employed. If an infinite timestep is chosen to solve for steady state, as in the present study, the number of iterations required to achieve convergence will be very large. Consequently, some linearized versions of implicit TVD schemes have been devised (Yee *et al.* 1985, Yee 1986, 1987a). We describe two of these versions next. Both versions will be adopted to solve the one–dimensional model problems.

6.4.2 Linearized Non–Conservative Implicit (LNI) Form

Using definitions (6.28a) and (6.28b), Eq. (6.27) can be written as

$$w_i^{n+1} + \frac{\lambda\theta}{2}\begin{bmatrix} f_{i+1} + f_i + g_{i+1} + g_i - Q\left(a_{i+1/2} + \gamma_{i+1/2}\right)\Delta_{i+1/2}w \\ -f_i - f_{i-1} - g_i - g_{i-1} + Q\left(a_{i-1/2} + \gamma_{i-1/2}\right)\Delta_{i-1/2}w \end{bmatrix}^{n+1} \tag{6.35}$$
$$= w_i^n - \lambda(1-\theta)\left(\tilde{f}_{i-1/2}^n - \tilde{f}_{i-1/2}^n\right) \equiv \mathcal{G}^n$$

Now using definitions (6.33c) and (6.33d), we can write the above as

$$w_i^{n+1} + \frac{\lambda\theta}{2}\left[\left(a_{i+1/2} + \gamma_{i+1/2}\right) - Q\left(a_{i+1/2} + \gamma_{i+1/2}\right)\right]^{n+1}\Delta_{i+1/2}w^{n+1}$$
$$- \frac{\lambda\theta}{2}\left[-\left(a_{i-1/2} + \gamma_{i-1/2}\right) - Q\left(a_{i-1/2} + \gamma_{i-1/2}\right)\right]^{n+1}\Delta_{i-1/2}w^{n+1}$$
$$= w_i^n - \lambda(1-\theta)\left(\tilde{f}_{i-1/2}^n - \tilde{f}_{i-1/2}^n\right) \equiv \mathcal{G}^n \tag{6.36}$$

This form can be shown to be TVD (Harten 1984, Yee 1986). However, this form of the implicit scheme cannot be expressed in conservation form and thus it is non–conservative except at steady state where it has been shown that it does reduce to a conservative form (Harten 1984). Thus, this linearized version is suitable only for steady state calculations.

For our purpose in the present study, we choose the fully implicit scheme, i.e., $\theta = 1$, and we employ the approach of taking an infinitely large timestep to make the numerical fluxes completely independent of the timestep. Thus we get the following:

$$\left(w_i^{n+1} - w_i^n\right)\frac{\Delta x}{\Delta t}$$
$$+ \frac{1}{2}\left[\left(a_{i+1/2} + \gamma_{i+1/2}\right) - Q\left(a_{i+1/2} + \gamma_{i+1/2}\right)\right]^{n+1}\Delta_{i+1/2}w^{n+1} \tag{6.37}$$
$$- \frac{1}{2}\left[-\left(a_{i-1/2} + \gamma_{i-1/2}\right) - Q\left(a_{i-1/2} + \gamma_{i-1/2}\right)\right]^{n+1}\Delta_{i-1/2}w^{n+1} = 0$$

where the first term vanishes due to the infinitely large timestep Δt. The superscripts n and

$n+1$ signify the previous and current iteration levels at steady state, respectively. If we linearize the above nonlinear equation by dropping the superscripts of the coefficients of $\varDelta_{i\pm1/2}w^{n+1}$ from $n+1$ to n, we get the following:

$$E_1 w_{i-1}^{n+1} + E_2 w_i^{n+1} + E_3 w_{i+1}^{n+1} = 0 \tag{6.38}$$

where

$$E_1 = \frac{1}{2}\left[-\left(a_{i-1/2} + \gamma_{i-1/2}\right) - Q\left(a_{i-1/2} + \gamma_{i-1/2}\right)\right]^n \tag{6.39a}$$

$$E_3 = \frac{1}{2}\left[\left(a_{i+1/2} + \gamma_{i+1/2}\right) - Q\left(a_{i+1/2} + \gamma_{i+1/2}\right)\right]^n \tag{6.39b}$$

$$E_2 = -\left(E_1 + E_2\right) \tag{6.39c}$$

It can be observed that the above form is spatially a five–point scheme with a tridiagonal system of linear equations and has a dominant diagonal.

6.4.3 Linearized Conservative Implicit (LCI) Form

A linearized conservative version of the implicit TVD scheme can also be obtained. First, we write Eq. (6.28a) as

$$\tilde{f}_{i+1/2}^{n+1} = \frac{1}{2}\left[f_i^{n+1} + f_{i+1}^{n+1}\right] + \left[\left[\frac{g_i + g_{i+1}}{\varDelta_{i+1/2}w}\right] - Q(a_{i+1/2} + \gamma_{i+1/2})\right]^{n+1} \varDelta_{i+1/2}w^{n+1} \tag{6.40}$$

Then using a local Taylor series expansion about w^n given by

$$f^{n+1} - f^n = a^n\left(w^{n+1} - w^n\right) + O\left(\varDelta t^2\right) \tag{6.41}$$

where $a = \partial f/\partial w$ and adding and subtracting f_{i+1}^n to f_{i+1}^{n+1} we get

$$\left(f_{i+1}^{n+1} - f_{i+1}^n\right) + f_{i+1}^n = a_{i+1}^n\left(w_{i+1}^{n+1} - w_{i+1}^n\right) + f_{i+1}^n \equiv a_{i+1}^n w_{i+1}^{n+1} \tag{6.42}$$

Performing a similar manipulation in the expression for $\tilde{f}_{i-1/2}$, we obtain a linearized conservative implicit (LCI) form which preserves the conservative property of the differencing scheme and is thus applicable for both steady and unsteady problems.

For our present applications, we choose the fully implicit scheme with an infinitely large timestep and linearize by dropping the superscripts of the coefficients of $\varDelta_{i\pm1/2}w^{n+1}$ from $n+1$ to n; we get the following form:

$$E_1 w_{i-1}^{n+1} + E_2 w_i^{n+1} + E_3 w_{i+1}^{n+1} = 0 \tag{6.43}$$

where

$$E_1 = \frac{1}{2}\left[-a_{i-1} + \tau_{i-1/2} - Q\left(a_{i-1/2} + \gamma_{i-1/2}\right)\right]^n \tag{6.44a}$$

$$E_3 = \frac{1}{2}\left[a_{i+1} + \tau_{i+1/2} - Q\left(a_{i+1/2} + \gamma_{i+1/2}\right)\right]^n \tag{6.44b}$$

$$E_2 = \frac{1}{2}\left[-\tau_{i+1/2} + Q\left(a_{i+1/2} + \gamma_{i+1/2}\right) - \tau_{i-1/2} + Q\left(a_{i-1/2} + \gamma_{i-1/2}\right) \right]^n \quad (6.44c)$$

with

$$\tau_{i+1/2}^n = \left[\frac{g_i + g_{i+1}}{\Delta_{i+1/2}w} \right]^n \qquad\qquad \tau_{i-1/2}^n = \left[\frac{g_i + g_{i-1}}{\Delta_{i-1/2}w} \right]^n \qquad (6.44d)$$

Like the LNI form, this too is a five–point scheme with a tridiagonal system of linear equations but it does not maintain diagonal dominance because in general $a_{i+1} - a_{i-1} \neq 0$. A drawback of the LCI form is that it may not be unconditionally TVD for $\theta = 1$; no proof exists yet (Yee 1986, 1987b). However, in practice, the LCI form has been found to be stable and yields high accuracy for fairly large CFL numbers (Yee 1986).

6.4.4 Model Problem I: Linear Convection–Diffusion Equation

The first model problem studied here is the one–dimensional, steady state linear Burgers' equation:

$$a\phi_x = \beta\phi_{xx} , \qquad a,\beta = constants > 0 \qquad (6.45)$$

with the following boundary conditions:

$$\phi(0) = 0 , \qquad \phi(1) = 1$$

The exact solution to this problem is given by

$$\phi(x) = \frac{1 - e^{Rx}}{1 - e^R} \qquad (6.46)$$

where $R = a/\beta$. For small values of R, the solution varies smoothly throughout the domain. But for large values of R, $\phi(x)$ is virtually zero except in a region near $x=1$ which has the thickness of the order of $1/R$. Thus, the solution is of boundary layer type since it is dominated by convection in the majority of the domain and the entire variation in the solution is contained in the thin region next to the right boundary.

Numerically, the diffusion term is conveniently and effectively approximated by second–order central differencing. If we use central differencing for the convection term, then the numerical approximation to Eq. (6.45) takes the form

$$(2 - P)\phi_{i+1} + (-4)\phi_i + (2 + P)\phi_{i-1} = 0 \qquad (6.47)$$

where $P = ah/\beta$ and is called the cell Peclet number. Here h is the grid spacing, assumed to be uniform. Eq. (6.47) indicates that if $|P| > 2$, then oscillations in the solution profile may result due to a lack of necessary damping.

If the CVS is employed for the convection term, Eq. (6.45) is approximated by

$$\left(\tilde{f}_{i+1/2} - \tilde{f}_{i-1/2}\right) = \frac{1}{P}\left(\phi_{i+1} - 2\phi_i + \phi_{i-1}\right) \qquad (6.48)$$

where $\tilde{f}_{i\pm1/2}$ are defined in Eq. (6.28); here $f \equiv \phi$ and $a_{i+1/2} = a_{i-1/2} = 1$. Using these definitions, we can write the numerical scheme in the form

$$\left(\phi_{i+1} - \phi_{i-1}\right) = \frac{2}{P}\left[\left(\phi_{i+1} - 2\phi_i + \phi_{i-1}\right) + S \cdot P\right] \tag{6.49}$$

where S consists of the net flux over and above the central difference flux, i.e.,

$$S = -\frac{1}{2}\left[g_{i+1} - g_{i-1} - Q(a_{i+1/2} + \gamma_{i+1/2})\Delta_{i+1/2}w\right.$$
$$\left. + Q(a_{i-1/2} + \gamma_{i-1/2})\Delta_{i-1/2}w\right] \tag{6.50}$$

From this viewpoint, the controlled variation scheme can be interpreted as nothing but the basic central difference scheme accompanied by a nonlinear numerical dissipation term which suppresses the dispersive characteristic of the central difference scheme at high cell Peclet numbers. In the next section, we discuss the issue of incorporation of this nonlinear term into the numerical scheme which can lead to different versions of the CVS.

For the linear convection–diffusion equation studied here, we choose $\alpha = 1000$ and $\beta = 1$. The length of the domain is 1 with the grid spacing $h = 0.02$. Thus, the cell Peclet number for the flow is 20. The central difference scheme, as expected, yields oscillations as seen in Fig. 18(a). The CVS operates by injecting a nonlinear numerical dissipation into the central difference scheme to suppress its dispersive effect. This is evident from Fig. 18(a) where it can be seen that the CVS yields an oscillation–free solution profile.

To compare the performance of the CVS with other convection schemes, the solution profile obtained with the QUICK scheme is also plotted in Fig. 18(a) and those obtained with first–order upwind and second–order upwind schemes are plotted in Fig. 18(b) along with that obtained with the CVS, in the vicinity of the right boundary. The critical value of cell Peclet number for the model problem, beyond which oscillations occur, is ∞ for the first– and second–order upwind schemes, 2 for the central difference scheme and 8/3 for the QUICK scheme (Shyy 1985a). As can be seen from Fig. 18(a), the QUICK scheme indeed yields oscillations, albeit smaller in magnitude and extent than those obtained with the central difference scheme. First– and second–order upwind schemes yield oscillation–free solution profiles similar to the profile obtained by the CVS. First–order upwinding, which has a high level of numerical dissipation, smears out the solution profile the most. The profile obtained with the CVS turns out to be more smeared than that with the second–order upwind scheme, and is comparable to that with the first–order upwind scheme. Nevertheless, suffice it to say here that all three numerical solutions shown in Fig. 18(b) are very acceptable, and the basic goal of the CVS, namely, to help improve the solution accuracy based on the central difference scheme for a case with high cell Peclet number, is realized. It should be pointed out that the linear Burgers' equation without source terms is a more straightforward problem to solve. The relative performance of all these schemes is more distinguishable for a very high cell Peclet number flow with a source term.

Next we demonstrate the mechanism by which the CVS operates. For the present model problem, if we use the central difference scheme for the convection term, wiggles are produced as seen in Fig. 18(a). Here, it can be easily shown that the wiggles are

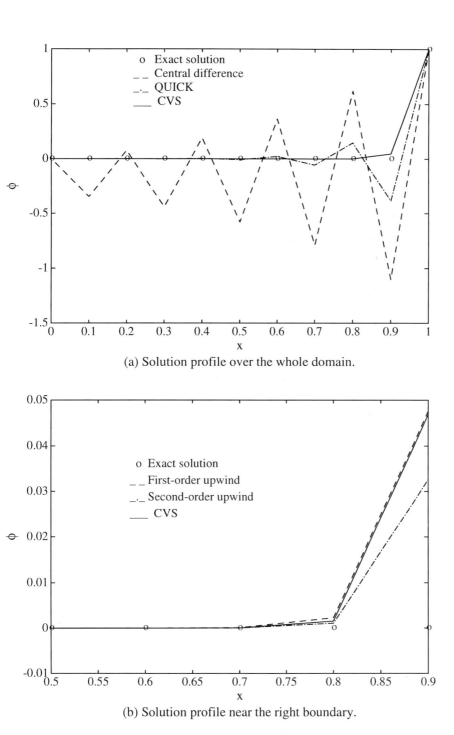

(a) Solution profile over the whole domain.

(b) Solution profile near the right boundary.

Figure 18. Solution profiles for model problem I (without source) using various schemes.

produced due to the right boundary condition. For example, for a high cell Peclet number such as the one used in the present case, the exact solution profile consists of zeros (virtually) everywhere except at the right boundary point. With the central difference scheme, at the grid point adjacent to the boundary, i.e., i=10, in order that Eq. (6.47) holds for initially given values of ϕ_{11} (=1) and ϕ_9 (=0) with $P > 2$, ϕ_{10} must assume a negative value to increase the magnitude of the diffusion term. Thus, the absence of any numerical dissipation in central differencing to augment the physical diffusion causes the solution profile to exhibit wiggles in order to supply an "artificial diffusion" to balance the convection term in the region where the important length scale can not be resolved appropriately, as quite lucidly illustrated by Roache (1972). From a different point of view, one of the roots of the characteristic equation corresponding to the finite difference equation resulting from the central difference scheme becomes negative when $|P| > 2$ and according to the standard theory of finite difference equations, oscillations in the solution might result. This, of course, is a necessary condition; for instance, if, instead of the Dirichlet condition specified presently for the right boundary, we specify a Neumann boundary condition of the type $\partial\phi/\partial x=0$, the central difference scheme will yield the oscillation–free exact solution which is zero everywhere, for any value of cell Peclet number. This is due to the fact that with such a Neumann boundary condition, the governing equation itself reduces to a first–order equation, thus eliminating the root of the finite difference equation whose sign is dependent on the magnitude of the cell Peclet number.

For the model problem under investigation, the CVS, which is based on the central difference scheme, eliminates the wiggles discussed above by injecting numerical dissipation where necessary. In order to illustrate the relation between the balancing terms in the CVS, we write the fully nonlinear form as follows

$$\frac{a}{2}\left(\phi_{i+1} + \phi_i\right) - \frac{a}{2}\left(\phi_i + \phi_{i-1}\right) = \tag{6.51}$$

$$\left[\frac{\beta}{h} + \frac{a\left(\overline{Q}_{i+1/2} - \overline{G}_{i+1/2}\right)}{\Delta_{i+1/2}\phi}\right]\left(\phi_{i+1} - \phi_i\right) - \left[\frac{\beta}{h} + \frac{a\left(\overline{Q}_{i-1/2} - \overline{G}_{i-1/2}\right)}{\Delta_{i-1/2}\phi}\right]\left(\phi_i - \phi_{i-1}\right)$$

In the above, a is the coefficient of effective convection; the first term within the square brackets on the right hand side of the equation, namely β/h, is the coefficient of physical diffusion and the second term in the square brackets is the nonlinear dissipation of the CVS. The effective cell Peclet number of the CVS for the model problem can thus be defined as the ratio of the convective coefficient and the coefficient of net dissipation:

$$P_{cvs} = \frac{a}{\left[\frac{\beta}{h} + a\left(\frac{\overline{Q}_{i+1/2} - \overline{G}_{i+1/2}}{\Delta_{i+1/2}\phi}\right)\right]} \tag{6.52}$$

While computing the coefficient of the nonlinear dissipation (second term in the denominator of the above expression), one can avoid the singularity that occurs when

$\Delta_{i+1/2}\phi = 0$ by prescribing this coefficient to be zero when $\left|\Delta_{i+1/2}\phi\right| \leq \varepsilon$ where ε is a small positive number. For the present case we use $\varepsilon=10^{-4}$. This is equivalent to saying that when $\left|\Delta_{i+1/2}\phi\right|$ is negligibly small, no extra numerical dissipation is required. It must be pointed out that this artificial cutoff is enforced only for the estimation of the effective cell Peclet numbers and not during the computation of the numerical fluxes used by the CVS. Note that the original cell Peclet number (i.e., with respect to the central difference scheme) is given by

$$P \equiv P_{central} = \frac{a}{\beta/h} \tag{6.53}$$

A comparison of P_{cvs} and $P_{central}$ shows that the CVS reduces the effective cell Peclet number of the flow, where necessary, by augmenting the physical diffusion with numerical dissipation.

Table III shows the values of the relevant quantities associated with the model problem. As presented in Eqs. (6.29)–(6.31), the $\overline{F}$ flux represents the central difference operator, $\overline{Q}$ the numerical dissipation and $\overline{G}$ the numerical anti–dissipation. The $\overline{Q}$ flux is responsible for the TVD property of the original scheme developed by Harten, whereby spurious oscillations near discontinuities are suppressed. The $\overline{G}$ flux serves to improve the accuracy by sharpening the resolution of the scheme via anti–diffusion in regions away from the discontinuities in the flow field. Thus $(\overline{Q}-\overline{G})$ represents the net numerical dissipation present in the scheme and this combination provides the higher–order accuracy in the smooth regions of the flow field along with the capability to suppress spurious oscillations near sharp gradients in the solution profiles. Using an artificial cutoff for the computation of effective cell Peclet numbers by assigning $\varepsilon=10^{-4}$, it can be observed that

Table III. Values of some variables for model problem I; $\alpha=200$, $\beta=1$, $P=20$ *and* $\varepsilon=10^{-4}$.

i	ϕ	$\Delta_{i+1/2}\phi$	Q	G	Q–G	P_{cvs}
1	0.0	–	–	–	–	–
2	8.50×10^{-14}	1.78×10^{-12}	0.0	0.0	0.0	20.0
3	1.87×10^{-12}	5.44×10^{-11}	0.0	0.0	0.0	20.0
4	5.63×10^{-11}	1.67×10^{-9}	0.0	0.0	0.0	20.0
5	1.72×10^{-9}	5.12×10^{-8}	0.0	0.0	0.0	20.0
6	5.29×10^{-8}	1.57×10^{-6}	0.0	0.0	0.0	20.0
7	1.62×10^{-6}	4.82×10^{-5}	0.0	0.0	0.0	20.0
8	4.98×10^{-5}	1.47×10^{-3}	0.7418	0.2581	0.4836	1.8737
9	1.52×10^{-3}	4.53×10^{-2}	0.7418	0.2581	0.4836	1.8737
10	4.69×10^{-2}	9.53×10^{-1}	0.5	0.0	0.5	1.8182
11	1.0	–	–	–	–	–

the CVS injects the dissipation $(\overline{Q} - \overline{G})$ near the right boundary where the wiggles originate for the central difference scheme. In the vicinity of the right boundary, the amount of numerical dissipation augmenting the physical diffusion is high enough to reduce the effective cell Peclet number to a value less than two. However, in the region away from the right boundary, the cell Peclet numbers remain high, resulting in a varying distribution of P_{cvs}. This result has clearly established that the cell Peclet number does not need to be uniformly lower than 2 for a central difference based scheme in order to yield a solution free from spurious oscillations. It should be noted that adjacent to the right boundary, i.e., at the grid point i=10, the first–order version of the CVS has to be employed for which $\overline{G}$=0.

6.4.5 Model Problem II: Linear Burgers' Equation with Source

As the next test case, we choose the model problem

$$\alpha\phi_x = \beta\phi_{xx} + \psi(x) , \qquad \alpha,\beta = constants > 0 \qquad (6.54a)$$
$$\phi(0) = 0$$

with two different choices of conditions at the right boundary, namely,

(i) Neumann boundary condition: $(\phi_x)_{x=L} = 0$ \qquad (6.54b)

(ii) Dirichlet boundary condition: $\phi(L) = 0$

to study the performance of the CVS for convection dominated flows in the presence of source terms. The form of the source term used here is the following:

$$\psi(x) = \begin{cases} ax + b , & 0 \le x \le x_1 \\ -\dfrac{(ax_1 + b)}{x_2}x + \dfrac{(x_1 + x_2)}{x_2}(ax_1 + b) , & x_1 \le x \le (x_1 + x_2) \end{cases} \qquad (6.55)$$

By varying the parameters in the source term, one can investigate the performance of the various schemes under different distributions of the source. Of the various cases suggested by Shyy (1985a), we choose a source spanning through several mesh lengths. The parameters used are: $P=10^8$ or 10^2, $L=15$, $N=35$, $a=-0.5$, $b=1.76$, $x_1=12h$ and $x_2=8h$. As shown in Eq. (6.54b), two different boundary conditions for the right boundary are investigated. It is of interest to compare the performance of the CVS with the other convection schemes mentioned earlier with the prescribed source distribution and two sets of boundary conditions.

In the previous model problem, the mechanism that generated the spurious oscillations with central differencing was the boundary condition. The oscillations there could also be explained from the viewpoint of the roots of the characteristic equation associated with the central difference scheme. In the presence of a source term in the linear Burgers' equation, the solution of the characteristic equation consists of the original general solution for the source–free equation and a particular solution due to the presence of the source term. The general solution here is the trivial one, namely, zero everywhere. Thus, the nature of the solution profile, i.e., the presence and the extent of wiggles for high cell Peclet number, using central differencing now depends on the nature of the source term. When the Neumann boundary condition on the right boundary is used in the presence of the source term with the central difference scheme, the solution exhibits very slight oscillations about the exact solution, as shown in Fig. 19(a). If the Dirichlet boundary condition is used instead of the zero gradient Neumann condition, oscillations, resulting from an interaction with the Dirichlet boundary condition similar to model problem I, appear even for much lower values of *P*. To demonstrate this, the solution profile using the central difference scheme for cell Peclet number 10^2 is shown in Fig. 19(b).

Next we present the results for model problem II obtained using the CVS. For both sets of boundary conditions, CVS yields oscillation–free profiles, as shown in Fig. 20. Thus, it is able to suppress the wiggles in the central difference scheme resulting from the interaction with both source terms and boundary conditions. Moreover, profiles obtained by the CVS are quite comparable to those by the second–order upwind scheme; it is noted that for the Dirichlet boundary condition used here, the QUICK scheme yields oscillations near the right boundary as discussed by Leonard (1979). The first–order upwind, as expected, exhibits an excessive amount of dissipation for both types of boundary conditions as shown in Fig. 20.

It is also illustrative to look at the distribution of the effective cell Peclet numbers, defined in Eq. (6.52). Table IV lists the effective cell Peclet numbers obtained using the CVS for the present model problem with $P=10^2$ and using the Dirichlet boundary condition, i.e., $\phi_{(x=L)}=0$. Fig. 21 shows the distribution of the effective cell Peclet number, P_{cvs}, over the whole domain. The solution profile is also plotted in the same figure (not to scale) to correlate the cell Peclet numbers with the regions of variation in the solution profile. Again, an artificial cutoff, similar to that for the previous model problem, is used for the estimation of effective cell Peclet numbers by prescribing $\varepsilon=10^{-4}$. Similar to that model problem, the CVS operates by introducing an appropriate amount of numerical dissipation, where necessary, to effectively reduce the cell Peclet numbers, thus balancing the convection and diffusion terms so as not to produce any spurious oscillations in the solution profile. For instance, it can be seen from Table IV that, near the left and right boundaries (i=2 and 35), and in the "downhill" part of the solution profile (i=9–13), the effective cell Peclet numbers are much smaller than in the rest of the domain. In fact, adjacent to the left boundary (i=2) and at the extremum (i=9) of the solution profile where the profile undergoes a change of sign in slope, the effective cell Peclet number is less than 2. Thus, the CVS effectively reduces the cell Peclet numbers in regions where spurious oscillations might occur with central differencing. In the other regions of the domain, the effective cell Peclet numbers can be quite high, without generating any wiggles.

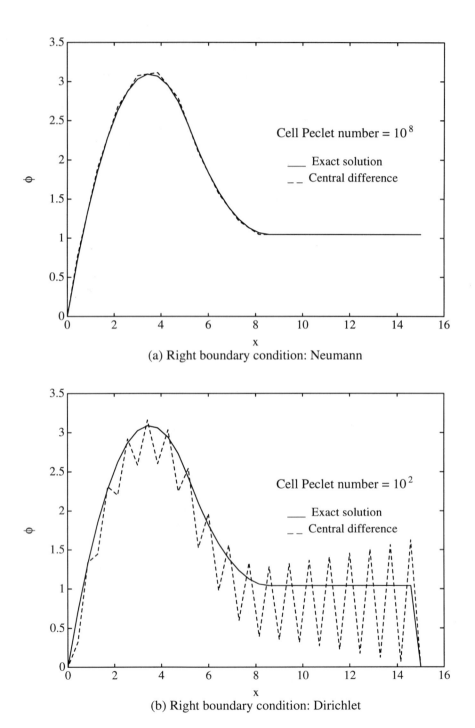

Figure 19. Solution profiles for model problem II (with source) using central differencing.

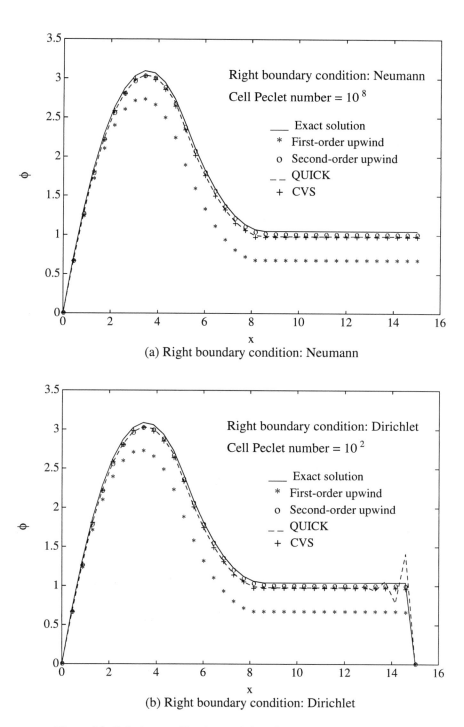

(a) Right boundary condition: Neumann

(b) Right boundary condition: Dirichlet

Figure 20. Solution profiles for model problem II using various schemes.

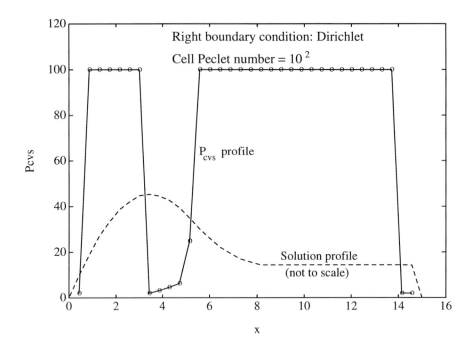

Figure 21. Effective cell Peclet numbers for the CVS for model problem II
with P=100.

Table IV.   Effective Cell Peclet Numbers for the CVS for Model Problem II; $P=10^2$
Right Boundary Condition: Dirichlet

i	x	P_{cvs}
2	0.42	1.96
9	3.42	1.96
10	3.85	3.15
11	4.28	4.66
12	4.71	6.32
13	5.14	24.85
34	14.14	1.97
35	14.57	1.96
Remaining i	Remaining x	100.00

6.4.6 Incompressible, Laminar Navier–Stokes Equations

For a two–dimensional fluid flow problem, the momentum equations can be written as follows:

$$\frac{\partial(\varrho\phi)}{\partial t} + \frac{\partial J_x}{\partial x} + \frac{\partial J_y}{\partial y} = S^* \tag{6.56}$$

where the J's are the net fluxes along the two directions and are composed of convective and diffusive fluxes, as follows:

$$J_x = (\varrho u\phi) - \mu\frac{\partial\phi}{\partial x} \tag{6.57a}$$

$$J_y = (\varrho v\phi) - \mu\frac{\partial\phi}{\partial y} \tag{6.57b}$$

where $\phi \equiv u$ or v, μ is the physical viscosity of the fluid, and S^* is the source term consisting of pressure gradient and any other external source terms. Integrating the above over the control volume, we get

$$\left[(\varrho\phi)_P - (\varrho\phi)_P^0\right]\frac{\Delta x\Delta y}{\Delta t} + [(\varrho u\phi)_e - (\varrho u\phi)_w]\Delta y + [(\varrho v\phi)_n - (\varrho v\phi)_s]\Delta x$$
$$- \left[\left(\mu\frac{\partial\phi}{\partial x}\right)_e - \left(\mu\frac{\partial\phi}{\partial x}\right)_w\right]\Delta y - \left[\left(\mu\frac{\partial\phi}{\partial y}\right)_n - \left(\mu\frac{\partial\phi}{\partial y}\right)_s\right]\Delta x = S^{**} \tag{6.58}$$

where the superscript 0 represents the previous time level and S^{**} is the source term integrated over the control volume.

We now formulate the controlled variation scheme (CVS) by formally extending the TVD scheme from one dimension to the present two–dimensional case. Similar to the model problems described in the previous sections, we use the fully implicit scheme. Steady state analogs of the four versions discussed for model problem I have been investigated for the two–dimensional SIMPLE algorithm and conclusions similar to those for the two model problems have been reached. Consequently, we focus on just the analog of version C of the CVS for model problem I, namely, the linearized version (LNI) of the fully implicit TVD scheme.

Using the LNI form of the fluxes, presented earlier, independently along the x– and y–directions, Eq. (6.58) can be expressed as

$$\varrho\,\Delta y\,C_{i+1/2}^n\left(\phi_E - \phi_P\right)^{n+1} - \varrho\,\Delta y\,C_{i-1/2}^n\left(\phi_P - \phi_W\right)^{n+1}$$
$$+ \varrho\,\Delta x\,C_{j+1/2}^n\left(\phi_N - \phi_P\right)^{n+1} - \varrho\,\Delta x\,C_{j-1/2}^n\left(\phi_P - \phi_S\right)^{n+1}$$
$$= D_{i+1/2}^n\left(\phi_E - \phi_P\right)^{n+1} - D_{i-1/2}^n\left(\phi_P - \phi_W\right)^{n+1} \tag{6.59}$$
$$+ D_{j+1/2}^n\left(\phi_N - \phi_P\right)^{n+1} - D_{j-1/2}^n\left(\phi_P - \phi_S\right)^{n+1} + S^{**}$$

where the various coefficients are given by

$$C_{i\pm1/2}^n = \frac{1}{2}\left[-\varrho\left(a_{i-1/2} + \gamma_{i-1/2}\right) \pm \left(a_{i-1/2} + \gamma_{i-1/2}\right)\right]^n \tag{6.60a}$$

$$D^n_{i+1/2} = \mu_e \frac{\Delta y}{\delta x_e} = \left(\frac{\mu_P + \mu_E}{2}\right)\frac{\Delta y}{\delta x_e} \text{ , etc.} \qquad (6.60b)$$

Note that the subscript i is used to denote the x–direction and j to denote the y–direction.

The local convection speeds are used as the local characteristic speeds for the whole system:

$$a_{i+1/2} \equiv a_{i+1/2,j} = \frac{1}{2}\left(u_{i,j} + u_{i+1,j}\right) \qquad (6.61a)$$

$$a_{j+1/2} \equiv a_{i,j+1/2} = \frac{1}{2}\left(v_{i,j} + v_{i,j+1}\right) \qquad (6.61b)$$

It should be noted that the above interpolations required for the local characteristic speeds result from the staggered nature of the grids employed for the velocity components. Eq. (6.59) can be written as

$$A_P\phi_P = A_E\phi_E + A_W\phi_W + A_N\phi_N + A_S\phi_S + b \qquad (6.62a)$$

The coefficients used in Eq. (6.62a) can be expressed as follows:

$$A_W = D^n_{i-1/2} - C^n_{i-1/2}\,\varrho\,\Delta y$$

$$A_E = D^n_{i+1/2} - C^n_{i+1/2}\,\varrho\,\Delta y$$

$$A_S = D^n_{j-1/2} - C^n_{j-1/2}\,\varrho\,\Delta x \qquad (6.62b)$$

$$A_N = D^n_{j+1/2} - C^n_{j+1/2}\,\varrho\,\Delta x$$

$$A_P = A_W + A_E + A_S + A_N$$

For the present two–dimensional steady state case, the fully nonlinear version of the CVS, similar to Eq. (6.51) for the one–dimensional model problem I, can be written as

$$\frac{F_e}{2}\left(\phi_E + \phi_P\right) - \frac{F_w}{2}\left(\phi_P + \phi_W\right) + \frac{F_n}{2}\left(\phi_N + \phi_P\right) - \frac{F_s}{2}\left(\phi_P + \phi_S\right) = \qquad (6.63)$$

$$\left[D_e + \varrho\Delta y\frac{\left(\overline{Q}_{i+\frac{1}{2}} - \overline{G}_{i+\frac{1}{2}}\right)}{\Delta_{i+\frac{1}{2}}\phi}\right]\left(\phi_E - \phi_P\right) - \left[D_w + \varrho\Delta y\frac{\left(\overline{Q}_{i-\frac{1}{2}} - \overline{G}_{i-\frac{1}{2}}\right)}{\Delta_{i-\frac{1}{2}}\phi}\right]\left(\phi_P - \phi_W\right)$$

$$+ \left[D_n + \varrho\Delta x\frac{\left(\overline{Q}_{j+\frac{1}{2}} - \overline{G}_{j+\frac{1}{2}}\right)}{\Delta_{j+\frac{1}{2}}\phi}\right]\left(\phi_N - \phi_P\right) - \left[D_s + \varrho\Delta x\frac{\left(\overline{Q}_{j-\frac{1}{2}} - \overline{G}_{j-\frac{1}{2}}\right)}{\Delta_{j-\frac{1}{2}}\phi}\right]\left(\phi_P - \phi_S\right)$$

where the subscripts e, w, n and s are used interchangeably with the subscripts $i+1/2, i–1/2,$ $j+1/2$ and $j–1/2$ to denote the east, west, north and the south face, respectively, of the control volume. In the above equation, F_e, for example, represents the mass flux (or the strength of convective effect) across the east face of the control volume and is commonly called the convective coefficient; it is defined as

$$F_e = (\varrho u)_e \, \Delta y \tag{6.64}$$

The coefficient D_e ($\equiv D_{i+1/2}$ given by Eq. 6.60(b)), called the diffusion coefficient, is an indicator of the magnitude of the physical diffusive effect across the east face. Likewise, the terms involving $(\overline{Q} - \overline{G})$ represent the numerical dissipation of the CVS at the four faces of a control volume and can be termed as nonlinear dissipation coefficients. We can define an effective cell Peclet number of the CVS on the east face (i.e., along the x–direction), as

$$(P_e)_{cvs} = \frac{F_e}{\left[D_e + \varrho \Delta y \left(\frac{\overline{Q}_{i+1/2} - \overline{G}_{i+1/2}}{\Delta_{i+1/2}\phi} \right) \right]} \tag{6.65}$$

Likewise, one can define an effective cell Peclet number for the north face, i.e., along the y–direction. The above definition is very similar to the definition given in Eq. (6.52) for the one–dimensional convection–diffusion equation and indicates the relative strength of local convection and the net local diffusion (physical plus numerical). Similar to model problem I, while computing the coefficient of the nonlinear dissipation (second term in the denominator of the above expression), one can avoid the singularity that occurs when $\Delta_{i+1/2}\phi = 0$ by prescribing this coefficient to be zero when

$$\frac{|\Delta_{i+1/2}\phi|}{\phi_{characteristic}} \leq \varepsilon$$

where ε is a small positive number. For the present case we use $\varepsilon = 10^{-4}$. Note that the cell Peclet number for the central difference scheme is defined as

$$(P_e)_{central} = \frac{F_e}{D_e}$$

In order to assess the performance of the controlled variation scheme (CVS) for recirculating flows, flows in a lid–driven cavity and over a backward facing step are chosen as test cases. As discussed earlier, the controlled variation scheme presented here can be viewed as the central difference scheme with an appropriate amount of nonlinear numerical dissipation. The $\overline{Q}$ flux illustrates the physical mechanism by which the CVS operates. It is repeated here to illustrate this mechanism:

$$\overline{Q}_{i+1/2} = \frac{1}{2} Q \left(a_{i+1/2} + \gamma_{i+1/2} \right) \Delta_{i+1/2}\phi$$

Here Q, given by Eq. (6.34), is the numerical dissipation and is a function of the local numerical characteristic speed. Since the local characteristic speed is nothing but the local convection speed, a high local convection speed, which implies a high local cell Peclet number, automatically selects a higher amount of numerical dissipation. However, if the local gradient ($\Delta_{i+1/2}\phi$) is negligible, this damping term is also negligible. A high value of $\Delta_{i+1/2}\phi$ which represents the local gradient of velocity and thus the local cell Peclet number, also prompts the use of a high (though controlled) amount of dissipation by the CVS. The $\overline{G}$ flux balances the $\overline{Q}$ flux in regions where numerical dissipation is not necessary and assumes a zero value in regions where $\Delta_{i+1/2}\phi$ and $\Delta_{i-1/2}\phi$ have an opposite sign. Thus high cell Peclet number situations are handled by the CVS by automatically triggering a variation control in the regions where required.

The results for the driven cavity flow presented here correspond to a Reynolds number 1000 based on the length of the cavity and the velocity of the moving lid. The relative significance of the net numerical dissipation and the physical diffusion can be estimated by defining the normalized viscosity as follows:

$$
Q^* - G^* = \frac{\left[\varrho \Delta y \dfrac{(\overline{Q}_{i+1/2} - \overline{G}_{i+1/2})}{\Delta_{i+1/2}\phi} \right]}{D_{i+1/2}} \tag{6.66}
$$

As the grid is refined, the magnitude of (Q^*-G^*) will diminish since the local cell Peclet numbers, based on the velocity differences across the computational cells, decrease with grid refinement. This means that the effective dissipation injected by the CVS should decrease monotonically as the grid is refined. This is clearly evident from Fig. 22 where the (Q^*-G^*) flux distribution is plotted on four grids with nodes varying from 21x21 to 161x161. It is observed that the (Q^*-G^*) flux decreases monotonically in magnitude as the grid is refined. This serves as an illustration of the mechanism of the CVS. From the viewpoint of the central difference scheme, for a high Reynolds number flow, the local cell Peclet numbers will monotonically decrease with the refinement of the grid; so will the wiggles in the solution and the amount of diffusion selected by the controlled variation scheme is regulated accordingly.

In summary, three key elements have been addressed in an attempt to utilize some of the modern concepts of convection treatment for solving low speed recirculating flows, namely, flux coordination via local convection speed, controlled numerical viscosity based on the formalism of Harten's TVD scheme, and coupling of source term and convection treatment. More work is needed to further advance this line of approach.

7 CONVERGENCE AND MATRIX SOLVER

7.1 General Characteristics

As discussed in Chapter III, efficient solvers are needed to handle large number of equations resulting from discretization based on finite mesh sizes. Among the many choices available for such a task, some methods are more popular; they include the standard LSOR–TDMA already mentioned, a five–point strongly implicit procedure (SIP) originally proposed by Stone (1968), a modified SIP (MSIP) developed by Schneider and Zedan (1981), and the preconditioned conjugate gradient method (Chen *et al.* 1991). It is well known that for the LSOR–TDMA type of method, the number of

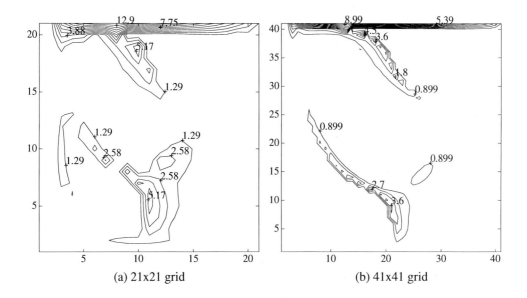

(a) 21x21 grid

(b) 41x41 grid

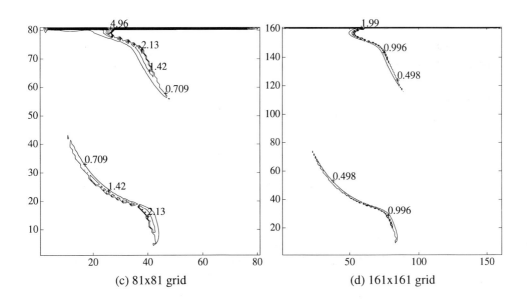

(c) 81x81 grid

(d) 161x161 grid

Figure 22. Distribution of normalized effective numerical viscosity, $Q^*–G^*$, with respect to grid size for lid–driven cavity flow with Re=1000. (Lid moves from left to right).

iterations required typically increases with the number of grid points employed. To demonstrate this fact, Fig. 23 shows the convergence paths of the two–dimensional lid–driven flow in a square cavity with three different grid sizes, 41×41, 81×81 and 161×161 uniformly distributed nodes. For a typical Reynolds number, say 100, and discretization scheme, say first or second–order upwind for convection and central difference for all other terms, it is clear that the number of iterations needed for the computation to yield the same degree of convergence increases as the number of grid points increase. Since the CPU time on a per iteration basis scales with the number of grid points, the demand on the computing resources grows quickly as the grid resolution is improved. Of course, it is desirable to have the convergence rate of a matrix solver to be as independent of the size of the coefficient matrix as possible. In this aspect the multigrid technique is very useful as will be discussed later.

There is another aspect of the solution procedure that should be pointed out. Generally speaking, for recirculating flows, two methods have been developed for obtaining the solution field, namely, the decoupled method (Braaten and Shyy 1987, Arakawa *et al.* 1988, Hortmann *et al.* 1990), and the coupled method (Vanka 1986, Thompson *et al.* 1988, Arakawa *et al.* 1988, Bruneau and Jouron 1990, Sathyamurthy and Patankar 1990). In the decoupled method, a two–level iterative procedure is usually employed; one is the outer iteration needed to progressively update different partial differential equations, and the other is the inner iteration devised to solve the system of linear algebraic equations resulting from the discretization procedures of each partial differential equation with other variables remaining unchanged. Within the outer iteration, for a two–dimensional flow with u, v, and p as the dependent variables for example, a cyclic outer iterative procedure is designed to sequentially solve, say, the linearized x–momentum equation first, the linearized y–momentum equation next, and the pressure (correction) equation last. After sweeping through all three equations to obtain partially converged solutions, the x–momentum equation is again invoked to initiate a new cycle, until all three equations are satisfactorily solved. Within the inner iteration, say, the x–momentum equation is discretized and linearized, and the resulting set of linear equations is then solved by an iterative procedure such as the LSOR method till the prescribed number of iterations or the convergence criterion has been reached. For a decoupled algorithm, the treatment of coupling among the dependent variables, such as velocity and pressure, is critical to the overall convergence (Patankar 1980, 1988, Raithby and Schneider 1979, Issa 1986, Braaten and Shyy 1987). Hence, the performance of this type of method is sensitive to factors such as the Reynolds number and the distribution and skewness of the grid (Vanka 1985, Braaten 1985, Braaten and Shyy 1986b).

In contrast, a coupled method, which solves the velocity vector and the scalar variables at a point, line, or plane simultaneously, usually shows robust performance with respect to parameters such as the Reynolds number. In the context of Cartesian coordinates, the coupled method is also found to be relatively insensitive to the number of the grid points employed (Vanka 1985, Braaten 1985). With the use of curvilinear

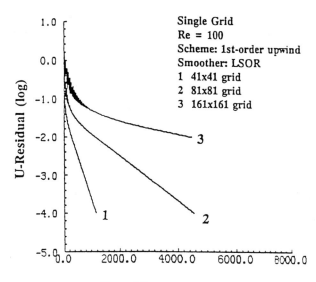

No. of finest grid iterations

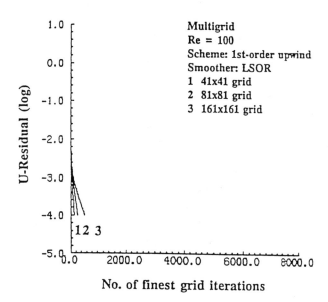

No. of finest grid iterations

first-order upwind scheme

Figure 23. Effect of grid size on convergence rate for the lid–driven cavity with Re=100, using the SIMPLE algorithm and line SOR method for solving the linearized equations for each partial differential equation.

coordinates, however, the situation is not as favorable, since, depending on the characteristics of the grid skewness, one either has to treat the cross–derivative terms explicitly as source term or solve equations whose coefficient matrix is no longer sparse (Braaten and Shyy 1986b). Furthermore, from the viewpoint of developing generic computational capabilities for flows involving different physical mechanisms such as turbulence, heat transfer, combustion, and phase change, it is preferable that one does not have to redo the algorithm for a different number of partial differential equations. In this regard, the decoupled method has a clear advantage since it can handle a different number of equations in more a flexible manner.

Regarding the inner iteration procedure, in a decoupled algorithm, for the convection–dominated flows, the pressure and pressure–correction equations are usually slower to converge. The pressure–correction equation is critical in determining the overall rate of convergence. It is essentially a Poisson–like equation, with anisotropic coefficients. The success of multigrid methods in solving Poisson equations has led to their application in the pressure correction equation (Rhie 1986; Phillips and Schmidt 1984). The effectiveness of line and point SSOR (Symmetric Successive Over–Relaxation) methods using various number of grid levels is evident from the inner convergence paths of the transport and pressure correction equations, as shown by Fig. 24 for a typical 3–D high Reynolds number incompressible flow calculation with $45 \times 21 \times 21$ nodes. Being strongly elliptic, the pressure correction equation requires more iterative steps for convergence than do the momentum and scalar transport equations, which are convection dominated. Figure 24(a) suggests that a more elaborate solution method is needed for the pressure correction equation only, whereas the point–SSOR method is sufficient for the velocity and scalar transport equations. Differences between line and point SSOR methods for the pressure correction equation, as well as differences between various levels of grid utilization are shown in Fig. 24(b). It is clear that the line solver is more effective than the point solver. This is consistent with the results of Stuben and Trottenberg (1982) who found that the point solver is not an effective smoother for highly anisotropic problems. More importantly, this observation demonstrates that the smoothing characteristics of a given equation solver is important.

In order to gain insight into the different smoothing properties possessed by the conventional matrix solvers, a simple model problem is used as the test case:

$$\phi_x = \phi_{xx} + \phi_{yy} + x(1 - x)\sin(\pi y), \qquad 0 \le x, \qquad y \le 1 \qquad (7.1)$$

with the boundary condition: $\phi = (Ax^2 + Bx + C)\sin(\pi y)$

where $A = -\dfrac{1}{\pi^2}$, $B = \dfrac{(1 - 2A)}{\pi^2}$, and $C = \dfrac{(2A - B)}{\pi^2}$

As to the discretization scheme, the first derivative term is represented by the second–order central difference scheme and the second derivative terms are represented by a second–order five–point operator resulting in finite difference equations of a five point stencil. Two sets of computations are made for identical problems, based on two

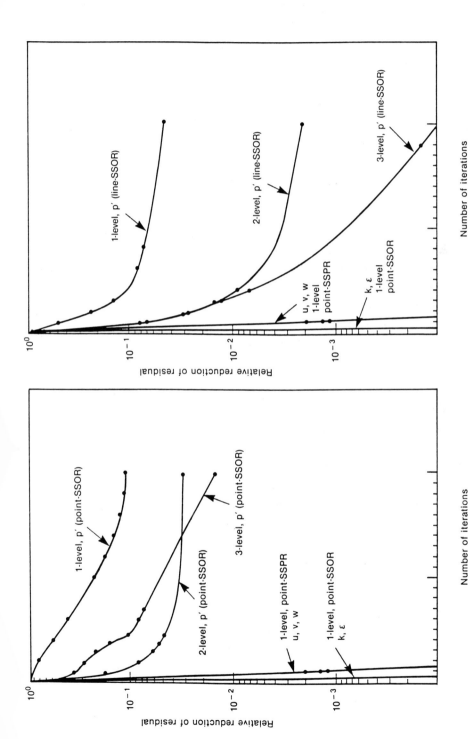

Number of iterations

Figure 24. Convergence characteristics of the inner iteration procedure of the individual differential equations. The multigrid method is applied only to the inner loop for the results shown in this figure.

different initial guesses. The two different initial guesses are: (i) zero initial guess, $\phi_{i,j} = 0$ (this is the starting condition for the results shown previously), and (ii) oscillating initial guess, $\phi_{i,j} = 0.015(-1^j+1)$. These two different initial guesses represent, respectively, the smallest and the largest wave numbers that a given grid system can possibly resolve. The case chosen is for a 51×51 grid system. Figure 25(a) compares the residual profiles resulting from the case of the oscillatory initial guess after 2 and 4 iterations. For the short–wave errors, the LSOR method has a better performance than the SIP.

For the case of zero initial guess, Fig. 25(b) compares the residual profiles of LSOR, and SIP along the central vertical grid line, $i = 26$, after 2 and 27 iterations. SIP exhibits faster convergence rates for the smooth initial profile. However, more importantly, Figs. 25(a) and (b) demonstrate that in general, the iterative methods are far more effective in eliminating the short–wave errors than the long–wave ones. Hence, if one can devise a computation strategy with varying grid spacings, each aimed at eliminating a range of wave lengths efficiently, then a good convergence rate can be achieved independent of either the wave length of the errors contained in the initial guesses, or the number of grid points employed. This is the basic motivation behind the multigrid method, as will be reviewed in the following section.

7.2 Multigrid Methodology

As already discussed in Section 6 of Chapter III, in the MG method the computation is carried out on a series of grids, G_k, with the corresponding solution vector $\{\vec{\psi}_k\}$ where $k = 1,2,3,...M$, with $k=M$ representing the finest mesh, and the meshes become coarser as the value of k becomes smaller. The exact solution for any variable ψ_k on grid G_k satisfies the equation:

$$[a_k]\{\vec{\psi}_k\} = \{\vec{\zeta}_k\} \tag{7.2}$$

where $[a_k]$ and $\{\vec{\zeta}_k\}$ are, respectively, the coefficient matrix and the source term vector derived directly from the discretization procedure adopted. Hence, at convergence, $[a_k]$ and $\{\vec{\zeta}_k\}$ are based on the exact solutions on grid G_k of the coupled variables. Before convergence, however, they are estimated based on the intermediate solution, and they are continuously updated in the course of iteration.

There are several algorithms for implementing the MG idea, each with several possible variations. One of the simplest is the correction storage (CS) scheme which is most useful for the linear problems. In this scheme, the calculation starts on the finest grid G_M, and an approximate solution to equation (7.2) is computed by a relaxation method. Unless the approximate solution $\{\vec{\phi}_k\}$ satisfies equation (7.2) and the boundary conditions, there will be a residual vector $\{\vec{R}_k\}$ given by

$$[A_k]\{\vec{\phi}_k\} = \{\vec{S}_k\} - \{\vec{R}_k\} \tag{7.3}$$

where $[A_k]$ and $\{\vec{S}_k\}$ are based on the intermediate solution $\{\vec{\phi}_k\}$ without being updated along with $\{\vec{\phi}_k\}$. A few iterations are performed on grid G_k until the rate of reduction of

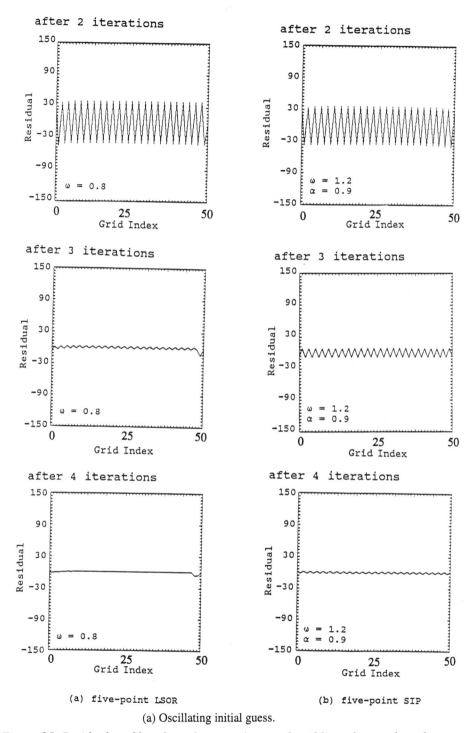

(a) five-point LSOR   (b) five-point SIP

(a) Oscillating initial guess.

Figure 25. Residual profiles along the central vertical grid line after a selected number of iterations with two different initial conditions for the convection–diffusion–source model equation.

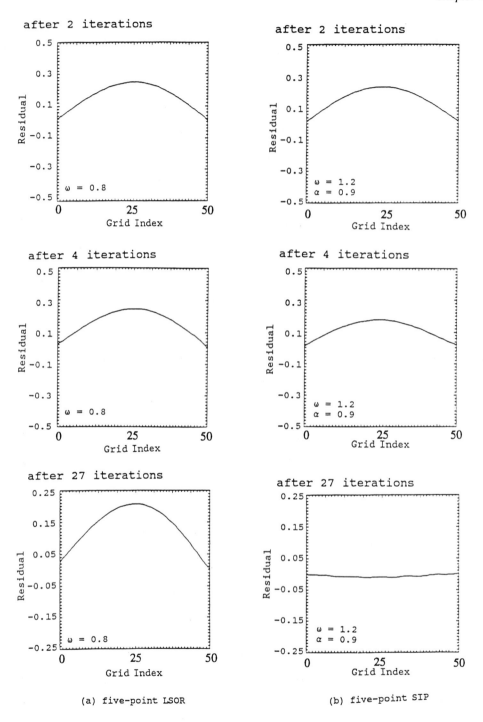

(a) five-point LSOR   (b) five-point SIP

(b) Zero initial guess.

Figure 25. continued.

the residuals falls below a desired level. The residuals are then transferred by a "restriction (or injection)" operator to the next coarser grid G_{k-1} and a correction vector $\{\delta\vec{\phi}_k\}$ is obtained by solving the equation:

$$[A_{k-1}] \ \{\delta\vec{\phi}_{k-1}\} \ = \ \{I_k^{k-1} \vec{R}_k\} \tag{7.4}$$

where I_k^{k-1} is the restriction operator that performs the task of transmitting the information from a fine grid to a coarse grid. Once equation (7.4) is solved, the correction vector $\{\delta\vec{\phi}_{k-1}\}$ is "prolongated (or interpolated)" to grid G_k and $\{\vec{\phi}_k\}$ is subsequently corrected as

$$\{\vec{\phi}_k^{new}\} \ = \ \{\vec{\phi}_k^{old}\} \ + \ I_{k-1}^k \ \{\delta\vec{\phi}_{k-1}\} \tag{7.5}$$

where I_k^{k-1} is the prolongation operator. This process of relaxation, restriction and prolongation is repeated until the desired accuracy on the finest grid G_M is achieved.

The scheme described above serves to illustrate the idea of the multigrid procedure, but is inadequate for non–linear problems such as the Navier–Stokes equations. Since it depends on the linearity of $[a]$ to derive the residual for computing the correction $\{\delta\vec{\phi}\}$ on the coarse grid, it does not work well for flow problems with strong nonlinear convection effects. In order to remedy this deficiency, the so–called Full Approximation Storage (FAS) scheme has been developed (Shyy and Sun 1993), as will be briefly described below.

Without assuming linearity, the fine–grid residual equation

$$[a_k](\vec{\phi}_k + \delta\vec{\phi}_k) \ - \ [A_k]\{\vec{\phi}_k\} \ = \ \{\vec{\zeta}_k\} \ - \ \{\vec{S}_k\} \ + \ \{\vec{R}_k\} \tag{7.6}$$

can be written on the coarse grid by restricting the residual to form a corresponding equation

$$[a_{k-1}] \ \hat{I}_k^{k-1}\left(\vec{\phi}_k + \delta\vec{\phi}_{k-1}\right) - [A_{k-1}]\{\hat{I}_k^{k-1} \ \vec{\phi}_k\} = \{\vec{\zeta}_{k-1}\} - \{\vec{S}_{k-1}\} + \{I_k^{k-1}\vec{R}_k\} \tag{7.7}$$

putting the known quantities on the right hand side reduces equation (7.7) to

$$[a_{k-1}] \ \{\vec{\psi}_{k-1}\} \ = \ \{\vec{\xi}_{k-1}\} \tag{7.8}$$

where

$$\{\vec{\psi}_{k-1}\} \ = \ \hat{I}_k^{k-1}\{\vec{\phi}_k\} \ + \ \{\delta\vec{\phi}_{k-1}\} \tag{7.9}$$

and

$$\{\vec{\xi}_{k-1}\} \ = \ \{\vec{\zeta}_{k-1}\} \ + \ [A_{k-1}] \ \{\hat{I}_k^{k-1} \ \vec{\phi}_k\} \ - \ \{\vec{S}_{k-1}\} \ + \ \{I_k^{k-1}\vec{R}_k\} \tag{7.10}$$

Here, I_k^{k-1} for $\{\vec{\phi}_k\}$ may be different from the residual transfer operator I_k^{k-1} for $\{\vec{R}_k\}$ although they are usually taken to be the same. This is known as the Full Approximation Storage (FAS) scheme, since from Equation (7.8) on the coarse grid the complete solution $\{\vec{\psi}_{k-1}\}$, not just the correction $\{\delta\vec{\phi}_{k-1}\}$, is computed. After solving Equation (7.8), the fine grid solution is updated via

$$\{\vec{\phi}_k^{new}\} \;=\; \{\vec{\phi}_k^{old}\} \;+\; I_{k-1}^k [\{\vec{\phi}_{k-1}^{new}\} \;-\; I_k^{k-1}\{\vec{\phi}_k^{old}\}] \tag{7.11}$$

which represents interpolating the approximate correction $\{\delta\vec{\phi}_{k-1}\}$ and adding to the intermediate solution $\{\vec{\phi}_k\}$ on fine grid G_k. Since the starting values for $\{\vec{\phi}_{k-1}\}$ are I_k^{k-1} $\{\vec{\phi}_k\}$, the source terms $\{\vec{\zeta}_{k-1}\}$ and $\{\vec{S}_{k-1}\}$ are identical in the first coarse grid iteration. All the right–hand side terms in Equation (7.10), except $\{\vec{\zeta}_{k-1}\}$, are calculated once and then kept unchanged during subsequent iterations on the coarse grid; they appear as extra explicit source terms with prescribed values. With FAS, the source term and the coefficient matrix on coarse grid can be continuously updated to reflect progress made to the dependent variables. Since both $[a]$ and $\{\vec{\zeta}\}$ can be strongly dependent on $\{\vec{\psi}\}$ in nonlinear manners, this feature is very helpful.

When using multigrid cycles as a solution method one starts with an initial approximation $\{\vec{\phi}_k\}$ on the finest grid G_M. If a good initial approximation can be made then fewer cycles will be required to solve the problem. To get a better first approximation, one can interpolate an approximate solution from a coarser grid through a coarse–to–fine grid transfer operator, namely

$$\{\vec{\phi}_k\}_{initial} \;=\; \Pi_{k-1}^k \{\vec{\phi}_{k-1}\}_{initial} \tag{7.12}$$

where Π_k^{k-1} is a coarse–to–fine grid transfer operator which need not be the same as I_k^{k-1} in Equation (7.5). As schematically depicted in Fig. 26, starting with an approximate solution on the coarsest grid G_1, a sufficiently large number of relaxation sweeps can be taken to give an accurate solution there. This solution is then interpolated onto the next finer grid G_2 by the prolongation operator and used as a starting guess for the first V–cycle multigrid procedure; this two–grid V–cycle procedure continues till it meets the local convergence criterion on G_2. After the convergence on G_2 is obtained, the solution $\{\vec{\phi}_2\}$ is then interpolated onto next finer grid, G_3, initiating a new three–grid V–cycle relaxation involving G_1, G_2, and G_3, until convergence is attained on G_3. This process is repeated by interpolating the solution from G_{k-1} to G_k as a first approximation, and solving the problem by a V–cycle iteration, until the final converged solution is obtained on the finest grid G_M. The above described solution process is termed the full multigrid (FMG) procedure, which is important for obtaining fast convergence.

With the FMG–FAS algorithm, multigrid codes differ in their cycling procedures, which can be either fixed or adaptive, that is, the decisions of when to switch grids and which direction to go (i.e., to a coarser or finer grid) can be either prescribed in advance or made internally as the numerical solution develops.

With a coupled nonlinear system of equations, the rate of convergence of the various equations may be substantially different. For example, as already demonstrated, for a high Reynolds number flow, the pressure equation is much harder to converge than other linearized transport equations due to its stronger diffusion characteristics. In a decoupled algorithm where all the variables are sequentially updated, an adaptive cycling

Full Multigrid (FMG) V-cycle

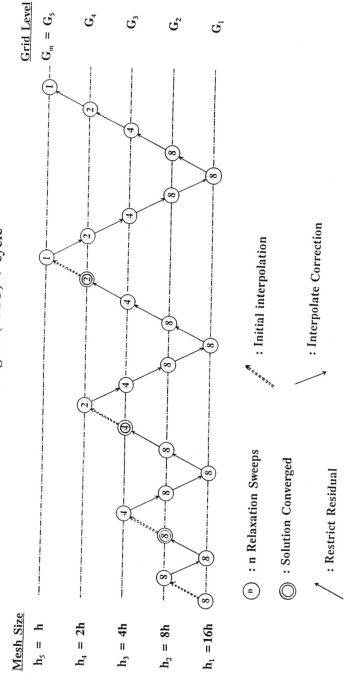

Figure 26. Schematic of a full multigrid (FMG) procedure with fixed V–cycles using five grid levels. Here the whole system of equations is solved on each grid level; both outer and inner iterations are conducted using the multigrid technique.

algorithm is not necessary for any of these convection–dominated equations because of their very good convergence rates.

7.3 Examples

As an illustrative example, the flow in a channel with multiple bumps, as shown in Fig. 27, is considered. There is a circular bump on both the top and bottom walls. The maximum height of both bumps is 0.3. The bump on the top wall is located between $x = 0.5$ and 1.5, and the one on the bottom wall is between $x = 1$ and 2. The length to inlet height ratio of the modeled domain is 6. Laminar flows with Reynolds numbers of 10^2 and 10^3, based on the inlet velocity and channel height, are considered. The inlet velocity profile is uniform and the value is 1. The density of the fluid is taken as 1, and the viscosity is the inverse of the Reynolds number. The outflow boundary condition of the velocity field is conducted by enforcing a zero value of the velocity gradients along the streamwise (ξ–) direction; this practice can help satisfy the continuity equation there. Since the streamwise momentum transport is dominated by convective effects, the outflow treatment does not affect the velocity field significantly. However, as demonstrated in Shyy (1988), for the channel flow with massive recirculation zones, depending on the grid resolution, and discretization schemes utilized, difficulties may be encountered in the course of numerical computation and an adaptive grid method can be critical for eliminating numerical instabilities.

In the present case, two grid sizes have been used, one with 193×97 nodes and the other with 97×49 nodes. The solution characteristics including both the streamlines and the static pressure contours for $Re = 10^2$ and 10^3 are exhibited in Fig. 28. Substantial complexities exist in those flowfields. At $Re = 10^2$, two closed recirculating bubbles exist behind the two bumps. For the flow over the lower bump, the incoming streamlines are more tilted toward the top wall due to the area expansion there, resulting in a much longer reattachment length on the bottom wall. The trend becomes much more pronounced as the Reynolds number is increased from 10^2 to 10^3; as exhibited in Fig. 28(b), within the domain considered, the recirculating flow behind the lower bump never reattaches on the

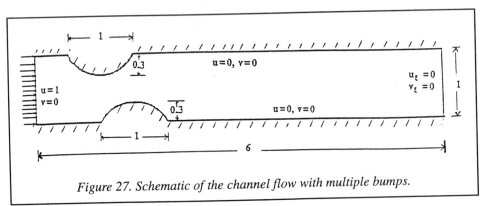

Figure 27. Schematic of the channel flow with multiple bumps.

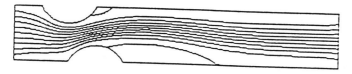

streamlines ($\Delta\psi = 0.1$)

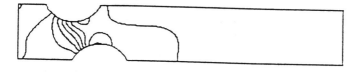

pressure contours ($\Delta p = 0.5$)

(a) Re = 100

streamlines ($\Delta\psi = 0.1$)

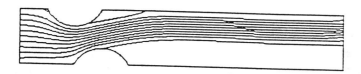

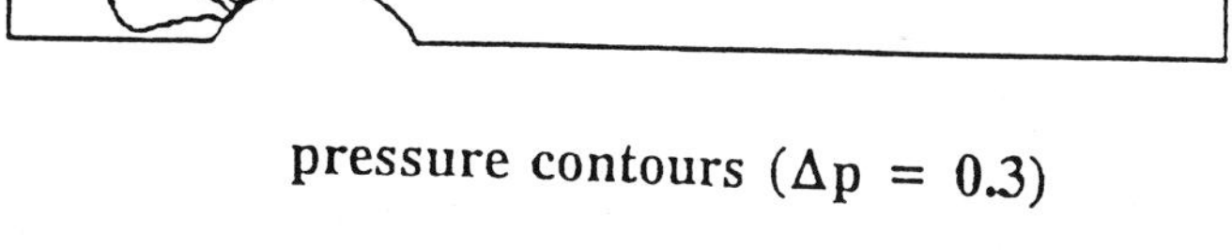

pressure contours ($\Delta p = 0.3$)

(b) Re = 1000

Figure 28. Solution characteristics of the channel flow with two different Reynolds numbers.

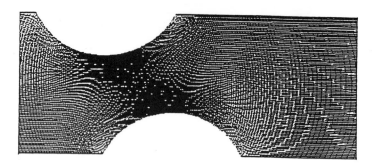

(i) original grid

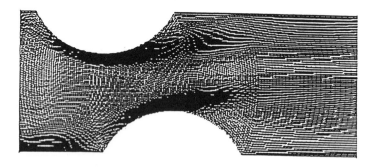

(ii) adaptive grid with Re = 100

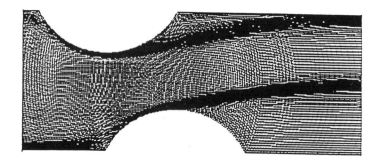

(iii) adaptive grid with Re = 1000

(a) Close–up view of the grids in the vicinity of the bumps.

Figure 29. Grid layout and convergence characteristics of the channel flows with adaptive and multiple grid techniques for different Reynolds numbers.

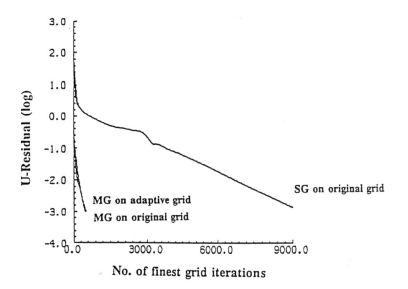

(a) Re = 100

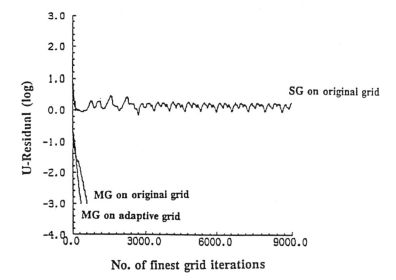

(b) Re = 1000

(b) Performance of the SG and MG methods for the channel flow.

Figure 29. continued.

bottom wall. Hence, both incoming and outgoing fluids now move across the so–called outflow boundary.

Fig. 29 compares the original and adaptive grids, as well as the convergence paths of the u–momentum equation yielded by the SG and MG methods. With $Re = 10^2$, the number of finest grid iterations needed by the MG method is an order of magnitude fewer than that by the SG method. The use of the adaptive grid technique does not affect the convergence rates very much, indicating that the qualitative changes of the grid characteristics at $Re = 10^2$ are modest despite noticeable adjustment made in the detailed grid distribution. At $Re = 10^3$, the situation becomes extremely interesting. On the original grid, the SG method fails to yield a fully converged steady state solution. It appears that the interaction of the lower recirculating eddy and the outflow boundary treatment is so strong that there is no adequate numerical damping to yield a converged solution (Shyy 1988). With the MG method, however, the coarser grids intrinsically contain higher levels of numerical dissipation, enabling the convergence path to change from a persistently oscillatory pattern to a fast convergent one. Hence, for $Re = 10^3$ the present MG method actually qualitatively alters the whole convergence characteristics.

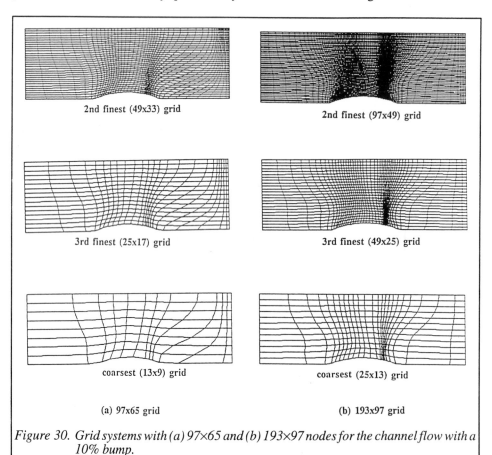

2nd finest (49x33) grid   2nd finest (97x49) grid

3rd finest (25x17) grid   3rd finest (49x25) grid

coarsest (13x9) grid   coarsest (25x13) grid

(a) 97x65 grid   (b) 193x97 grid

Figure 30. Grid systems with (a) 97×65 and (b) 193×97 nodes for the channel flow with a 10% bump.

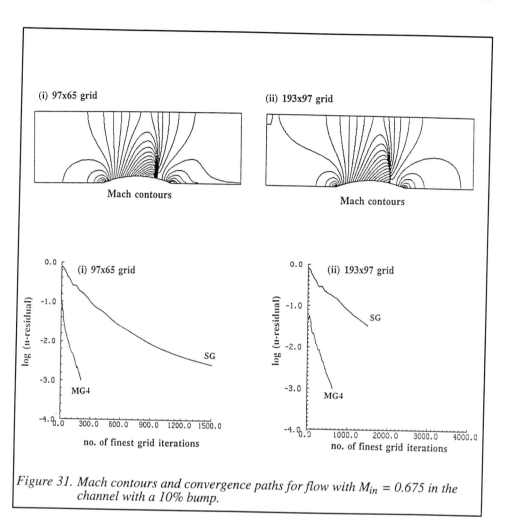

(i) 97x65 grid

Mach contours

(ii) 193x97 grid

Mach contours

Figure 31. Mach contours and convergence paths for flow with $M_{in} = 0.675$ in the channel with a 10% bump.

Furthermore, as shown in Fig. 29(b), it appears that the use of the adaptive grid method not only improves the numerical resolution but also results in a faster convergence rate.

Finally, a transonic case with an inlet Mach number of 0.675 over a 10% bump is presented in the following. For this flow, since the outflow condition is subsonic, outlet pressure must be specified, and the elliptic nature of the subsonic outflow condition causes a nonuniform pressure distribution near the outlet. Fig. 30 shows adaptive grid systems of 97×65 and 193×97 nodes at different multigrid levels utilized by the MG scheme. For the present case, the MG procedure can yield solutions with significant improvement in convergent rate. The Mach number contours and the convergence paths of the SG and MG procedures computed with both grid systems are depicted in Fig. 31. Again, as the grid is refined along with adaptation, the solutions are improved while the relative advantage of the MG procedure in convergence rate becomes higher.

7.4 Turbulent Flow Treatment

Most papers found in the multigrid literature deal with laminar or inviscid flows. For turbulent flows, however, relatively few successes have been reported, and all of the successes are quite recent, for example, Braaten and Shyy (1987), Joshi and Vanka (1991), Lien and Leschziner (1991), Mavriplis and Martinelli (1991), Shyy *et al.* (1993a). It should be noted that in Braaten and Shyy (1987), only the pressure equation is solved by the MG method, resulting in a decoupled procedure with no substantial speed–up in terms of the convergence rate of the whole set of equations. In the references listed above, the complete set of governing equations is solved in the context of the FMG method. With regard to the methodology adopted, they differ significantly both in terms of the basic formulation, pressure based (Joshi and Vanka 1991, Lien and Leschziner 1991, Shyy *et al.* 1992d) versus density based (Mavriplis and Martinelli 1991), grid layout, staggered (Joshi and Vanka 1991, Shyy *et al.* 1992d), collocated (Lien and Leschziner 1991), and unstructured (Mavriplis and Martinelli 1991), choice of the dependent velocity variables, Cartesian (Lien and Leschziner 1991, Mavriplis and Martinelli 1991, Shyy *et al.* 1992d) versus contravariant components (Joshi and Vanka 1991), and the discretization schemes.

For turbulent flow, the constraint of the physical possibility imposed by the turbulence model calls for some special treatments. Some treatments developed by Shyy *et al.* (1993a) are briefly discussed in the following.

Treatment 1

Since both, the turbulent kinetic energy, k and the rate of turbulent energy dissipation, must always be positive, it is critical to check that these requirements are met during the prolongation step. The task of the prolongation operator transferring the solution correction from grid level $k - 1$ to grid level k can be expressed by:

$$\{\vec{\phi}_k^{new}\} = \{\vec{\phi}_k^{old}\} + I_{k-1}^k\{\delta\vec{\phi}_{k-1}\} \tag{7.13}$$

where $\{\vec{\phi}_k^{new}\}$ is the solution vector after prolongation

$\{\vec{\phi}_k^{old}\}$ is the solution vector before prolongation

$\{\delta\vec{\phi}_{k-1}\}$ is the solution correction at the coarse grid level after FAS procedure.

The variable ϕ designates either k or ε. However, before the above equation is invoked, one must ensure that $\{\vec{\phi}_k^{new}\}$ is positive. If the prolongation procedure violates this requirement, then in the present algorithm, one simply chooses not to prolong the solution correction from the coarse grid to the fine grid at that iterative step.

Treatment 2

The eddy viscosity μ_e is determined according to

$$\mu_e = C_\mu \frac{\varrho \; k^2}{\varepsilon} \tag{7.14}$$

hence depending on the intermediate paths that k and ε evolve in the course of the MG procedure, an upper limit of say, 10^4 times the molecular viscosity can be imposed to avoid generating unrealistic values; this practice does not affect the final accuracy but can help stabilize the computation. The above two procedures can effectively prevent the MG procedure from generating nonphysical intermediate solutions, when solving the k–ε two–equation closure (Launder and Spalding 1974), through the prolongation procedure, without affecting the final accuracy.

Treatment 3

Another feature that affects overall convergence rate is the way the grid restriction is conducted. The prolongation and the restriction operators can substantially affect the near–wall solutions yielded by the wall function between grid levels. It is demonstrated in Shyy *et al.* (1992e) that if the grid line next to the solid boundary is retained during the grid restriction procedure, an improved convergence rate is obtained.

With these treatments, the multigrid solution technique yields better performance. It is clear that more work is needed to further utilize the effectiveness of this method.

8 COMPOSITE GRID METHOD

8.1 Introduction

In the following, we describe some recent advances made in the field of composite gridding for treating complex geometries. For many problems of engineering interest, the generation of a single grid which discretizes the domain adequately for resolving the various flow features is very difficult or even impractical. This problem can be overcome to a limited extent by applying sophisticated grid generation schemes to construct a single grid with suitable characteristics, however, the degree of satisfaction achieved with such a process is highly problem dependent. To resolve such difficulties unstructured grids utilizing triangular meshes have been developed (Mavriplis and Martinelli 1991). An alternative approach is to partition the domain into a number of distinct blocks, each block being topologically simple. Grids can then be independently generated for each block with little difficulty. Furthermore, the grid resolution in each sub–domain can be better controlled according to the fluid physics there. For example, the solution of heat transfer problems requires the estimation of solution variation in thin regions next to extended solid surfaces, and accordingly the length scale disparity between different directions needs to be accounted for. This need can be more satisfactorily met by the use of a composite grid algorithm. Further efforts are needed before a robust and accurate procedure can be developed.

For such a composite (or multiple block) solution procedure in which the governing equations are solved in each block, one of the main issues of importance concerns the transfer of information between different blocks in the system. Strategies for transferring information between blocks can be generally classified as either conservative or non–conservative. For many multiple–block solution techniques, information is transferred between blocks strictly via interpolation. This is the case for many multiple–block schemes using overlaid grids (Steger and Benek 1987, Moon and Liou 1989, Chesshire and Henshaw 1990), or abutting (patched) grids (Yang 1990, Hsu and Lee 1991). For some techniques employing the so–called chimera grids, information transfer via interpolation is the best approach (Benek *et al.* 1985, Rimlinger *et al.* 1992, Chyu *et al.* 1993), as implementation of conservative transfer schemes is very difficult. For some applications, interpolation proves to be satisfactory, however, in many instances in which large solution gradients, or elliptic features such as recirculation exist in the vicinity of block boundaries, significant errors can be introduced into the solution. In addition, for schemes based upon a control volume formulation, the use of a non–conservative interface treatment, such as direct interpolation, may result in incompatible boundary conditions which can violate mass conservation and hence prevent convergence (Wright and Shyy 1992). Hence, a conservative approach is preferable for transferring information between the grid blocks in the system. Much work has been done in this area, mainly in the context of density–based methods for compressible fluid flows. Rai (1986a,b) for example, has successfully developed and implemented conservative boundary schemes for Euler equations on composite patched grids in the framework of general curvilinear coordinate systems, for both explicit and implicit time integration schemes. Reggio *et al.* (1990) have also developed conservative multiple–block strategies for the Euler equations using overlaid grids. However, further work is needed, especially for the case involving grid discontinuities between two adjacent blocks. Some efforts aiming at improved accuracy of the interface treatment have been made by Moon and Liou (1989) and by Thompson and Ferziger (1989). In a work by Perng and Street (1991), a composite grid technique was developed for solving the incompressible Navier–Stokes equations using a pressure–based method with the staggered grid arrangement, however, the interface treatment used for transferring information between the blocks is non–conservative. Davis and Thompson (1992) have conducted a study to investigate several interpolation schemes for the grid interface treatment. Their results indicate that a quadratic interpolation scheme aided by a gradient weighted correction procedure is a better choice among several different candidate schemes tested. The relative merits between an interpolation procedure and a conservative procedure still awaits further clarification. The flexible grid computation is an active research area at the present time, as evidenced by some interesting papers published recently (e.g., Bayyuk et al 1993, Connell and Holmes 1993, Kao et al 1993, Melton et al 1993, Pember et al 1993, Weatherill et al 1993). It is clear that while rapid progress is being made, the two key difficulties, namely, conservation treatment across grid interface and handling of disparate length scales of high Reynolds number flows, have yet been adequately resolved.

    In this section, a detailed discussion of the pressure–based composite grid method is presented. More illustrations and details can be found in Wright (1993). A pressure correction algorithm is used in conjunction with a staggered grid system to solve the continuity and momentum equations in a sequential manner, within each of the blocks of the domain, and a conservative interface treatment is used for transferring information between the blocks. Specifically, mass and tangential momentum flux conservation across the interface is always enforced and used as the basis to conduct the interpolation necessary to yield the velocity distribution along the boundary of each block. The pressure distribution, on the other hand, is naturally computed in each block with the merit of the staggered grid. The only extra procedure that needs to be incorporated into the composite grid algorithm for the pressure calculation is the explicit balance of the total force on each block interface, which eliminates the arbitrary constant of the pressure field within each block. The first point of discussion involves the organization of complex composite grid systems involving an arbitrary number of overlapping subgrids. A generic organizational scheme is first developed which will allow the conservative internal boundary treatment to be implemented in a straightforward manner for arbitrary configurations. After the development of the organizational scheme, the global conservation conditions for the staggered grid are presented, followed by a discussion of the global conservation strategy for the pressure–based method. Next, the explicit local conservation procedure is detailed, followed by a discussion of the conditions under which both local and global conservation can be simultaneously maintained. Following the development of the basic methodology, a sample calculation is presented which demonstrates the capability of the current method. Finally, some recent developments in the area of composite multigrid are addressed.

8.2 Organization of Composite Grids

One of the primary difficulties in dealing with composite grids is the organizational task of determining the information flow from block to block. For a composite grid consisting of only two overlapping grid blocks, or for any composite grid in which no more than two blocks in the system overlap at any single point in the domain, the channel of information flow is already determined, since there is only one neighboring block for each block in the system which can provide the required internal boundary data. However, for cases in which three, or even more grid blocks in the composite grid system overlap, determination of the internal boundary information flow paths becomes more difficult. In these cases, several grid blocks may be available to provide information across some parts of an internal boundary, while for other parts of the boundary, information may only be obtained from one other neighboring block. An example of this may be seen from the simple composite grid system shown in Fig. 32. Here, for the lower side of block two, which is entirely an internal boundary, for the right portion of this boundary, information can be obtained only from block three, however, for the left portion of this boundary, information may be obtained from either block one or block three.

The problem of handling composite grids consisting of multiple overlapping blocks, at least in terms of information transfer, is equivalent to determining what information flows across particular segments of each side of each block in the system. Each block side can have two types of segments associated with it, which will be designated as boundary condition segments and internal boundary segments. Boundary condition segments, which include both Dirichlet and outflow (i.e. gradient condition)

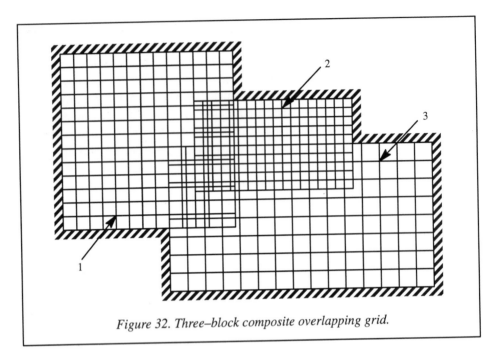

Figure 32. Three–block composite overlapping grid.

segments are specified at the outset of the problem, and in terms of information flow, all required information is specified via the particular boundary condition for each segment. Internal boundary segments, on the other hand, are created by the overlappings of the blocks in the composite grid. Each side of each block in the system is spanned by a combination of internal boundary segments and/or boundary condition segments. Across each of the internal boundary segments, information is obtained from a single neighboring grid block. While the internal boundary segments are essentially determined by the arrangement of the composite grid system, for composite grids containing regions in which more than two blocks overlap, the set of internal boundary segments is not unique. This again relates to the fact that information may be obtained across portions of the internal boundaries, from more than one of the neighboring blocks. Thus, some strategy must also be developed for selecting a unique set of internal boundary segments for each of the blocks. Once the internal boundary segments for each block in the composite grid system have been determined, then along with the boundary condition segments of the blocks, the complete paths of information flow into each block are specified, and the conservative internal boundary scheme can be implemented.

The procedure used for determining a unique set of internal boundary segments for each block in the composite grid system can be summarized as follows. First an intersection test is performed for each block in the composite grid with every other block to arrive at a preliminary set of internal boundary segments. For general composite grids, the internal boundary segments may overlap along portions of the boundaries, resulting in an ambiguity in terms of information transfer there. In order to resolve this ambiguity, each block in the composite grid system is given an overlapping priority, so that when two internal boundary segments overlap, the segment with the highest priority takes precedence. In this way, a unique set of internal boundary segments can be obtained. This two–step process is best illustrated by example.

Consider again, the composite grid system shown in Fig. 32. We will determine a unique set of internal boundary segments for block two of this three block composite grid. First, an intersection of block two independently with each of blocks one and three is performed. The preliminary internal boundary segments obtained from these intersection tests are shown in Fig. 33(a). The number by each of the segments indicates that that segment was created by the intersection of block two with the block of that number. For the lower internal boundary as well as the left internal boundary of block two, the internal boundary segments are seen to overlap at certain locations. Now, the following overlapping priorities are further specified for the blocks in the composite grid. Block number one is given the highest priority, followed by block two, and then block three. After using this priority information to eliminate internal boundary segment overlaps, the final

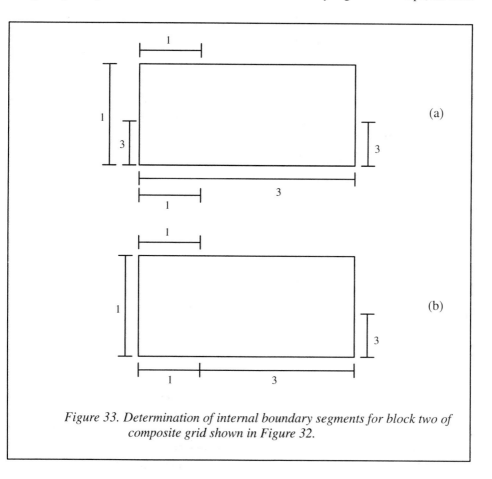

Figure 33. Determination of internal boundary segments for block two of composite grid shown in Figure 32.

set of internal boundary segments appear as shown in Fig. 33(b). With this set of internal boundary segments, the information channels through the internal boundaries are precisely specified. Along with the boundary condition segments, a complete specification of the information flow into block two is obtained. It should be noted that this procedure for determining the exchange of information within a composite grid arrangement is quite general, and can be applied to composite grids created from any number of grid blocks overlaid in an arbitrary fashion.

8.3 Global Conservation Conditions for the Staggered Grid

Consider the u–momentum equation written in conservative form as follows:

$$E_x + F_y = 0 \qquad (8.1a)$$

where the terms E and F represent the local fluxes of u momentum in the x and y directions respectively and are written as follows:

$$E = \varrho u u - \mu \frac{\partial u}{\partial x} + p \qquad\qquad F = \varrho u v - \mu \frac{\partial u}{\partial y} \qquad (8.1b)$$

Integrating this equation over a typical u control volume yields

$$(E_e - E_w)\varDelta y + (F_n - F_s)\varDelta x = 0 \qquad (8.2)$$

The terms $E_e\varDelta y$ and $E_w\varDelta y$ are respectively the total fluxes of u–momentum through the east and west control volume faces, and the terms $F_n\varDelta x$ and $F_s\varDelta x$ are respectively the total fluxes of u–momentum through the north and south control volume faces.

Now consider the single grid domain shown in Fig. 34 with uniform spacing in both coordinate directions. Summing the above equation for a typical u control volume over all the u control volumes in the domain yields

$$S = \sum_{j=2}^{nj} \sum_{i=3}^{ni} \{(E_e - E_w)\varDelta y + (F_n - F_s)\varDelta x\}$$

$$= \sum_{j=2}^{nj} (E_e|_{i=ni} - E_w|_{i=3})\varDelta y + \sum_{i=3}^{ni} (F_n|_{j=nj} - F_s|_{j=2})\varDelta x = 0 \qquad (8.3)$$

Note that based on the staggered grid convention adopted in this work, this procedure, which is taken over all the internal unknown u control volumes, results in a summation over the indices $i=3, ni$ and $j=2, nj$. From the second part of Equation (8.3) it is clear that S is only a summation of the boundary fluxes, since the interior fluxes for neighboring control volumes cancel each other out. This equation represents the global conservation property for any control–volume based scheme as expressed in Equation (8.2). Notice, however, for the staggered grid, the boundary in question is not the physical boundary of the domain as is usually the case when dealing with a nonstaggered grid, when the unknowns are located at the cell centers, but the boundary formed by the boundary sides of the u control volumes along the physical boundary. This boundary is hereafter referred to as the global

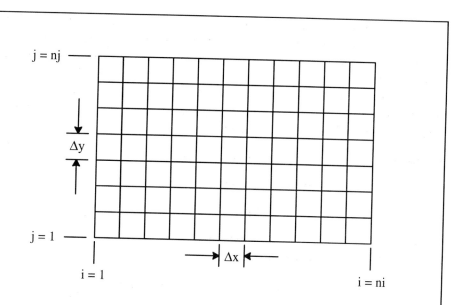

Figure 34. Single grid used in deriving the global conservation conditions for the staggered grid.

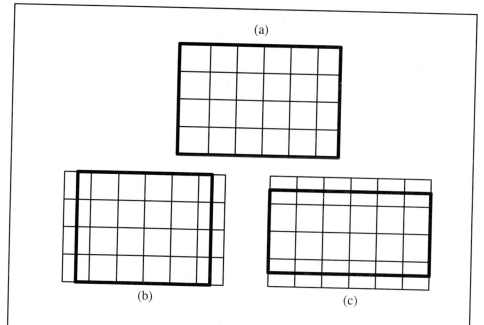

Figure 35. Global conservation boundaries for the staggered grid. a) Continuity; b) u–momentum; c) v–momentum.

conservation boundary. For the staggered grid system, three different global conservation boundaries exist, one for the continuity (pressure correction) equation, one for the u–momentum equation, and one for the v–momentum equation. The three global conservation boundaries for the staggered grid covered by the control volumes of the interior unknowns are shown in Fig. 35.

For composite grids, the global conservation property also requires that a summation of the u control volume equations over all the u control volumes results in only a summation of the u momentum fluxes across the global conservation boundary for the composite grid. Note that for such a summation, the total area of the summation should be equal to the area formed by an exclusive summation of the areas enclosed by the global conservation boundaries of each of the blocks forming the composite grid. By an exclusive summation, we mean that in an overlap zone created by the intersection of the global conservation boundaries of two blocks, the overlap area should only be included in the control volume summation of one of the two blocks.

For example, consider the composite grid shown in Fig. 36(a), constructed from two blocks with a horizontal overlap region. The global conservation boundary for the u–momentum equation is highlighted in bold. If in the overlap region, we sum over u control volumes in the upper block only, then the exclusive overlapping appears as shown

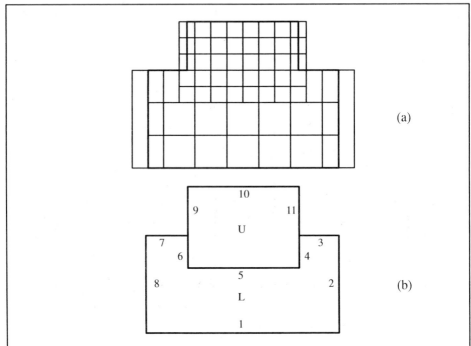

(a)

(b)

Figure 36. Two–block composite grid with horizontal overlap region. a) Composite
grid with global conservation boundary in bold; b) Exclusive
overlapping used in control volume summation.

in Fig. 36(b). Numbers have been given to the various boundary segments which will be used in the ensuing derivation of the conditions for global u–momentum conservation. Summing over all the u control volumes using the exclusive overlapping shown in Fig. 36(b), yields

$$\int_2 Edy - \int_8 Edy + \int_7 Fdx + \int_3 Fdx - \int_1 Fdx + \int_{11} Edy - \int_9 Edy + \int_{10} Fdx$$

$$= \left[\int_{6_U} Edy - \int_{6_L} Edy \right] + \left[\int_{4_L} Edy - \int_{4_U} Edy \right] + \left[\int_{5_U} Fdx - \int_{5_L} Fdx \right]$$ (8.4)

The left–hand side of this expression represents the summation of the boundary fluxes through the u–momentum global conservation boundary for the composite grid. The right–hand side of the expression represents the difference in the summation of the fluxes along the internal boundaries of the upper block as estimated from the two different blocks in the composite grid system. Now, for global conservation of u–momentum, the summation of the u–momentum flux through the u–momentum global conservation boundary should vanish. Accordingly, the right–hand side of Equation (8.4) should vanish, resulting in the following necessary condition for global u–momentum conservation

$$\left[\int_{6_U} Edy - \int_{6_L} Edy \right] + \left[\int_{4_L} Edy - \int_{4_U} Edy \right] + \left[\int_{5_U} Fdx - \int_{5_L} Fdx \right] = 0$$ (8.5)

A more restrictive condition can be imposed which requires that each of the terms in Equation (130) vanish individually, resulting in

$$\int_{6_U} Edy = \int_{6_L} Edy \qquad \int_{4_U} Edy = \int_{4_L} Edy \qquad \int_{5_U} Fdx = \int_{5_L} Fdx$$ (8.6)

These relations imply that the *u*–momentum flux through any internal boundary must be identical when estimated from the blocks on either side of the internal boundary. It should be noted, however, that these are only sufficient conditions for global *u*–momentum conservation, since Equation (8.5) represents the only requirement for global *u*–momentum conservation. Equation (8.5) states that for global conservation of *u*–momentum, no net *u*–momentum can be created at the internal boundaries. Similar necessary conditions apply for the *v*–momentum and continuity equations.

8.4 Conservation Strategy for Pressure–Based Method

Because we are using a pressure correction technique for solving the governing equations, the global conservation procedure involves a different strategy than that used with density–based compressible flow techniques employing nonstaggered grids. This is due in part to the nature and type of boundary condition used in solving the pressure correction equation. Two types of boundary conditions in general can be used for the pressure correction equation. If the pressure is known at the boundary, then the value of the pressure correction at the boundary can be set to zero. If the pressure is not known at the boundary, then the velocity component normal to the boundary must be specified. Since, in general, the boundary pressure values are not known, especially for blocks which are located completely interior to the physical boundaries of the domain, we have adopted the use of the normal velocity component boundary condition in this work exclusively. Normal velocity component boundary conditions, in fact, can be used quite generally throughout, even at boundaries where the normal velocity component is not initially known (such as internal boundaries and outflow boundaries), but which evolves to a known value as the iterative solution process converges. By using a normal velocity component boundary condition for the pressure correction equation, the pressure field is only obtained to within an arbitrary constant, however, this represents no problem in terms of the solution technique, since the density is unaffected by the magnitude of the pressure, and for single grid solutions, if a pressure value is known at a certain point in the field, then the entire pressure field can be adjusted accordingly, after the solution has been obtained. However, for composite grids, in which the governing equations are solved independently within each block, in a block to block iterative fashion, each block in the system will generate a pressure field independently from those created in neighboring blocks. It is this characteristic that requires a different global conservation strategy than that used in density–based compressible flow solvers applied to composite grids.

Solution of the continuity (pressure correction) equation requires the specification of the mass flux into each of the pressure correction control volumes located along the boundaries of the grid. Similarly, solution of the u– and v–momentum equations requires the specification of the u and v–momentum fluxes into each of the respective u and v control volumes located along the boundaries of the grid. Each of these control volume fluxes may involve contributions from fluxes due to specified external boundary condition segments, or fluxes entering through internal boundaries of the grid. For the case of internal boundaries, flux contributions must be calculated in such a way that the global conservation conditions previously outlined are satisfied. A discussion of the exact procedure for achieving this will be undertaken in the following sections. This section deals with issues concerning the global conservation strategy that has been developed for the pressure correction algorithm with a staggered grid system.

The global conservation strategy can be summarized as follows. First, explicit conservation of mass across the horizontal and vertical sides of the global conservation boundary for the pressure correction equation is used to determine the mass fluxes into each of the pressure correction control volumes along the boundaries of the grid block in question. By an explicit conservation procedure, we mean that the mass flux through the boundaries is computed directly from the imposed boundary condition segments and using information from the neighboring blocks corresponding to the internal boundary segments. Once the mass fluxes have been determined, then the normal component of the velocity profile along the horizontal and vertical sides of the grid block can be determined. From these velocity values, we can compute the contributions to the u–momentum fluxes into each of the u control volumes along the vertical boundaries of the grid block which do not involve pressure (e.g. the term $\varrho u u - \mu \ (\partial u / \partial x \ )$), as well as the contributions to the v–momentum fluxes into each of the v control volumes along the horizontal boundaries which do not involve pressure (e.g. the term $\varrho v v - \mu \ (\partial v / \partial y \ )$). The pressure values required for calculating the rest of these fluxes are already specified via the staggered grid arrangement. Since the u–momentum fluxes into the u control volumes along the horizontal boundaries do not contain a pressure term (i.e. $\varrho u v - \mu \ (\partial u / \partial y \ )$), they can be obtained via explicit conservation, as was done for the mass fluxes. Similarly, this can be done for the v–momentum fluxes into the v control volumes along the vertical boundaries. In this way, all the required boundary fluxes for the various control volumes along the boundaries of the staggered grid block in question are obtained.

For a converged solution to a composite grid problem, since the pressure fields within each of the blocks of the system have developed independently from the others, and are only determined to within a constant within each grid block, the pressure fields must be corrected in some manner so that they are compatible with each other. In order to determine the compatibility constant between two blocks sharing an internal boundary, a normal momentum balance is performed across the internal boundary interface, and the pressure field in one of the blocks sharing that interface is adjusted by a constant value so that normal momentum flux across that segment is identical when computed from either of the two blocks. For the general case in which the grid lines from neighboring blocks across the interface are discontinuous, this post–processing procedure is dependent upon the choice of the segment over which the normal momentum flux balance is performed, however, one can compute the appropriate compatibility constants for ensuring global conservation of the normal momentum flux by performing the balance described above along an appropriate length of the internal boundary.

Since the present interface procedure is designed to satisfy the conservation laws without artificially imposing the continuity of solution variables, extra care must be taken in ensuring that no spurious multiple solutions occur for configurations that contain solid obstacles. For the staggered grid arrangement adopted here, there is no need to specify boundary conditions for pressure; this basic merit remains the same for any multi–block configuration. However, for multiply connected domains, it is known that extra boundary constraints need to be prescribed, otherwise the solution may not be physically correct. An earlier study conducted by Shyy (1985c) has established this point clearly. By solving a channel flow with two outlet branches, besides the normal extrapolating outflow conditions, an additional regulatory mechanism is needed. In Shyy (1985c), it is the ratio of the mass flux between the two branches that helps the numerical computation reach a unique and physically correct solution. This extra condition is a mathematically correct procedure, and not a numerical artifact. For example, Milne–Thompson (1979) has studied a potential flow in a branched canal. In that analysis, the stagnation point of the flowfield is given a priori, amounting to an equivalent specification of the mass split ratio as well.

In the present formulation, for configurations with internal obstacles, there is no need to explicitly assign any extra boundary conditions to facilitate the numerical computation. The problem arises from the fact that since one does not have to prescribe any pressure boundary conditions within each block, potentially, a pressure jump can occur across a block interface. This jump is physically incorrect and can result in multiple spurious solutions. To circumvent this potential pitfall, one can simply enforce a pressure continuity constraint at the interfaces shared by adjacent blocks. The practice devised here is that in the course of iteration, the pressure continuity at interfaces common to adjacent blocks is enforced via explicit conservation of the normal momentum through these interfaces. For cases with internal obstacles, individual grid blocks need this continuous enforcement of pressure continuity at internal interfaces to prevent unphysical pressure jumps from appearing. As can be observed in the test cases to be presented later, the present procedure can maintain a physically correct solution. One should emphasize that the present treatment of pressure is completely consistent with the spirit of the staggered grid, i.e. no artificial pressure boundary condition is used to affect the final solution; the pressure continuity is enforced strictly via a normal momentum balance between adjacent blocks to avoid creating a nonphysical flowfield.

8.5 Explicit Conservation Procedure

In order to update the values of the dependent variables in a particular block of a composite grid, the fluxes of mass and momentum through the boundaries of the block into the various boundary control volumes must be specified. According to the global conservation strategy previously outlined, explicit conservation of the mass and u–momentum flux through the horizontal boundaries and the mass and v–momentum flux through the vertical boundaries, provides all of the information required for calculating the fluxes into all of the boundary control volumes. In order to illustrate the explicit conservation procedure used in this work, the explicit conservation of mass and u–momentum through horizontal internal boundaries is presented in detail here. The same concepts apply for explicitly conserving mass and v–momentum through vertical internal boundaries.

Consider a composite grid comprised of two blocks with a horizontal overlap region. Fig. 37 shows a blowup of part of the overlap region, where the upper block (whose dependent variables are to be updated), is drawn with dotted lines, and the neighboring lower block is drawn with solid lines. The upper block will hereafter be denoted as block one, and the lower block as block two. In the figure, a pressure correction control volume along the lower boundary of block one has been shaded. In order to formulate the discrete form of the pressure correction equation for this control volume, the mass flux through the lower boundary of the control volume (segment AB) must be obtained. Since segment AB lies completely within block two, all the information required for calculating the mass flux through the segment is obtained from this block. Segment AB is comprised of a number of sub–segments, formed by the intersection of the vertical grid lines of block two with the

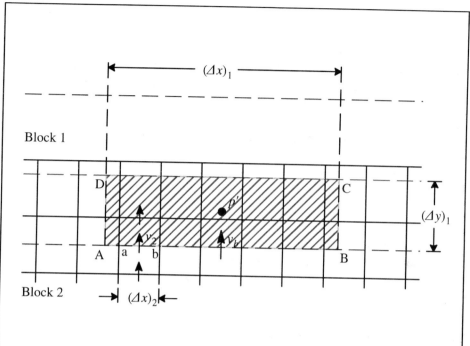

Figure 37. Blowup of horizontal overlap region for demonstrating explicit conservation of mass. Pressure correction control volume is shaded.

segment. In this case, segment AB is comprised of five complete sub–segments (forming the center portion of the segment), and two partial segments (on the left and right ends of segment AB). Each of these sub–segments contributes a portion to the total mass flux through AB.

The mass flux contribution for a typical sub–segment in block two is obtained in the following manner. The constant normal velocity component along the segment is obtained using a linear distance–weighted interpolation from the velocity components just above and below, located within block two. For example, consider the second sub–segment, denoted as ab, of segment AB, shown in Fig. 37. The normal velocity component along the entire sub–segment is taken to be that at the center of the sub–segment, and is denoted v_2. The value of v_2 is obtained based on a linear interpolation within block two from the staggered grid values located directly above and below, which are also shown in the figure. Once the segment velocity has been found, the mass flux contribution is obtained by multiplying this value by the density and the segment length. This calculation procedure results in a piecewise constant mass flux distribution over the segment AB. With the total mass flux from block two denoted as $mflux_2$, and assuming that the north, west, and east pressure correction neighbors are internal unknowns of block one, then the discretized form of the pressure correction equation for the pressure correction control volume ABCD becomes

$$A_P \, p'_P \; = \; A_E \, p'_E + A_W \, p'_W \; + \; A_N \, p'_N \; + \; b \qquad (8.7a)$$

$$A_P \; = \; A_E + A_W + A_N \qquad (8.7b)$$

where the coefficients A_i are as given before (except for A_s which has been set to zero since we consider the mass flux obtained to be a known quantity), and the source term b takes the form,

$$b \; = \; [(\varrho u^*)_w \; - \; (\varrho u^*)_e](\varDelta y)_1 \; + \; mflux_2 \; - \; (\varrho v^*)_n \, (\varDelta x)_1 \qquad (8.7c)$$

Once the mass flux into the pressure correction control volume has been calculated, the velocity component at the control volume boundary for block one, denoted v_b in the figure, is given by the following

$$v_b \; = \; \frac{mflux_2}{\varrho(\varDelta x)_1} \qquad (8.8)$$

Thus, as previously stated, conservation of mass along the horizontal boundaries provides both the boundary condition for each of the pressure correction control volumes along the boundary, as well as the normal velocity component profile along the boundary. Once this profile is determined, then all the information required for the calculation of the contributions to the v–momentum fluxes not involving pressure (i.e. $\varrho vv \; - \; \mu \, (\partial v/\partial y)$) is available.

The procedure for explicitly conserving the u–momentum fluxes into each of the u control volumes along horizontal boundaries is similar to that for explicitly conserving mass. Consider again, a composite grid comprised of two blocks with a horizontal overlap region. Fig. 38 shows a blowup of part of the overlap region, where the upper block (whose dependent variables are to be updated) is again denoted as block one, and the lower neighboring block as block two. In the figure, a u control volume, labelled ABCD, along the lower boundary of the block has been shaded. In order to formulate the discrete form of the u–momentum equation for this control volume, the u–momentum flux through the lower boundary of the control volume (segment AB) must be obtained. Again, segment AB is comprised of a number of sub–segments, however, in this case the sub–segments are defined by the u control volumes in block two. Consider the sub–segment ab shown in the figure. The u–momentum flux, including the convective and diffusive components, through this sub–segment, denoted $uflux_{ab}$ is calculated as follows:

$$uflux_{ab} \; = \; [D(1 - 0.5|F/D|) \; + \; [\![\, F, 0.0]\!] \;] \, u_L \; -$$
$$\qquad\qquad [D(1 - 0.5|F/D|) \; + \; [\![- F, 0.0]\!] \;] \, u_U \qquad (8.9a)$$

where the terms u_U and u_L are respectively the u velocity components located on the staggered grid of block two just above and below the segment. In this expression, central differences have been used for both the convection and diffusion terms for illustration

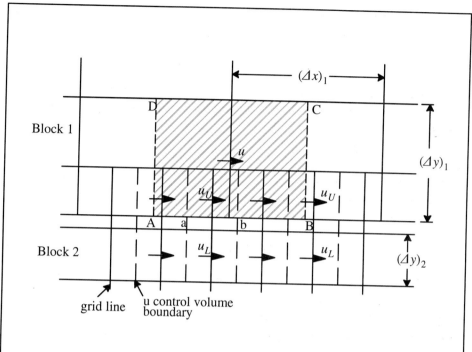

Figure 38. *Blowup of horizontal overlap for demonstrating explicit conservation of u–momentum. U control volume is shaded.*

purposes. The terms F and D represent the convection and diffusion fluxes through the segment, and are given by

$$D = \frac{\mu \Delta x}{(\Delta y)_2} \qquad F = \varrho v_s(\Delta x) \tag{8.9b}$$

where Δx represents the length of the sub–segment, and $(\Delta y)_2$ is the vertical distance between u_L and u_U which for this case is equal to the constant vertical grid spacing in block two. The quantity v_s is the velocity component calculated via bi–linear interpolation from the neighboring v component values of block two at the center of the segment. With the total flux contribution from block two denoted as $uflux_2$, and assuming that the north, east, and west neighbors of the u boundary control volume in block one are internal unknowns, the discretized form of the u–momentum equation for the control volume becomes

$$A_P u_P = A_E u_E + A_W u_W + A_N u_N - (F_n + F_e - F_w)u_P +$$
$$(p_W - p_E)(\Delta y)_1 + uflux_2 \tag{8.10a}$$

$$A_P = A_E + A_W + A_N \tag{8.10b}$$

In the original formulation for the u control volume in Patankar (1980), the term $(F_n - F_s + F_e - F_w)$ in the coefficient A_P is identically zero for a continuity satisfying velocity field and can be dropped. The term $(F_n + F_w - F_e)u_P$ remains in Equation (8.10a) because A_P is now computed without the contribution from A_S. Due to the interface treatment, A_S and its corresponding component in A_P are now explicitly given in the form of $uflux_2$. The term $(F_n + F_w - F_e)$ can be replaced by F_s and computed using the normal velocity component profile at the boundary obtained from explicitly conserving mass.

In the examples above, illustrating the explicit procedure for conserving mass and u–momentum across horizontal boundaries, all the required information was obtained from a single neighboring block. It should be noted here, that in a general composite grid, for any control volume located along a boundary, contributions to the total flux into the control volume may be required from any number of neighboring blocks and/or externally imposed boundary conditions. Since the path of information flow into each block is entirely specified via boundary segments, we can determine which neighboring block or boundary condition is to provide the required information across a particular portion of a control volume boundary. Once this is known, then if a neighboring block is required to provide the information, the flux through that portion can be calculated as described above, and if a boundary condition is required, then the flux across that portion can be calculated accordingly.

8.6 Conditions for Local and Global Conservation

In this work, as previously mentioned, the only restriction placed on the overlap of grid blocks is that internal block boundaries must fall within a constant column or row index of any neighboring blocks. For such configurations, the question arises as to whether both local and global conservation can be maintained simultaneously. Alternatively posed, does the local conservation procedure detailed in the previous section result in a globally conservative formulation for general configurations? In accordance with Patankar's requirement that acceptable formulations apply equally well to fine and coarse grid discretizations, a similar requirement is placed here on the conservative interface treatment, namely that we be able to drive global mass and momentum residuals to machine accuracy for arbitrary configurations satisfying the single constraint mentioned above.

To begin the discussion, consider the following two–block cavity flow configuration shown in Fig. 39. For the purpose of clarity, the two blocks are shown separated, however, physically they overlap. The dotted vertical line shown within the left block (block one), represents the physical location where the left boundary of the right block (block two) falls within block one. Suppose that the mass fluxes into the control volumes along the left boundary of block two are determined from block one using the explicit conservation procedure previously detailed. In order to satisfy global mass conservation for block two, the mass fluxes must sum to zero, since no mass can enter

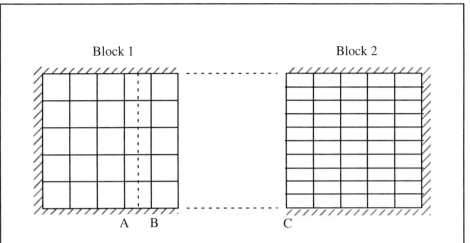

Figure 39. Two–block cavity flow problem. Explicit conservation of mass flux through line C results in a globally conservative formulation for block two.

through the upper, lower, and right solid boundaries. From the explicit conservation of mass into block two, the total mass flux through line C, denoted as M_C, can be written in terms of the mass fluxes through lines A and B as

$$M_C = \left(\frac{x_B - x}{x_B - x_A}\right) M_A + \left(\frac{x - x_A}{x_B - x_A}\right) M_B \tag{8.11}$$

where x_A and x_B represent the horizontal coordinates of lines A and B respectively, and x the horizontal coordinate of the dotted line. Since we are using a control volume formulation, both M_A and M_B are identically zero (to machine precision), and thus $M_C = 0$ holds. Therefore, for this case, the explicit local conservation procedure also results in a globally conservative formulation for mass. One important thing to note here is that due to the piecewise constant local conservation procedure, the total fluxes M_A and M_B are consistent with the internal mass summations along the lines A and B, and are thus by definition zero. If a piecewise linear local conservation (in the vertical direction) is used, this is not the case, and in general a global mass correction must be distributed along line C to enforce global conservation, which destroys any concept of local conservation. Another interesting thing to note is that as far as v–momentum is concerned, for this case, global conservation is always maintained, regardless of the explicit conservation procedure used along line C, due to the presence of the solid boundaries, which will adjust accordingly to produce a net v–momentum flux of zero from block two. In this regard, solid walls, in general, provide an additional degree of freedom for ensuring global conservation of momentum, however, they allow no such flexibility as far as mass is concerned. Outflow boundaries, on the other hand, can provide relief for both mass and momentum. For grid blocks which are located entirely internal to a domain, however, a global imbalance of momentum can occur as with mass above, since no additional degree of freedom is provided in this case via either a solid wall or outflow condition.

Now consider the three–block cavity flow configuration shown in Fig. 40. Here we are concerned with the total mass flux entering block three. For this case, the overlapping priorities are set with block one having the highest priority, followed by block two, and then block three. With this arrangement, the internal boundaries of block three are divided into three segments, labelled A, B, and C at the center of the segments, as shown.. The total mass flux entering through the internal boundaries of block three is given as

$$mflux_3 \; = \; M_A \; + \; M_B \; + \; M_C \qquad (8.12)$$

For global conservation, $mflux_3$ must be identically zero, since no mass can enter block three through the lower or right solid walls. In this case, however, unlike the previous one, no control volume–based argument can be made to prove that this summation is zero, and in general, the summation is not zero. Therefore, a global mass imbalance for block three results, preventing the continuity (pressure correction) equation from driving its

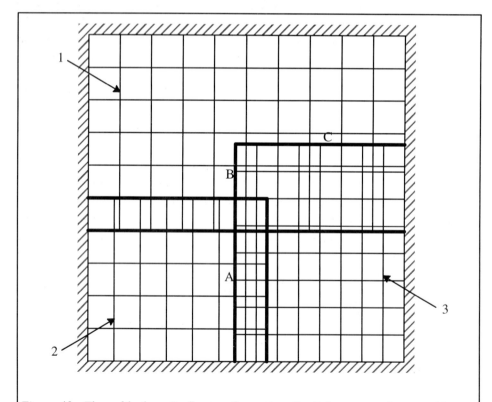

Figure 40. Three–block cavity flow configuration. Explicit conservation procedure does not guarantee global conservation of mass for block three. Internal boundaries are highlighted in bold.

residual to machine zero. The level of mass imbalance for this scenario is also highly dependent on both the discretization as well as the solution gradients within the domain, since both of these factors influence the accuracy of the local interpolation. For a sufficiently fine discretization, global mass conservation may be closely satisfied, but only in the limit of infinite grid resolution is it guaranteed to be exactly satisfied. Again though, no global conservation problem exists for the u–momentum and v–momentum equations, due the presence of the solid walls.

In general, in order to achieve a globally conservative formulation via the piecewise constant explicit local conservation, all locally conserved fluxes must be based strictly upon the interior fluxes of neighboring blocks as they are represented within the control volume formulation, without resorting to interpolation among the fluxes. For the global conservation of mass, this implies that the internal boundaries of every block in the composite grid must fall identically upon the grid lines of the neighboring blocks. Since mass and tangential momentum are the only quantities which are explicitly conserved, and since they share the same physical internal boundary on every side of a block, this, in fact, is the only condition required to guarantee global conservation of mass and both momentum components for general configurations. This overlap condition for ensuring global conservation is similar to that imposed by Rai (1986a) in his development of a conservative transfer scheme for patched grid, compressible flows, where artificial grid lines are extended into the neighboring blocks until they intersect the nearest interior control volume face at the boundary, which in effect results in a direct usage of the neighboring block fluxes (with the piecewise constant local conservation) without resorting to interpolation.

8.7 Composite Grid Calculation

A demonstrative example is given here to show the capability of this composite grid method in treating complex geometry. Fig. 41(a) shows the configuration studied, where we have the internal obstacle CFD inside a cavity with sliding upper and lower walls and entry and exit jets on the left and right walls respectively. Fig. 41(b) shows the multi–block configuration used to solve the problem. The grid consists of eleven individual grid blocks, ten of which are used to form the interior portion of the cavity, and one which is used on the interior of the letter D. The problem prescribed on the block forming the interior of D is actually a separate lid–driven cavity flow running simultaneously with the problem prescribed in the main cavity. Both flows are computed for a Reynolds number of 1000 based on the lid velocity and cavity width of each problem. Fig. 41(c) displays the streamfunction contours.

8.8 Composite Multigrid

8.8.1 Implementation Issues Associated with Composite Multigrid

In this section, we examine two important implementation issues associated with the composite multigrid solution technique. The first issue involves the interaction

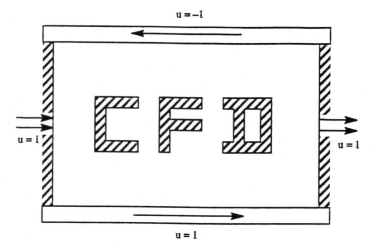

(a) Physical problem with boundary conditions.

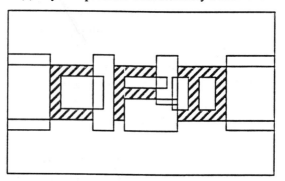

(b) Grid block arrangement.

(c) Streamfunction contours for Re=1000.

*Figure 41.   Example of the composite grid technique for a CFD–shaped body
placed inside a cavity.*

between the composite grid and multigrid components. From the discussion, a methodology is chosen which is considered most suitable for solving highly nonlinear equations, such as the Navier–Stokes equations, on multilevel composite grids. The second issue concerns the coarsening strategy which is adopted for multilevel overlapping grids. This issue in considered from the standpoint of both the numerical efficiency achieved in obtaining solution to complex flows requiring composite grids, as well as the flexibility in discretizing the domains for such problems.

For multilevel composite grids, one area of primary interest involves the procedural relationship between the composite grid component and the multigrid component. The question as to which component should reside as the outermost shell in solution algorithms for highly nonlinear equations such as the Navier–Stokes equations is still largely unresolved and in need of further study. The two main composite multigrid approaches which have been studied in the literature can be summarized as follows. In the first approach (hereafter called method I), the composite grid component acts as the outermost shell. For any block in the composite grid, the appropriate boundary conditions are obtained at the finest grid level and a multigrid cycle is then used to solve the governing equations within the block. This procedure is carried out for each block in the composite grid in some predetermined cyclic order. Thus, for each multigrid cycle for the composite grid, the boundary conditions for each block are obtained only at the finest grid level. In the second approach (method II), the multigrid component acts as the outermost shell. With this approach, for each level in the multigrid cycle, the appropriate composite grid problem is solved, with internal boundary conditions for each block being determined from the neighboring blocks within the same level. Therefore, within each multigrid cycle for the composite grid, internal boundary information is exchanged multiple times, the exact number of which depends on the type of cycle being executed.

Model problem analysis was performed by Hinatsu and Ferziger (1991) to compare the convergence rates obtained with the above two methods. As test problems, they used a 1–D linear ODE and a 2–D Poisson equation. For the 1–D model problem, two subgrids were used, and both Dirichlet and Neumann internal boundary conditions (IBC's) were employed to transfer information between the subgrids. The influence of the domain boundary conditions (DBC's) was also considered. In some instances they found that method I outperformed method II in convergence rate, however, this was highly dependent on both the IBC's as well as the DBC's. In addition, the relative performance of the two methods was found to be largely dependent on the IBC relaxation parameter, which was used to accelerate the variation of the IBC's. Overall, method II was seen to be less sensitive to the IBC relaxation parameter, and for no relaxation, slightly outperformed method I. The 2–D Poisson equation was solved on a domain composed of two diagonally shifted squares. Neumann conditions were imposed for both the IBC's and DBC's. Overall, the fastest convergence was obtained using method I with slight over–relaxation of the IBC's.

While method I was deemed more favorable than method II in the above study, others have chosen to employ method II for composite grids. Henshaw and Chesshire (1987) used a correction–storage (CS) strategy with method II for solving linear, variable–coefficient PDE's. Tu and Fuchs (1992) used method II within a FAS scheme based upon a V cycle, for solving the incompressible Navier–Stokes equations on staggered grid arrangements. In their work, multigrid was used for both the momentum and pressure equations, unlike other works where only the pressure equation is solved using multigrid (Perng and Street 1991). Their results show a substantial speedup with the FAS scheme, which is largely maintained when applied to composite grids. While the model problem analysis performed by Hinatsu and Ferziger indicates a slight preference for method I, they were employing a CS strategy for solving individual linear equations. Since we are solving a system of nonlinear equations with a FAS strategy, it is doubtful that their conclusions can be extended here.

Bearing the above considerations in mind, we have chosen to employ method II in this work exclusively. For nonlinear problems, method II allows changes in the solution variables to propagate freely to neighboring blocks within the multigrid cycle, thus providing improved estimates of the boundary conditions at each grid level. This feature is consistent with the original FAS concept and allows the boundary values at each grid level to be treated in the same manner as the interior points. In our opinion, this should enhance the convergence rate when solving nonlinear equations on composite grids, however, further study is required to determine if this is indeed the case, and if so, to what extent convergence is enhanced. Since the original composite grid methodology was developed for general systems, little difficulty is encountered in extending the original framework to handle multilevel composite grids. In this regard, the original composite grid procedure operates as a module within the outer framework of the FMG–FAS solution process.

Another issue which arises when attempting to apply multigrid to composite overlapping grids concerns the minimum overlapping requirement. The minimum overlapping is determined by the requirement that the sides of control volumes located along internal boundaries must fall within interior known, calculated values of the neighboring grid blocks. Consider the two–block case shown in Fig. 42, where the left and right blocks have been horizontally separated for visualization purposes, but are assumed to overlap along some vertical boundary. Now, assume that the boundary conditions for block 2 are to be determined. In this case the required overlap is determined only by the left boundary for block 2 and the right boundary for block 1, since the surrounding boundaries are Dirichlet conditions. For a vertical internal boundary between the adjacent blocks, the two types of boundary control volumes of interest are the continuity and v–velocity component control volumes, as detailed previously. These two types of control volumes along the left internal boundary of block 2 are shown in the figure. For the continuity control volumes along the left internal boundary of block 2, the inlet u velocity component profile along the line denoted by A_2 is the required boundary condition. This line must fall within interior, calculated u velocity component values of block 1, and with the staggered

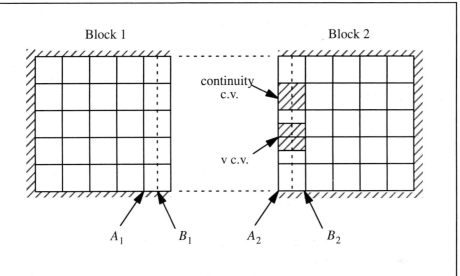

Block 1   Block 2

continuity c.v.

v c.v.

A_1 B_1 A_2 B_2

Figure 42. Two–block composite grid used to demonstrate minimum overlapping requirement.

grid convention means that line A_2 must fall on or to the left of line A_1 in block 1. Boundary conditions for the v control volumes require the calculation of the horizontal v–momentum flux from block 1 which requires both u and v velocity component values. Now, the rightmost interior, calculated v–component values in block 1 are located along line B_1. Since u component values are also required, the minimum overlapping again requires that line A_2 of block 2 fall on or to the left of line A_1 in block 1, identical to the case of the control volume used to compute the continuity equation. Thus, based on these requirements, line A_1 in block 1 defines the minimum overlapping requirement for block 2. Similarly, line B_2 defines the minimum overlapping requirement for block 1. For a valid overlap, both overlapping constraints must be satisfied. Similar constraints also exist for horizontal overlapping zones.

With the multigrid procedure adopted in Shyy and Sun (1993), as already discussed, successively coarser grids are created by removing every other interior grid line in both coordinate directions from the grid at the previous (finer) level. Letting M denote the total number of multigrid levels and n_k the number of grid lines in any coordinate direction for grid level k, then $n_k = (n_{k+1} + 1)/2$ for $1 \le k \le M - 1$. With this notation, level M is again the finest grid and coarser grids have lower k values. For composite overlapping grids, this type of coarsening strategy quickly leads to a violation of the overlapping requirements just discussed unless appropriate measures are taken. Consider a two–block continuous grid line case as shown below in Fig. 43(a). The two blocks combine to form a vertical single cell overlap region which satisfies the overlapping requirements for the finest level. However, at the next multigrid level shown in Fig. 43(b),

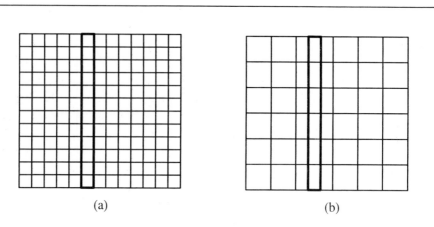

(a)   (b)

*Figure 43.   Two–block composite overlapping grid used to demonstrate violation of
overlap constraint with grid coarsening. a) Valid overlap at finest level;
b) Invalid overlap after coarsening.*

the overlapping requirements are not met and thus the overlap is not valid. In general, for a composite grid with uniform grid spacing and continuous grid lines across internal boundaries, in order to ensure a valid overlap on the coarsest multigrid level, an overlap thickness of 2^{M-1} cells must be used on the finest grid level. In the FMG procedure, for $M=6$ an overlap thickness of 32 cells is required. Clearly, this strategy is very wasteful of computational time and physical memory, as many redundant grid nodes are required just to maintain the appropriate overlap.

One possible remedy to this situation hinges on the fact that with the above strategy, the location of the physical boundaries of each grid block were maintained during the coarsening process. For some simple composite grid configurations, such as a two–block cavity flow problem, it is possible to adjust the physical location of certain internal block boundaries at each multigrid level to ensure that the appropriate overlap is maintained. Thus for each multigrid level, the physical locations of the internal boundaries are different. Algorithmically, this type of strategy is not difficult to implement for simple configurations such as the one above, however, for configurations in which multiple (>3) blocks overlap at the same location, or in which some physical boundaries are composed of both boundary condition segments and internal boundary segments, this strategy becomes very difficult to implement, and in some cases is impossible due to the fact that physical boundary constraints may be violated by moving sides with internal boundaries.

A general solution to this problem lies in modifying the coarsening method used during the restriction process. With the previously mentioned coarsening method, the grid at level $k-1$ is created from level k by removing every other interior grid line in both coordinate directions. As a result of this, the rows of cells directly adjacent to the boundaries of the grid thicken as $k \rightarrow 1$ until eventually the off–boundary side of the row leaves the overlap region established by the finest grid level, thus creating a non–valid overlap. In order to prevent this, a row or rows of fine grid cells can be retained next to the boundaries at each multi–grid level. For composite grid schemes utilizing standard 5–point stencils (such as the hybrid scheme for the convection terms and the central difference scheme for diffusion terms), it is sufficient to maintain a single layer of fine grid cells at the boundaries for all multigrid levels. For 9–point stencils commonly arising from the use of higher–order convection schemes, at least two layers of fine grid cells must be maintained. In this work, we will retain two rows of fine grid cells at each multigrid level. With this coarsening strategy, when employing 5–point stencil schemes, now for a composite grid with uniform grid spacing and continuous grid lines across internal boundaries, a single–cell overlap thickness on the finest grid level is sufficient to ensure a valid overlap at all multigrid levels, while for 9–point stencils a two–cell overlap is sufficient. Letting n_k again denote the number of grid lines in any coordinate direction for multigrid level k, then $n_k = (n_{k+1} + 5)/2$ for $1 \leq k \leq M - 1$. As an example of this grid coarsening strategy, consider the four–level multigrid for a square lid–driven cavity shown in Fig. 44. The grid shown in Fig. 44(a) is the finest grid, and Figs. 44(b)–(d) represent successive coarsenings from previous levels. The finest grid contains 21×21 uniformly–spaced nodes, while successive grids contain 13×13, 9×9, and 7×7 nodes respectively. It should be noted that this type of uneven grid coarsening procedure has been developed previously in a single block multigrid procedure (Shyy *et al.* 1993a). The primary motivation there was to maintain adequate grid resolution in the region close to the solid boundary. For high Reynolds number flows, especially those in the turbulent regime, it has been found that this type of grid coarsening procedure can help maintain the necessary length scale resolution requirement, resulting in improved convergence rates with the multigrid solver. Here, this aspect will still be useful.

8.8.2 *Evaluation of Composite Multigrid Technique*

In this section, three test cases are used to examine the effectiveness of the composite multigrid technique which has been developed. First we consider a simple geometry in order to investigate the interaction between the composite gridding procedure and the multigrid technique. The well–known lid–driven cavity flow is used as the test case. For the next two cases, we compute two complex flows, namely a channel with internal baffles and the flow about the previously shown CFD configuration. These cases are used to assess the ability of the current procedure to efficiently compute fluid flows with complex geometry.

For composite grid methods, two factors, namely, the convergence rate within the individual subgrids, and the presence of the internal boundaries, interact to determine the net effectiveness of the method. For a fixed physical domain and grid resolution, as the domain is divided into a larger number of smaller subgrids, the convergence rate within each of the subgrids increases. At the same time, however, the increased presence of the internal boundaries reduces the coupling between the unknowns in the different subgrids, and eliminates the implicit connection to boundary conditions for some subgrids. Thus,

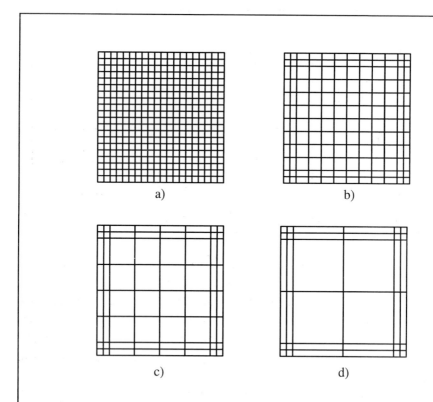

*Figure 44.   Four–level multigrid for square lid–driven cavity showing new
             coarsening strategy for composite multigrid. a) Level four; b) Level three;
             c) Level two; d) Level one.*

while the convergence rate of the linearized equations within individual subgrids may be improved, that of the overall system may not. The net effect of composite gridding is certainly problem dependent, however, for the current procedure, it has been shown (Shyy *et al.* 1993b) that for a square, lid–driven cavity flow, the convergence rate is virtually unaffected by the way in which the domain is partitioned, or the number of subgrids used. At the same time, however, for a high aspect ratio channel flow, division of the domain into several subgrids was found to significantly increase the rate of convergence of the overall calculation (by almost an order of magnitude for large aspect ratio).

With the incorporation of the composite grid method into the outer framework of a multigrid strategy, the interaction of the composite grid and multigrid components becomes a relevant issue of concern. To examine this interaction, and the overall effect of composite gridding on multigrid efficiency, a single–block, lid–driven cavity flow is computed, and compared with corresponding multi–block configurations for the same flow. The sequence of grids used is shown in Fig. 45. Grid A consists of a single block of 69×69 nodes, while grids B and C consist of two 69×37 blocks and four 69×21 blocks

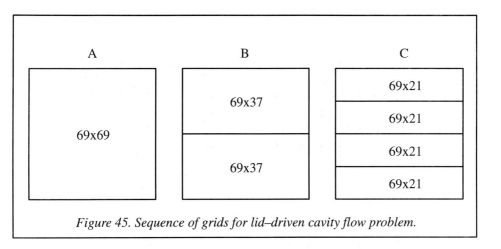

Figure 45. Sequence of grids for lid–driven cavity flow problem.

respectively. These grid resolutions were chosen, because with a 69×69 initial multigrid level, five coarsenings can be achieved with the new coarsening procedure, resulting in successive grids of $37 \times 37, 21 \times 21, 13 \times 13, 9 \times 9$, and 7×7 nodes. For cases B and C, the resulting configurations have overlaps of five cells and four cells respectively on the finest grid level, however, as stated previously, for this case the convergence rate is essentially unaffected by the degree of overlap, and thus any overlap satisfying the minimum overlapping requirements is sufficient for this case.

Since the multi–block grids contain more control volumes than the single grid, equivalent convergence tolerances must be used. For the following computations, a normalized residual of 1.0×10^{-4} is used for the single grid case. The corresponding normalized residuals for cases B and C are 1.07×10^{-4} and 1.21×10^{-4} respectively. These normalized global residuals are used to give an indication when the various cases have reached the same level of convergence. Another note needs to be made regarding the relative amount of work performed between successive multigrid levels. In the following results, the amount of work required to achieve convergence is stated in work units. A work unit is here defined as one sweep through all of the blocks on the finest grid. Since we are not using the standard coarsening procedure, whereby the number of grid points at any multigrid level k is half that at level k+1, the work per iteration on grid level k is not one fourth of that on level k+1 per iteration. For the new coarsening procedure, the work required per iteration on level k is greater than one fourth of that on level k+1, and is dependent on the configuration. For the three cases to be considered here, the work per iteration on each of the possible multigrid levels with respect to the finest grid level is shown in Table V.

During the course of the FMG cycle the current level is taken to complete convergence, i.e. the equivalent convergence tolerance of 1.0×10^{-4}, before the next multigrid level is initiated. While this requirement is wasteful of iterations on coarser grid

Table V. *Relative work for successive multigrid levels compared to finest grid level for three cavity flow cases. Note: Only five multigrid levels are possible for case B and four for case C.*

Level Number	Relative Work Compared to Finest Level		
	Case A	Case B	Case C
6	1.0000	——	——
5	0.2875	1.0000	——
4	0.0926	0.3043	1.0000
3	0.0355	01069	0.3320
2	0.0170	0.0458	0.1304
1	0.0103	0.0247	0.0628

levels during the initial process of obtaining an approximate solution for the finest grid V cycle, it does not affect the results to be presented.

Results for cases A, B, and C are presented in Table VI. For each case, the number of work units required to achieve convergence and the corresponding speedup with respect to the single level calculation, in terms of a work unit ratio, are tabulated for each possible number of multigrid levels. Reynolds numbers of both 100 and 1000 were computed in order to evaluate the flow dependency of the results as well as to provide a database for the following complex flow computations. In all of the calculations, the hybrid scheme of Spalding (1972) was used for the convection terms while standard central differences were employed for the diffusion terms. The hybrid scheme was chosen in order to have the same scheme at all multigrid levels, since with the central difference scheme, divergence of the solution was found to occur at some coarser grid levels for Re=1000. In any case, the hybrid scheme defaults to the central difference scheme for a majority of the flow field for both Re=100 and Re=1000, since the local cell Peclet number exceeds the value of 2 on the finest grid only near the upper corners of the domain (Shyy and Thakur 1993).

From the tabulated values shown in Table VI, several observations can be made. Firstly, we see that the number of work units required to achieve convergence for the single–level computations for cases A, B, and C are virtually the same, for both Re=100 and Re=1000, indicating the relative insensitivity of the computations to the number of blocks used to partition the domain. This is consistent with previous findings of Shyy *et al.* (1993b) for the same flow, but with different grid resolutions and domain partitioning. Secondly, for this configuration, the solution for Re=1000 converges much quicker than that for Re=100. This surprising result was also observed by Shyy and Sun (1993), and is inconsistent with other configurations whereby the number of iterations for the single–level computation increases as the Reynolds number is increased.

Table VI. Multigrid computation results for cavity flow. a) Case A; b) Case B; c) Case C.

a)

MG Levels	Re = 100		Re = 1000	
	Work Units	Speedup	Work Units	Speedup
1	4978	1.00	1950	1.00
2	1022	4.87	1527	1.28
3	616	8.08	1253	1.56
4	524	9.50	1113	1.75
5	548	9.09	1145	1.70
6	580	8.58	1203	1.62

b)

MG Levels	Re = 100		Re = 1000	
	Work Units	Speedup	Work Units	Speedup
1	4976	1.00	2099	1.00
2	1076	4.62	1611	1.30
3	673	7.40	1335	1.57
4	596	8.35	1197	1.75
5	618	8.06	1164	1.80

c)

MG Levels	Re = 100		Re = 1000	
	Work Units	Speedup	Work Units	Speedup
1	5030	1.0	2152	1.00
2	1188	4.24	1696	1.27
3	911	5.52	1449	1.49
4	824	6.10	oscillatory	

Regarding the effect of composite gridding on multigrid performance, we see that as the number of subgrids increases, multigrid effectiveness decreases. For two multigrid levels, the reduction in effectiveness is only slight, however, as the number of multigrid levels increases, the effectiveness is further reduced. This can be explained as follows. For the single grid case A, the finest grid level contains 69×69 nodes, resulting in successive multigrid levels of 37×37, 21×21, 13×13, 9×9, and 7×7 nodes. As the number of multigrid levels is increased for this case, we see a continual improvement in convergence rate up to five multigrid levels, at which point the convergence rate decreases. This clearly results from the relative resolutions of multigrid levels three, two, and one, which are 13×13, 9×9, and 7×7, respectively. Since these resolutions are comparable, the errors introduced via interpolation during the prolongation and restriction process are not efficiently eliminated since the wavelengths of the errors are comparable to the grid spacing onto which they are being interpolated. In other words, the additional speedup obtained with the introduction of a new multigrid level will be little, or none if the resolution of the new level is comparable to that of the previous level. For case B, which starts with an initial subgrid resolution of 69×37, this limitation is reached with fewer multigrid levels than case A, due to the resolution of only 37 nodes in the vertical direction on each of the finest subgrids. Case C, which starts with an initial subgrid resolution of 69×21, approaches this limitation even sooner. Thus, for the three grid configurations A, B, and C, the difference in resolution between any two multigrid levels is the least for case C, followed by case B, and then case A. Therefore, a reduction in multigrid effectiveness should occur as the domain is partitioned into two blocks, as in case B, and then four blocks, as in case C.

The situation described above, in which newly introduced multigrid levels have roughly the same resolution as their predecessors, is a direct consequence of the new coarsening strategy, and introduces a tradeoff between domain partitioning and the computational efficiency which can be obtained with composite multigrid. For the cases computed above, however, the grid resolutions employed on the finest levels are still relatively coarse, and thus the ratio of successive multigrid level resolutions quickly decreases below a factor of two (which is always maintained for the standard coarsening procedure) as the number of multigrid levels is increased. For much finer grid resolution cases, for which the multigrid technique is most useful, little difference exists between the new coarsening procedure and the standard procedure, and the tradeoff between domain partitioning and multigrid efficiency is minimal. In terms of this tradeoff, one last observation from the cases presented above is noted. While the overall multigrid effectiveness is much less for all cases with $Re=1000$, the sensitivity of the effectiveness to domain partitioning is less for $Re=1000$ than for $Re=100$. For two multigrid levels the degradation in multigrid performance between cases A and C is 13% for $Re=100$, and 1% for $Re=1000$, while for three multigrid levels a degradation of 32% occurs for $Re=100$, and only 5% occurs for $Re=1000$.

8.8.3 Composite Multigrid for Complex Flows

In order to test the current composite multigrid procedure for complex flows, two multiply connected flows are computed here. The first case considered is a channel with internal baffles. The flow is computed for Reynolds numbers of 100 and 800 based on the inlet velocity and configuration width. The convection scheme of Spalding (1972) is again employed here. Due to the x–coordinate resolution of 41 nodes in block four, at most three multigrid levels can be used for the computation. Fig. 46 shows the finest grid level along with the two successive multigrid coarsenings obtained using the new coarsening strategy. We note here that grid lines within some blocks of the coarsest multigrid level have been adjusted so as to guarantee global mass conservation for each of the those blocks, as discussed previously.

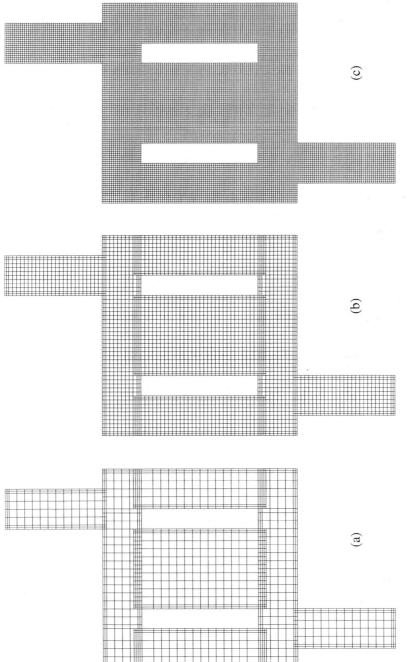

Figure 46. Three multigrid levels used for baffle flow computation. a) Level one; b) Level two; c) Level three.

Results from the multigrid computations for both *Re*=100 and *Re*=800 are presented in Table VII, in terms of number of fine grid iterations required to achieve convergence on the finest grid level, the total work performed during the FMG cycle, and the speedup in terms of a work unit ratio. Fig. 47 shows a comparison of the convergence paths of the single–level and multi–level computations for both Reynolds numbers. The speedup ratios obtained for this configuration demonstrate that even for complex composite grid configurations, multigrid effectiveness can be largely maintained. In general, for any composite grid configuration, the maximum attainable speedup ratio will be determined roughly be the lowest resolution block of the composite grid. For this configuration, the lowest resolution grid blocks are blocks one and seven, each having a resolution of 21×53 nodes. Therefore, one would expect the speedup ratio to fall close to that for a 21×21 single block cavity flow, which is consistent with the findings of Shyy and Sun (1993). Finally, the multigrid effectiveness obtained for this configuration is again

Table VII. Multigrid computations for baffle flow.

# levels	Re = 100			Re = 800		
	# f. g. iter.[*]	# work u.[**]	speedup	# f. g. iter.[*]	# work u.[**]	speedup
1	1977	1977	1.00	1288	1288	1.00
2	720	1099	1.80	660	1010	1.28
3	242	612	3.23	340	835	1.54

[*] indicates number of fine grid iterations   [**] indicates number of work units

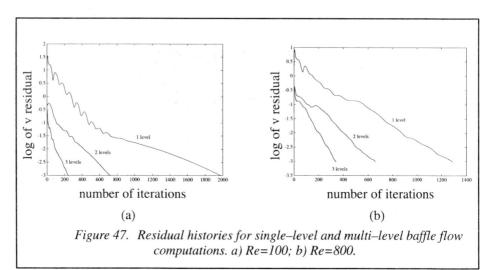

Figure 47. Residual histories for single–level and multi–level baffle flow computations. a) Re=100; b) Re=800.

dependent upon Reynolds number, and follows the same trends observed for the single–block, lid–driven cavity flow previously presented.

For the CFD configuration, a new grid block arrangement is required to apply the multigrid procedure. The physical configuration and boundary conditions are identical to those used previously, however, the grid has been constructed in a different manner in order to provide not only a sufficient number of multigrid levels for each subgrid, but subgrids which do not violate the global mass conservation property upon coarsening. Thirty one subgrids are used to discretize the domain. Fig. 48 shows the configuration and subgrid partitioning. Overlap regions have not been shown for clarity. Each of the subgrids is composed of either 21×21, 21×37, 37×21, or 37×37 nodes, so that four multigrid levels are possible for this configuration. The different multigrid levels are shown in Fig. 49.

The flow in the main portion of the enclosure is computed for Re=100 based on the sliding lid velocity and configuration width. For the enclosed lid driven cavity in the letter D, the lid velocity is the same as the main configuration, resulting in Re=6 based on the enclosed cavity width. Results for the various multigrid computations are presented in Table VIII in terms of the number of fine grid iterations required to reach convergence, the corresponding number of work units for the entire FMG cycle, and the effective speedup in terms of a total work unit ratio. Since the resolution for this configuration, based on the total number of grid nodes, is equivalent to a 171×171 square grid, convergence was taken

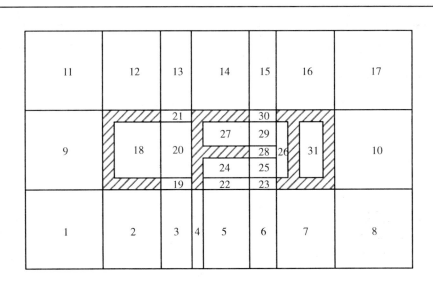

Figure 48. Subgrid partitioning for multigrid CFD configuration.

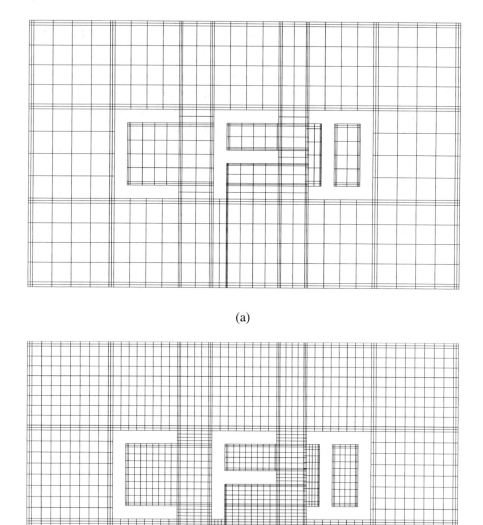

(a)

(b)

Figure 49. Multigrid levels for CFD configuration. a) Level one; b) Level two; c) Level three; d) Level four.

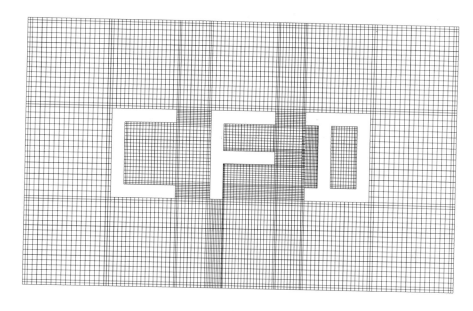

(c)

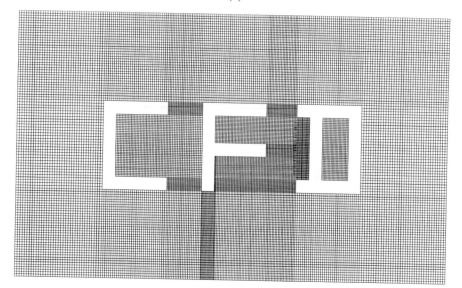

(d)

Figure 49 — continued

Table VIII. Multigrid computations for CFD configuration.

# levels	# fine grid it.	# work units	speedup
1	2735	2735	1.00
2	1020	1740	1.57
3	468	1284	2.13
4	228	782	3.50

to a normalized tolerance of 10^{-3}. From the speedup ratio values shown in Table VIII, we see that the multigrid procedure is still very effective in reducing the total amount of work required to achieve convergence on the finest grid level. In fact, a larger speedup ratio is obtained for this configuration than for the baffle flow, due primarily to the finer discretization on the finest grid level. These results indicate that for high–resolution grids (much higher than computed here) we can anticipate an order of magnitude speedup even for very complex multiply–connected configurations.

9 CONCLUDING REMARKS

In the previous sections, various aspects of the pressure–based computational algorithm relevant to complex fluid flow and heat/mass transfer problems have been presented. The point stressed here is that in order to conduct a successful calculation for problems of engineering interest, all these various aspects need to be adequately handled. That is, one must consider the grid layout and adjustment within the flow domain, the choice of the primary dependent variables and their geometric relationships due to the computational mesh, the form of the governing differential equations and the corresponding discrete form, the discretization and interpolation schemes for all terms, in particular, the convection terms, the treatment of solid and open boundaries, and solution techniques for the set of algebraic equations resulting from the discretization procedure. These issues need to be tackled individually; but, more importantly, the treatment of these elements must also be mutually compatible so that they can collectively form an effective computational tool for the practical problems of heat and mass transfer.

CHAPTER VI
PRACTICAL APPLICATIONS

In the following, several topics of different characteristics and flavor have been chosen to facilitate discussion. These include simulations of three–dimensional flow through a combustor, heat transfer and fluid flow characteristics in a modern high pressure discharge lamp, two–phase thermocapillary flow under normal and microgravity conditions and flow in a complex duct–airfoil assembly.

1 THREE–DIMENSIONAL COMBUSTOR FLOW SIMULATION

1.1 Background

In general, in a combustor flowfield, there exists a range of complex, interacting physical and chemical phenomena. Included are fuel spray atomization and vaporization, turbulent transport, finite–rate chemistry of combustion and pollutant formation, radiation and particle behavior, and recirculation zones involving multiple flow streams. Rigorous description of these phenomena are, however, either not available or require mathematical models which are too complex for computation, when taken together in the context of multi–dimensional flows. For these reasons, models of varying degrees of sophistication have been used depending on the particular application. Over the last three decades, the level of sophistication of the predictive models has been continuously increased with improvements in numerical methods, computer capabilities and physical understanding.

An important factor in the selection of sub–models is "computational tractability", which means that the differential or other equations needed to describe a sub–model should not be so computationally intensive as to preclude their practical application in three–dimensional Navier–Stokes calculations. For views on the current status of turbulence modelling, the interested reader is referred to a recent book edited by Lumley (1990).

1.2 Combustion Model

Basic material on combustion theory and models can be found in Spalding (1955), Libby and Williams (1980), Williams (1985) and Rosner (1986). Previous studies (Jones and Whitelaw 1984) have established that the "mixed–is–burned" model (equilibrium chemistry) along with the k–ε eddy viscosity turbulence model and an assumed shape pdf/moment equations for scalar fluctuations, are computationally tractable in complex flows and have replaced the eddy break–up models (Spalding, 1970). Numerical procedures used to be based on the SIMPLE algorithm (Patankar 1980) and rectangular grid systems, as exemplified by the highly popular TEACH code (Gosman and Iderish 1976) and its derivatives (Jones and Priddin 1979). It may be noted that other techniques are available (Oran and Boris 1981, Butler *et al.* 1981, Pope 1985); however these have not found widespread use in combustor simulation because of lack of flexibility, robustness and experience. Only the broad features of such flowfield have been simulated with TEACH–based codes (Kenworthy *et al.* 1983, Mongia *et al.* 1979). It is not trivial, however, to improve the fidelity of these simulations. The models for turbulence and chemistry and the numerical procedures are together responsible for the problems encountered.

First, on the experimental side, turbulent jet diffusion flames have been important for assessing combustion models due to the advances in experimental non–intrusive laser–based techniques and their application to such flows (Drake *et al.* 1986a,b; Penner *et al.* 1984, Eckbreth 1988, Chen and Goss 1989, Masri *et al.* 1992). Detailed and carefully chosen comparisons of data and calculations in jet flames have led to the improvement of models describing some of the fundamental mechanisms. However, jet flames involve a highly simplified fluid dynamics model and hence the much needed database for recirculating flows is not available from such experiments. In this regard, Switzer *et al.* (1985) and Goss and Switzer (1986) applied the CARS technique to an axisymmetric bluff–body stabilized flame. Recently, Correa and Gulati (1992) have conducted a study to probe the combustion characteristics in a bluff body stabilized burner based on a partial equilibrium model and Raman scattering technique.

Finite–rate chemistry models for turbulent combustion have been developed and assessed in jet flames (Janicka and Kollman 1979, Correa *et al.* 1985, Bilger 1980, Peters 1984, Chen and Kollman 1990, Smooke 1991, Peters and Rogg 1993). These models have been used in calculations of simple flow field when kinetically–influenced behavior is of interest. For example, the emission of thermal and prompt nitric oxides (NO_x)is a very prominent issue in the design of stationary gas–turbine systems. The formation of nitric oxides is limited kinetically and so the equilibrium models are not adequate. The model must also recognize the fluctuations in the NO_x production rate due to turbulence. Extensions of the "fast" chemistry model have been proposed (Kent and Bilger 1976) but have difficulties in accounting for superequilibium free radicals, which increase NO_x levels (as demonstrated by Drake *et al.* 1987). Prompt NO_x is more problematic as it involves the chemistry of hydrocarbon fragments (Iverach *et al.* 1972); the challenge is to describe the chemical kinetics with a computationally tractable yet realistic scheme. Other pollutants such as unburned hydrocarbons and CO also detract from the applicability of "fast" chemistry models. Efforts are being made (e.g., Dasgupta *et al.* 1993) to predict droplet evaporation and NO_x formation in a gas turbine combustor environment.

Flame extinction is another important non–equilibrium phenomenon. Reductions in fuel flow rate (to reduce power in engines) or the inlet air density (because of increasing altitude) may cause combustors to reach their blowout limits. The nature of turbulent flame extinction is currently under scrutiny in both premixed and non–premixed flames (Liew *et al.* 1981, 1984; Peters 1984, Miller *et al.* 1984, Chen *et al.* 1989). These phenomena further emphasize the importance of non–equilibrium chemistry and provide incentives to improve the models. The knowledge acquired by the application of such codes to the development of new engines will help expand blowout and re–ignition limits and will increase the performance envelopes of practical combustors. Here, however, the nonequilibrium effects will not be addressed.

1.3 Numerical Algorithm

To capture the underlying physics, computational algorithms with increased mathematical sophistication are being developed. The numerical prediction of shear layer flows – to within the accuracy of the physical models is now well established (Anderson *et al.* 1984). Many numerical procedures have been shown to be successful and most of the difficulties associated with such predictions relate to a lack of physical understanding and consequent inadequacies in the various models used. Complex (recirculating) flows, however, fail to satisfy the shear layer approximation. The accuracy of numerical predictions for this type of problem depends not only on the accuracy of the physical and chemical models, but also on the accuracy of the numerical techniques used to solve the equations.

In the following, a methodology for computing steady turbulent combusting flow in combustors of complex shape (Shyy and Braaten 1986, Shyy *et al.* 1988, 1989), is briefly outlined. The approach taken here attempts to strike a reasonable balance in handling two competing aspects of the modelling work, namely, the complicated physical and chemical interactions in the flowfield, and the requirements in resolving the three–dimensional geometrical constraints of the combustor contours, film cooling slots, and circular dilution holes. In order to handle these features, the CONCERT code has been developed and applied.

The key elements of the numerical algorithm and turbulent combustion models embodied in the CONCERT code are listed in the following:

(1)   Conserved scalar (with assumed pdf to account for variance effect) and fast chemistry approach for turbulence/chemistry interaction (Williams 1985). A partial equilibrium model has recently been incorporated into the two–dimensional version of CONCERT (Correa and Gulati 1992)

(2)   The Jones–Launder k–ε two–equation model with wall function treatment (Launder and Spalding 1974) for turbulence effects.

(3)   Zonal method for three–dimensional non–orthogonal grid generation which yields good control on the local geometrical variations (holes, slots) and produces a grid system with a unified index notation.

(4)   Semi–implicit iterative algorithm solving strong conservation form of transport
        equations (mass, momentum, and scalar fields) in general non–orthogonal
        curvilinear coordinates.

(5)   Second–order finite difference operators for all terms including convection,
        pressure and diffusion effects

(6)   Multi–step predictor–corrector method for the pressure correction equation.

(7)   Multi–grid method (with either line or point method) for solving the system of
        linear equations resulting from the discretization procedure, and for handling the
        whole system of equations.

The contribution of the CONCERT algorithm is that, for the first time, a gas–turbine combustor flow calculation can be conducted with a reasonable combustion model on the one hand, and a satisfactory numerical procedure on the other. Along with the development of CONCERT, there have been efforts devoted in this direction in the research community (Priddin and Coupland 1986, Joos and Simm 1987, Lee *et al.* 1990, McGuirk and Palma 1992). Priddin and Coupland (1986) used the orthogonal curvilinear coordinate system which fails either to resolve all of the geometrical complexities, such as circular dilution holes, or to adjust the grid according to the flow characteristics to reduce the numerical errors. Many other reported three–dimensional studies essentially are all still in the stage of conducting the reacting flow calculation based on Cartesian/polar coordinate systems, which is even less adequate.

1.4 Case Studies

As an illustration, the flowfield in the GE CF6–80C2 turbofan engine combustor (Shyy *et al.* 1989) is presented first. The high pressure core system is currently in use on several aircraft, including the Boeing 747–400. The combustor is annular in geometry and the calculation was performed for a single swirl–cup sector of 12–degree with periodic boundary conditions being imposed on the two side planes. A schematic of the combustor side view, the grid system (with $65 \times 25 \times 21$ grid points) and a comparison, between prediction and measurement, of the exit temperature profile are shown in Fig. 1, where T_4 and T_3 designate the overall averaged temperature in the exit and inlet of the combustor, respectively and $T_4(r)$ designates the circumferentially averaged temperatures at each local radial position. The profiles shown in Fig. 1 have been averaged along the circumferential direction. The measured data were obtained from four arrays of seven thermocouple measurements, between the top and the bottom liners, rotated around the entire combustor annulus at $1.5°$ interval. Good agreement has been obtained for this extremely complicated flow.

For a flow with this level of complexity, it is extremely demanding to make measurements within the flow domain. Some attempts have been made by Heitor and Whitelaw (1986) and Jones and Toral (1983) where assessments of experimental accuracies have also been reported. For example, the issues of undesirable influences from the catalysis effects and flow disturbance effects of the available probing techniques need

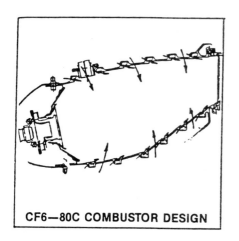

CF6—80C COMBUSTOR DESIGN

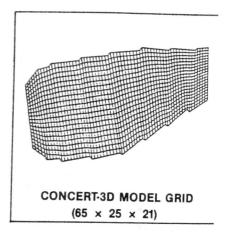

CONCERT-3D MODEL GRID
(65 × 25 × 21)

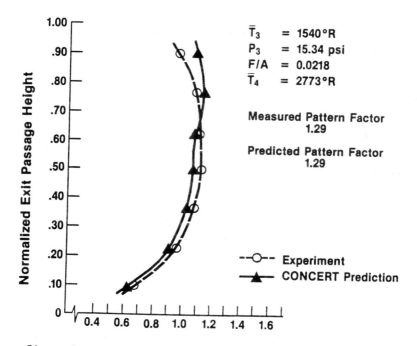

$\overline{T}_3$ = 1540°R
P_3 = 15.34 psi
F/A = 0.0218
$\overline{T}_4$ = 2773°R

Measured Pattern Factor
1.29

Predicted Pattern Factor
1.29

--○-- Experiment
▲ CONCERT Prediction

Circumferentially-averaged Normalized Temperature

$$\frac{\overline{T}_4(r) - \overline{T}_4}{\overline{T}_4 - \overline{T}_3} + 1$$

Figure 1. Exit temperature profile of CF6 combustor.

to be addressed. Hence, due care must be taken in the interpretation of probe measurements. With this background in mind, it is at least instructive and informative to use the computational tool to analyze the flow characteristics within the combustor. Figure 2 shows velocity and temperature fields in a selected side–view plane obtained from the numerical solution of a single cup sector of the CFM 56 engine combustor. The strong interactions of the jets and main flow and the resulting impact on the scalar transport process and temperature distribution can be clearly observed. Complicated flow pattern is always present in the combustor as a result of the need to enhance the mixing process. To illustrate the extremely complicated mixing process within typical combustors, Fig. 3 depicts some representative fluid particle trajectories of the mean velocity field. In Fig. 3(a), the trajectories are issued from the top and lower regions in the inlet plane. The pattern is complicated and is not easily revealed by experimental techniques in the presence of turbulent combustion and the liners. For example, it is evident that a large scale mixing process is created by the primary jets which can quickly exchange the fluid particles between the top and bottom wall regions. It is also evident that there is fluid exchange between adjacent sectors through the side boundary planes. Figure 3(b) shows the particle trajectories of the mean flow field issued from the centerline of the main inlet plane, where a recirculating zone located in the middle of the main dome can be clearly observed. Figure 3(c) shows the mean fluid particle trajectories issued from the primary and secondary dilution holes on the top wall. The differences between the incoming momentum of the two primary jets cause the depth of jet penetration to be considerably different. Collectively, Fig. 3, can help understand in a visually vivid manner, the evolution of the key flow processes, such as fuel–air mixing, jet–main flow interactions, and the jet signatures in the exit plane. It is this kind of detailed presentation of flow characteristics that a design engineer has been so far in dire need of, but usually unable to obtain. The advanced computational tool developed here has been making a highly useful contribution to design improvement. For instance, the information on the impact of the change of dilution hole pattern on the exit temperature profile has proven to be very helpful to improve combustor performance.

2 HIGH PRESSURE DISCHARGE LAMP

2.1 Background

The discharge lamp is an important light source for general illumination (Elenbass 1951, Waymouth 1971). It usually utilizes two electrodes, contained in a small quartz tube which is, in turn, embedded in a much larger glass jacket. The quartz tube is filled with mercury and often with metal halide additives. Within the tube, the electric discharge vaporizes these species to form an arc of high temperature, up to 7000 K, and high pressure, of several atmospheres. Figure 4 illustrates a modern design which depicts a curved arctube contained in an outer jacket filled with nitrogen. Due to its technical and commercial importance, and the intrinsic physical and chemical complexities, considerable research has been conducted into various areas relevant to the design of the

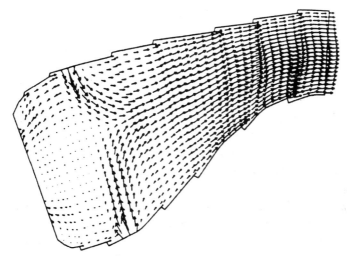

Velocity field in plane No. 9 from right side

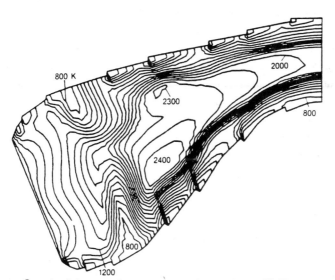

Constant temperature contours (interval = 100 K).

Figure 2. Representative velocity and temperature fields.

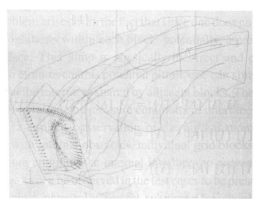

**(a) streamlines issued from top and bottom
of main inlet**

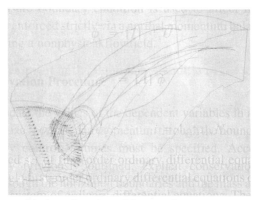

**(b) streamlines issued from center
of main inlet**

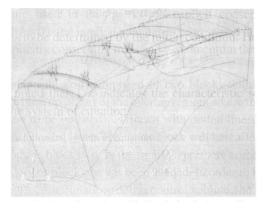

(c) streamlines from dilution holes in top wall

Figure 3. Fluid particle trajectories highlighting mixing pattern.

discharge lamp. Among many important issues, natural convection has received increasing emphasis since a critical requirement for designing a high quality discharge lamp is to distribute its wall temperature as evenly as possible in order to maximize metal halide vapor pressure, which is largely determined by the temperature of the coldest region of the arctube. The degree of uniformity of temperature distribution, both within the arctube and on its solid wall, can greatly affect the light quality as well as life expectancy of the lamp.

Kenty (1938) conducted the first experimental research into natural convection inside a discharge arctube. Zollweg (1978) and Lowke (1979) were the first to conduct the analytical and computational studies of natural convection in high pressure mercury arcs. In their works, the fluid flow and energy equations in an axisymmetric domain were solved. Dakin and Shyy (1989) developed a more comprehensive and detailed model to study the vertical metal halide discharge by including two–dimensional transport of the convection, the electric field, demixing of different species and radiation.

More recently, progress has been made in three–dimensional computation of the transport processes in the discharge lamp (Shyy and Dakin 1988, Chang *et al.* 1990, Shyy and Chang 1990a,b, Chang and Shyy 1991a,b) with a capability of handling the complex geometry of the lamp, including the curved surface and electrode insertion and the jacket. The model solves the combined momentum, mass continuity, energy and electric field equations, based on the fundamental conservation laws, and a simplified radiation heat transfer model (Lovin and Lubkowitz 1969). A finite volume algorithm based on general non–orthogonal curvilinear coordinates has been adopted in numerical computation. The transport processes within the arctube as well as between the arctube and the jacket walls have been carefully analyzed. Good agreement between experimental measurements and theoretical prediction has been obtained in terms of mounting angle, curvature effect and the wall temperature distribution. The results appear to be the first successful theoretical modeling of the very complicated three–dimensional transport phenomena in a discharge lamp. Much new insight has been gained from these studies.

2.2 Case Studies

The geometrical definition of a particular modern discharge lamp, including both the arctube and jacket, is given in Fig. 4. In this design, the arctube is bowed to accommodate the upward biased temperature profiles caused by natural convection (Chang *et al.* 1990). Accordingly, the mounting orientation of a lamp has a definite impact on its performance and life expectancy (Shyy and Chang 1990a). The lamp shown in Fig. 4 is specially designed for horizontal mounting and is often used in turnpike, street or billboard lighting. The outer jacket approximates an ellipsoid, with a cylindrical neck region. Between the arctube and the jacket, there are wiring and supporting structures to position the arctube as well as to supply electric power. Since these auxiliary structures are of negligible physical dimensions in comparison to the arctube and outer jacket, their presence is not accounted for in the present model. A three–dimensional illustration of the grid system generated and utilized in our computations is shown in Fig. 5.

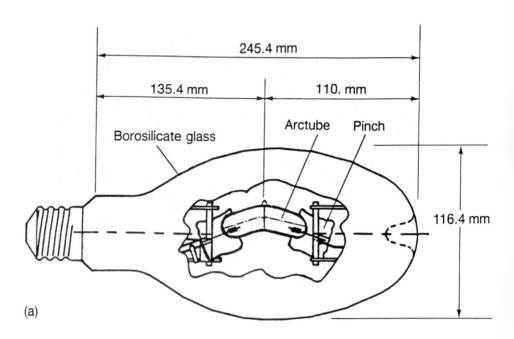

(a)

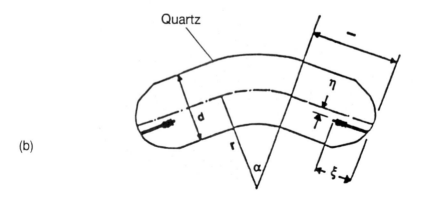

(b)

Figure 4. (a) Jacket geometry (b) Arctube geometry

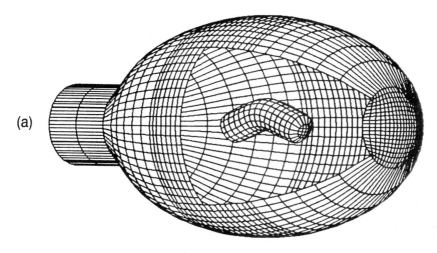

(a)

Mesh of jacket – 31 × 19 × 19

Mesh of arctube – 17 × 5 × 5

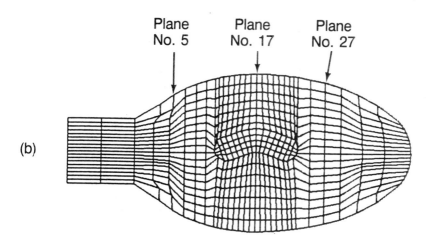

(b)

Figure 5. (a) Computer generated geometry and mesh (in 3–D perspective).
(b) Middle side–view plane (plane of symmetry).

Conduction, convection, radiation and electric discharge effects are included in regions where these effects are important. For the general fluid flow field, the Navier–Stokes equations are adopted.

2.2.1 Arctube

Direct photographs of operating arctube, as well as measurements of wall temperature have been made to facilitate assessment of the predictions of the present model (Shyy and Chang 1990a). The quartz wall temperature measurements were made with an Ircon–series–700 radiation pyrometer, an optical device which can quantitatively measure the radiance of a blackbody or other radiating object. Measurements were made for several mercury filled arctubes with vacuum outer jackets. Aquadag was applied on the locations where the temperature measurements were intended. Each aquadag spot was about 2 mm in size and was opaque. To take measurements, the optical head position was adjusted to sight on an aquadag spot at normal incidence. Then the radiant measurements obtained from the radiation pyrometer were converted to temperature.

Based on the computed flowfield and temperature contours, it is estimated that the Grashof number is of the order of 10^5, and hence the flow is in the laminar regime but contains a substantial convective effect. Figure 6(a) indicates the locations where the calculated and measured data of the wall temperature are compared. Figure 6(b) compares the calculated and measured top wall temperatures while Fig. 6(c) compares the calculated and measured bottom wall temperatures. The unsymmetric profiles of the measurement indicate the degree of experimental inaccuracy.

The calculated top wall temperature profile agrees well with the measured data. It is observed that the curving of the arctube causes the profile to show a double peak. For the bottom wall temperature, on the other hand, the curving causes the profile to show a single peak at the center. These characteristics result from the interaction of the upward buoyancy effects of the temperature contours and the bending of the arctube shape. It is noted that in the arctube, the highest temperature appears in regions near the two electrode tips. Between the electrodes, there is a current strip along the axial direction where the temperature is higher than in the other regions. The curvature of the strip is a function of the balance of the buoyancy effect, which tends to push the higher temperature gas upward, and the electric field effect, which tends to minimize the length of the strip to maintain the continuity of the electric current. It is clear that, in order to yield the uniform wall temperature distribution, the optimal design of an arctube with a given power input should accommodate the wall curvature to the curvature of the high temperature strip between the electrodes. The curvature of the high temperature strip results from the balance of the competing effects of the buoyancy and the electric field. With this physical picture in mind, the results presented in Fig. 6 suggest that in view of the double–peak temperature profile, the curvature along the top wall of the present arctube may be too high.

The calculated bottom wall temperature agrees well with the measurement in the center region but there is quite a discrepancy outside this region. The discrepancy is possibly caused by the following factors:

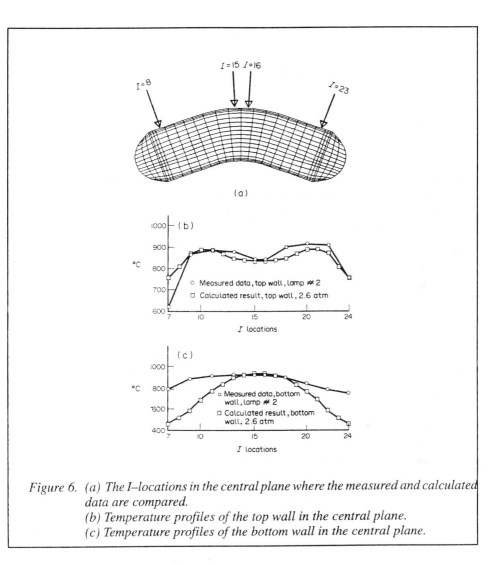

Figure 6. (a) *The I–locations in the central plane where the measured and calculated data are compared.*
(b) *Temperature profiles of the top wall in the central plane.*
(c) *Temperature profiles of the bottom wall in the central plane.*

(1)   The tungsten electrodes are much hotter than the inner quartz wall; therefore, conduction from the electrodes will heat up the wall regions. However, this is not taken into account.

(2)   The surfaces of the tungsten electrodes are much hotter than the inner quartz wall; therefore, the radiant energy exchanges between these two surfaces would increase the quartz inner wall temperature, especially the regions around the electrodes. This is rather difficult to account for in the context of the simplified treatment of radiation heat transfer adopted here.

(3)   The steady–state temperature distributions show that the bottom wall would absorb more radiant energy from the hot arc than the upper wall. But the

approximate treatment of radiant energy absorbed by the wall makes no distinction of positions around the circumferential boundary.

2.2.2 *Jacket*

In a different case, Figure 7 compares the predicted and measured top and bottom outer jacket wall temperatures where the predictions were based on three operating pressures inside the jacket, namely, 500, 600 and 760 Torr. It is clear that as the pressure increases, the strength of natural convection increases. With increasing pressure, convection causes the gas temperature to be more uniformly distributed inside the jacket as shown by Figs. 8(a) and (b). Consequently, in the region directly above the arctube, the wall temperature decreases as the pressure increases, while in the region close to the end zone, the wall temperature increases with increasing pressure. For the bottom wall temperature, on the other hand, increasing jacket pressure causes it to increase monotonically, but only to a modest extent. Figure 8(b) also shows that in the cross–view planes, the temperature distribution is symmetric at 500 and 600 Torr, but becomes asymmetric at 760 Torr.

Figures 9 and 10 depict selected plots of the velocity field on several side and cross view planes to illustrate more detailed information regarding the convection field induced by the temperature nonuniformities. A pair of contra–rotating convection cells above the arctube can be clearly identified in the middle side–view plane. Figure 9 shows that in the side–view plane of symmetry and above the middle of the arctube, high temperature combined with upward curving of the tube wall creates a thermal plume type of flow which moves to the top surface of the jacket, then turns to opposite directions to form two large scale recirculating eddies. The downward entrainment in regions close to the two ends of the arctube induced by the aforementioned contra–rotating cells are large enough to have the fluid penetrate into the domain around and beneath the arctube, where multiple recirculating eddies are also present.

Figure 10 shows a series of velocity vector plots in cross–view planes, from plane No. 5 to plane No. 27. It is noted that plane No. 17 is the one cutting through the cross–view plane of symmetry of the arctube. The flow structures are obviously three–dimensional and complicated. The top wall temperature of the arctube, which is around 800°C, is consistently much higher than the top wall temperature of the jacket, which is around 200°C. Furthermore, according to Figs. 4 and 5, their difference is higher than those along the top wall of the arctube. However, as far as convection is concerned, the flow development in the side–view planes seems to dominate that in the cross–view planes. For example, despite the fact that all the gas particles above the arctube are heated from below, they do not always move upward. Figure 10 shows that the gas particles above the arctube, but close to the end regions of the arctube, as in cross–view planes No. 9, 11 and 25, actually move downward. On the other hand, the gas particles above the middle of the arctube, as in cross–view planes No. 15, 17 and 19, move upward. This phenomenon is apparently caused by the eddy structure developed in the side–view plane already illustrated in Fig. 9. Due to the bending curvature of the arctube, the gap between the top

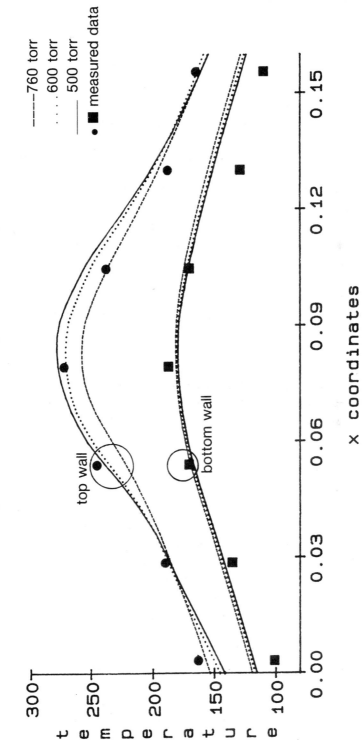

Figure 7. Comparison of predicted and measured outer jacket wall temperatures at different operating pressures.

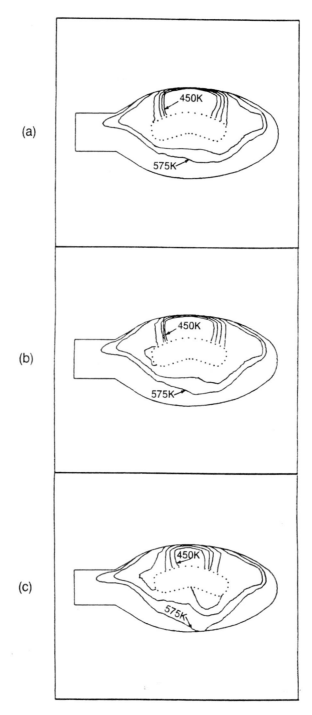

Figure 8. (a)Temperature contours in middle side–view plane corresponding to three operating pressures inside the jacket: (i) 500 Torr (ii) 600 Torr (iii) 760 Torr   (contour interval: 25 K).

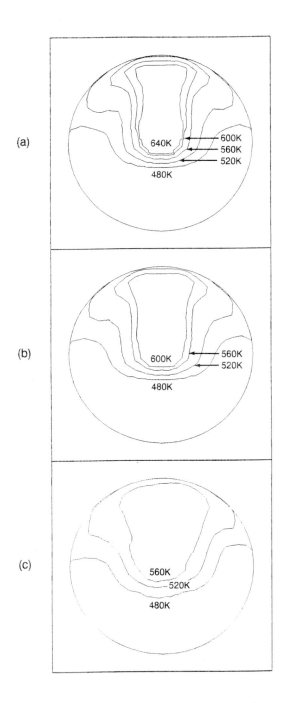

Figure 8. (b) Temperature contours in middle cross–view plane corresponding to three operating pressures inside the jacket: (i) 500 Torr (ii) 600 Torr (iii) 760 Torr (contour interval: 25 K).

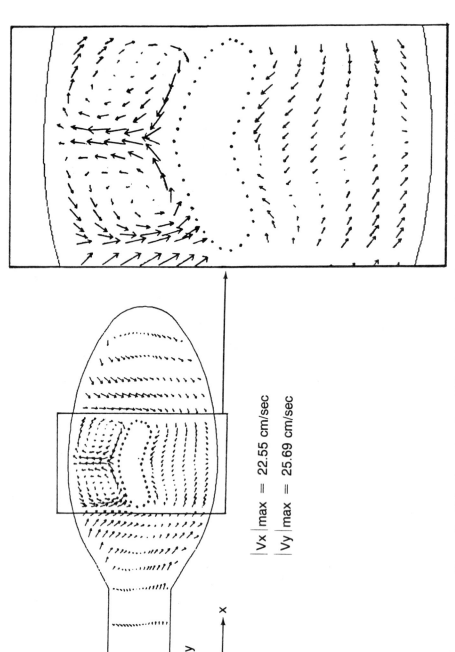

$|V_x|\ max\ =\ 22.55\ cm/sec$

$|V_y|\ max\ =\ 25.69\ cm/sec$

Figure 9.   Velocity vectors in middle side–view plane for operating pressure of 600 Torr.

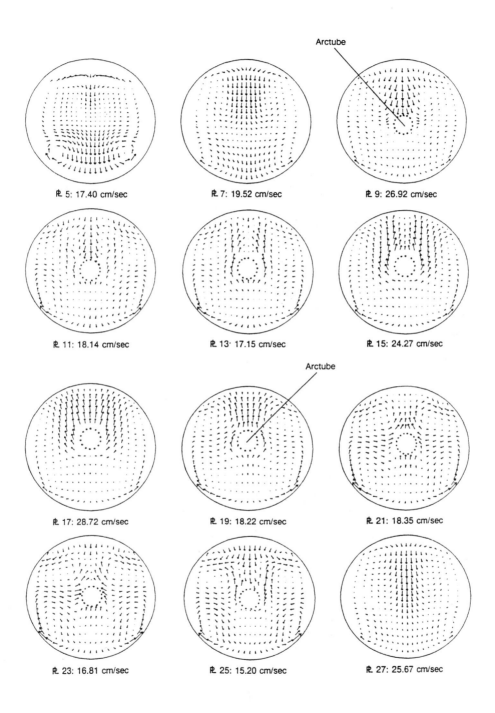

*Figure 10. Velocity vectors and maximum vertical component values on cross view
planes from plane no. 5 to 27 (see Fig. 5(b)) for operating pressure of
600 Torr.*

walls of the arctube and the jacket is minimum at the middle and increases towards the end regions. Consequently, the vertical temperature gradient is the highest in the middle region, and a thermal plume type of flow appears here. This structure dictates flow development in the other regions. In summary, the top wall temperature differences between the arctube and the jacket, combined with the variation of distance between them are responsible for the overall convection structure depicted in Figs. 9 and 10. The other notable feature observable in Fig. 10. is that in all cross–view planes, two eddies appear in the two lower corner regions.

3 TWO–PHASE THERMOCAPILLARY FLOW UNDER NORMAL AND MICROGRAVITY CONDITIONS

3.1 Background

Two–phase flow technology has the potential to significantly improve spacecraft heat acquisition, transport, and control. Using the latent heat of fluid, ammonia in particular, orders of magnitude more energy can be transferred than is possible using the sensible heat of single phase fluids. During the past several years, two–phase heat transport systems (in which surface tension forces established in a fine–pore capillary wick circulate the working fluid) have demonstrated performance potentials compatible with the thermal requirements on future advanced space–based programs such as the Earth Observatory System (EOS) Platforms and the Space Station. One concept, the capillary pumped loop (CPL) (Kroliczeck *et al.* 1984, Anderson and Beam 1992) has been developed to a high state of technology readiness and demonstrated via extensive ground testing (Ku *et al.* 1986a) and a limited flight test program (Ku *et al.* 1986b). This proof–of–concept breadboard development and advanced engineering model testing clearly demonstrated the potential of CPL as a reliable and versatile transport system capable of developing heat transport capabilities in the tens of kilowatts over long distances (Chalmers *et al.* 1988).

The "heart" of any capillary pumped loop system is the reservoir. The reservoir maintains the operating conditions within the CPL system during heat load variations by increasing or decreasing the system liquid mass at constant pressure and temperature conditions. The maintenance of operating conditions is accomplished by simultaneously having energy radiated from the reservoir's outer surface to a cold sink in conjunction with the variation of the heater voltage to maintain the reservoir's fluid temperature. The maintenance of system temperature and pressure is important to insure instrument operation within a predetermined temperature range.

It is well known that due to the different transport mechanisms prevailing under different conditions, substantial uncertainties exist when one tries to apply the information obtained from the normal gravity condition to the zero gravity condition. Since it is very difficult and expensive to conduct tests in actual operating conditions under microgravity, most of the design and experiments have been confined to earth–bound conditions, where the gravitational force exerts a much higher level of influence on performance through buoyancy induced transport. In order to help facilitate such an extrapolation of knowledge, computational modelling can play a very useful role. Shyy *et al.* (1991) and Shyy and Gingrich (1992) have made efforts to develop a computational capability to fulfill this need.

3.2 Case Studies

To give an overview of the work in progress, a reservoir flow and heat transfer simulation will be presented. The reservoir chosen is axisymmetric with distributed thermal boundary conditions, as schematically illustrated in Fig. 11. Along part of the cylinder walls, a porous wick made of aluminum or stainless steel screen is placed to create a temperature dependent surface tension. Under microgravity conditions, and in the absence of forced convection, the reservoir regulates the thermal field of the loop largely via thermocapillary convection and phase change. The present model has incorporated a treatment to allow a time dependent mass exchange between the reservoir and the outside, under the constraint of a constant thermodynamic pressure. With heat added to the reservoir at a constant rate, the ammonia, originally in the liquidus state, tends to remove the temperature non–uniformities through the transport processes of convection and conduction. If these processes are not effective enough, the temperature of ammonia will increase continuously and cross a critical value, causing a phase change from liquid to vapor.

Due to the phase change from liquid to vapor, the density of ammonia reduces by a factor of about 25, resulting in a substantial volumetric expansion. Since the reservoir has a constant volume and is at a constant thermodynamic pressure, the volume change of the fluid inside causes a mass exchange with the fluid outside the reservoir. The outlet of the reservoir is located at the center of one end, as shown in Fig. 11, with a diameter of 0.2 times the reservoir diameter. The whole mass exchange process is complicated in nature since it is influenced by the distribution of the moving phase boundaries and the thermal field within the reservoir, which affects the convection and conduction processes.

Shyy and Gingrich (1992) have utilized the time dependent Navier–Stokes equations, including mass continuity, momentum and energy conservation. To summarize the findings of the different cases, involving different gravity levels and power input levels, Figs. 12 and 13, compare the stream functions and the phase boundaries in the whole reservoir at t = 50 s at the various conditions. At 0–g and 120 W/m^2, the wick is partially covered by vapor; however, the evaporation rate is relatively modest, resulting in a weak convection field dictated by the thermocapillary effect alone. Also, under 0–g, but with 1200 W/m^2, the ammonia evaporation takes place at much higher rates, causing the wick to be entirely covered by vapor; in absence of buoyancy and thermocapillary effects, convection is of the rapid–expansion mode, which is quite strong compared to the thermocapillary mode. Under 1–g and 120 W/m^2, although evaporation takes place at an earlier time, once the temperature non–uniformity within the liquid is high, the buoyancy induced convection becomes correspondingly strong and is able to remove enough heat to

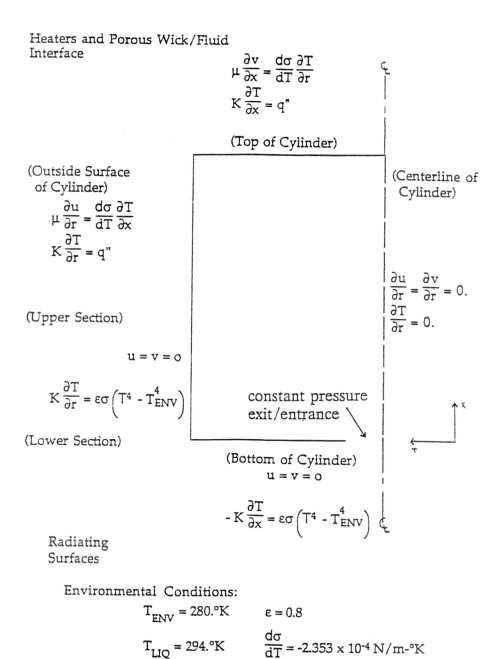

Heaters and Porous Wick/Fluid
Interface

$$\mu \frac{\partial v}{\partial x} = \frac{d\sigma}{dT} \frac{\partial T}{\partial r}$$

$$K \frac{\partial T}{\partial x} = q''$$

(Top of Cylinder)

(Outside Surface
of Cylinder)

$$\mu \frac{\partial u}{\partial r} = \frac{d\sigma}{dT} \frac{\partial T}{\partial x}$$

$$K \frac{\partial T}{\partial r} = q''$$

(Centerline of
Cylinder)

$$\frac{\partial u}{\partial r} = \frac{\partial v}{\partial r} = 0.$$

$$\frac{\partial T}{\partial r} = 0.$$

(Upper Section)

$$u = v = o$$

$$K \frac{\partial T}{\partial r} = \varepsilon\sigma \left(T^4 - T_{ENV}^4 \right)$$

constant pressure
exit/entrance

(Lower Section)

(Bottom of Cylinder)

$$u = v = o$$

$$- K \frac{\partial T}{\partial x} = \varepsilon\sigma \left(T^4 - T_{ENV}^4 \right)$$

Radiating
Surfaces

Environmental Conditions:

$$T_{ENV} = 280.°K \qquad \varepsilon = 0.8$$

$$T_{LIQ} = 294.°K \qquad \frac{d\sigma}{dT} = -2.353 \times 10^{-4} \, N/m\text{-}°K$$

$$T_{VAP} = 296.°K$$

*Figure 11.   Computational model and boundary conditions for a capillary pumped
loop reservoir.*

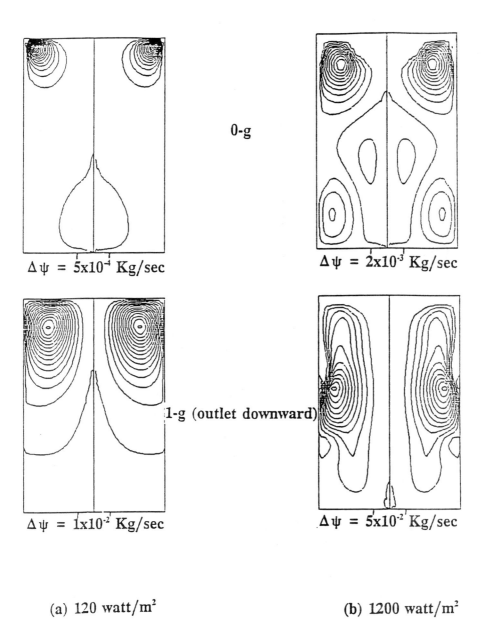

Figure 12. Comparison of stream function in the whole reservoir at t = 50 s under different conditions and heat input levels.

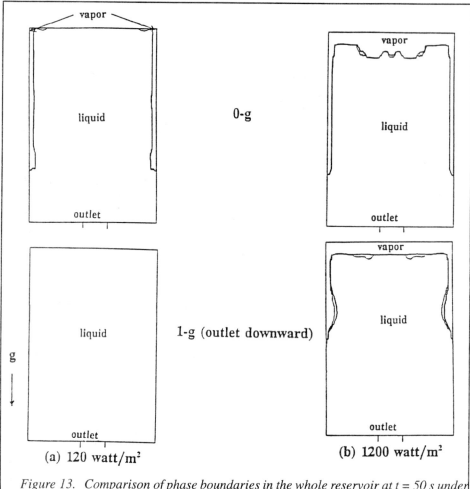

Figure 13. Comparison of phase boundaries in the whole reservoir at t = 50 s under different conditions and heat input levels.

convert all the ammonia vapor back to the liquidus state at t = 50 s. At 1–g and 1200 W/m², on the other hand, the rate of heat input into the reservoir is so high that the buoyancy effects can no longer suppress the evaporation process and consequently, the convection field results from the combined buoyancy and rapid–expansion (due to phase change) modes.

        The cases modelled here have exhibited substantial differences in the two–phase heat transfer behavior due to the three convection modes identified. The relative role each mode plays depends on the system configuration, operating conditions and gravity level. The degree of satisfaction a given design can achieve is strongly affected by these factors and needs to be considered carefully. A related example of fluid flow under reduced gravity can be found in Chapter IX, Section 8.4.

4 FLOW IN A 360° PASSAGE WITH MULTIPLE AIRFOILS

4.1 Background and Problem Formulation

In this section, we report an attempt to predict the complicated three–dimensional turbulent flows within a device containing large curvature and a number of solid elements. The flow device investigated here is a spiral casing of a hydraulic turbine power plant. As schematically illustrated in Fig. 14, both in three–dimensional perspective and in two–dimensional projections, the spiral casing is a passage of 360° turning that directs water, coming from the dam, to enter from one end, and to exit circumferentially along the radially inward direction. In the exit region, there are twenty–four pairs of airfoils, forming the distributor, which serve to control the mass–flux distribution and angle of the water. The water, after leaving the distributor, enters the runner to drive the shaft and produce usable power. Representative geometries as well as body–fitted grids of both the spiral casing and the distributor are shown in Fig. 15. The spiral casing considered here is a Piquet type distributor with 360° turning and decreasing cross–sectional areas. The inlet is of a circular shape, and the outlet occupies the whole inner circumference. The distributor, consisting of a cascade of stay vanes and wicket gates, is housed in the outlet region of casing.

The flow in a combined spiral casing and distributor is difficult to model and predict. Since the spiral casing and the distributor closely affect each other, both must be treated in a coupled manner. We have found that in the computational model, without taking an appropriate account of the presence of the distributor, the flow in a spiral casing can be persistently oscillatory due to the lack of dissipation of the flow kinetic energy in the exit region. However, from the viewpoint of aiding the design practice for such devices on a routine basis, it is not feasible to model the whole spiral casing and distributor combination simultaneously with good grid resolution. In view of these considerations, instead of solving the whole problem all at once, an alternative method is devised utilizing different grid resolutions to satisfy the different needs of physical modeling in casing and in distributor. To predict the overall characteristics of the flow in the spiral casing, whilst accounting for the geometry of the casing in detail, it is not necessary to consider each individual airfoil element of the distributor. On the other hand, from the viewpoint of the distributor flow analysis, only the inlet condition of the distributor is needed from the global spiral casing flow analysis. Hence, a two–level approach is proposed: 1) To predict the flow characteristics inside the spiral casing, the distributor region is treated as a porous medium with the details of individual airfoils being smeared out; 2) Within the distributor region, analysis is conducted for flow between each pair of wicket gates and stay vanes to determine the local distributor performance and to provide input to the porous medium treatment. The information is exchanged and coupled between the two levels. We believe that the present type of approach can be useful for aiding a designer to understand the performance as well as the flow behavior of a given casing/distributor configuration without having to resort to exceedingly large computing resources.

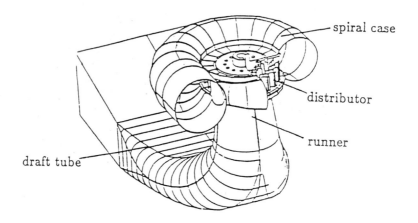

(a) Schematic representation of a hydraulic turbine

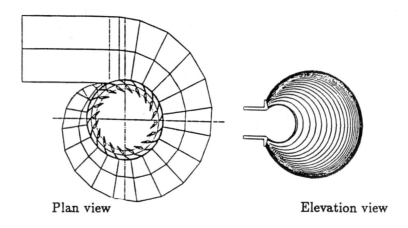

Plan view   **Elevation view**

(b) Plan view and elevation view of a spiral case.

Figure 14. Schematic of turbine and spiral casing.

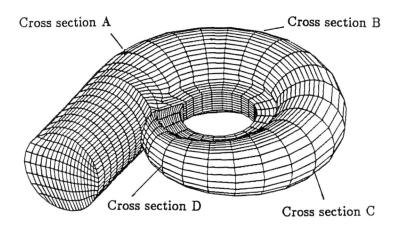

Cross section A

Cross section B

Cross section D

Cross section C

(a) Flow domain and body–fitted grid system of a spiral case.

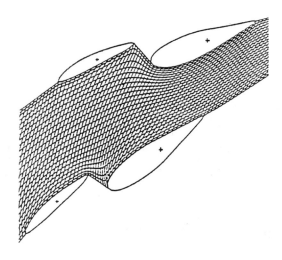

(b) Flow domain and body–fitted grid system of a distributor.

Figure 15. Representative grids of spiral casing and distributor.

The equations adopted are the Reynolds–averaged three–dimensional Navier–Stokes equations with an extra inclusion of the porous medium treatment based on Darcy's law. The equations for mass continuity and momentum conservation of the incompressible fluid are:

$$\nabla \cdot \vec{q} = 0 \tag{4.1}$$

$$\vec{q} \cdot \nabla \vec{q} = -\frac{1}{\varrho}\nabla p + v_{eff}\nabla^2\vec{q} - K\vec{q} \tag{4.2}$$

where $\vec{q}$ is the velocity vector with three velocity components, v_{eff} is the effective viscosity, and K is the Darcy's coefficient whose magnitude regulates the resistance exerted by the presence of porous medium in the distributor region. With regard to the turbulence closure, the k–ε two–equation model with the wall function treatment proposed by Launder and Spalding (1974) is adopted to supply the information of eddy viscosity. By the nature of the Reynolds–averaging procedure, the mean flow quantities are solved.

As K becomes large, both convection and viscous terms in the momentum equations become negligible due to the small values of velocity, and the pressure term balances the Darcy's term. As K approaches zero, the original Navier–Stokes equations are recovered. In the present case, K is zero everywhere except in the distributor region. Within the distributor region, K varies according to the nature (e.g., number, shape, and thickness) of the airfoils, and the local angles of attack of the incoming flow. In order to determine K, the distributor flow analysis (Vu and Shyy 1988) is conducted for flow in a passage bounded by a cascade of airfoils, as shown in Fig. 15. Since the inlet flow angles of the distributor vary along the circumferential direction, individual analysis for the flow between each pair of wicket gates and stay vanes is made with the inlet flow condition supplied by the spiral casing flow analysis. The total pressure loss profile of the distributor flow along the circumference of the spiral casing is then used to correlate the Darcy's coefficient, K. In the present work, the value of K is taken to be proportional to the total pressure loss to the second power. This choice is made according to the relative agreement observed between the computation and the measurement for a base geometry and flow condition. The proportional constant is fixed by trying to match the prediction with the experimental measurement at one operating point. Once the correlation formula is determined, calculation is made to model other operating conditions with no change of any constants. In summary, the global casing model supplies the inlet condition to the local distributor flow analysis; the total pressure loss predicted by the distributor model, in turn yields guidance to determine the distribution of K in the porous medium treatment. Such a global–local iterative procedure continues till convergence is reached simultaneously at both levels.

Figure 16 depicts the top–view of the geometry and boundary conditions in both the physical and transformed computational domains. The velocity at the inlet, AB, is of a uniform profile, with a Reynolds number, based on the inlet diameter, of 10^6. The dotted region is where the distributors are located and the only zone that the Darcy's coefficient, K, is nonzero. Other boundaries, including BC, CD, and AE, are all solid wall. The results

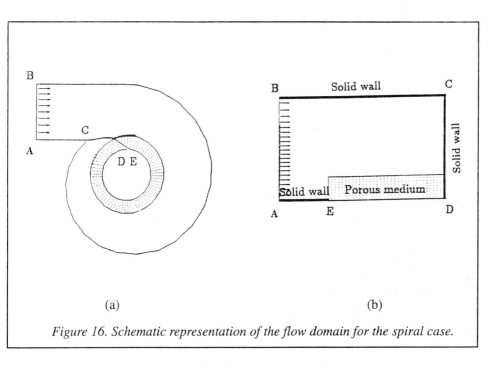

Figure 16. Schematic representation of the flow domain for the spiral case.

reported in the following are based on a 3–D grid of $99 \times 21 \times 15$ nodes for the spiral casing, and a 2–D grid of 25×85 nodes for the distributor. Typical CPU time of a complete calculation required for such a grid system is about 20 hours on a Silicon Graphics 4D/120. However, in general, much less time is needed to assess a design change since the calculation can be initiated from a solution obtained on an existing geometry.

4.2 Results and Discussion

Figure 17 shows three calculated velocity fields based on inlet flow angles (between the inlet velocity vector and the tangent of the cascade), of 10–degree, 15–degree, and 25–degree. The flow characteristics are qualitatively different among the three cases. Under the favorable condition (25–degree), the flow is attached, resulting in a small amount of total pressure loss. As the inlet flow angle is reduced from 25–degree to 15–degree, flow is no longer attached and separation appears in the region above the stay vane. Hence the flow under such a condition produces a larger total pressure loss. With the inlet flow angle further reduced to 10–degree, massive separation results from the larger adverse pressure gradient, which also produces noticeable streamline curvatures downstream of the separation zone. The loss is obviously the highest for the lowest flow angle. The total pressure loss of the distributor flow, normalized by the fluid kinetic energy at the inlet is presented in Fig. 18. At very low angle of attack, where the pressure distribution is unfavorable, large total pressure loss, up to about 45, is observed. As the flow angle increases, the loss quickly decreases and the flow stabilizes as the inlet flow angle exceeds 25–degree.

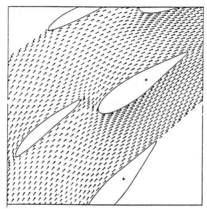

25° Flow Attack Angle

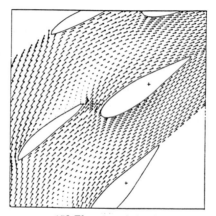

15° Flow Attack Angle

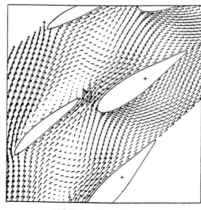

10° Flow Attack Angle

Figure 17. Three representative solutions of distributor flow with different angles of attack.

One of the key features in the present work is the supply of the inlet flow profiles of the distributor computation by the spiral casing analysis. The distributor flow analysis, in turn, yields information of the total losses within the distributor region. Based on this information, Darcy's resistance term is devised to represent the presence of the distributor. It is noted that, throughout the whole casing, the inlet angle of the distributor flow varies substantially along the circumferential direction. Figure 19 shows two profiles of inlet angle of the distributor flow, one at the mid–height location of the casing and the other averaged over the whole height of each cross–section of the casing. The variation of the angle of attack of the distributor flow results from the change of flow rate along the circumferential direction of the casing, and the nonuniform total pressure loss within the distributor region. Information related to experimental measurement can also be found in Vu and Shyy (1988, 1990).

As already stated, in the present model, Darcy's coefficient is determined from the local total pressure loss in the distributor region. Figure 20 exhibits three K profiles at different heights, from bottom to mid–height of the casing, along the circumferential direction. In the upstream portion of the casing, the inlet flow angles are low, resulting in high values of K for the distributor region. In the middle circumferential portion of the casing, as indicated by Fig. 19, the inlet angles of the distributor flow produce favorable conditions, resulting in lower total pressure losses and, accordingly, lower values of K. Toward the end circumferential portion of the casing, the inlet angles of the distributor flow vary substantially from bottom (high angle) to mid–height (low angle), and hence causes large differences in K to appear there.

Based on the combined casing/distributor flow analysis, a complete solution can be obtained for the whole passage. The numerical solutions have been compared to the in–house experimental measurements made with pitot tubes for various aspects of flow characteristics. We first compare the numerical prediction with the experimental measurement for the distributions of the static pressure, radial– and tangential–velocity components on two cross–sectional planes. The data shown in Fig. 21 are for physical quantities predicted and measured on the cross–sectional planes A and B of Fig. 15; the plots are shown along the radial direction, from the outside wall (left end of figure) to the outlet (right end of figure) of the casing along the mid–height line. Figure 21 illustrates the ranges of agreements between the numerical prediction and the experimental measurement for these variables. Qualitatively, the static pressure generally decreases from the outside wall (upstream) toward the casing/distributor outlet (downstream). With regard to the velocity components, they also increase in magnitude along the same direction until reaching the distributor region. Within the distributor region, the radial velocity component increases in magnitude while the tangential velocity component becomes smaller, indicating that the flow is evolving toward a two–dimensional behavior due to the flow acceleration caused by the distributors.

Figure 22 compares the predicted and measured circumferential distribution of the static pressure at three locations, all on the mid–height plane. These three locations are,

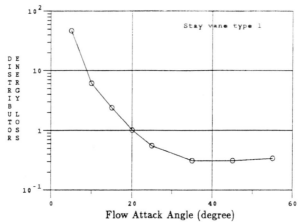

Figure 18. Total pressure loss in the distributor as a function of angle of attack.

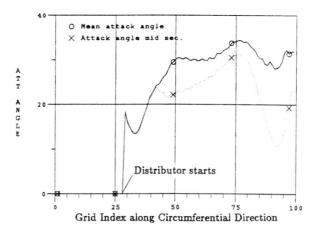

Figure 19. Variation of flow angle of attack in upstream of distributor.

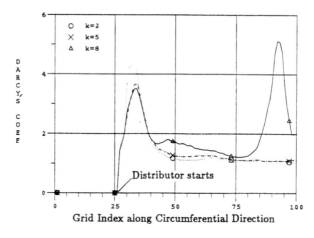

Figure 20. Distribution of Darcy's coefficient along circumferential direction.

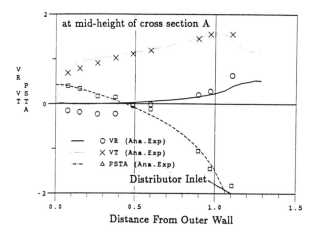

(a) At mid–height of cross section A.

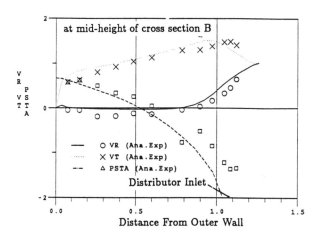

(b) At mid–height of cross section B.

Figure 21. Theory / Data comparison of static pressure, radial and tangential velocity components along radial direction. (symbols: measurements; lines: prediction)

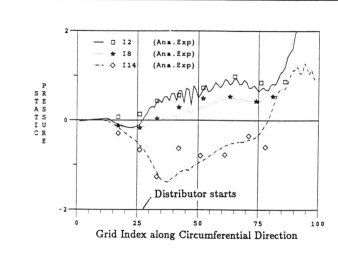

Figure 22.   Theory / Data comparison of static pressure along circumferential direction. (symbols: measurements; lines: prediction)

respectively, close to the outside wall, at the middle of the cross–section, and slight
upstream of the distributor.   While there are quantitative differences between t
measurements and the predictions, the overall characteristics are in agreement.   The fi
portion of the spiral casing is just a circular duct, and hence the pressure there decreas
along the streamwise direction.   As the spiral casing starts to turn, the wall curvature caus
a pressure gradient within each cross–section.   Furthermore, the presence of the distribut
produces more resistance to the flow field, resulting in an increasing pressure along t
circumferential direction.   In the end–wall region, the numerical solution predicts a mo
rapid increase of the static pressure, indicating that less fluid is predicted to be left in t
end zone of the spiral casing.   One can also clearly notice the pressure oscillatio
exhibited by the numerical solution in region close to the outside wall.   As discussed I
Shyy and Vu (1991), this phenomenon is not a numerical phenomenon; it is caused by t
way that the casing is manufactured.   Due to its large size, the casing is made by joini
many pieces of straight segments together, resulting in discontinuous slopes of the wa
contours.   This geometric characteristic is responsible for the apparent pressu
oscillations observed along much of the outside wall.   Shyy and Vu (1991) demonstrate
that with a smooth wall contour, pressure oscillations no longer appear.

Figure 23 shows both predicted and measured velocity fields on the midd
top–view plane of the casing.   Very close resemblance can be observed between the tw
Since the Reynolds number is high, the pressure variation appears mainly along the radi
rather than the circumferential, direction in order to balance the curvature effe
Qualitatively, the overall convection field is of combined characteristics of a free vort
flow superimposed upon a sink flow.   Three–dimensional illustrations of the flu
trajectories from inlet toward outlet in the form of the "ribbon" plots are shown in Fig. 2

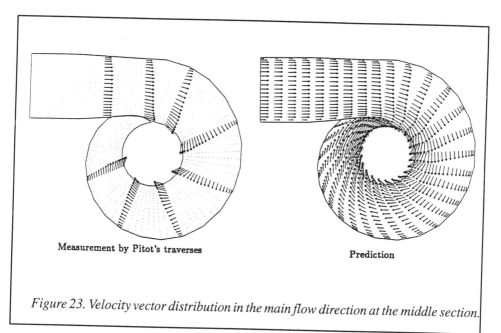

Measurement by Pitot's traverses   Prediction

Figure 23. Velocity vector distribution in the main flow direction at the middle section.

which depicts the ribbons initiated along the middle line of the inlet section horizontally and vertically. As evidenced by the complicated mixing pattern exhibited in Fig. 24, the flow is highly three–dimensional.

Figure 25 gives representative views of the secondary flow patterns on four cross–sectional planes from upstream to downstream of the spiral casing, as indicated in Fig. 15. Several factors influence the development of the flow behavior. Since the outlet is located along the inner circumference of the spiral casing, a strong tendency exists to push the fluid to move radially inward. Furthermore, along the circumferential direction, the continuous reduction of cross–sectional area competes with the continuous loss of mass through the outlet; the balance of these factors determines whether the flow accelerates or decelerates in the streamwise direction, and interacts with the curvature of the spiral casing to produce the velocity pattern observed in Fig. 25. The strength of the secondary flow varies from cross–section to cross–section. At the beginning of the spiral casing, the flow is accelerated evenly along the casing height. As flow moves along the circumferential direction, double vortices are generated by the interaction between the casing curvature and the presence of the wall. From cross–sectional plane A to B (see Fig. 15) the secondary flow strengthens. The radial velocity component is very strong near the top and bottom of the casing and much weaker near the center. After the 180–degree turning, the core of the double swirls becomes smaller, as illustrated on cross–section C. Near the end of the casing where the flow continues accelerating, the double vortices completely disappear on cross–section D. The qualitative development of the secondary flow shown here is the same as the experimental observation made by Kurokawa and Nagahara (1986).

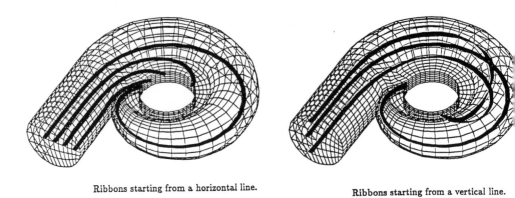

Ribbons starting from a horizontal line.   **Ribbons starting from a vertical line.**

Figure 24. Spiral case flow characteristics with ribbon representations.

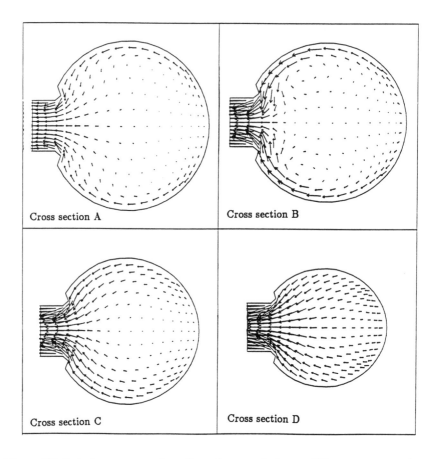

Figure 25. Spiral case secondary flow charcateristics at different cross sections.

5 CONCLUDING REMARKS

This chapter reviews the application of computational fluid dynamics, with the use of body–fitted coordinates, in the prediction of macroscopic transport phenomena associated with several distinctly different areas. Assessment has been made regarding the current capability of handling complex geometrical configurations and physical processes. Besides presenting quantitative comparisons between prediction and measurement, the various modeling issues that arise in the course of computation have also been briefly discussed. Many more examples involving fluid flows driven by and interacting with body force, surface tension, phase change and turbulence will be given in Chapter VIII (Section 4) and Chapter IX (Sections 5, 7 and 8). It is demonstrated here and the aforementioned sections that substantial insight can be gained with the aid of computational tools, thereby enabling a better understanding of the physical mechanisms responsible for the observed phenomena. Based on the improved knowledge, the predictive models can be refined, which can in turn produce new information to further advance the state–of–the–art. This kind of mutual guidance and enhancement is the most fruitful way to conduct research and to make practical impact on engineering practice.

PART III

INTERFACIAL TRANSPORT

This part discusses transport processes involving interfacial dynamics. Specifically, transport processes influenced by phase change, gravity, and capillarity are emphasized, and both the macroscopic and morphological (microscopic) scales are dealt with. Basic concepts of interface dynamics, capillarity, and phase change processes are first summarized, to help clarify the physical mechanisms involved. Since the existing books on theoretical and computational fluid dynamics usually do not emphasize interfacial transport, the presentation in this part starts by summarizing fundamental concepts and basic analyses, then addresses scaling procedures and phenomena of complex systems. Recent developments in computational modeling related to interfacial transport are also presented in some detail.

In many physical systems, the dimension of the interfaces belong in the range of a few mean free paths. Physically, there is a series of cascading processes which transfer the information from very small submicron level, all the way to the global system level. Rigorous modeling of interfacial behaviour, at the macroscopic level, needs to account for the presence of all these scales, from small to large, in a continuum framework. Much of the submicron range of interface behaviour is not well established to date. The author does not presume to undertake the task of presenting all the issues involved. However, due consideration is given to properly account for the conditions across the interface from a continuum point of view. The focus here is on addressing those aspects of interfacial phenomena that are relevant to computations of the macroscopic and morphological behavior of physical systems containing interfaces.

Much of the material in this part is drawn from the research efforts made recently. A variety of numerical solutions are also discussed to illustrate the salient features of the thermofluid transport and interfacial dynamics. Solutions obtained by using the schemes discussed in Part II and this part are presented to discuss the issues arising from both macroscopic and morphological scales. Since virtually the whole scope of Part III is within the incompressible flow regime, the pressure–based algorithms presented in Part II are particularly suited for solving these problems. Examples are mainly selected from ongoing research conducted by the author and his collaborators in general areas of buoyancy–induced convection and materials processing. However, the computational schemes and the scaling procedures developed in this part can be applied to other problems as well.

The order of presentation in this part is as follows. Due to the realization that some of the background material may not be familiar to many readers, basic thermodynamics concepts, particularly those concerning interfaces, are first summarized in Chapter VII. To aid appreciation of the physical mechanisms involved in the problems encountered in Part III, several examples are discussed in Chapter VIII; they illustrate both equilibrium and nonequilibrium aspects of gravity and capillarity effects, including basic analyses of the behavior of liquid menisci, jet breakup, and natural and Marangoni convection. Finally, in Chapter IX, a relatively detailed account is given, with materials processing as the primary physical motivation, to address the issues of scaling, phase change dynamics and thermofluid transport.

CHAPTER VII

BASIC CONCEPTS OF THERMODYNAMICS

In this chapter, we will focus on the thermodynamics of the interface separating two or more phases in a system. The importance of this topic is growing with the technological progress in many areas involving multiphase phenomena, such as the processing of modern materials and microgravity operations. Since some of the basic concepts may not be familiar to some readers, a brief summary of thermodynamics will be presented in this chapter.

1. BASIC CONCEPTS

A thermodynamic system is distinguished from the surroundings by a clearly defined boundary. The state of the thermodynamic system is then completely specified by thermodynamic variables. Any function determined completely by such parameters is a function of state. For example, consider a quantity Z, which may be a function of pressure P, volume V, and number of moles n_i of its constituent substances. Then the quantity Z will be a function of state if the total derivative dZ is given by:

$$dZ = \left(\frac{\partial Z}{\partial P}\right)_{V,n_i} dP + \left(\frac{\partial Z}{\partial V}\right)_{P,n_i} dV + \sum_{i=1}^{k} \left(\frac{\partial Z}{\partial n_i}\right)_{P,V,n_{j,j\neq i}} dn_i \qquad (1.1)$$

This means that the change in Z accompanying the transition of the system from one state to another depends only on the initial and the final states, specified by the variables P, V, and n_i, and is independent of the path taken by the system to reach the final state. Thermodynamic variables may be divided into two classes:

(i)   Extensive parameters, such as volume and mass, which depend on the size of the system and on the amount of the various substances present in the system, and

(ii)   Intensive parameters, such as pressure, temperature and concentration, which do not depend on the size of the system.

It may be noted that any extensive parameter can be converted to an intensive parameter by normalizing with a property which characterizes the size of the system, such as mass or number of moles.

1.1 Thermodynamic Equilibrium and Basic Laws

Thermodynamic equilibrium is the state reached by the system when it is left undisturbed for a long time in the absence of external constraints. An important property of the system at equilibrium is that its internal parameters, such as temperature and pressure are fully determined by the external parameters. The external parameters are functions of the generalized coordinates such as the ones defining the geometric dimension, the volume of the system and the external pressure exerted by the surroundings on the walls of the system.

The first law is a statement of energy conservation. For any thermodynamic system,

$$dU = \delta Q - \delta W \tag{1.2}$$

where U is the internal energy and is a state variable. Q and W are the heat input and work output of the system respectively, and δ indicates an inexact differential since Q and W are path dependent and not functions of state.

For reversible processes, we have,

$$TdS = \delta Q \tag{1.3}$$

Combining the first and the second laws, we have, for reversible processes

$$dU = TdS - \delta W \tag{1.4}$$

1.2 The Gibbs Equation

The Gibbs equation generalizes the concept of work. In thermodynamics, the most commonly used expression for work is the mechanical work, i.e.,

$$\delta W = P dV \tag{1.5}$$

however, numerous other forms of work are possible, e.g., electrical work. For a system that loses an electric charge, de, under a potential difference of ψ volts, the work done is $- \psi de$. For multiphase systems the concept of chemical work assumes importance. Consider a system having n components. For $- dn_i$ moles of the i'th component transported from the system to the surroundings, the work done is $- \mu_i \, dn_i$ where the intensive quantity μ_i is the chemical potential of the i'th component. Hence, if a system can perform both mechanical and chemical work, Eq. (1.5) should be modified to read,

$$\delta W = PdV - \sum_{i=1}^{k} \mu_i dn_i \tag{1.6}$$

in which the summation is carried out over all chemical components that are transported between the system and its surroundings. Combining the above equation with the first and second laws, we obtain the Gibbs equation,

$$dU = TdS - PdV + \sum_{i=1}^{k} \mu_i dn_i \tag{1.7}$$

The Gibbs equation takes into account all possible changes in extensive properties (dS, dV, dn_i, and dU) and relates the total change in internal energy to the sum of the products of the intensive properties (T, P, and μ_i) and the change in the corresponding capacities which are extensive properties.

Utilizing the notion of the state function, we can write $U=U(S, V, n_i, ...)$. The total differential,

$$dU = \left.\frac{\partial U}{\partial S}\right|_{V,n_i} dS + \left.\frac{\partial U}{\partial V}\right|_{S,n_i} dV + \sum_{i=1}^{k} \left.\frac{\partial U}{\partial n_i}\right|_{S,V,n_j,j \neq i} dn_i + \dots \qquad (1.8)$$

and by comparing with the Gibbs equation, we obtain,

$$\left.\frac{\partial U}{\partial S}\right|_{V,n_i} = T$$

$$\left.\frac{\partial U}{\partial V}\right|_{S,n_i} = -P \qquad (1.9)$$

$$\left.\frac{\partial U}{\partial n_i}\right|_{S,V,n_j,j \neq i} = \mu_i$$

An integrated form of the Gibbs equation can also be obtained. However, the intensive properties are functions of all the independent parameters of the system, e.g., P is a function of V and also T, n_i, etc. Hence, PdV cannot be integrated in a straightforward manner. The same holds true of the other terms also. In order to facilitate such an integration, we introduce the concept of homogeneous functions originated by Euler.

Homogeneous Equations:

An equation $\phi(x, y, z) = 0$ is a homogeneous equation of degree n if

$$\phi(\lambda x, \lambda y, \lambda z) = \lambda^n \phi(x, y, z)$$

For example, $ax + by + cz = 0$ is a homogenous equation of degree 1

$ax^2 + bxy + czy = 0$ is a homogenous equation of degree 2

$ax^2 + bx + czy = 0$ is not a homogenous equation.

Now, if we differentiate $\phi(\lambda x, \lambda y, \lambda z)$ with respect to λ, keeping x, y and z unchanged, we get

$$\frac{\partial \phi(\lambda x, \lambda y, \lambda z)}{\partial \lambda} = x\frac{\partial \phi}{\partial(\lambda x)} + y\frac{\partial \phi}{\partial(\lambda y)} + z\frac{\partial \phi}{\partial(\lambda z)} \qquad (1.10)$$

However, by the definition of the homogeneous equation,

$$\frac{\partial \phi(\lambda x, \lambda y, \lambda z)}{\partial \lambda} = \frac{\partial(\lambda^n \phi)}{\partial \lambda} = n\lambda^{n-1}\phi(x, y, z) \qquad (1.11)$$

and for the special case of $\lambda = 1$, by combining Eq. (1.10) and Eq. (1.11) we obtain Euler's relation:

$$n\phi = x\left.\frac{\partial \phi}{\partial x}\right|_{y,z} + y\left.\frac{\partial \phi}{\partial y}\right|_{x,z} + z\left.\frac{\partial \phi}{\partial z}\right|_{x,y} \qquad (1.12)$$

It may be noted that all extensive properties of a thermodynamic system are homogeneous functions of first order, when λ is taken to be the mass of the system. For example, the internal energy of the system, $U = U(S, V, n_i)$ is doubled by doubling the mass of the system, i.e., $U(2S, 2V, 2n_i) = 2U(S, V, n_i)$. Thus, by Euler's relation we have,

$$U = S \left.\frac{\partial U}{\partial S}\right|_{V,n_i} + V \left.\frac{\partial U}{\partial V}\right|_{S,n_i} + \sum_{i=1}^{k} n_i \left.\frac{\partial U}{\partial n_i}\right|_{S,V,n_j,j \neq i} \tag{1.13}$$

where we restrict the independent variables to S, V and n_i. Utilizing the set of relations for S, V and n_i derived earlier in Eq. (1.9), we obtain the integrated form of the Gibbs equation,

$$U = TS - PV + \sum_{i=1}^{k} n_i \mu_i \tag{1.14}$$

1.3 The Gibbs – Duhem Equation

We can start from the integrated form of the Gibbs equation, Eq. (1.14), to obtain an expression for the total differential dU,

$$dU = TdS + SdT - PdV - VdP + \sum_{i=1}^{k} \mu_i \, dn_i + \sum_{i=1}^{k} n_i \, d\mu_i \tag{1.15}$$

Now we subtract the differential form of Gibbs equation, Eq. (1.7) from the above expression to give,

$$SdT - VdP + \sum_{i=1}^{k} n_i \, d\mu_i = 0 \tag{1.16}$$

which is the Gibbs–Duhem equation. The Gibbs–Duhem equation plays an important role in our understanding of phase–change processes because it conveniently connects changes in temperature and pressure with changes in chemical potential. Furthermore, it also tells us that for isothermal and isobaric systems, $\sum_{i=1}^{k} n_i \, d\mu_i = 0$.

1.4 Interpretation of Free Energy and Chemical Potential under Phase Equilibrium

The temperature of a system may be considered a measure of the tendency for energy to be transferred from the system to its surroundings. The heat flux is governed by Fourier's law, $\vec{q} = -k\nabla T$. We say that a system is in thermal equilibrium with its surroundings when both the system and the surroundings are at the same temperature. Thus thermal equilibrium exists in any combination of systems when all the systems involved have the same tendency to lose energy thereby yielding no net heat exchange between them.

Analogously, we can think of thermodynamic equilibrium of two or more phases of a given substance under given conditions. Consider steam and liquid water in contact with each other at a pressure of 1 atm and temperature of 100°C. Water molecules will tend to escape from the gas phase and enter the liquid phase. Simultaneously, molecules will leave the liquid phase and enter the gas phase. Unless these two tendencies establish a dynamic balance, one of the phases will gradually disappear over time in favor of the other. Thus, thermodynamic equilibrium can exist only if such a dynamic balance can be established, i.e., the tendency of the molecules to leave a given phase must be exactly balanced by the tendency to enter that phase.

It is now useful to introduce the concept of free energy which serves as an indicator of thermodynamic equilibrium. Thermodynamically, any isolated system has a tendency to pass from a state of higher free energy to a state of lower free energy. An appropriate indicator of the tendency of phase change is the chemical potential, μ, which is the intensive quantity associated with the Gibbs free energy.

2. THERMODYNAMIC POTENTIALS

2.1 Gibbs Free Energy

At equilibrium, the extensive parameters U, S, V, and n_i and linear combinations of these, are functions of state. Such combinations are often found more useful than the internal energy, U, for describing systems under some conditions. These combinations are called thermodynamic potentials. They are extensive parameters. For example, the enthalpy of a system can be defined as follows,

$$H = U + PV \qquad (2.1a)$$

Then
$$dH = dU + PdV + VdP \qquad (2.1b)$$

and at constant pressure (isobaric process), $dH = dU + P\,dV$. If other forms of work, such as electrical work, are absent, then, from the first law, $dH = \delta Q$ for an isobaric process. Thus enthalpy is a measure of heat exchanged by a closed system with the surroundings during an isobaric process.

Gibbs free energy is defined as

$$G = U - TS + PV = H - TS \qquad (2.2)$$

At constant temperature and pressure,

$$dG|_{T,P} = dU - TdS + PdV \qquad (2.3a)$$
$$= -\delta W + PdV \qquad (2.3b)$$
$$-dG|_{T,P} = \delta W - PdV \qquad (2.3c)$$

i.e., the decrease in G is due to the sum of all work contributions excluding the component of work due to mechanical compression. It is the part of work done due to chemical reactions and by transport of electricity. Thus, in systems where mechanical compression (PdV) does not contribute significantly to the total work W, the Gibbs free energy makes for a good choice as the characteristic potential.

The general expression for dG is given by:

$$dG = dU - TdS - SdT + PdV + VdP \qquad (2.4a)$$

and noting Eq. (1.7)

$$dU = TdS - PdV + \sum_{i=1}^{k} \mu_i \, dn_i$$

$$\therefore \quad dG = -SdT + VdP + \sum_{i=1}^{k} \mu_i \, dn_i \tag{2.4b}$$

Thus the chemical potential can be defined as:

$$\mu_i = \frac{\partial G}{\partial n_i}\bigg|_{T,P,n_j, j \neq i} \tag{2.5}$$

2.2 Condition for Equilibrium

From the definition of G and for constant T and P, we have Eq. (2.3a). Note that from the second law:

$$\delta Q - TdS \leq 0 \tag{2.6a}$$

$$\therefore \quad dG\big|_{T,P} \leq dU + PdV - \delta Q \tag{2.6b}$$

From the first law, if the only work performed is mechanical compression, i.e., $\delta W = PdV$, then

$$dG\big|_{T,P} \leq 0 \tag{2.7}$$

in which the equality applies to reversible processes and the inequality to irreversible processes.

We can specialize the above reasoning to a phase change process. Considering the system to be composed of k phases, we have

$$
\begin{aligned}
dG &= (dU) + PdV + VdP - TdS - SdT \\
&= (TdS - PdV + \sum_{i=1}^{k} \mu_i \, dn_i) + PdV - TdS + VdP - SdT \\
&= -SdT + VdP + \sum_{i=1}^{k} \mu_i \, dn_i
\end{aligned}
\tag{2.8}
$$

$$\therefore \quad dG\big|_{T,P} = \sum_{i=1}^{k} \mu_i \, dn_i \tag{2.9}$$

We can now generalize the above results to the case where there are contributions to the work term, apart from that due to mechanical compression (PdV). We utilize the Gibbs equation, Eq. (1.7),

$$dU = TdS - PdV + \sum_{i=1}^{k} \mu_i \, dn_i$$

and by the second law, $dS \geq 0$, and under equilibrium conditions, i.e., $dU = dV = 0$, we obtain,

$$\sum_{i=1}^{k} \mu_i \, dn_i \leq 0 \tag{2.10}$$

and again we have, $\qquad\qquad dG\big|_{T,P} \leq 0 \tag{2.7}$

2.3 Implications of Equilibrium Conditions during Phase Change

Consider a closed system of two phases, say 1 and 2, in which the chemical potentials of the i'th component are $\mu_i^{(1)}$ and $\mu_i^{(2)}$ respectively. If there is a change in the quantity (measured in number of moles) of the i'th component in phase 1, given by $dn_i^{(1)}$, and a corresponding change in phase 2 given by $dn_i^{(2)}$, then the total change in free energy of the system will be,

$$dG = \sum_{i=1}^{k} \left(\mu_i^{(1)} \, dn_i^{(1)} + \mu_i^{(2)} \, dn_i^{(2)} \right) \tag{2.11}$$

Furthermore, if there is no chemical reaction to change the i'th component, then the total number of moles of any given component is conserved, $dn_i^{(1)} + dn_i^{(2)} = 0$.

$$dG = \sum_{i=1}^{k} \left(\mu_i^{(1)} - \mu_i^{(2)} \right) \, dn_i^{(1)} \tag{2.12}$$

Since the dn_i's are arbitrary, the condition for equilibrium, i.e., when $dG = 0$, can be deduced as,

$$\mu_i^{(1)} = \mu_i^{(2)} \tag{2.13}$$

Now, consider a system with only one component, i.e., $k = 1$. If the condition for equilibrium is violated, i.e.,

$$\mu^{(1)} > \mu^{(2)} \tag{2.14}$$

and since $dG \leq 0$, we can conclude that $dn^{(1)} \leq 0$. Thus, in a nonequilibrium situation, the system tends to minimize its free energy through the mechanism of phase change.

3. THERMODYNAMICS OF INTERFACES

An important aspect of phase–change is the existence of a boundary between the phases. A macroscopic treatment of the boundary or interface between two phases necessarily assumes a sharp discontinuity in density and other material properties and may involve composition discontinuities across the interface. This is analogous to the modeling of shocks in inviscid gas dynamics and of flame fronts in combustion processes. In reality, the rapid change in material and other properties actually occurs over a finite thickness centered about the interface, and whose characteristic length is of the order of the mean free path of the molecules.

3.1 A Heuristic Account of Interface Formation

The interaction between two molecules may be repulsive or attractive and is usually characterized in terms of a potential function $\phi(r)$, which is defined as the energy input required to bring two molecules from an initially infinite separation to some finite spacing, r. This field can be attractive or repulsive. In general, the forces between two molecules are attractive at long range and repulsive at short range. Typically, potentials are modeled as inverse powers of the intermolecular spacing. A widely used model is the Lennard–Jones model,

$$\phi(r) = 4\varepsilon \left[\left(\frac{r_0}{r} \right)^{12} - \left(\frac{r_0}{r} \right)^{6} \right] \tag{3.1}$$

where the first term describes the repulsive potential and the second term the attractive part of the potential. The interaction force between the two molecules is then obtained from the gradient of the potential function.

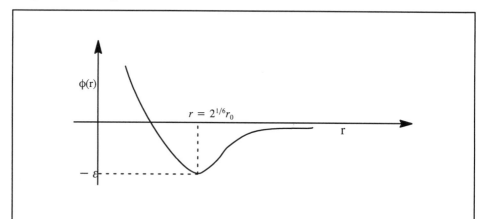

Figure 1   Lennard–Jones potential for interatomic separation. Note that work needs to be done to increase the interatomic spacing from the equilibrium value.

3.2 Origins of the Forces

Repulsive forces:

These usually result from the overlapping of electron orbits belonging to the two atoms causing mutual repulsion. According to the Pauli exclusion principle, no two electrons can occupy the same energy state. Thus overlapping electronic orbitals result in repulsive forces.

Attractive forces:

These may be of several types.

(i)   Ionic bonding: This is due to electrostatic attraction between oppositely charged ions, as in NaCl.

(ii)   Covalent bonding: This occurs when atoms share electron orbitals as in diatomic molecules like O_2, N_2, etc. This type of bonding is also called homopolar bonding.

(iii)   Metallic bonding: This occurs in metals where the free electrons can move about and hold the atoms together. Many atoms may thus be bonded together.

(iv)   Van der Waals force: This is a weak attraction between atoms or molecules without sharing or exchange of electrons. Thus there is no chemical bonding between the atoms or molecules. These forces are important in phase change phenomena. An example is the condensation of inert gases. The following table gives an idea of the typical bond energies involved.

Table 3.2.1 Typical bond energies

	Ionic	Covalent	Metallic	Van der Waals
ε/eV	3	3	2	10^{-2}

Note that $1\,eV = 1.6 \times 10^{-19}\,J$. Also, gravitational attraction, although definitely present, is negligibly small compared to the electrostatic forces. For example, for Ar, the gravitational potential is 10^{30} times smaller than the Van der Waals forces.

3.3 Latent Heat

Latent heat is the thermal energy absorbed/released at constant temperature when a material changes phase. When a substance changes state from, say liquid to vapor, energy input is required for two reasons,

(1)   The interatomic or intermolecular spacing increases substantially. Consequently, work has to be done to effect the large increase in potential energy. This energy has to be supplied.

(2)   As the substance evaporates, the system boundary moves and displaces the surrounding atmosphere with the vapor phase. Energy has to be supplied for this expansion.

Usually, the energy required for (2) is small compared to (1). As an example, the latent heat of vaporization of water at 100°C is $4.06 \times 10^4\,J/mol$. The molar volume of an ideal gas at S.T.P. is $2.24 \times 10^{-2}\,m^3$. The molar volume of water at 100°C and 1 atm can be estimated from the ideal gas law as $3.06 \times 10^{-2}\,m^3$. The mechanical (PdV) work done due to the volumetric change is $P(V_{vapor} - V_{liquid}) = 3.09 \times 10^3\,J/mol$. This is only 8% of the total latent heat of vaporization.

3.4 Latent Heat of Vaporization: ΔH_{l-g}

An approximate estimate of the latent heat may be obtained from a consideration of the potential energy of separation of molecules. For liquids, the intermolecular distance will be approximately r_o. Thus, from Eq. (3.1), the potential energy associated with each bond between neighboring molecules will be approximately ε (Figure 1). Now, further suppose that each molecule is associated with Z nearest neighbors, where Z is the coordination number. Then the total number of bonds in one mole of the substance will be $ZN/2$ where N is the Avogadro number and the factor $1/2$ is due to the fact that there is one bond for every two molecules. Next, we assume that the interatomic spacing for gases is very large so that the potential energy is nearly zero and neglect the mechanical work due to volumetric expansion against the surrounding atmosphere. Thus the molar latent heat of vaporization is,

$$\Delta H_{l-g} \cong \frac{1}{2} ZN\varepsilon \qquad (3.2)$$

where $\varepsilon = \phi(r_o)$

3.5 Solidification and Melting

As a background to the solidification and melting process, we take note of the following facts:

(1)   The volume change is usually very small, so the mechanical work performed against the surrounding atmosphere is negligible.

(2)   In the solid, the molecules are arranged regularly in fixed positions. In the liquid, the molecules are more mobile, indicating that each molecule has a lower number of nearest neighbors associated with it.

Therefore, we can conclude that latent heat consists of the energy needed to,

(a)   adjust the intermolecular spacing,

(b)   reduce the effective number of bonds.

Since the molecules in the liquid are still associated together, the latent heat of fusion (ΔH_{s-l}) is substantially less than the latent heat of vaporization, since the molecular interaction is nearly reduced to zero during phase change from liquid to vapor. Thus, for water, the latent heat of fusion is about 6×10^3 J/mol which is about 15% of the latent heat of vaporization of water.

3.6 Enthalpy as a Function of Temperature

Figure 2 shows the variation of enthalpy with temperature for a typical phase change process, which is also mathematically represented by Eq. (3.3). In Eq. (3.3), the subscripts s, l and g designate the solid, liquid and gas phases respectively. As already explained, $\Delta H_{s-l} < \Delta H_{l-g}$; furthermore, in general, the sensitivity of C_p with respect to temperature increases from solid to liquid to gas as shown by the different slopes of the enthalpy–temperature curve shown in Figure 2.

$$H^{(g)}(T) = \left\{ \begin{array}{l} H^{(s)}(T_{ref}) + \displaystyle\int_{T_{ref}}^{T_m} C_p^{(s)}\, dT \\[2ex] + \Delta H_{s-l} + \displaystyle\int_{T_m}^{T_v} C_p^{(l)}\, dT \\[2ex] + \Delta H_{l-g} + \displaystyle\int_{T_v}^{T} C_p^{(g)}\, dT \end{array} \right\} \tag{3.3}$$

4. PHASE EQUILIBRIA

As already derived, the general condition for mass transfer equilibrium is the basis for phase change equilibrium, i.e., $\mu_i^{(1)} = \mu_i^{(2)}$ for all species i, where the superscripts (1) and (2) indicate the different phases. Thus two phases can coexist stably only if the chemical potential of each of their chemical components is equal in all the phases present. In this condition, the Gibbs free energy of the system is a minimum.

Thus, to grow solid material from the melt, the chemical potential of the solid needs to be lower than that of other phases present. Consider a single component system. Utilizing the Gibbs–Duhem equation for this system, we have, for each phase,

$$[\; SdT - VdP + nd\mu = 0\;]^{(k)} \quad , \qquad k = l, s \qquad (4.1)$$

or for one mole of material,

$$[\; d\mu = -sdT + vdP\;]^{(k)} \quad , \qquad k = l, s \qquad (4.2)$$

where $s = S/n$ and $v = V/n$, where n is the number of moles in the system. Based on this equation, one can, in principle, find different ways to manipulate the system to obtain the desired material.

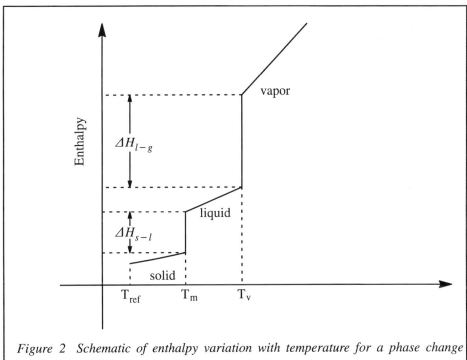

Figure 2 Schematic of enthalpy variation with temperature for a phase change process.

(i)   *Temperature Control:*

The most common method employed to lower the chemical potential of the incipient solid below that of the melt or the nutrient phase is to lower the temperature of the system. Note that

$$[\; d\mu\,|_P = -sdT\;]^{(k)} , \qquad k = l, s \qquad (4.3a)$$

$$\text{also knowing that} \quad s_l > s_s \qquad\qquad\qquad (4.3b)$$

Based on Eqs. (4.3a) and (4.3b) the following plot of chemical potential vs temperature can be constructed for both phases.

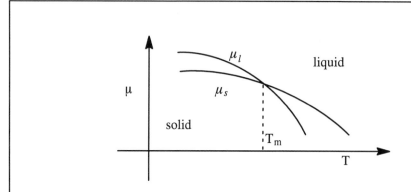

Figure 3 Plot of chemical potential vs temperature for both phases. T_m is the melting temperature where the solid and the liquid phases coexist in equilibrium. The system tends to minimize its free energy, G. Thus, if heat is extracted from the system at the melting temperature, the solid phase will grow at the expense of the liquid phase.

(ii)   Pressure Control:
The pressure can also be varied to enhance the solidification process. However, whether the pressure should be increased or decreased depends on the material properties. Noting that,

$$\left[\, d\mu|_T = v\,dP \, \right]^{(k)} \quad , \qquad k = l, s \tag{4.4}$$

If $v_l > v_s$, as is the case for most materials, then increasing the pressure favors the formation of the solid phase; if $v_l < v_s$, as in the case for water, Bi, Ge, Si, Hg, and III – V compounds, decreasing the pressure promotes solidification. Figure 4 shows the different possibilities.

Usually, temperature control is favored over pressure control in industrial applications. Typically the density difference between the condensed phases, solid and liquid, is quite small, so it may require a substantial pressure increase to create significant changes in the chemical potential to drive the phase change process at a desirable rate. This is generally inefficient.

4.1 The Clausius–Clapeyron Equation

When two phases are in equilibrium, then for each component, $d\mu^{(1)} = d\mu^{(2)}$. Keeping all other intensive parameters constant, the above equation can be expanded in a Taylor series for small variations of pressure and temperature as,

$$\left.\frac{\partial\mu}{\partial T}\right|_P^{(1)} dT + \left.\frac{\partial\mu}{\partial P}\right|_T^{(1)} dP = \left.\frac{\partial\mu}{\partial T}\right|_P^{(2)} dT + \left.\frac{\partial\mu}{\partial P}\right|_T^{(2)} dP \qquad (4.5)$$

According to the Gibbs–Duhem equation, Eq. (4.2),

$$(\ d\mu = - sdT + vdP \ )^{(k)} \quad , \qquad k = 1, 2$$

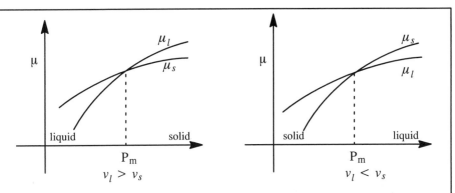

Figure 4 *Plot of chemical potential vs system pressure at a given temperature illustrating both the possibilities. If the liquid has a higher specific volume than the solid, then increasing the system pressure favors the formation of the solid. If the liquid has a lower specific volume than the solid, as is the case for water, Bi, Ge, Si, Hg, and III – V compounds, decreasing the pressure promotes solidification.*

Hence,

$$s^{(k)} = - \left.\frac{\partial\mu}{\partial T}\right|_P^{(k)} \qquad (4.6a)$$

$$v^{(k)} = + \left.\frac{\partial\mu}{\partial P}\right|_T^{(k)} \qquad (4.6b)$$

Substituting these equations into Eq. (4.5), we get,

$$- s^{(1)} dT + v^{(1)} dP = - s^{(2)} dT + v^{(2)} dP \qquad (4.7)$$

which can be rearranged as

$$\left(s^{(2)} - s^{(1)} \right) dT = \left(v^{(2)} - v^{(1)} \right) dP \qquad (4.8)$$

which results in the Clausius–Clapeyron equation

$$\frac{dP}{dT} = \frac{\Delta s}{\Delta v}$$

$$\text{where } \Delta s = s^{(2)} - s^{(1)} \qquad (4.9)$$

$$\Delta v = v^{(2)} - v^{(1)}$$

4.2 Interpretation of the Clausius–Clapeyron Equation

(i)   The Clausius–Clapeyron equation relates the *slope of the phase–equilibrium line at a given point of the line* to the entropy change and the volume change of the substance *across the line* at this point, i.e., the entropy change and the volume change in undergoing phase change at the given pressure and temperature.

(ii)   The left hand side is a function of intensive parameters while the right hand side is a function of the extensive parameters.

(iii)   Since there is an entropy change associated with the phase transformation, heat must also be absorbed or released. This fact provides a macroscopic viewpoint supplementing the microscopic viewpoint offered earlier for the latent heat.

(iv)   Since this process takes place at a constant transition temperature, T, the corresponding entropy change is $\Delta s = \dfrac{\Delta H_{1-2}}{T}$, where ΔH_{1-2} is the latent heat of phase change from phase 1 to phase 2.

(v)   Hence the Clausius–Clapeyron equation can also be written as,

$$\frac{dP}{dT} = \frac{\Delta s}{\Delta v} = \frac{\Delta H_{1-2}}{T \, \Delta v} \tag{4.10}$$

It may be noted that since the right hand side is based on extensive parameters, the choice of units will not affect the left hand side which is solely dependent on intensive parameters.

Shown below in Figure 5 is the P–T diagram for a typical liquid–vapor phase change process. In undergoing phase change from solid to liquid the entropy of the substance (degree of disorder) almost always increases (an exception being solid He^3). Thus the corresponding latent heat, ΔH, is positive and heat is absorbed in the phase transformation. For most materials, the volume change, $\Delta v > 0$ for a solid to liquid phase change and hence $\dfrac{\partial P}{\partial T} > 0$. However, an exception is water for which $\Delta v < 0$ and hence $\dfrac{\partial P}{\partial T} < 0$. Furthermore, $v^{(l)} - v^{(s)}$ is usually much smaller than $v^{(g)} - v^{(l)}$. Hence in a triple–phase diagram in P–T coordinates, the slopes of the different phase boundaries are quite different as shown in Figure 6.

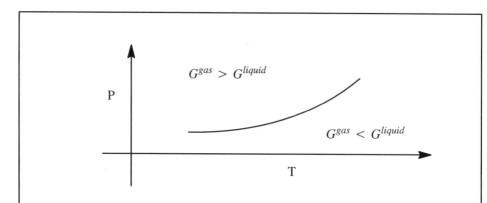

$G^{gas} > G^{liquid}$

P

$G^{gas} < G^{liquid}$

T

Figure 5 The curve shown is the equilibrium line between the two phases. The slope of the P–T curve is positive because both Δs and Δv are positive for phase change from liquid to gas.

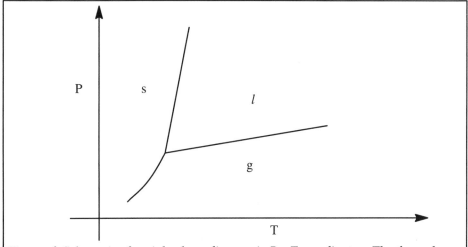

Figure 6 Schematic of a triple phase diagram in P – T coordinates. The three phases, solid, liquid and gas coexist in equilibrium at the triple point.

5. SURFACE TENSION AND THE GIBBS–THOMSON EFFECT

The terms surface tension and surface energy are used interchangeably for liquids that do not support residual stresses, e.g., Newtonian fluids. For solids, the surface tension or the stress needed to stretch a surface can result in residual stresses as well as in elastic and plastic deformations and in general can be different from the surface energy.

We need to extend the definition of work to take into account the energy required to form the interface between the phases. Gibbs (1931) discussed the thermodynamics of interfaces in terms of a conceptual *dividing surface* between the phases. The work term now becomes,

$$- dW = - PdV + \sum_{i=1}^{k} \mu_i dn_i + \sigma dA \tag{5.1}$$

where σ is the surface tension (or surface energy per unit area) and A is the surface area of the interface.

Thus,
$$dU = TdS - PdV + \sigma dA + \sum_{i=1}^{k} \mu_i dn_i \tag{5.2}$$

and
$$dG = - SdT + VdP + \sigma dA + \sum_{i=1}^{k} \mu_i dn_i \tag{5.3}$$

Therefore, the surface energy is the change in the Gibbs free energy per unit area, i.e.,

$$\sigma = \frac{\partial G}{\partial A}\bigg|_{T,P,n_i} \tag{5.4}$$

We can now state that for equilibrium under conditions of constant temperature, and pressure

$$dG\big|_{T,P} = \sigma dA + \sum_{i=1}^{k} \mu_i dn_i \leq 0 \tag{5.5}$$

and if chemical equilibrium is guaranteed, $\mu_i^{(1)} = \mu_i^{(2)}$.

$$\therefore \; \sigma dA \leq 0 \quad \Rightarrow \quad dA \leq 0 \tag{5.6}$$

Thus a system in equilibrium tends to minimize the interface area.

5.1 A Simple Kinetic Estimation

The existence of surface energy implies that it takes a finite amount of energy to move an atom or molecule from the interior of the phase to the phase boundary (interface). Consider a vaporization process where the liquid is separated from the vapor by an interface. In the interior of the liquid, a molecule is surrounded by m neighboring molecules on all sides, where usually $m \sim 10$. However, a molecule at the surface has neighbors towards the bulk of the liquid but none on the other. Consequently it experiences an asymmetric distribution of force. It also has about half the number of neighbors as a molecule in the interior.

As an example, consider a system having N molecules per mole and a volume V per mole. Then the number of molecules per unit volume is N/V. Hence, at the surface, the number of molecules per unit area will be roughly $(N/V)^{2/3}$. Also, the energy required to move one molecule from the interior to the surface will be approximately half the latent heat of vaporization per molecule since only half the bonds need to be broken, i.e., $\left(\Delta H_{l-g}/2\right)\big/N$. Therefore, the surface energy per unit area is the product of the number of molecules per unit area times the work required per molecule, i.e.,

$\sigma \sim \left(\dfrac{N}{V}\right)^{2/3} \left(\dfrac{\Delta H_{l-g}}{2N}\right)$ where as already derived, $\Delta H_{l-g} \approx ZN\varepsilon/2$. From this we can

obtain an estimate for the surface energy as, $\sigma \sim \left(\dfrac{N}{V}\right)^{2/3}\left(\dfrac{Z\varepsilon}{4}\right)$.

5.2 The Gibbs–Thomson Effect

In our previous discussion of phase change, we only accounted for the effect of the chemical potential, μ, and neglected the effect of the surface energy required to form the boundary between the phases. This deficiency will now be corrected.

Consider a crystal with a chemical potential μ^s solidified in the form of a sphere of radius r, out of a liquid phase with a chemical potential of μ^l. For simplicity we only consider a pure material so that both phases have exactly the same composition. Also consider the solid phase to be the bulk phase and the thermodynamically preferred phase, i.e., $\mu^s < \mu^l$. Then the surface energy required to form the phase boundary in such a system is $4\pi r^2\sigma$, where σ is the specific free energy of the interface.

Now, we transfer dN molecules from the the liquid phase to the solid phase. This gives rise to the following effects:

(i) The thermodynamic potential of the system is reduced by $(\mu_l - \mu_s) \, dN$.

(ii) The surface energy of the system will increase by σdA where $A = 4\pi R^2$. Thus the surface energy will increase by $8\pi\sigma R dR$. We can arrive at an estimate of dR as follows: Suppose Ω is the volume occupied by a molecule of the material in the liquid phase, then the total volume change due to transfer of dN molecules from the liquid phase to the solid phase will be $dV = \Omega \, dN$ where $V = 4/3\pi R^3$ i.e., $\Omega \, dN = 4\pi R^2 dR$. Hence, $dR = \Omega \, dN/4\pi R^2$. Thus the net increase in surface energy due to the formation of the interface is $2\sigma\Omega dN/R$.

We know that for any spontaneous process to occur, there must be an accompanying decrease in the Gibbs free energy. In the above case the Gibbs free energy is mainly affected by the change in the thermodynamic potential and the increase in surface energy. The solidification process can continue only if the decrease in the thermodynamic potential is greater in magnitude than the increase in the surface energy of the system. Otherwise the solidified sphere will melt back into the liquid phase. The new particles solidified do not have to be uniformly distributed over the surface of the sphere; however such a uniform increase of the radius maintaining the spherical shape of the crystal causes the minimum increase in surface area and hence the free energy.

From the information given above, we can now derive the Gibbs–Thomson equation. At equilibrium, the decrease of the thermodynamic potential must be exactly balanced by the increase in surface energy due to the increase in the interface area, i.e.,

$$\mu_l - \mu_s = \frac{2\sigma\Omega}{R} \tag{5.7}$$

This is the Gibbs–Thomson equation which defines the shift of phase equilibrium at the spherical interface due to the surface energy. The shift can be expressed in terms of the deviation in the temperature, pressure or concentration from their equilibrium values obtained by neglecting the contribution of surface energy. The temperature form is the most useful for solidification processes and will be derived in the following section.

We can generalize this concept to a non–spherical interface. The phase equilibrium at any point on the interface is governed by

$$\mu_l - \mu_s = \Omega\sigma \left(\frac{1}{R_1} + \frac{1}{R_2} \right) \tag{5.8}$$

where R_1 and R_2 are the principal radii of curvature of the interface at the given point. This formula is based on the assumption that the surface energy is isotropic. For anisotropic surface energy, this equation needs to be modified to account for the variation of surface energy with crystallographic orientation. It may also be noted that this derivation is based on the assumption of constant volume and temperature within the system during the phase change process. If these conditions are not satisfied, other components of work will have to be taken into account.

For a constant volume and isothermal process, it is convenient to use the Helmholtz free energy defined as $F = U - TS$. Thus $dF = dU - TdS - SdT$. Noting that from the second law, $dQ \leq TdS$ and from the first law, $dU = dQ - PdV \leq TdS - PdV$, and since $dT = dV = 0$, we have

$$dF|_{T,V} \leq 0 \tag{5.9}$$

which is a general requirement for equilibrium at constant temperature and volume.

5.3 The Young–Laplace Equation

Consider a closed system composed of two phases of a pure material separated by an interface A. The total volume of the system is V with the phases I and II occupying V_I and V_{II} respectively. The energy balance for each phase, I and II, as well as the interface, A, can be written as

$$dU_I = (TdS - PdV + \mu dN)_I \tag{5.10a}$$

$$dU_{II} = (TdS - PdV + \mu dN)_{II} \tag{5.10b}$$

$$dU_A = (TdS + \sigma dA + \mu dN)_A \tag{5.10c}$$

subject to the following constraints

(i) $V = V_I + V_{II} = \text{const} \Rightarrow dV_I = - dV_{II}$

(ii) $U = U_I + U_{II} + U_A = \text{const} \Rightarrow dU = 0$

(iii) $N = N_I + N_{II} + N_A = \text{const} \Rightarrow dN = 0$

and at equilibrium, we also have,

(iv) $S = S_I + S_{II} + S_A$ reaches a maximum implying that $dS = 0$.

From these we obtain,

$$
\begin{aligned}
dU = 0 &= d(U_I + U_{II} + U_{III}) \\
&= (T_I - T_A)dS_I + (T_{II} - T_A)dS_{II} - (P_I - P_{II})dV_I + \sigma dA \\
&\quad + (\mu_I - \mu_A)dN_I + (\mu_{II} - \mu_A)dN_{II}
\end{aligned} \tag{5.11}
$$

We note that S_I, N_I, V_I are independent variables, but that dA is not independent of dV_I. Thus it is evident that $T_I = T_{II} = T_A$ and also that $\mu_I = \mu_{II} = \mu_A$. Then equating the remaining terms to zero gives,

$$P_I - P_{II} - \sigma \frac{dA}{dV_I} = 0 \tag{5.12a}$$

or

$$P_I - P_{II} = \sigma \left(\frac{1}{R_1} + \frac{1}{R_2} \right) \tag{5.12b}$$

where R_1 and R_2 are the principal radii of curvature of the interface; they are positive if measured from the side of phase I. This is the Young–Laplace equation. The Young–Laplace equation can also be derived from considerations of a force balance at the interface (Gibbs 1931).

5.4 The Temperature Form of the Gibbs–Thomson Equation

Recall that Eq. (5.7) states that $\mu_l - \mu_s = \frac{2\sigma\Omega}{R}$, implying that when the interface has non–zero curvature, the chemical potential $\mu_l \neq \mu_s$. Whereas it was also shown earlier, in Eq. (5.11), that under equilibrium conditions $\mu_l - \mu_s = 0$. It is appropriate to remark on the cause of this inconsistency. An implicit assumption made in the derivation of Eq. (5.7) is that there is no pressure difference across the phase boundary implying that the phase change is isobaric. Figure 7 shows the shift in the equilibrium point as the chemical potential curves shift to accommodate the change in pressure across the curved interface. Thus, the above mentioned inconsistency is caused by artificially constraining the pressure to be continuous across the curved interface and defining the chemical potential at that pressure.

We can also develop the Gibbs–Thomson relation in a more general manner. We start with the Gibbs–Duhem relation inclusive of surface tension effects and written on a unit mole basis as,

$$d\mu = -sdT + vdP + \sigma dA \tag{5.13}$$

and the condition for equilibrium as

$$\mu_l(P_l, T_m) = \mu_s(P_s, T_m) \tag{5.14}$$

We now expand the terms in Eq. (5.14) in a Taylor series about P_s and T_m^* which are the equilibrium values in the absence of interface curvature. Under equilibrium conditions $d\mu_l = d\mu_s$.
Thus

$$\mu_l(P_l, T_m) - \mu_l(P_s, T_m^*) = \mu_s(P_s, T_m) - \mu_s(P_s, T_m^*) \tag{5.15}$$

and substituting Eq. (5.15) into Eq. (5.13) and neglecting the variation of surface tension, we get,

$$v_l(P_l - P_s) - s_l(T_m - T_m^*) = -s_s(T_m - T_m^*) \tag{5.16}$$

$$\Rightarrow (s_l - s_s)(\Delta T_m) = v_l P_l - v_l P_s \tag{5.17}$$

$$\Rightarrow \frac{(s_l - s_s)}{v_l}(\Delta T_m) = P_l - P_s \tag{5.18}$$

where $\Delta T_m = T_m^* - T_m$ and v is the specific volume. From the Young–Laplace equation, Eq. (5.12) we get,

$$\frac{(s_l - s_s)}{v_l}(\Delta T_m) = \sigma\left(\frac{1}{R_1} + \frac{1}{R_2}\right) \tag{5.19}$$

At equilibrium conditions, the heat transfer during phase change is the latent heat, $\Delta h_{l-s} = T_m\, ds/v_l$. Thus we obtain the Gibbs–Thomson equation,

$$\frac{\Delta h_{s-l}}{T_m}(\Delta T_m) = \sigma\left(\frac{1}{R_1} + \frac{1}{R_2}\right) \tag{5.20}$$

which relates ΔT_m to the latent heat, surface tension and interface curvature as,

$$\frac{\Delta T_m}{T_m} = \frac{1}{\Delta h_{l-s}}\, \sigma\left(\frac{1}{R_1} + \frac{1}{R_2}\right) \tag{5.21}$$

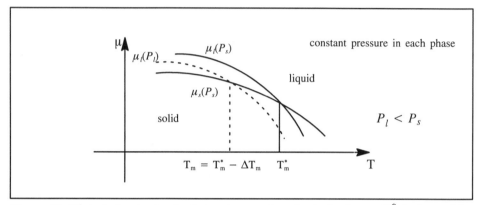

Figure 7. Plot of chemical potential vs temperature for both phases. T_m^* is the melting temperature for a planar phase boundary, and $T_m = T_m^* - \Delta T_m$ is the the melting temperature with ΔT_m being the change in the melting temperature due to the effects of surface tension and curvature. The equilibrium point shifts because the two phases are not at the same pressure. The dotted lines show that the chemical potential is a function of the pressure also. Hence the curve shifts and a new equilibrium point is attained.

A few general remarks may be made about surface energy. During the phase change process the equilibrium shape of the material is a configuration of minimum surface energy. Thus for isotropic materials the equilibrium shape is a sphere. However, for certain types of crystalline solids the surface energy is not an isotropic property. Low index faces of the crystal have low surface energy. One would expect a crystal at equilibrium to have low index planes as its surfaces, but the exact combination of these planes that give the lowest surface energy is not straightforward to deduce. The surface energy for such crystals is given by $\sum\limits_{i=1}^{m} A_i \sigma_i$ which is minimized at equilibrium, where A_i is the area of face i and σ_i is the surface energy of face i.

6. FURTHER READING

More detailed information of the various aspects covered here can be found in Adamson (1990), Gibbs (1931), Guggenheim (1957), Hatsopoulos and Keenan (1965), Hirschfelder, Curtiss and Bird (1954), Kestin (1979), Morse (1969), Reif (1965), and Rosenberger (1979).

CHAPTER VIII
THERMOFLUID PHENOMENA INVOLVING
CAPILLARITY AND GRAVITY

Interfacial phenomena appearing at phase boundaries have a profound influence on the behavior of multiphase systems. Capillarity is one of the most important aspects of interfacial phenomena. Surface–controlled phenomena, such as capillary effect, often occur in conjunction with gravity and in many cases these may be mutually counteracting. Thus, equilibrium of such systems occurs by the compound minimization of the free energy due to both gravity and capillary effects.

Capillary and gravitational effects act over different length scales. As a consequence, capillary effects are usually neglected for large scale phenomena. However, for small length scales, or under microgravity conditions, both phenomena need to be accounted for. Under microgravity, the influence of gravity is substantially reduced and capillary effects often become dominant. For example, not only do taps not drip, but suspensions do not experience sedimentation, and surface tension gradients which, on earth may be of negligible effect, can significantly contribute to fluid flow. This is one of the most important aspects of microgravity fluid mechanics in space related applications.

In the following, three different phenomena will be used to illustrate the underlying physics; they are: (1) Meniscus formation in an isothermal tri–phase system, i.e., with gas, liquid and solid simultaneously present, (2) jet breakup due to Rayleigh instability and (3) Benard convection and illustrations of the interaction of surface tension induced convection and natural convection in complicated manners.

1. MENISCUS FORMATION AND CONTACT ANGLE

It is now appropriate to discuss technologically important applications which serve to illustrate and highlight the physical principles described so far. The physical phenomena involved include phase change, free surface phenomena, multiphase flow, heat and mass transfer. Keeping in mind the increasing interest in space based manufacturing facilities, issues relating to the effects of a low gravity environment as opposed to a 1–g environment are discussed in the context of the physical phenomena mentioned above. A good understanding of these issues will be necessary prior to the establishment of commercial manufacturing operations in outer space.

1.1 Edge–defined Fibre Growth Process

In this section we discuss the fluid flow and transport phenomena which play an important role in the solidification and growth of crystals from the liquid melts. The edge–defined film–fed growth (EFG) technique is chosen to illustrate the interplay of solidification, meniscus behavior, heat and solute transport, convection in the melt and gravitational effects in the growth process. Figure 1 shows a schematic of the meniscus corresponding to the edge defined fibre growth process.

An important aspect of this problem is the equilibrium meniscus shape as characterized by the Bond number which will be defined later, the reservoir pressurization, the aspect ratio and the contact condition at the trijunction point where the melt, solid and the ambient gas meet and interact. Control of the crystal diameter involves control of the meniscus shape and the contact angle, while the growth rate of the crystal is controlled by the heat transfer characteristics at the crystal–melt interface. Neglecting the effect of convection in the melt, the meniscus shape is governed by the Young–Laplace equation.

Assuming an axisymmetric configuration, we can now derive the governing equation for the shape of the meniscus as follows. The Young–Laplace equation, Eq.(VII–5.12b), relates the pressure difference across the free surface and the free surface curvature, as follows:

$$P_G - P_L = \sigma\left(\frac{1}{R_1} + \frac{1}{R_2}\right) \tag{1.1}$$

where $P_G = P_g - \varrho_g gy$ and $P_L = P_l - \varrho_l gy$ are the respective static pressures on the gas and the liquid sides of the interface.

Now, we define $\Delta P = P_l - P_g$ and $\Delta\varrho = \varrho_l - \varrho_g$, where ΔP is an imposed pressure difference,

$$\Delta\varrho gy - \Delta P = \sigma\left(\frac{1}{R_1} + \frac{1}{R_2}\right) \tag{1.2}$$

and assuming an axisymmetric configuration, we get,

$$\Delta\varrho gy - \Delta P = \sigma\left[\frac{f_{yy}}{(1 + f_y^2)^{3/2}} - \frac{1}{f(1 + f_y^2)^{1/2}}\right] \tag{1.3}$$

which is the governing equation for the equilibrium meniscus shape.

We now proceed to non–dimensionalize this equation. The relevant parameters are: g, $\Delta\varrho$ and σ, and the geometric parameters r_b and h_c, where r_b is the radius of the die tip and h_c is the height of the trijunction point measured from the die tip. Thus we have five parameters and three fundamental units, in which case we deduce from the Buckingham Pi theorem that this problem is characterized by two non–dimensional parameters.

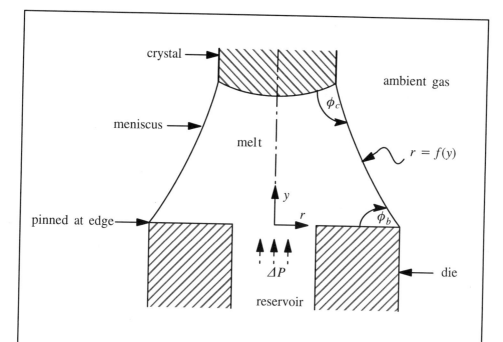

Figure 1. Schematic of the edge–defined film–fedgrowth (EFG) process. The radii of curvature are taken as positive on the concave side of the interface. ϕ_c is the contact angle of the meniscus with the crystal. It is observed in practice that the meniscus is pinned to the sharp edge of the die. The crystal grows when the melt solidifies at the solid–liquid interface and is continuously pulled out. The melt is supported by the surface tension forces between the melt and the ambient gas. ΔP corresponds to the pressurization of the reservoir. The equilibrium shape of the meniscus is dictated by the competing effects of gravity (hydrostatic forces), reservoir pressurization and the surface tension forces between the ambient gas and the melt.

They are the aspect ratio,

$$AR = \frac{r_b}{h_c} \tag{1.4}$$

and the Bond number

$$Bo \equiv \frac{\Delta\varrho g r_b^2}{\sigma} = \left(\frac{r_b}{l_\sigma}\right)^2 \tag{1.5}$$

where $l_\sigma = \sqrt{\sigma/\Delta\varrho g}$ is the capillary length scale. The Bond number characterizes the relative strength of the hydrostatic pressure force and the capillary force. The behavior of a body of fluid is dominated by the capillary force if any of $r_b, \Delta\varrho$ or g is small. The surface tension is usually in a range that does not vary too much. With this information, the Young–Laplace equation can be non–dimensionalized using l_σ as the reference length scale, $\Delta\varrho$ as the density scale and a pressure scale of $\Delta\varrho g l_\sigma$ as follows:

$$\left[\frac{f^*_{YY}}{(1 + f^{*2}_Y)^{3/2}} - \frac{1}{f^*(1 + f^{*2}_Y)^{1/2}} \right] = Y - \Delta P^* \qquad (1.6)$$

where $f^*(Y)$ and Y are the non–dimensional meniscus shape and vertical coordinate. The starred quantities are non–dimensionalized variables. This nondimensionalized equation is apparently invariant with Bond number, but it should be recalled that the vertical dimension has been scaled by the capillary length. Therefore the meniscus profile will still be dependent on the Bond number. The choice of any other length scale, say r_b, will again bring in *Bo* as a parameter in the equation.

1.2 Boundary Condition: the Contact Angle

When a liquid drop is formed on a plane solid surface or when a liquid bridge is formed between two solid planes, it will usually be in contact with both solid and ambient gas. The contact line is the line defining the perimeter of the wetted area of the solid. The contact angle is the angle formed by the liquid and the solid at the contact line.

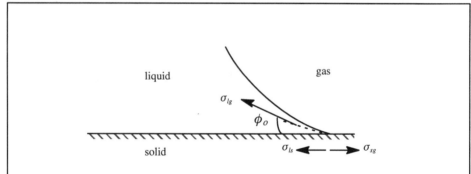

Figure 2. *Schematic of the trijunction point and the static contact angle, ϕ_o , for a liquid drop in contact with a flat surface.*

The existence of the contact angle implies that
(i)   There must be a force component associated with the liquid–gas surface tension σ_{lg} that acts parallel to the solid surface and whose magnitude is $\sigma_{lg} \cos \phi_o$;
(ii)   If the liquid is to remain in static equilibrium without moving along the solid surface, the above force component must be balanced by other forces that act at the contact line. These other forces have been assumed to be σ_{ls} and σ_{sg}.
Then by force balance at the contact line we have the Young's contact condition,

$$\sigma_{lg} \cos \phi_o = \sigma_{sg} - \sigma_{ls} \qquad (1.7)$$

indicating that ϕ_o, along with the surface tension, σ's, are a material property. The above contact condition can also be derived on the basis of free energy minimization (Johnson 1959).

Young's contact condition can also be written as

$$k = \frac{\sigma_{sg} - \sigma_{ls}}{\sigma_{lg}} = \cos\phi_o \qquad (1.8)$$

if $k = 1$ then $\phi_o = 0$ and we have a completely wetted surface.

In reality, not all of the surface energies are easily measurable. At the small scales, the liquid surface can still be viewed as homogeneous and smooth; however it is virtually impossible to obtain a purely planar solid surface. Also solid surfaces are almost always contaminated, which can substantially affect the surface tension. Thus, in practice, it is difficult to obtain the magnitude and even the "sign" of the surface tension forces between the solid and the liquid. Experimentally, any direct measurement yields two values of the contact angle, the advancing contact angle, ϕ_A and the receding contact angle, ϕ_R. Figure 3 illustrates this schematically. It is to be noted that in general the two angles are not the same. This difference is called contact angle hysteresis. Figure 4 illustrates contact angle hysteresis.

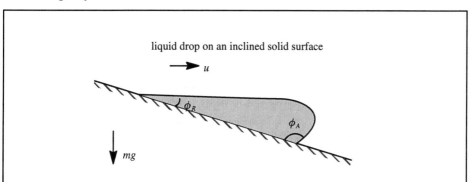

liquid drop on an inclined solid surface

Figure 3. Schematic of a liquid drop on a tilted plate showing the advancing and the receding contact angle.

The sources of contact angle hysteresis are thought to be the following (Carey 1992, deGennes 1985 and Dussan V. 1979):

(i)   surface inhomogeneity
(ii)   surface roughness
(iii)   surface contamination.

In the EFG process, the liquid meniscus is formed between the crystal on top and the die beneath. The following characteristics are noteworthy.

(i)   The shape of the meniscus is governed by the Young–Laplace equation,
(ii)   The lower contact line is usually anchored at the outside sharp edge of the die tip as illustrated in Figure 1.
(iii)   The contact angle at the lower contact line varies with the meniscus shape, the reason being that at the outside edge the die tip is not flat, but sharp and hence the contact line can remain fixed as long as the contact angle there is bound by, say, the values of ϕ_A and ϕ_R.
(iv)   The upper contact line usually moves due to the diameter variation of the crystal.

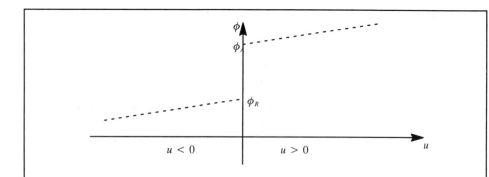

Figure 4. Schematic showing the hysteresis effect of interface motion on the contact angle. According to the above model, the contact angle shows a hysteresis effect at u = 0 and varies linearly with u thereafter.

(v)   If a fixed contact angle is to be maintained at the upper contact line, then the radius of the solid–liquid interface there must change with changes in the height of the liquid bridge. On the other hand, if contact angle hysteresis is allowed, then the interface radius can experience non–smooth variations since the radius of the contact line does not always change when the interface height varies.

Let us now consider the situation of thin fibre growth by solving the Young–Laplace equation subject to (i) a fixed radius at the lower boundary and (ii) a fixed angle at the top boundary. A schematic illustration of a model with no phase change is given in Figure 5.

The problem is that of solving a non–linear ordinary differential equation subject to the boundary conditions at the two end points. Such a two point boundary value problem does not yield a unique solution. Thus there arises the problem of selecting, out of the multiple solutions that are mathematically permissible, the one that corresponds to the physical equilibrium condition that exists in reality. Therefore, we now invoke the thermodynamic condition that at equilibrium the free energy is a minimum and select the solution that minimizes the free energy (Shyy *et. al.* 1992g).

For simplicity, consider only the isothermal condition with no phase change and examine the Helmholtz free energy, $F = U - TS$ which contains three contributions:

(i)   The potential energy from the effective head

$$U_1 = \int \pi f^2 \left(\Delta \varrho g y - \Delta P \right) dy \qquad (1.9)$$

(ii)   The surface energy of the meniscus forming the gas–liquid interface

$$U_2 = \int 2\pi \, \sigma_{lg} \, f \left(1 + f_y^2 \right)^{1/2} dy \qquad (1.10)$$

(iii)   The surface energy needed to wet the solid–liquid wetted area

$$U_3 = \left(\sigma_{ls} - \sigma_{sg} \right) \pi \, r_c^2 \qquad (1.11a)$$

From Young's contact condition, Eq. (1.8), this becomes,

$$U_3 = -\sigma_{lg}\cos\phi_0\,\pi\,r_c^2 \tag{1.11b}$$

where ϕ_0 is the static contact angle. The equilibrium shape is the one that minimizes

$$E = U_1 + U_2 + U_3 \tag{1.12}$$

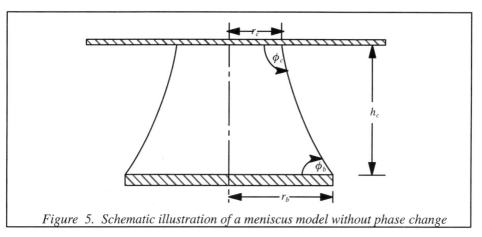

Figure 5. Schematic illustration of a meniscus model without phase change

1.3 Basic Meniscus Behavior with a Static Contact Angle

In this section, a series of cases will be discussed to illustrate the characteristics of the meniscus formation under both equilibrium and quasiequilibrium conditions. Consider a typical case (Case 1) shown in Figure 6. The parameters are $Bo = 4 \times 10^{-5}$, h_c/r_b (aspect ratio) = 0.5, $\Delta p = 0$. The meniscus is maintained by pulling the plate upward as illustrated in the curves corresponding to ϕ_c versus ϕ_b, and r_c/r_b versus ϕ_c clearly show the multiple solution behavior. Thus, there is only an interval of possible contact angles ϕ_c that can be achieved at the top of the meniscus. Outside this range of angles no solutions exist. This is not due to the instability of possible equilibrium profiles, but because it is not even possible to obtain profiles satisfying the Young–Laplace equation. Also shown in Figure 6 are sample meniscus profiles, corresponding to ϕ_0 of 60° and 90°, respectively.

As shown in Figure 6 when the value of ϕ_0 is specified, and E is calculated by fixing the value of ϕ_o in Eq. (1.11b) and scanning through the whole range of meniscus profiles obtained by fixing the lower trijunction location (with varying angle) and the height of the upper trijunction point (with varying locations and angles), distinct extrema are obtained at specified values of ϕ_0. Among the multiple solutions, the one corresponding to a point of minimum on the curve is the physically realizable stable meniscus profile. When either maxima or non–extrema arise, we may surmise that the corresponding solutions belong to the unstable branch.

That the Young's condition is in fact the condition for stability is straightforward to show, as say in the case of a sessile drop. In such a (constant volume) case, when gravity is ignored, the drop assumes the shape of the arc of a circle (Dyson 1978). Then the minimization result becomes merely a geometric condition compatible with minimum surface area. When gravity is considered, the drop shape no longer remains the element of an arc, since minimization of surface free energy is not sufficient, it being necessary to minimize the total energy including the gravitational potential. In the case of menisci above infinite baths, Pitts (1976) , and others have shown the minimization principle via variational methods. It is the conventionally held view that Young's contact condition can be obtained both from the consideration of forces as well as energy (however, see Johnson (1959) and discussion therein for criticism of this view). However, it is interesting to see that in the present configuration, multiple shapes corresponding to an energy minimum, maximum and even non–extremum may all be observed at the given contact angle ϕ_0. Furthermore, solutions to the Young–Laplace equation, which is after all a static force balance across the interface, are all equilibrium solutions. From the standpoint of local energy minimization, there may be more than one stable shape. Thus, the end condition and the path of development of the meniscus selects the stable solution. However, in a process in which the aspect ratio is varied, an unstable meniscus may form first. This can happen if the new trijunction location corresponding to this shape is closer to the initial trijunction location. Then, in achieving the stable solution profile for the given height, one may pass through an unstable solution profile.

   As discussed in connection with the non–dimensional form of the Young–Laplace equation, for low Bond numbers, the term corresponding to gravity is small compared to the curvature terms on the right hand side of Eq. (1.3). Thus, the Bond number serves principally to rescale the dimensions of the meniscus according to the capillary length. However, for higher Bond numbers, the hydrostatic term on the left hand side of Eq. (1.3) becomes comparable to the curvature terms on the right hand side, and gravity does significantly impact the meniscus shape, as will be illustrated in the results to follow.

   For taller menisci, i.e., higher aspect ratios, the solutions offer a wider variety of shapes and range of angles. There is also a greater tendency for multiple solutions. Retaining $Bo = 5 \times 10^{-5}$ and $\Delta p=0$ as in case 1 we can increase the height of the meniscus. For example, for $h_c/r_b = 2$, which represents a taller meniscus, the available range of contact angles will shrink. Thus, as expected, when the height of the meniscus is increased, in the absence of pressurization a large number of solutions to the Young–Laplace equation fail to exist.

   Cases 2 and 3 (Figure 7 and Figure 8 respectively) are illustrative of the effects of pressure and Bond number on menisci with a higher aspect ratio. For Case 2, Bo = 2.5×10^{-6}, $\Delta p=5\%$, $h_c/r_b = 2$. For this low Bond number case, from Figure 7, we notice that for certain contact angles it is possible to obtain three solutions. However, with an aspect ratio of 2, the range of permissible contact angles is narrow. Meniscus shapes obtained are not significantly different from those for $\Delta p=0$, all other parameters remaining the same. It is noticed that for a contact angle of 16°, three equilibrium profiles could be captured. The possible contact angles at the top are in the range of 18° to 26°. These profiles may not be seen in a physical situation, since the contact angles corresponding to the materials of the system may not lie in this range at all. Thus, ultimately, the existence of menisci will be dictated by the range of contact angles ϕ_c at the top permitted by the materials, the range of ϕ_b for which the pinning condition is maintained at the edge of the die, and the stability of the equilibrium solution (of the Young–Laplace equation) obtained.

The contrasting situation to Case 2 emerges in Case 3 (Figure 8). Here the Bo = 2.5×10^{-3} (h_c/r_b =2, and Δp=5% as in Case 2) which is 10^3 times that of Case 2. The available range of contact angles is considerably widened, and a greater variety of meniscus profiles than for the lower Bo in Case 2 can be obtained. Comparison of profiles with those in Figure 7 shows the difference that Bond number makes to the effect of pressure. Changes in curvature arise in the profile. This implies that for the given parameters, on the right hand side of Eq. (1.3), the pressure and gravity terms must compete to determine the shape of the meniscus. Thus, as mentioned before, for $\Delta P \neq 0$, the Bond number can significantly influence meniscus shapes and permissible solutions. This can be seen from the scaling that we have employed in non–dimensionalization. With fixed h_c/r_b, the height of the meniscus is scaled with respect to Bond number as $h_c \propto \sqrt{Bo}$. Hence, for given $\Delta \varrho$, g and a, the meniscus height increases with Bond number. For higher Bo, by definition, the relative importance of gravity ($\Delta \varrho g y$) becomes more significant compared with the pressure term. Thus, the competition between gravity and pressure in the force balance leads to a variety of curvatures for higher Bond numbers. For lower Bo, a relatively small Δp can overwhelm the gravity term, and the curvature is less sensitive to location along the y–axis.

Another aspect to note in Figure 7 and Figure 8 is that in the E vs. ϕ_c curves both maxima and minima are obtained at specified ϕ_o. As a contrast, however, consider Case 4 (Figure 9). Here, we have $Bo = 4 \times 10^{-4}$, Δp=5% and with the h_c/r_b increased to 5. While the other properties appear to follow the general trend of Case 3, the behavior of energy is distinctive. Here, for ϕ_o =90°, both minimum and maximum are obtained. However, the minimum energy state so obtained is only local. There appears to be a global minimum at 130°. For ϕ_o =130°, Figure 9(c) and (d), the local minimum occurs at around 100° and there appears to be a non–extremum point corresponding to the specified ϕ_o indicating that although only one meniscus profile exists, it is not a stable one. Thus, for this case the decision regarding the stable meniscus profile becomes more involved.

In summary, the numerical simulation reveals that depending on the range of parameters a variety of solutions are possible for a given contact angle. Multiple energy minima, implying multiple stable shapes, or no minima, implying unstable shapes may be obtained for a given contact angle.

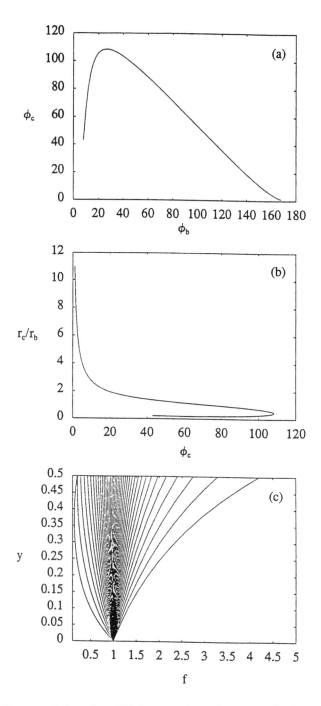

Figure 6. Characteristics of equilibrium meniscus formation for Bo = 4 × 10⁻⁵,
ΔP = 0, h_c/r_b=0.5, and upward pulling.
(a) Variation of angle ϕ_c with ϕ_b. (b) permissible range of r_c/r_b with respect
to ϕ_c. (c) profile shapes for different angles, ϕ_b.

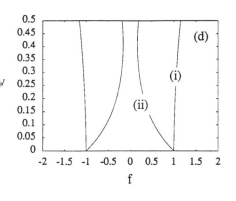

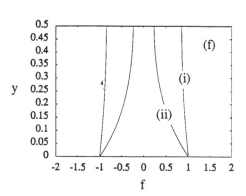

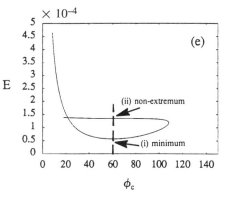

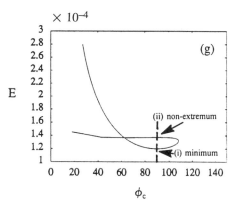

Figure 6(continued) Characteristics of equilibrium meniscus formation for
Bo = 4 × 10⁻⁵, ΔP = 0, h_c/r_b=0.5, and upward pulling
(d) profiles corresponding to ϕ_o = 60°. (e) E profile versus ϕ_c for ϕ_o = 60°.
(f) profiles corresponding to ϕ_o = 90°. (g) E profile versus ϕ_c for ϕ_o = 90°.

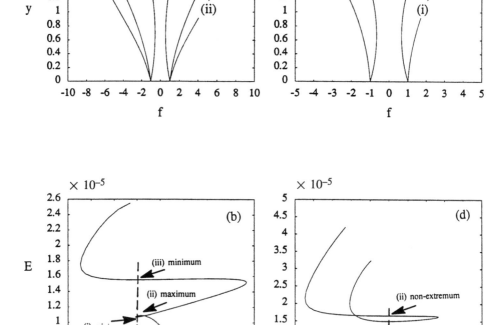

*Figure 7. Characteristics of equilibrium meniscus formation for Bo = 2.5 × 10⁻⁶,
ΔP = 5%, h_c/r_b = 2, and upward pulling
(a) profiles corresponding to φ_o = 16°. (b) E profile versus φ_c for φ_o = 16°.
(c) profiles corresponding to φ_o = 20°. (d) E profile versus φ_c for φ_o = 20°.*

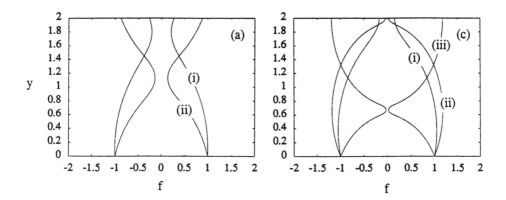

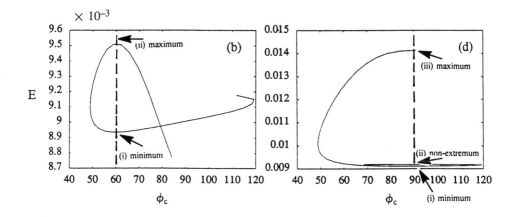

Figure 8. Characteristics of equilibrium meniscus formation for Bo = 2.5 × 10⁻³,
$$\Delta P = 5\%, \ h_c/r_b = 2, \ and \ upward \ pulling$$
(a) profiles corresponding to $\phi_o = 60^o$. (b) E profile versus ϕ_c for $\phi_o = 60^o$.
(c) profiles corresponding to $\phi_o = 90^o$. (d) E profile versus ϕ_c for $\phi_o = 90^o$.

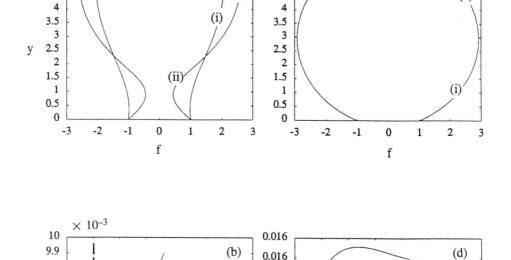

Figure 9. Characteristics of equilibrium meniscus formation for Bo = 4 × 10⁻⁴,
ΔP = 5%, hᴄ/hᵦ = 5, and upward pulling
(a) profiles corresponding to φₒ = 90°. (b) E profile versus φᴄ for φₒ = 90°.
(c) profiles corresponding to φₒ = 130°. (d) E profile versus φᴄ for φₒ = 130°.

1.4 Quasi–Equilibrium Motion with Hysteresis

In this third set of experiments we investigate the quasi–equilibrium path negotiated by the meniscus, by incorporating a simplified hysteresis model at the top trijunction, i.e., on our modeled flat plate. Specifically, the top plate is oscillated about a mean height h_{c*} and the assumption is made that the time scale of these oscillations is such that a succession of equilibrium states is achieved. In other words, the Young–Laplace equation is solved for each position $h_c(t)$ of the oscillating plate. The contact line at the die is allowed to remain pinned at the edge, as long as the Gibbs inequality is not violated. At the top, the boundary condition is modified by the hysteresis model adopted for the top trijunction, which is defined by the following generally observed criteria:

(i) The contact line does not move, i.e., $r_c(t)$ is a constant when the contact angle varies between ϕ_A and ϕ_R, where ϕ_A and ϕ_R are respectively the advancing and receding contact angles (Dussan V. 1979).

(ii) When the contact angle ϕ_A or ϕ_R is reached, contact line motion ensues. To simplify the situation, we assume that the contact angle does not vary with contact line motion, i.e., $\phi_c = \phi_A$ or ϕ_R depending on the direction of motion.

(iii) It is imposed that an advancing contact line cannot recede without passing through the hysteresis range ϕ_A to ϕ_R. Thus, if the oscillatory motion of the plate induces an impending recession of an advancing contact line, the model would perforce pin the radius $r_c(t)$ at such a location and require ϕ_c to travel from ϕ_A to ϕ_R before allowing recession. Similar conditions hold for the receding angle, ϕ_R.

An additional simplification resulting in the model is relevant in the presence of multiple solutions. When the contact line is advancing, for instance, we choose that solution for which $r_c(t)$ is closest in the direction of impending motion. So also for a receding contact line. Thus, irrespective of the stability of the equilibrium profiles, proximity of the contact line is made the criterion for choice of a solution from those available. Actually, in the situation where dynamic contact angles exist, the criterion in terms of the existence of a minimum of the energy curve (E versus ϕ_c) are no longer applicable. In particular, the wetting component of energy E_{wetted}, namely $\pi r_c{}^2(\sigma_{ls}-\sigma_{sg})$ cannot be replaced by an expression such as $\pi r_c^2 \sigma_{lg} \cos(\phi_c)$, for otherwise $(\sigma_{ls}-\sigma_{sg})$ would cease to be a material property. Thus, in passing through the successive equilibrium states, the stability of menisci cannot be deduced in terms of an energy minimization principle. From a microscopic viewpoint, if we consider that the hysteresis range ϕ_A to ϕ_R merely represents the apparent contact angle behavior, and that, in actuality, the static contact angle is ϕ_0 (which is the Young's contact angle, not related in any straightforward way to ϕ_A or ϕ_R), then we are justified in replacing $(\sigma_{ls}-\sigma_{sg})$ by $\sigma_{lg} \cos(\phi_0)$. However, on a microscopic scale, the phenomenon of hysteresis is occasioned by the presence of asperities (Eick 1975) and surface chemical inhomogeneities (Neumann and Good 1972) or by several other effects which are still a matter of investigation. Thus, the wetted energy expression will suffer modification to $\sigma_{lg} \cos\phi_0 A_{real}$ where A_{real} will incorporate the effects of roughness etc. Thus, no tacit assumption regarding E_{wetted} may be made either from a microscopic or macroscopic view, and for this reason we will refrain from drawing any conclusions regarding stability of the quasi–equilibrium path simulated hereinunder. It may be pointed out that the situations presented in this section and the two previous ones represent two limiting types of behavior. The boundary conditions and stability criterion, in terms of the minimization of energy are applicable to a situation where the time scale of motion of the top contact line is greater than that required for the meniscus to achieve equilibrium, as discussed in the previous section. Here, on the other hand, it is supposed that the time scale of contact line motion is much shorter than that required for reaching an equilibrium state, the latter being determinable only by incorporation of the dynamics of the interface as well as the fluid contained in the meniscus.

Traditionally, the following formula has been employed in models describing growth of fibres via the EFG process. The variation of the radius at the top of the meniscus with pull rate is given by the following equation,

$$\frac{dr_c}{dt} = (u_p - \frac{dh_c}{dt}) \tan(\phi - \phi_o) \qquad (1.13)$$

where $\phi-\phi_o$ is the deviation in contact angle from the value that leads to fibre growth with uniform diameter (Tatarchenko and Brenner 1980, Thomas *et. al.* 1986). This equation is a geometric statement describing a constant contact angle condition. We attempt, in what follows, to evaluate the applicability of this expression in the light of hysteresis. In the event of ϕ being a constant, it is obvious that the variation of the radius with time follows the same waveform but has an opposite phase with respect to the variation of meniscus height, $h_c(t)$. In this respect, the presence of hysteresis introduces a non–linearity in the above expression, in that the value of ϕ in turn depends on $r_c(t)$ and its rate of change. However, if a pinning condition arises, i.e., r_c=constant, then the above formula indicates that, provided $u_p \neq dh/dt$, $\phi=\phi_o$, i.e., that the contact angle remains fixed. But this is neither necessary nor true when hysteresis is included, because it can happen that ϕ_c assumes values between ϕ_A and ϕ_R, while the contact line remains stationary. The results of our calculations, under the restriction of equilibrium, and the conditions 1 to 3, prescribed by the hysteresis model above, will call to question the implications of Eq. (1.13).

Under the present model, the oscillation of the plate is effected by specifying $h_c(t)=h_c{}^*+\Delta h_c \sin \pi t$, where in all the results presented, $\Delta h_c= 0.2h_c{}^*$ is given. The hysteresis range ϕ_A to ϕ_R is also specified, with the values corresponding roughly to the material properties of sapphire. We start with an initial condition at the top trijunction which facilitates the calculation of top radius $r_c(t)$, when the contact angle is $\phi_c=\phi_A+\Delta\phi_A$, i.e., the contact angle is assumed to have been displaced so that the contact line motion may ensue. But for the initial transients, the final periodic behavior of the meniscus is found to be independent of the specified initial state. The subsequent development is followed by plotting the time variation of meniscus profiles, top radius and contact angle. As in the previous sections, Bond numbers of 10^{-5} and 2.5×10^{-3} are used, and the meniscus is internally pressurized to a given ratio of the atmospheric pressure. The cases simulated below relate to downward pulling of the fibre.

First, a low Bond number case (Bo=10^{-5}) is considered, without hysteresis, the contact angle being fixed at 90°. In Figure 10, the sinusoidal motion of the plate given by $h_c(t)$, causes a corresponding sinusoidal motion of the top radius, $r_c(t)$. The r_c variation is here out of phase with $h_c(t)$ as represented by Eq. (1.13). It is noted that the radius variation in Figure 10 for the imposed amplitude Δh_c, is quite wide, assuming values from $0.62r_b$ to $0.83r_b$. Introduction of a hysteresis range gives rise, in Figure 11, to a square waveform for $r_c(t)$. For the hysteresis range imposed, ϕ_R=85° to ϕ_A=95°, the pinning condition comes into effect when ϕ_c lies in the hysteresis range. This can be seen from Figure 11 (c) and (d), in which, while ϕ_c varies between ϕ_A and ϕ_R, the radius remains a constant. It is not difficult to foresee that when the hysteresis range is further increased, for this amplitude of oscillation, the pinning condition is eventually enforced throughout the cycle of $h_c(t)$, while $\phi_c(t)$ assumes a sinusoidal form within the range ϕ_R to ϕ_A. In terms of the motion of the contact line, the variation of $r_c(t)$ displayed in Figure 10 can be interpreted as a stick–slip motion (Jansons 1986).

Figure 12 corresponds to a higher Bond number of 2.5×10^{-3}, where the contact angle is fixed at 100° with no hysteresis, mean height, h_c^*=0.5, and Δp=5%. Here the behavior of the meniscus registers a marked change. This is consistent with the difference in meniscus behavior arising due to the different Bond number regimes as seen in the previous sections. It is particularly noticed here that for the same relative variation in height Δh_c as in Figure 10, in this case the variation in radius r_c is less significant. Also the radius variation is in fact in phase with the h_c variation, being significantly in opposition to Figure 10, and formula, Eq. (1.13). In other words, the radius in this case increases with height instead of decreasing, while the angle remains constant. This is evidently due to the change in meniscus curvature from concave, as in Figure 10, to convex. For a higher meniscus, h_c^*=1.5, for the same ϕ_c and Δp shown in Figure 13, the radius variation is also not very substantial in magnitude and the waveform is not identical to $h_c(t)$, indicating the non–linearity that is introduced by $\phi(t)$. When hysteresis is present, and for the case of fairly large hysteresis, and tall meniscus, Figure 14 shows an interesting situation arising due to the presence of multiple solutions. Note that in the previous cases, only one solution existed in each case, while here 2 or 3 solutions are obtained for certain heights. The choice of solutions is made as explained above. The marked excursions of radius evident in the figure arise when the outer solutions seen in Figure 14(a), are allowed to be attained by the imposed conditions 1 to 3. The hysteresis model in this case causes a noticeable change in the contact angle behavior (Figure 14(d)) in comparison to the static model.

Finally, Figure 15 shows a case of shorter h_c and a smaller range of hysteresis, which again exhibits highly non–linear patterns different from the other configurations based on the same Bond number. Thus, there appears to be a Bond number and configuration dependence of the phase relationship between dr_c/dt and dh_c/dt, which is not reflected in the relation, Eq. (1.13).

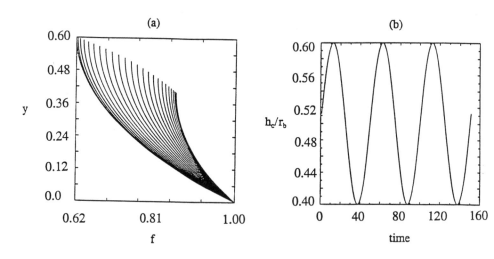

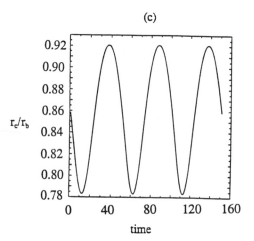

Figure 10. Meniscus behavior at lower Bond number, with no hysterisis.($Bo=10^{-5}$, $\Delta p=5\%$, $h_c{}^*/r_b=0.5$, $\phi_A=\phi_R=100^o$). (a) Meniscus profiles obtained for the duration of oscillation.(b) Variation of height h_c with time. (c) Computed value of top radius r_c versus time.

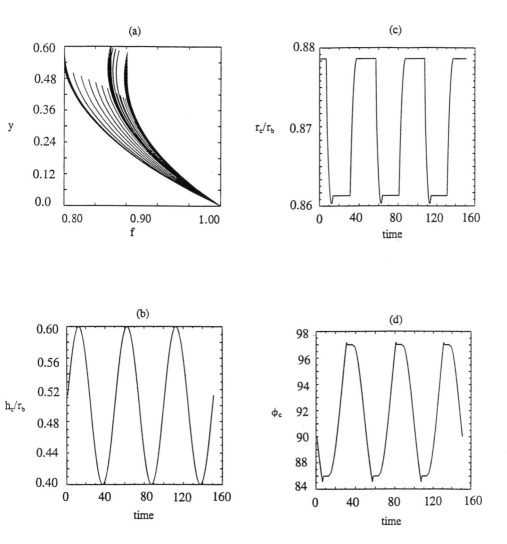

Figure 11. Meniscus behavior at lower Bond number, with small hysterisis range.
(Bo=10⁻⁵, Δp=5%, h_c^*/r_b=0.5, ϕ_A=95°,ϕ_R=85°). (a) Meniscus profiles
obtained for the duration of oscillation. (b) Variation of height h_c with time.
(c) Computed value of top radius r_c versus time. (d) Computed variation of
contact angle ϕ_c versus time.

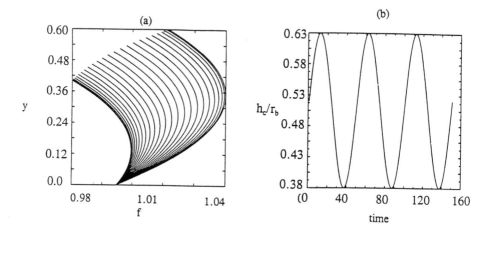

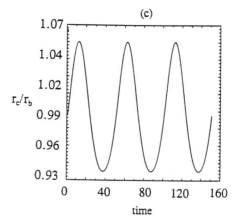

Figure 12. Meniscus behavior at higher Bond number, with no hysteresis.
(Bo=2.5 × 10⁻³, Δp=5%, h_c^*/r_b=0.5, ϕ_A=ϕ_R=100°). (a) Meniscus profiles
obtained for the duration of oscillation. (b) Variation of height h_c with time.
(c) Computed value of top radius r_c versus time.

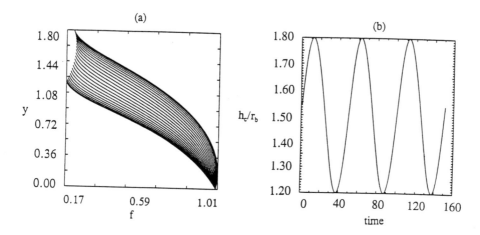

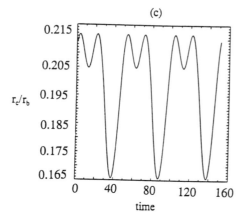

Figure 13. Meniscus behavior at higher Bond number, with no hysteresis.
($Bo=2.5 \times 10^{-3}, \Delta p=5\%, h_c*/r_b=1.5, \phi_A=\phi_R=100^\circ$). (a) Meniscus profiles
obtained for the duration of oscillation. (b) Variation of height h_c with time.
(c) Computed value of top radius r_c versus time.

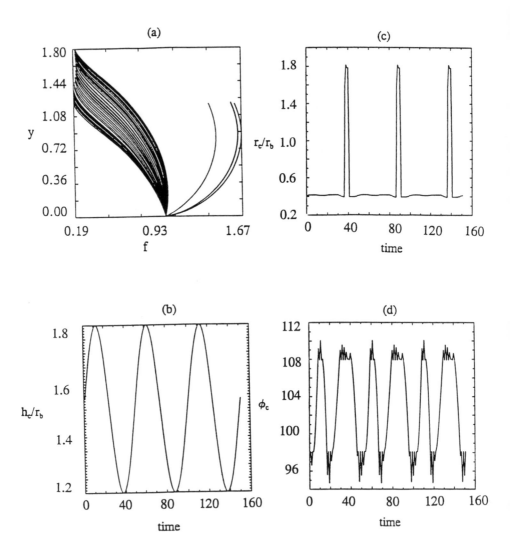

*Figure 14. Meniscus behavior at higher Bond number, with hysteresis. (Bo=2.5 × 10⁻³,
Δp=5%, h_c*/r_b=1.5, φ_A=110°,φ_R=95°). (a) Meniscus profiles obtained for
the duration of oscillation. (b) Variation of height h_c with time. (c) Computed
value of top radius r_c versus time. (d) Computed variation of contact angle
φ_c versus time.*

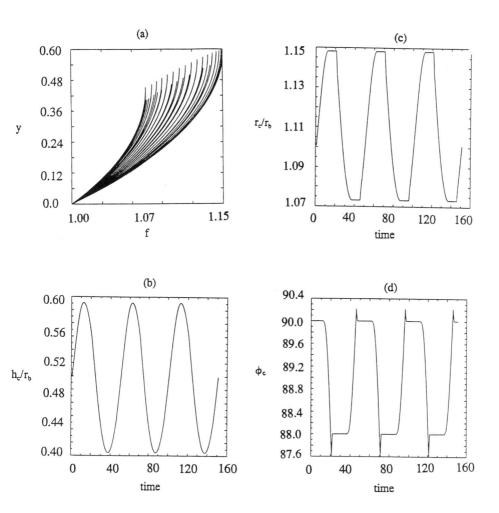

Figure 15. Meniscus behavior at higher Bond number, with small hysteresis range.
*(Bo=2.5 × 10⁻³, Δp=5%, h_c */r_b=0.5, ϕ_A=90°,ϕ_R=88°.) (a) Meniscus*
profiles obtained for the duration of oscillation. (b) Variation of height h_c
with time. (c) Computed value of top radius r_c versus time. (d) Computed
variation of contact angle ϕ_c versus time.

Thus, Eq. (1.13) relating the radius with height of the meniscus, seen in the light of a quasi–equilibrium path, does not appear to exhibit the appropriate features under contact angle hysteresis. In the absence of hysteresis, the Bond number dependence of the phase is not captured. When the Young–Laplace equation yields multiple solutions, the choice of profiles is not straightforward, and the particular selection procedure adopted indicates an abrupt dynamic behavior which may not be justifiably represented by a quasi–equilibrium situation, and the full fluid–dynamic consideration is required to assess the implication of such behavior. We also reiterate that no account has been taken of the stability of the computed equilibrium profiles.

1.5 Surface Tension and Capillary Flow

An easy and accurate method of determining the surface tension of a liquid is via the measurement of the rise or fall of the liquid in a circular capillary tube. Figure 16 is a schematic illustrating the process.

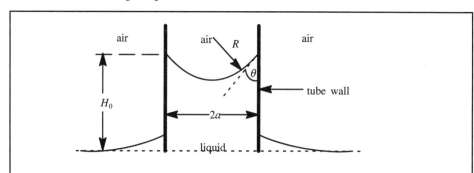

Figure 16. Schematic of the rise of a liquid in a capillary tube due to surface tension.

We can obtain an expression for the measurement of surface tension. If the pressure variation along the top interface is zero, then $R_1 = R_2 = \dfrac{a}{\cos\theta}$. Now,

$P_I - P_{II} = P_{liquid} - P_{air} = -\varrho g H_0$.

But $P_{air} = P_{atm}$ outside the top surface of the liquid.

$$\therefore \quad P_I - P_{II} = -2\frac{\sigma}{R} \tag{1.14}$$

$$\Rightarrow \varrho g H_0 = \frac{2\sigma\cos\theta}{a} \tag{1.15}$$

Thus we have,

$$H_0 = \frac{2\sigma\cos\theta}{\varrho g a} = 2\cos\theta \cdot \frac{l_\sigma^2}{a} \tag{1.16}$$

where l_σ is the capillary length defined in section 1.1. Note that:

(i) $H_0 \propto \left(\dfrac{l_\sigma^2}{a}\right)$

(ii)   if $\theta > 90^0$ then $H_0 < 0$. For example, Hg has $\theta = 140^0 \Rightarrow$ the liquid column will fall.

(iii)   Under microgravity, $H_0 \to \infty$ i.e., the capillary may continue to rise or fall (Levich 1962).

We next estimate the rate at which the capillary will rise to the height H_0 and the time it takes to attain it. To simplify the calculation, we assume that the velocity profile at any instant of time is given by the Poiseuille profile, i.e., fully developed profile. Under this assumption, the axial pressure gradient balances the shear stress.

$$\frac{\partial P}{\partial y} = \mu\left(\frac{\partial^2 u}{\partial r^2} + \frac{1}{r}\frac{\partial u}{\partial r}\right) = \beta \ , \quad \text{the applied pressure gradient} \qquad (1.17)$$

$$\therefore \ u = \frac{-\beta}{4\mu}\left(a^2 - r^2\right) \qquad (1.18)$$

and also the average velocity of the fluid

$$u_{avg} = \frac{u_{max}}{2} \qquad \text{where} \qquad u_{max} = -\frac{\beta a^2}{4\mu} \qquad (1.19)$$

$$\therefore \quad u_{avg} = \frac{dH}{dt} = -\frac{\beta a^2}{8\mu} \qquad (1.20)$$

Now, during the capillary rise, the pressure gradient in the rising column of liquid depends on the hydrostatic pressure due to the increasing height of the liquid column and the capillary pressure difference across the meniscus which separates the liquid and the ambient gas. The pressure gradient may be assumed constant along the capillary and equal to,

$$\frac{dP}{dy} = \beta = -\frac{\Delta P}{H} = -\frac{2\sigma\cos\theta/a - \varrho g H}{H} = \text{constant} \qquad (1.21)$$

and substituting Eq. (1.21) into Eq. (1.20), we get

$$u_{avg} = \frac{dH}{dt} = \frac{1}{8}\frac{\sigma}{\mu}\left[\left(2\frac{a}{H}\cos\theta\right) - \frac{\varrho g a^2}{\sigma}\right] \qquad (1.22)$$

Now, defining a Bond number,

$$Bo = \frac{\varrho g a^2}{\sigma} \qquad (1.23)$$

and a capillary number $Ca = \frac{\mu u}{\sigma} = \frac{\text{viscous force}}{\text{capillary force}}$

$$Ca = \frac{1}{8}\left[\left(2\frac{a}{H}\cos\theta\right) - Bo\right] \qquad (1.24)$$

Integrating Eq. (1.22) we obtain,

$$t = \frac{8\mu}{a^2\varrho g}\left[H_0\ln\left(\frac{H_0}{H_0 - H}\right) - H\right] \qquad (1.25)$$

Remarks:

(i)   The assumption of a fully developed velocity profile is reasonable only when the initial transient is over, i.e., $t > O\left(\dfrac{\varrho a^2}{\mu}\right)$. However, this time scale is small compared

to the overall rise time for capillary tube flow.

(ii) As $t \to \infty$, $H \to H_0$

(iii) Asymptotically, $\dfrac{H}{H_0} \to 1 - \exp\left(-\dfrac{t}{\tau}\right)$ where $\tau = \dfrac{8\mu H_0}{\varrho g a^2} = 8\,\dfrac{\mu}{\sigma}\,\dfrac{H_0}{Bo}$

2. JET BREAKUP AND DROP FORMATION

In this section we examine the breakup of liquid jets into discrete drops by the action of capillary and gravitational surface waves (Rayleigh 1964, Lamb 1945, Levich 1962, Yih 1965, Bogy 1977, Lefebvre 1989 and Probstein 1989). This subject is too vast to be completely covered here. The motivation of this section is to illustrate the interplay between gravity and capillarity using a well known example.

Lord Rayleigh was the first to demonstrate by a hydrodynamic stability analysis that a liquid jet is unstable to small perturbations and breaks up into segments which under the action of surface tension, form individual drops. Depending on the jet velocity four different regimes can be identified.

(i) At low jet velocity: drops form individually at the nozzle tip and grow in size until the weight overcomes the surface tension force and the drop is released.

(ii) At increased velocity: a continuous jet forms which tends to oscillate and form "necks" with subsequent breakup to produce discrete drops.

(iii) At higher velocity: the jet takes on a ruffled appearance immediately outside the nozzle and the drops formed are less uniform than in stage (ii).

(iv) At high velocities: the jet breakup point retreats to the nozzle tip and a nonuniform spray is issued.

Fundamentally, there are two types of surface waves from the hydrodynamic point of view.

(i) When the waves are driven by a balance between the fluid's inertia and the restoring force of gravity, they are termed surface gravity waves.

(ii) If surface tension, rather than gravity is the restoring force, the waves are called capillary waves

Both types of waves share some common characteristics:

(i) They are confined to a distance of about a wavelength from the free surface. Hence they are called surface waves.

(ii) The speed of propagation depends on the individual wavelengths; these waves are dispersive.

(iii) They are waves of small amplitude, and accordingly they can be usefully studied by linear analysis.

2.1 Basic Notions

We shall discuss situations where the jet does not carry a high momentum, making a linear analysis applicable. With the following definition:

time: angular frequency $\omega = \frac{2\pi}{\tau}$: where τ is the period of the wave

space: wave number $k = \frac{2\pi}{\lambda}$: where λ is the wavelength

wave speed: $c = \frac{\lambda}{\tau}$

Consider the perturbed momentum equations simplified by assuming (1) inviscid flow, (2) small disturbances, i.e., neglecting the nonlinear terms

$$\varrho \frac{\partial \vec{q}}{\partial t} = - \nabla P_e \tag{2.1}$$

where $\vec{q}$ is the velocity vector, and P_e is the excess pressure over the undisturbed value. Due to the small amplitude assumption, i.e., $a \ll \lambda$, $\frac{\partial P_e}{\partial x} \ll \frac{\partial P_e}{\partial y}$, thus the problem can be reduced to a one dimensional pressure imbalance. The undisturbed hydrostatic pressure distribution, P_o, defined by considering that the undisturbed liquid surface is located along $y = 0$, is $P_0 = P_a - \varrho g y$, where P_a is the undisturbed atmospheric pressure.

According to the standard linear analysis, the vertical displacement of the free surface, $f(x, t)$, is given by

$$f(x, t) = a \sin(\omega t - kx) \tag{2.2a}$$

$$\therefore \quad f_{xx} = - k^2 f \tag{2.2b}$$

The pressure at any point in the disturbed liquid can be split into the undisturbed hydrostatic part and an excess pressure as follows:

$$P = P_0 + (P_e)_g \tag{2.3}$$

and $P = P_a$, the ambient pressure, at $y = 0$. Then, from the Young–Laplace equation specialized to a 2–D planar case, we have:

$$(P_e)_\sigma = \frac{\sigma}{R} \tag{2.4}$$

The excess hydrostatic pressure: $\qquad (P_e)_g = \varrho g f \tag{2.5}$

and from the geometry of the free surface,

$$\frac{1}{R} = - \frac{f_{xx}}{\left(1 + f_x^2\right)^{3/2}} \approx - f_{xx} \text{ since } f_x \ll 1 \tag{2.6}$$

with the center of curvature on the concave side of the interface. Then the total excess pressure is $\qquad P_e = (P_e)_g + (P_e)_\sigma \tag{2.7}$

Combining the equations (2.2b) through (2.7) and defining an effective gravity, g_{eff}, we get,

$$P_e = (P_e)_g + (P_e)_\sigma = \varrho \left(g + \frac{k^2 \sigma}{\varrho} \right) f = \varrho \, g_{eff} \, f \tag{2.8}$$

Based on the relative magnitudes of g and $\frac{k^2 \sigma}{\varrho}$, we can decide the dominant physical mechanism.

(1) $g \gg \dfrac{k^2\sigma}{\varrho}$:   gravity dominated $\Rightarrow \lambda \gg 2\pi\left(\dfrac{\sigma}{\varrho g}\right)^{1/2}$:   long waves

(2) $g \ll \dfrac{k^2\sigma}{\varrho}$:   capillary dominated $\Rightarrow \lambda \ll 2\pi\left(\dfrac{\sigma}{\varrho g}\right)^{1/2}$: short waves

Now we can solve the linear stability problem of the 2–D planar jet. We start by assuming the flow to be irrotational. Then,

$$\nabla^2\phi = 0 \tag{2.9}$$

Hence from the x–momentum equation we obtain,

$$P_e = -\varrho\frac{\partial\phi}{\partial t} \tag{2.10}$$

and by applying this to the free surface we get,

$$\left(\frac{\partial\phi}{\partial t}\right)_{y=0} = -g_{eff}\, f \tag{2.11}$$

From the linearized boundary condition, $\dfrac{Df}{Dt} = v$ at $y = f(x)$, we get,

$$\frac{\partial f}{\partial t} = v = \left(\frac{\partial\phi}{\partial y}\right)_{y=f} \approx \left(\frac{\partial\phi}{\partial y}\right)_{y=0} \tag{2.12}$$

i.e.,

$$\frac{\partial f}{\partial t} = \left(\frac{\partial\phi}{\partial y}\right)_{y=0} \tag{2.13}$$

The overall mathematical statement is the Laplace equation along with the boundary conditions given by Eq. (2.11) and Eq. (2.13).

Now, with the surface displacement given by Eq. (2.2a), the velocity potential is of the form,

$$\phi = A\, e^{ky}\cos(\omega t - kx) \tag{2.14}$$

where A is a constant.

(1)   The above form of the velocity potential satisfies the boundary conditions.
(2)   A may be determined by letting ϕ satisfy the Laplace equation.
(3) $\phi \to 0$ as $y \to -\infty$ due to the exponential term exp(ky). In fact, the functional form of the velocity potential shows that the disturbance dies off exponentially with increasing depth from the free surface. This is a confirmation of the surface nature of the waves.
(4)   From the boundary conditions given by Eq. (2.11) and Eq. (2.13), we get

$$\left(\frac{\partial^2\phi}{\partial t^2}\right)_{y=0} = -g_{eff}\left(\frac{\partial\phi}{\partial y}\right)_{y=0} \Rightarrow \omega^2 = k\, g_{eff} \tag{2.15}$$

which shows that the wave speed is wave number dependent and thus the wave is dispersive.

The wave speed is $c = \dfrac{\lambda}{\tau} = \dfrac{2\pi/k}{\tau} = \dfrac{2\pi/\tau}{k} = \dfrac{\omega}{k}$. Noting that $\omega^2 = k\, g_{eff}$

$$\therefore \quad c = \sqrt{\frac{g_{eff}}{k}} = \sqrt{\frac{g + \sigma k^2/\varrho}{k}} = \left(\frac{g}{k} + \frac{\sigma k}{\varrho}\right)^{1/2} \tag{2.16}$$

$$\Rightarrow \quad c = \left(\frac{g\lambda}{2\pi} + \frac{2\pi\sigma}{\varrho\lambda}\right)^{1/2} \tag{2.17}$$

i.e., for gravity waves the wave speed increases with wavelength whereas for capillary waves the wave speed decreases with wavelength. It may also be noted that in the absence of viscosity and constant surface tension, a 2–D planar surface wave will propagate undamped and unamplified since the solution is a pure sinusoid along the x–direction.

2.2 Cylindrical Jets

The case of the instability of a cylindrical jet was first investigated by Lord Rayleigh. We now proceed to perform the same analysis as done by Rayleigh.

(i)   Within the context of irrotational flow, the velocity potential satisfies the Laplace equation in cylindrical coordinates,

$$\frac{1}{r}\frac{\partial}{\partial r}\left(r\frac{\partial\phi}{\partial r}\right) + \frac{1}{r^2}\frac{\partial^2\phi}{\partial\theta^2} + \frac{\partial^2\phi}{\partial z^2} = 0 \tag{2.18}$$

Also the radial displacement of the disturbed jet surface may be expressed functionally as

$$r = f(z, \theta, t) \tag{2.19}$$

(ii)   The appropriate boundary conditions can be derived from the linearized momentum equations and the kinematic boundary conditions. The linearized kinematic boundary condition reads,

$$\frac{Df}{Dt} \approx \frac{\partial f}{\partial t} = v_r|_{r=a} = \left.\frac{\partial\phi}{\partial r}\right|_{r=a} \tag{2.20}$$

where a is the jet diameter.

(iii) From the Young–Laplace equation, we get

$$(P_e)_\sigma = -\sigma\left(\frac{\partial^2 f}{\partial z^2} + \frac{1}{a^2}\frac{\partial^2 f}{\partial\theta^2} + \frac{f}{a^2}\right)_{r=a} \tag{2.21}$$

For the typical jet, the diameter is small and usually we don't have to consider the hydrostatic pressure $(P_e)_g$.

(iv)   From the streamwise momentum equation, after linearization, we obtain

$$P_e = -\varrho\frac{\partial\phi}{\partial t} \tag{2.22}$$

A general form of the solution is

$$\phi(r, \theta, z, t) = \psi(r)\, e^{\beta t} \cos(kz + n\theta) \tag{2.23}$$

where

> k : the real axial wave number
>
> n : an integer
>
> β : the amplification factor. If this is positive, the instability grows and if

this is purely imaginary we have a travelling wave with unchanging magnitude.

Substituting Eq. (2.23) into the Laplace equation, we get

$$\frac{d^2\psi}{dr^2} + \frac{1}{r}\frac{d\psi}{dr} - \left(\frac{n^2}{r^2} + k^2\right)\psi = 0 \tag{2.24}$$

with the general solution,

$$\psi(r) = A \text{ In }(kr) + B \text{ Kn }(kr) \tag{2.25}$$

where the In and Kn are modified Bessel functions of the first and the second kind respectively. Now applying the boundary conditions (2.20 – 2.21), we get, after some manipulation,

$$f = \frac{Ak \text{ In}'(ka)}{\beta} e^{\beta t} \cos(kz + n\theta) \tag{2.26}$$

also, it can be shown that $\beta^2 \approx \dfrac{\sigma}{\varrho a^3}$ which is similar to the plane capillary wave where $\omega^2 \approx \dfrac{\sigma}{\varrho\lambda^3}$.

Following the concept of free energy minimization discussed in Chapter VII, a fundamental reason behind the breakup mechanism studied here is that the jet, under the action of surface tension tends to minimize its surface area to drive the free energy to a minimum. To make a rough estimation, consider a cylindrical jet of radius a, and length l. Then its volume is $\pi a^2 l$ and its surface area is $\pi a^2 + 2\pi al$. If the jet breaks up into n spherical drops of radius R, then from mass conservation we have $n\left(\frac{4}{3}\pi R^3\right) = \pi a^2 l$. Minimization of surface energy requires that $n\left(4\pi R^2\right) < 2\pi al$. This implies that

$$R > \frac{3}{2}a \tag{2.27}$$

According to Rayleigh's analysis, if the interval between the drops is taken as the fastest growing wavelength (which can be shown to be $\lambda \approx 9a$), then $R \approx 1.89a$, which is quite close to the result from the thermodynamic considerations.

3. BENARD CONVECTION

In 1900 Henri Benard studied fluid motions in a shallow fluid layer (0.5 to 1 mm deep) uniformly heated from below. He observed cellular flows produced on the free surface of the liquid film and cellular deformation associated with the motion. The convection cells were hexagonal in shape. In 1916 Lord Rayleigh performed the stability analysis of a thin layer of fluid confined between two flat, rigid, horizontal plates (no free surfaces) heated uniformly from below. He found the existence of a minimum temperature gradient required for the onset of convection. This is why we call this type of convection, Rayleigh–Benard convection. However, it is observed that the critical temperature gradient for the onset of convection, as predicted by Rayleigh, far exceeds those observed by Benard. The prediction of Rayleigh is about 10^4 to 10^5 times higher than the experimentally observed values.

An explanation for this discrepancy was advanced by Block (1956). He did an experiment where a thin layer of hydrocarbon liquid is heated from below. Then he used a needle which was wetted with a silicone oil that was insoluble in the hydrocarbon, to touch the surface of the hydrocarbon. The Benard cells were observable until the needle touched the free surface. However, as soon as the needle touched the surface, the silicone oil floated off and as soon as the silicone oil passed over the Benard cell, the surface deformation disappeared and the convection stopped. Since the density of the original fluid could not be affected so rapidly by the silicone oil, the explanation was that surface tension plays a significant role in the formation of the hexagonal convection cells.

One of the first theoretical analyses of such a phenomenon was given by Pearson (1958). The reason advanced was that nonuniformity in surface tension causes convection to occur. This nonuniformity in surface tension is caused by nonuniformity of temperature and/or surfactant concentration. In the words of C.G.M. Marangoni (1871), "If for any reason, differences of surface tension exist along a free liquid surface, the liquid will flow towards the region of higher surface tension."

After Block's work, some notable early contributions in this area were the works of Young, Goldstein and Block (1959) and Scriven and Sterling (1964). A schematic illustration of the flow patterns associated with Benard cells is given in Figure 17, which depicts the various flow patterns formed under different conditions.

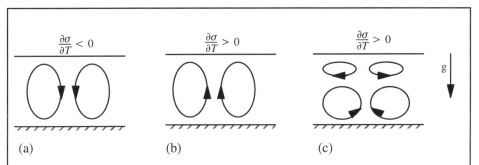

Figure 17. Sketch showing the various possibilities associated with Benard convection cells. A thin layer of liquid with a free surface at the top is heated from below, giving rise to a characteristic cellular convection pattern. In (a) the surface tension decreases with temperature. In (b) and (c) the surface tension increases with temperature. In (b) the surface tension effect is overwhelmed by the buoyancy driven convection, whereas in (c) the surface tension gradient results in a pair of contra–rotating convection cells near the free surface.

3.1 Surface Tension Induced Convection

The presence of an interface between two fluids can influence the motion of the fluids when

(1)   the interface has a finite, non–zero curvature that is different from that at equilibrium, or

(2)   when the surface tension varies from point to point.
The force balance at any point on the interface is

$$\left[P^{(2)} - P^{(1)} + \sigma\left(\frac{1}{R_1} + \frac{1}{R_2}\right)\right] n_i =$$

$$\left[\mu^{(2)}\left(\frac{\partial q_i^{(2)}}{\partial x_k} + \frac{\partial q_k^{(2)}}{\partial x_i}\right) - \mu^{(1)}\left(\frac{\partial q_i^{(1)}}{\partial x_k} + \frac{\partial q_k^{(1)}}{\partial x_i}\right)\right] n_k - \frac{\partial\sigma}{\partial x_i} \tag{3.1}$$

where n_i are the components of the unit vector normal to the surface and directed into the interior of fluid (1), P is the static pressure, σ is the surface tension and the q_i are the velocity components. With no fluid motion and a constant surface tension, this reduces to the Young–Laplace equation. In component form this equation becomes,
Normal component:

$$P^{(2)} - P^{(1)} + \sigma\left(\frac{1}{R_1} + \frac{1}{R_2}\right) = 2\mu^{(2)}\frac{\partial q_n^{(2)}}{\partial n} - 2\mu^{(1)}\frac{\partial q_n^{(1)}}{\partial n} \tag{3.2}$$

Tangential component:

$$\mu^{(2)}\left(\frac{\partial q_n^{(2)}}{\partial s} + \frac{\partial q_s^{(2)}}{\partial n}\right) - \mu^{(1)}\left(\frac{\partial q_n^{(1)}}{\partial s} + \frac{\partial q_s^{(1)}}{\partial n}\right) = \frac{\partial\sigma}{\partial s} \tag{3.3}$$

where s is the direction tangential to the interface.
        The normal component is the Young–Laplace equation modified to account for the normal stress terms due to the velocity gradient. This normal stress component of the force balance equation is always present. On the other hand unless there is a surface tension gradient along the interface, the tangential component of the force balance equation is not meaningful.
        We know that the surface tension varies with
(1)   The surface concentration of the substances absorbed on the boundary
(2)   The temperature distribution on the boundary.
Thus we need to know the concentration and the temperature distribution on the surface in order to solve the fluid dynamics problem. These distributions are determined by fluid dynamic considerations. Thus we need to solve a coupled problem of heat, mass and momentum transfer.

3.2 Linear Stability Analysis

        A linear stability analysis (Chandrasekar 1961, Drazin and Reid 1981), is a useful first step in obtaining a better understanding of the various flow phenomena that we will encounter in this and in subsequent sections. We first consider the case of (thermal) buoyancy driven flow without surface tension effects. The governing equations, including the Bousinnesq approximation, are as follows:

continuity: $\dfrac{\partial q_i}{\partial x_i} = 0$ (3.4)

momentum:
$$\frac{\partial q_i}{\partial t} + q_j \frac{\partial q_i}{\partial x_j} = -\frac{1}{\varrho_m} \frac{\partial P}{\partial x_i} + \frac{\mu}{\varrho_m} \frac{\partial^2 q_i}{\partial x_j^2} + \frac{\varrho}{\varrho_m} g_i \tag{3.5}$$

energy:
$$\frac{\partial T}{\partial t} + q_j \frac{\partial T}{\partial x_j} = a \frac{\partial^2 T}{\partial x_j^2} \tag{3.6}$$

where q_i are the velocity components, x_i are the coordinate directions, P is the static pressure, T is the temperature, μ is the dynamic viscosity and a is the thermal diffusivity. A rest state designated by an overbar is permissible if

$$\frac{\partial \overline{P}}{\partial x} = \frac{\partial \overline{P}}{\partial y} = \frac{\partial^2 \overline{T}}{\partial x_j^2} = 0 \text{ and } \frac{\partial \overline{P}}{\partial z} = -\overline{\varrho} \, g \tag{3.7}$$

The coefficient of thermal expansion is defined as

$$\beta = -\frac{1}{\varrho_m} \left(\frac{\partial \varrho}{\partial T} \right)_{T_m} \tag{3.8}$$

Note that the density $\varrho_m = \varrho(T_m)$ and the temperature $T_m = 0.5 \, (T_1 + T_2)$. Thus

$$\overline{\varrho} = \varrho_m - \varrho(z) = \varrho_m (1 - \beta \Delta T) \tag{3.9}$$

where $\Delta T = T_1 - T_2$ and β is the coefficient of thermal expansion. At the rest state, the temperature distribution is linear, i.e.,

$$\overline{T} = T_1 - \frac{\Delta T}{h} z \tag{3.10}$$

Hence,
$$\frac{\partial \overline{P}}{\partial z} = -\varrho_m \, g \left[1 - \beta \left(T_1 - \frac{\Delta T}{h} z - T_m \right) \right] \tag{3.11a}$$

and
$$\overline{P} = -\varrho_m \, g \, z + \varrho_m \, g \, \beta \, (T_1 - T_m) \, z - \frac{1}{2} \varrho_m \, g \, \beta \, \frac{\Delta T}{h} z^2 + \text{constant} \tag{3.11b}$$

We start the perturbation analysis by first assuming that

$$q_i = 0 + q_i' \tag{3.12a}$$
$$P = \overline{P} + P' \tag{3.12b}$$
$$T = \overline{T} + T' \tag{3.12c}$$

Assuming these perturbations to be small, a set of linearized equations result,

continuity:
$$\frac{\partial q_i'}{\partial x_i} = 0 \tag{3.13}$$

momentum:
$$\frac{\partial q_i'}{\partial t} = -\frac{1}{\varrho_m} \frac{\partial P'}{\partial x_i} + \frac{\mu}{\varrho_m} \frac{\partial^2 q_i'}{\partial x_j^2} \tag{3.14}$$

energy:
$$\frac{\partial T'}{\partial t} + w' \frac{\partial \overline{T}}{\partial z} = a \frac{\partial^2 T'}{\partial x_j^2} \tag{3.15}$$

where w' is the velocity component in the z direction. The convective terms are not retained in the momentum equations as they are small compared to the rest of the terms, being products of the perturbation velocities, which are by definition small quantities. In the energy equation, only the term involving w' survives the linearization process.

3.3 Normal Mode Analysis

We define

$$T' = \Delta T \hat{T}(Z) \exp\left(i\gamma_1 X + i\gamma_2 Y - i\gamma_3\left(\frac{\mu}{\varrho_m h^2}\right) \tau \right) \tag{3.16}$$

$$w' = \left(\frac{a}{h}\right) \hat{w}(Z) \exp\left(i\gamma_1 X + i\gamma_2 Y - i\gamma_3\left(\frac{\mu}{\varrho_m h^2}\right) \tau \right) \tag{3.17}$$

$$u' = \left(\frac{i\gamma_1}{k^2}\right) \left(\frac{a}{h}\right) \frac{d\hat{w}(Z)}{dZ} \exp\left(i\gamma_1 X + i\gamma_2 Y - i\gamma_3\left(\frac{\mu}{\varrho_m h^2}\right) \tau \right) \tag{3.18}$$

$$v' = \left(\frac{i\gamma_2}{k^2}\right) \left(\frac{a}{h}\right) \frac{d\hat{w}(Z)}{dZ} \exp\left(i\gamma_1 X + i\gamma_2 Y - i\gamma_3\left(\frac{\mu}{\varrho_m h^2}\right) \tau \right) \tag{3.19}$$

$$P' = \left(\frac{\varrho_m a \mu}{\varrho_m h^2}\right) \hat{P}(Z) \exp\left(i\gamma_1 X + i\gamma_2 Y - i\gamma_3\left(\frac{\mu}{\varrho_m h^2}\right) \tau \right) \tag{3.20}$$

where $k^2 = \gamma_1^2 + \gamma_2^2$ and x, y, z have been non–dimensionalized by h to yield X, Y, and Z and time, t, has been non–dimensionalized by $\varrho_m h^2/\mu$ to yield τ. Note that u', v' and w' have been chosen so as to satisfy the continuity equation, Eq. (3.13). Then the Eqs. (3.13–3.15) become

$$k^2 \hat{P} = (D^2 - k^2 + \gamma_3) D\hat{w} \tag{3.21}$$

$$D\hat{P} = (D^2 - k^2 + \gamma_3) \hat{w} + Ra \, \hat{T} \tag{3.22}$$

$$\hat{w} + (D^2 - k^2 + \gamma_3 Pr) \hat{T} = 0 \tag{3.23}$$

where $D \equiv \dfrac{d}{dz}$, and

$$Ra \equiv \frac{\varrho_m g \beta \Delta T h^3}{a\mu} \tag{3.24a}$$

is the thermal Rayleigh number and

$$Pr \equiv \frac{\mu}{\varrho_m a} \tag{3.24b}$$

is the Prandtl number.

Eliminating $\hat{P}$ and $\hat{T}$ from the above equations, we get

$$(D^2 - k^2 + \gamma_3 Pr) (D^2 - k^2 + \gamma_3) (D^2 - k^2) \hat{w}(Z) = - k^2 Ra \, \hat{w}(Z) \tag{3.25}$$

Different stability limits result from different types of boundary conditions. Here we consider the case of two free and perfectly conducting boundaries on the top and bottom. At a boundary we have $\hat{T} = \hat{w} = 0$. Then from the momentum and the energy equations we also obtain

$$D^2 \hat{w} = D^2 \hat{T} = 0 \tag{3.26}$$

at the boundary. Repeated differentiation of the momentum equations and application of the boundary conditions gives

$$D^{2n} \hat{w} = 0 \tag{3.27}$$

at an interface. Eq. (3.25) along with boundary conditions given by Eq. (3.26) and Eq. (3.27) at $z = 0$ and 1 has an eigensolution

$$\hat{w}(Z) = A \, \sin n\pi Z \quad , \qquad n \neq 0 \tag{3.28}$$

Thus we obtain,

$$(n^2\pi^2 + k^2 - \gamma_3 Pr)(n^2\pi^2 + k^2 - \gamma_3)(n^2\pi^2 + k^2) = k^2 \, Ra \tag{3.29}$$

For neutral stability, $\gamma_3 = 0$. Hence, $Ra = \dfrac{(n^2\pi^2 + k^2)^3}{k^2}$.

From the minimum condition that $\dfrac{dRa}{dk^2} = 0$, we get the critical Rayleigh number,

$$Ra = \frac{27\pi^4}{4} = 657.5 \tag{3.30}$$

It is noted that the critical Rayleigh number is independent of the Prandtl number.

3.4 Buoyancy with Surface Tension

With surface tension on the top boundary, local shear stresses are generated by the variation of the surface tension due to the thermal gradient on the liquid surface. The BC's now become,

(i)   at $z = 0$, a rigid and conducting boundary.,

$$\hat{w}(0) = \frac{d\hat{w}(0)}{dz} = \hat{T}(0) = 0 \tag{3.31}$$

(ii)   at $z = h$, the upper boundary is free and we have the thermal flux balance,

$$\mathcal{K} \frac{\partial T}{\partial z}\bigg|_{z=h} = \mathcal{H}\left(T_{liq} - T_\infty\right) \tag{3.32}$$

where $\mathcal{H}$ is the heat transfer coefficient. In non–dimensional form and after the normal mode analysis, this becomes,

$$D\hat{T} = -Bi \, \hat{T} \tag{3.33}$$

where

$$Bi \equiv \frac{\mathcal{H} \cdot h}{\mathcal{K}} \tag{3.34}$$

is the Biot number.

(iii)   the shear stress balance at the interface at $z = h$ is

$$\mu\left(\frac{\partial w}{\partial x} + \frac{\partial u}{\partial z}\right) = \frac{\partial \sigma}{\partial x} = \frac{d\sigma}{dT} \cdot \frac{\partial T}{\partial x} \tag{3.35}$$

In the perturbed form, this becomes

$$\mu\left(\frac{\partial w'}{\partial x} + \frac{\partial u'}{\partial z}\right) = \frac{d\sigma}{dT} \cdot \frac{\partial T'}{\partial x} \tag{3.36}$$

and after the normal mode analysis

$$(D^2 + k^2)\,\hat{w} = -\,k^2 Ma\,\hat{T} \tag{3.37}$$

where

$$Ma \equiv \frac{\frac{d\sigma}{dT}\cdot \Delta T \cdot h}{\mu\alpha} \tag{3.38}$$

is the thermal Marangoni number. Other boundary conditions are $\hat{w} = 0$ at $z = h$.

The solutions depend on Ma, Bi and Ra. But the neutral stability limit is independent of Pr. For a given Bi and Ra, there is a critical Ma. For Bi = Ra = 0, the critical Ma = 80. For the limiting case of Bi approaching infinity, the critical Ma also approaches infinity. Also it is interesting to note that Ma is proportional to h whereas Ra is proportional to h^3. Thus for thin liquid layers the surface tension effects predominate whereas for thick liquid layers the buoyancy effect predominates. An important implication is that a liquid "layer" heated from below may exhibit instabilities even for very small Ra, or even in the absence of gravity, if the Ma exceeds a critical value. It may be noted that the contribution of Ma and Ra to instability reinforce each other and are tightly coupled. As a general approximation (Nield 1958),

$$\frac{Ra_c}{Ra_c^0} + \frac{Ma_c}{Ma_c^0} \approx 1 \tag{3.39}$$

where Ma_c^0 is the critical Ma with Ra = 0 and Ra_c^0 is the critical Ra with Ma = 0. Figure 18 shows the insensitivity of this relation with Bi. More comprehensive discussions of this topic can be found in Chandrasekar (1961), Drazin and Reid (1981), Platten and Legros (1984) and in Davies (1987).

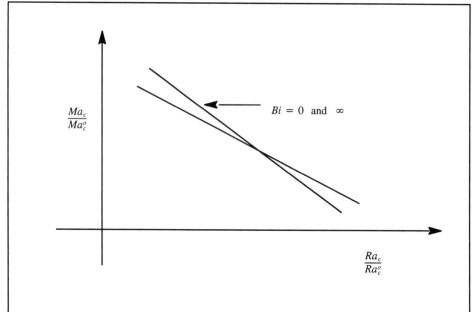

Figure 18. Schematic showing the insensitivity of the relation given by Eq. (3.39) to Bi.

3.5 Scaling Analysis

As has become clear, there are numerous dimensionless parameters involved in a flowfield induced by buoyancy and surface tension effects. To help gain insight into the physical mechanisms involved, we shall conduct some scaling analyses to characterize the relative sizes of the velocity and length scales involved. Some relevant information can be found from the works of Bejan (1984) and Ostrach (1982 and 1983). In the following, x, y, u, v and T designate respectively the dimensional Cartesian coordinates, the corresponding dimensional velocity components and the dimensional temperature. The characteristic length scales used for non–dimensionalization are h, the depth and l, the width, yielding the dimensionless coordinates X and Y. The characteristic velocity scales are U and V, yielding non–dimensional velocity components u^* and v^*. The characteristic temperature scale is ΔT, yielding the dimensionless temperature, T^*.

(I)　　For the case of a horizontal flat interface of nonuniform surface tension, bounding a liquid layer, the shear stress balancing the surface tension is

$$\mu \frac{\partial u}{\partial y} = \frac{\partial \sigma}{\partial x} \tag{3.40}$$

at the interface, assuming that the y–axis is normal to the interface. If the temperature varies along the interface, then the above expression can be written as

$$\mu \frac{\partial u}{\partial y} = \frac{d\sigma}{dT} \frac{\partial T}{\partial x} \tag{3.41}$$

A dimensional argument yields,

$$\mu \frac{U}{\delta_v} \sim \left(\frac{d\sigma}{dT}\right) \frac{\Delta T}{l} \tag{3.42}$$

$$\Rightarrow \quad U \sim \frac{\left(\frac{d\sigma}{dT}\right) \Delta T}{\mu} \cdot \frac{\delta_v}{l} \tag{3.43}$$

where U, l and δ_v are appropriate velocity, horizontal and vertical length scales respectively. The problem of determining the length scale δ_v still remains.

(i)　　For a flow where the viscous effects dominate, the convection effects are negligible, hence,

$$\frac{\partial^2 u}{\partial x^2} \sim \frac{\partial^2 u}{\partial y^2} \quad \Rightarrow \quad \delta_v \sim l \tag{3.44}$$

$$\therefore \quad U \sim \frac{\left(\frac{d\sigma}{dT}\right) \Delta T}{\mu} \tag{3.45}$$

(ii)　　For a highly convective flow, the concept of the boundary layer can be utilized, i.e., δ_v becomes the length scale characterizing the thin layer adjacent to the interface and within which the velocity profile varies

$$\Rightarrow \quad \frac{\delta_v}{l} \sim Re^{-1/2} = \left(\frac{\varrho U l}{\mu}\right)^{-1/2} \tag{3.46}$$

$$\therefore \quad U \sim \left\{ \frac{(\frac{d\sigma}{dT})^2 (\Delta T)^2}{\varrho \mu l} \right\}^{1/3} \tag{3.47}$$

A dimensional analysis carried out with the parameters $\left(\frac{d\sigma}{dT}\right)$, α, ϱ, μ, l and ΔT reveals that two dimensionless groups can be defined. A commonly used choice is the Prandtl number, Pr, and the Marangoni number, Ma, where l is used as the length scale, i.e.,

$$Ma = \frac{\left(\frac{d\sigma}{dT}\right) \Delta T \, l}{\mu \alpha} \tag{3.48}$$

In addition, based on the characteristic velocity scale, a Reynolds number can be defined as

$$Re = \frac{\varrho \, U l}{\mu} \tag{3.49}$$

and if we use the velocity scale derived from the viscous shear stress – surface tension gradient balance shown in Eq. (3.45), we obtain the Marangoni number as

$$Ma = Re \, Pr \tag{3.50}$$

whereas if we use the velocity scale defined in Eq. (3.47), we obtain

$$Ma = Re^{3/2} \, Pr \tag{3.51}$$

Furthermore,

$$\frac{\delta_v}{l} \sim \frac{1}{\sqrt{Re}} \sim \left(\frac{Ma}{Pr}\right)^{1/3} \tag{3.52}$$

(II)   It is instructive to consider flow induced by surface tension gradients inside a box filled with a liquid and open to the atmosphere at the top. A schematic is shown in Figure 19. In the thin layers adjacent to the boundaries, we can effect a balance between the convection and the shear stress terms, in nondimensional form, as follows,

$$u^* \frac{\partial u^*}{\partial X} \approx \mathcal{O}\left(\frac{1}{Re} \frac{\partial^2 u^*}{\partial Y^2}\right) \tag{3.53}$$

where $Re = \frac{\varrho \, U l}{\mu}$, where U and l are characteristic velocity and length scales respectively. The thickness of the thin layer can be defined by a length scale δ_v^* where

$$\delta_v^* = \mathcal{O}\left(\frac{1}{\sqrt{Re}}\right) \tag{3.54}$$

The key issue now is to appropriately select the reference velocity scale, U. It is noted that l is defined by the container geometry and ν is the kinematic viscosity of the liquid. Also, two types of thin–layers can be distinguished, wall layers and the free shear layer adjacent to the open boundary. The reference velocity scale must be appropriately chosen in each layer.

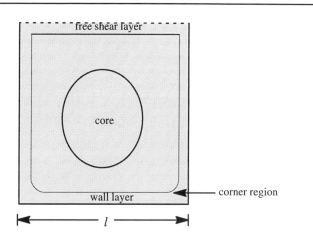

Figure 19. A schematic showing the fluid flow in a square box open to the ambient atmosphere at the top. The fluid flow is driven by the surface tension gradient at the open boundary. The schematic shows the development of thin layers close to the walls and the free shear layer at the open boundary. The corner regions are not included in the thin layer analysis. The velocity scale for the free shear layer is obtained from a shear stress – surface tension gradient balance. However, for the wall layers, the velocity scale is dictated by the velocity scale of the core region.

Region I: The free shear layer

The reference velocity scale can be obtained by the balance of the shear stress and the surface tension gradient at the open boundary,

$$\frac{\partial \sigma}{\partial T} \frac{\partial T}{\partial x} \approx \mu \frac{\partial u}{\partial y} \tag{3.55}$$

$$\Rightarrow \quad U = O\left(\frac{\partial \sigma}{\partial T} \cdot \frac{\Delta T}{l} \cdot \frac{\delta_v}{\mu}\right) \tag{3.56}$$

and using the value of δ_v from Eq. (3.54), we obtain,

$$U = \left\{\frac{(\partial \sigma / \partial T)^2 (\Delta T)^2}{\varrho \mu l}\right\}^{\frac{1}{3}} \tag{3.57}$$

With this choice of the reference velocity scale, we have the following dimensionless scales

$$u^* = O(1) \tag{3.58a}$$

$$v^* = O(\delta_v^*) = O\left(\frac{1}{\sqrt{Re}}\right) \tag{3.58b}$$

the Nusselt number

$$Nu = \mathcal{O}\left(\frac{1}{\delta_v^*}\right) = \mathcal{O}\left(\sqrt{Re}\right) \tag{3.58c}$$

and the stream function

$$\psi^* = \mathcal{O}\left(u^*\delta_v^*\right) = \mathcal{O}\left(\frac{1}{\sqrt{Re}}\right) \tag{3.58d}$$

Here the starred quantities refer to the nondimensional dependent variables

Region II: The wall layers

Formally, the relationships among the u–velocity, v–velocity, the Nusselt number and the stream function remain the same. However, the reference velocity scale will be different. An appropriate choice for the velocity scale may be effected by noting that for the wall layers, a "free stream" velocity may be defined by considering the flow inside the *whole cavity* as being driven by the surface tension gradient. Hence,

$$\frac{\partial \sigma}{\partial T}\frac{\partial T}{\partial x} \sim \mu \frac{\partial u}{\partial y} \tag{3.59}$$

where $x \sim \mathcal{O}(l)$ and $y \sim \mathcal{O}(l)$. Consequently, the velocity scale becomes,

$$U = \frac{\partial \sigma / \partial T \Delta T}{\mu} \tag{3.60}$$

With such a choice of the reference velocity, the nondimensional scales of u, v, Nu, and ψ along the wall layer will be identical to the forms given in Eq. (3.58).

On the other hand, if the reference velocity scale given by Eq. (3.60) is used to scale the whole domain, as is sometimes done in the literature (Zebib *et. al.* 1985), then the nominal orders of magnitude of the dependent quantities, in the free shear layer (region I), now become,

Region I

$$u^* = \mathcal{O}\left(\frac{1}{Re^{1/3}}\right) \tag{3.61a}$$

$$v^* = \mathcal{O}\left(\frac{1}{Re^{2/3}}\right) \tag{3.61b}$$

$$\delta_v^* = \mathcal{O}\left(\frac{1}{Re^{1/3}}\right) \tag{3.61c}$$

the Nusselt number

$$Nu = \mathcal{O}\left(Re^{1/3}\right) \tag{3.61d}$$

and the stream function

$$\psi^* = \mathcal{O}\left(\frac{1}{Re^{2/3}}\right) \tag{3.61e}$$

It is now evident that depending on the choice of the reference scales, different non–dimensional forms of the governing equations can result. However, it is also obvious that not all the choices yield appropriate normalization in the sense that the nondimensional terms are of order one in terms of magnitude. An appropriate choice of the normalization scales may facilitate simplification of the governing equations and improved understanding of the physics involved.

(3)   We can examine the length and the velocity scales of a laminar natural convection boundary layer adjacent to a heated vertical wall. We start by defining the scales. Let the thickness of the thermal boundary layer be δ_T. Then the reference scales for the various physical quantities can be defined as

$$u \sim U, \quad \mathrm{v} \sim V, \quad T \sim \Delta T, \quad x \sim \delta_T, \quad y \sim h \tag{3.62}$$

where ΔT is the temperature differential between the vertical plate and the ambient, and h is the height of the plate. Then we have,

Continuity: $\qquad\qquad\qquad\qquad U/\delta_T \sim V/h \tag{3.63}$

Heat Transfer: $\qquad\qquad U\dfrac{\Delta T}{\delta_T} \sim V\dfrac{\Delta T}{h} \sim a\dfrac{\Delta T}{\delta_T^2} \tag{3.64}$

and in the y–momentum equation,

Inertia terms : $\qquad\qquad \sim \dfrac{UV}{\delta_T}, \ \dfrac{V^2}{h} \tag{3.65a}$

Viscous terms : $\qquad\qquad \sim \dfrac{\mu V}{\varrho \delta_T^2} \tag{3.65b}$

Buoyancy term : $\qquad\qquad \sim g\beta\Delta T \tag{3.65c}$

where μ is the dynamic viscosity of the liquid, β is the coefficient of thermal expansion and a is the thermal diffusivity of the liquid.

For a high Prandtl number fluid, the viscous length scales dominate over the thermal length scales so that the buoyancy term is balanced by the viscous terms.

(i)   For a *high Pr fluid*, as shown in Figure 20, from v–momentum balance,

$$g\beta\Delta T \sim \mu V/\varrho\delta_T^2 \quad \Rightarrow \quad V \sim \varrho g\beta\Delta T\delta_T^2/\mu \tag{3.66}$$

Substituting into the energy balance, $V/h \sim a/\delta_T^2$, we get

$$\left(\dfrac{h}{\delta_T}\right)^4 \sim \dfrac{\varrho g\beta\Delta Th^3}{\mu a} \sim Ra \tag{3.67}$$

$$\Rightarrow \qquad\qquad \left(\dfrac{\delta_T}{h}\right) \sim Ra^{-1/4} \tag{3.68}$$

for the thermal boundary layer thickness where Ra is the Rayleigh number.

Since the buoyancy force is balanced by the shear stress, there will be an outer layer of entrained fluid whose length scale is greater than that of the thermal boundary layer. In this layer, the balance is between fluid inertia and the viscous terms in the momentum balance, yielding,

$$V^2/h \sim \mu V/\varrho \delta_v^2 \tag{3.69}$$

The velocity scale, V, is known from the energy balance. Thus

$$a/\delta_T^2 \sim \mu/\varrho \delta_v^2 \quad \Rightarrow \quad \left(\delta_v/\delta_T\right)^2 \sim \mu/\varrho a \;=\; Pr \tag{3.70}$$

Thus, for the entrained layer, we get

$$\left(\frac{\delta_v}{h}\right) \quad \sim \quad Ra^{-1/4}\; Pr^{1/2} \tag{3.71}$$

(ii)   For a *low Pr fluid*, as shown in Figure 21, the buoyancy term is balanced by the inertia term. Hence, the momentum balance becomes

$$g\beta\Delta T \sim V^2/h \quad \Rightarrow \quad V \sim \sqrt{g\beta h\Delta T} \tag{3.72}$$

where h is the characteristic length of the plate. Substituting this into the energy balance gives

$$V/h \sim a/\delta_T^2 \quad \Rightarrow \quad V \sim ah/\delta_T^2 \quad \Rightarrow \quad g\beta h^3 \Delta T/a^2 \sim (h/\delta_T)^4 \tag{3.73}$$

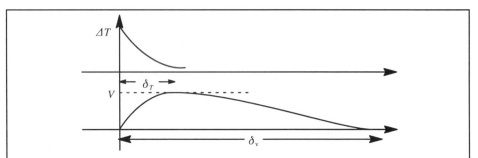

Figure 20. *Schematic of the two layer structure for a natural convection boundary layer (high Prandtl number fluid). δ_T is the length scale of the thermal boundary layer and δ_v is the length scale of the entrained layer.*

$$\Rightarrow \qquad \left(\frac{\delta_T}{h}\right)^4 \sim \frac{\varrho g\beta h^3 \Delta T}{a\mu} \cdot \frac{\mu}{\varrho a} \quad \sim \quad Ra \; Pr \tag{3.74}$$

$$\Rightarrow \qquad \left(\frac{\delta_T}{h}\right) \quad \sim \quad \left(Ra \; Pr\right)^{-1/4} \tag{3.75}$$

for the thermal boundary layer thickness where Pr is the Prandtl number.

At the wall, however, the flow has to meet the no–slip condition. Therefore, close to the wall, there is a thin layer where the viscous forces are of comparable order of magnitude. Effecting an inertia – viscous force balance, we get

$$a/\delta_T^2 \sim \mu/\varrho \delta_v^2 \quad \Rightarrow \quad \left(\delta_v/\delta_T\right)^2 \sim \mu/\varrho a \;=\; Pr \tag{3.76}$$

The thickness for the wall layer is

$$\left(\frac{\delta_v}{h}\right) \quad \sim \quad Ra^{-1/4} \; Pr^{1/4} \tag{3.77}$$

For assessment of the effect of disparate diffusivities between velocity and scalar transport processes, in a different condition, see Probstein (1989, Section 4.2).

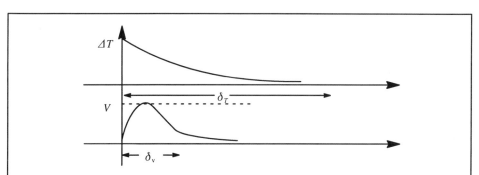

Figure 21. Schematic of the two layer structure for a natural convection boundary layer (low Prandtl number fluid). δ_T is the length scale of the thermal boundary layer and δ_v is the length scale of the wall layer.

We are frequently interested in the interaction of surface tension gradients and the buoyancy effect in fluid motion. One motivation is to be able to predict the differences in flow behavior under microgravity and 1–g conditions. For such problems it is frequently useful to devise alternate characteristic velocity scales as

(a)   Buoyancy scale $\Rightarrow$ $U \sim \sqrt{g\beta_T \Delta Tl}$

(b)   Diffusion scale $\Rightarrow$ $U \sim \dfrac{\alpha}{l}$

The steady Navier–Stokes and transport equations can be written based on the above length and velocity scales, along with consistently chosen density, pressure and temperature reference scales, in the following nondimensional form, subject to the Boussinesq approximation:

(i)   Based on the buoyancy velocity scale

u–momentum:

$$u^* \frac{\partial u^*}{\partial X} + v^* \frac{\partial u^*}{\partial Y} = -\frac{\partial P^*}{\partial X} + \frac{1}{\sqrt{(Gr)_T}} \nabla^2 u^* + T^* \pm \frac{(Gr)_S}{(Gr)_T} \phi^* \tag{3.78}$$

heat transport:

$$u^* \frac{\partial T^*}{\partial X} + v^* \frac{\partial T^*}{\partial Y} = \frac{1}{Pr \sqrt{(Gr)_T}} \nabla^2 T^* \tag{3.79}$$

solute transport:

$$u^* \frac{\partial \phi^*}{\partial X} + v^* \frac{\partial \phi^*}{\partial Y} = \frac{1}{Sc \sqrt{(Gr)_T}} \nabla^2 \phi^* \tag{3.80}$$

(ii)   Based on the diffusion velocity scale

u–momentum:

$$u^* \frac{\partial u^*}{\partial X} + v^* \frac{\partial u^*}{\partial Y} = -\frac{\partial P^*}{\partial X} + Pr \nabla^2 u^* + (Ra)_T \, Pr \, T^* \pm (Ra)_s \frac{Pr^2}{Sc} \phi^* \qquad (3.81)$$

heat transport:

$$u^* \frac{\partial T^*}{\partial X} + v^* \frac{\partial T^*}{\partial Y} = \nabla^2 T^* \qquad (3.82)$$

solute transport:

$$u^* \frac{\partial \phi^*}{\partial X} + v^* \frac{\partial \phi^*}{\partial Y} = \frac{Pr}{Sc} \nabla^2 \phi^* \qquad (3.83)$$

where the thermal Grashof number,

$$(Gr)_T = \frac{\varrho^2 g \beta_T l^3 \Delta T}{\mu^2} \qquad (3.84)$$

and the solute Grashof number

$$(Gr)_S = \frac{\varrho^2 g \beta_\phi l^3 \Delta \phi}{\mu^2} \qquad (3.85)$$

where β_T and β_ϕ are the magnitudes of the coefficient of expansion with respect to the temperature and the solute concentration respectively. The Rayleigh numbers are defined as $(Ra)_T = (Gr)_T \cdot Pr$ and $(Ra)_S = (Gr)_S \cdot Sc$. The Schmidt number is defined as

$$Sc = \frac{\mu}{\varrho \mathfrak{D}} \qquad (3.86)$$

where $\mathfrak{D}$ is the mass diffusivity of the binary mixture. The starred quantities indicate nondimensionalized variables. As indicated by Eqs. (3.78) and (3.81), depending on the properties of the material and the boundary conditions, the body force resulting from the thermal effect can either augment or counteract the solutal effect.

4. CASE STUDIES

It will be instructive to conclude this chapter with a few examples to illustrate fluid flow due to the combination of thermocapillary and buoyancy induced convection. A relatively simple geometry, the square cavity, will be considered so as to bring out the main features of such flows without introducing additional complications due to complex geometry. The solutions to be presented are obtained based on the algorithm described in Chapter V.

4.1 Buoyancy Induced Convection

In this section, two aspects of buoyancy induced flow and transport are illustrated. The first deals with the effect of Rayleigh number variation on the transport characteristics and the second with the effect of Prandtl number variation. Figure 22 shows a schematic of the domain. The Boussinesq approximation has been adopted to simplify the density treatment. The definition of the Grashof and Rayleigh numbers are based on the boundary conditions imposed at the two vertical walls.

1. *Effect of Rayleigh Number:*

The problem under consideration here is that of buoyancy induced convection in a square cavity. The equations solved are Eqs. (3.81) and (3.82) along with the continuity equation. The top and bottom walls are adiabatic while the side walls are held at constant temperatures. The temperature is non–dimensionalized with respect to the wall temperatures. The left wall is held at a temperature of T_1 and the right wall at a temperature

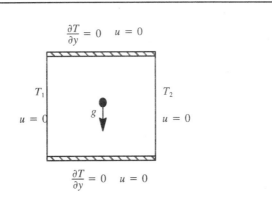

Figure 22. Schematic of the domain illustrating the geometry. The top and bottom boundaries are adiabatic whereas the left and the right boundaries are isothermal and are held at different temperatures. The sides of the box are no–slip walls.

of T_2. The determination of the velocity and temperature field was carried out using a numerical technique with adaptive gridding. Figure 23 shows the steady state computed results. An important feature of the solution is the skew–symmetry of the velocity and the temperature field with respect to the center of the box. Fluid is heated at the left wall, decreases in density and rises up whereas fluid adjacent to the cold right wall loses heat and increases in density. This establishes a clockwise rotating recirculating flow inside the cavity. Results are shown here for Rayleigh numbers ranging from 10^5 to 10^7. With increasing Rayleigh number, it may be observed that the convection strength increases and the isotherms are nearly horizontal in the core flow. The flow close to the walls is of the thin layer type and also a thermal boundary layer can be observed at the isothermal walls. At higher Rayleigh numbers, secondary vortices may be observed apart from the core recirculating flow at the center of the cavity. The heat transfer rate is observed to be enhanced with increasing Rayleigh number as can be deduced from the isotherm spacing at the left and right walls. Increasing the Rayleigh number thus increases the vorticity generation in the flowfield which in turn increases the convection strength and consequently enhances the overall heat transfer rate through the domain. More information can be found in Shyy (1988) and Gebhart *et. al.* (1988).

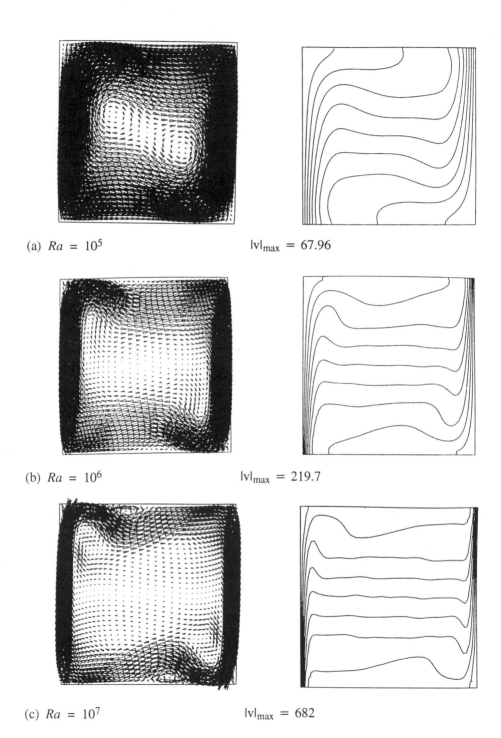

(a) $Ra = 10^5$ $|v|_{max} = 67.96$

(b) $Ra = 10^6$ $|v|_{max} = 219.7$

(c) $Ra = 10^7$ $|v|_{max} = 682$

Figure 23. Effect of Rayleigh number on the velocity field and isotherms at Pr = 0.7

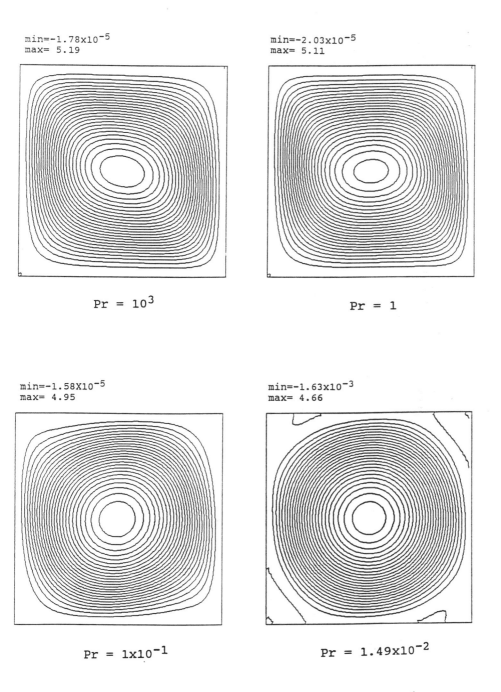

min=-1.78x10^{-5}
max= 5.19

min=-2.03x10^{-5}
max= 5.11

$Pr = 10^3$

$Pr = 1$

min=-1.58X10^{-5}
max= 4.95

min=-1.63x10^{-3}
max= 4.66

$Pr = 1x10^{-1}$

$Pr = 1.49x10^{-2}$

Figure 24. Effect of Prandtl number on the flow field at Ra = 10^4
(a) streamfunction.

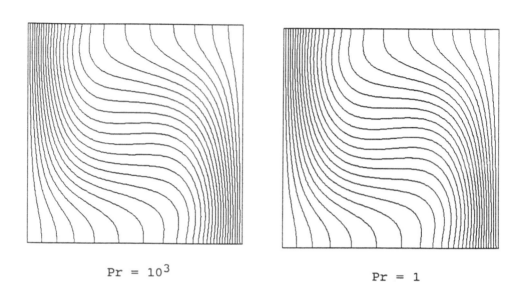

$$Pr = 10^3 \qquad\qquad\qquad Pr = 1$$

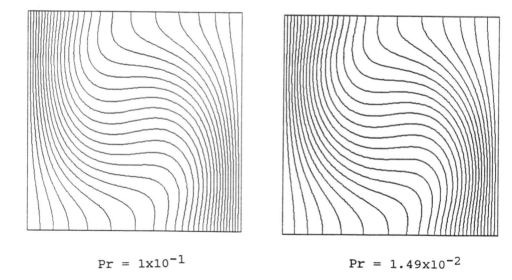

$$Pr = 1 \times 10^{-1} \qquad\qquad Pr = 1.49 \times 10^{-2}$$

Figure 24 (contd). Effect of Prandtl number on the flow field at Ra = 10⁴
  (b) isotherms.

2. *Effect of Prandtl number:*

The model problem to be considered in this section has a domain as depicted in Figure 22 Here, the top and bottom walls are adiabatic while the right wall is held at a non–dimensional temperature of unity and the left wall is held at a temperature of zero, i.e, $T_1 = 0$ and $T_2 = 1$. The objective is to examine the behavior of fluid flow and heat transfer with change in the Prandtl number. The equations solved are Eqs. (3.78) and (3.80) along with the continuity equation.

Figure 24 shows the streamfunction and the isotherms for various Prandtl numbers, at a constant Rayleigh number. The variation of overall convection strength with Prandtl number is quite modest when compared with the variation of the Rayleigh number.

A qualitative explanation of this phenomenon can be given as follows. At constant *Ra*, but varying *Pr*, the thermal diffusivity and kinematic viscosity of the fluid must simultaneously vary in opposing directions and at the same rate, other things remaining the same. Hence, for instance, as *Pr*, increases, the Grashof number, *Gr*, decreases, indicating that *if* the temperature field remains unchanged, then the convection field is weakened due to higher kinematic viscosity. However, a higher *Pr* also means a lower thermal diffusivity, which results in a more non–uniform and concentrated distribution of the thermal field *if* the convection strength remains the same. Ultimately, in this case, the weakened tendency of the convection effect and the strengthened tendency of the thermal buoyancy effect largely compensate each other, causing the overall transport characteristics to be insensitive to the Prandtl number. More information can be found in Shyy and Chen (1990b).

4.2 Surface Tension Driven Convection

The third model problem we consider is that of convection in a square cavity driven by surface tension gradients. Consider a geometry as shown in Figure 25. The top boundary is a free surface and the direction of gravity is so as to counteract the convection due to surface tension gradients. The right wall is heated and held at a constant temperature. The left wall is cooled and is also isothermal. The lower wall is adiabatic. In the cases described, the thermal Grashof number is 3.09×10^6, the Prandtl numbers considered are 0.0149 and 0.7. The Marangoni number is varied from 0 to 8.4×10^3. Figure 26 shows the streamlines, u–velocity contours and isotherms for the various cases considered. From (a) it may be observed that whereas the recirculating eddy in the core is driven by buoyancy, there is a contra–rotating eddy adjacent to the free surface that is driven by the Marangoni effect. In (b) and (c), due to the higher Prandtl number, the contra–rotating eddy moves towards the cold wall.

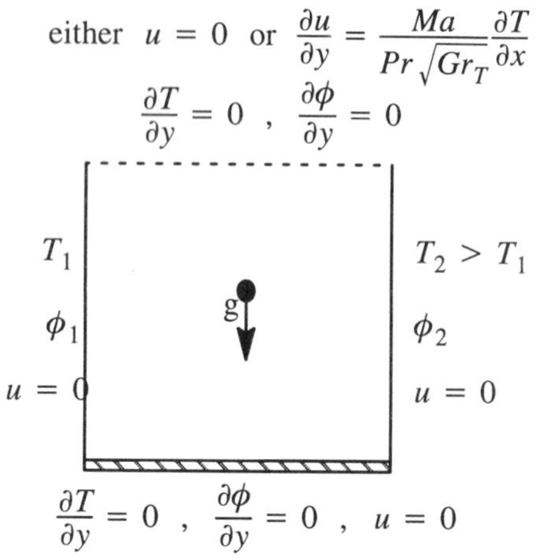

$$\text{either} \quad u = 0 \quad \text{or} \quad \frac{\partial u}{\partial y} = \frac{Ma}{Pr\sqrt{Gr_T}}\frac{\partial T}{\partial x}$$

$$\frac{\partial T}{\partial y} = 0 \quad, \quad \frac{\partial \phi}{\partial y} = 0$$

T_1

ϕ_1

$u = 0$

$T_2 > T_1$

ϕ_2

$u = 0$

g

$$\frac{\partial T}{\partial y} = 0 \quad, \quad \frac{\partial \phi}{\partial y} = 0 \quad, \quad u = 0$$

Figure 25. Schematic of the domain illustrating the geometry. The top and bottom boundaries are adiabatic whereas the left and the right boundaries are isothermal and are held at different temperatures. The sides of the box are no–slip walls except the top boundary which can be either a no–slip boundary or a free surface. In the second case, the temperature gradient gives rise to a surface tension gradient which gives rise to a shear stress on the fluid in the cavity. For this particular case, the Marangoni and the buoyancy effects counteract each other.

Also, the Prandtl number is almost unity, therefore the length scales of heat and momentum transport are comparable whereas in (a) the length scales differ by almost an order of magnitude. This is responsible for the qualitative differences between the solutions for the two different Prandtl numbers. At higher Marangoni numbers, due to the counteracting effect, the overall convection strength decreases as evidenced by the magnitudes of the minimum and maximum values of the streamfunction in each case. More information can be found in Shyy and Chen (1993).

4.3 Double–diffusive Convection in a Square Cavity

We now consider an example of double diffusive flow in which we have a temperature field as well as a solute field both of which contribute to the buoyancy effect. A detailed discussion of double diffusive effects has been given by Turner (1973). Figure 25 illustrates the geometry and boundary conditions that were used in the calculations based on Eqs. (3.78) – (3.80). It is emphasized that in this case the coefficient of solutal expansion is negative and thus the buoyancy effect due to the thermal expansion and that due to the solute counteract each other. In other words, the negative sign is chosen for the ϕ^* term in Eq. (3.78).

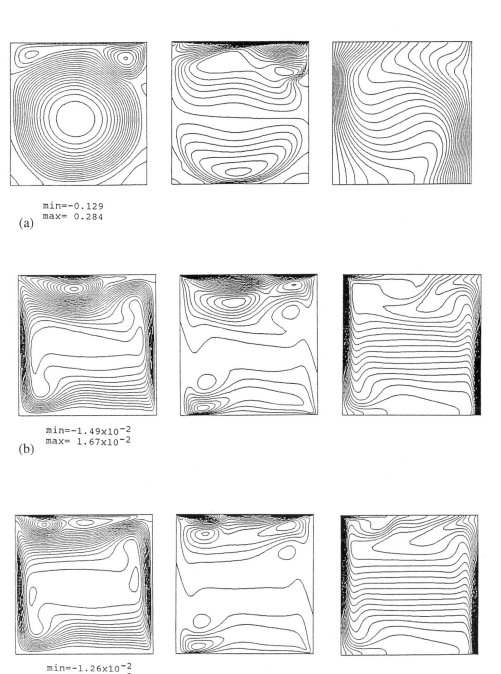

min=-0.129
max= 0.284
(a)

min=-1.49x10^{-2}
max= 1.67x10^{-2}
(b)

min=-1.26x10^{-2}
max= 1.60x10^{-2}
(c)

(left: streamfunction, center: u–velocity contours, right: isotherms)

Figure 26. *Flow in a square cavity under counteracting buoyancy and surface tension gradient. (a) Pr = 0.0149, Ma = 4.2 × 10^3, (b) Pr = 0.7, Ma = 4.2 × 10^3, (c) Pr = 0.7, Ma = 8.4 × 10^3. All cases are for (Gr)$_T$ = 3.09 × 10^6.*

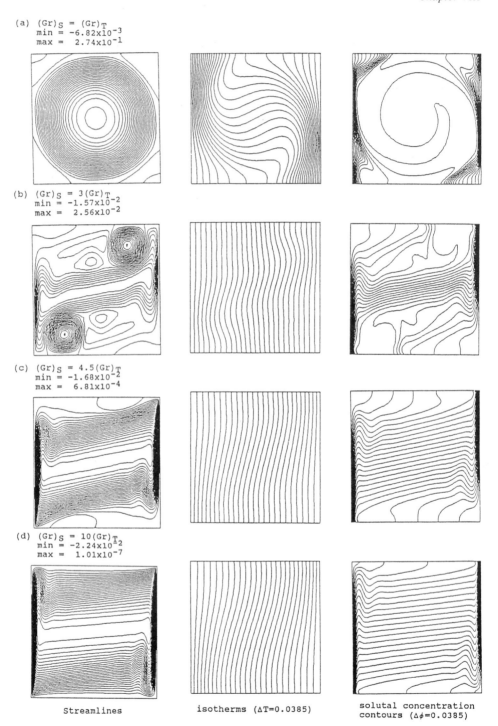

(a) $(Gr)_S = (Gr)_T$
min $= -6.82 \times 10^{-3}$
max $= 2.74 \times 10^{-1}$

(b) $(Gr)_S = 3(Gr)_T$
min $= -1.57 \times 10^{-2}$
max $= 2.56 \times 10^{-2}$

(c) $(Gr)_S = 4.5(Gr)_T$
min $= -1.68 \times 10^{-2}$
max $= 6.81 \times 10^{-4}$

(d) $(Gr)_S = 10(Gr)_T$
min $= -2.24 \times 10^{-12}$
max $= 1.01 \times 10^{-7}$

Streamlines   isotherms ($\Delta T=0.0385$)   solutal concentration
                                                                        contours ($\Delta\phi=0.0385$)

Figure 27. Effect of solutal Grashof number on double–diffusive flow in a square cavity. The Grashof number, $(Gr)_T = 3.09 \times 10^6$, the Prandtl number, $Pr = 1.49 \times 10^{-2}$ and the Schmidt number, $Sc = 1$, no surface tension effects.

Figure 25 shows the domain of the problem considered in this study of double–diffusive fluid flow. The Prandtl number is 1.49×10^{-2} and the Schmidt number is 1. Figure 27 shows the flow solutions without the influence of surface tension. The no–slip condition was imposed on all the four boundaries. The thermal Grashof number is 3.09×10^6. Hence the length scales of the velocity and the solute concentration fields are comparable and are substantially smaller than that of the thermal field. The solute Grashof number was varied and a rich variety of flow patterns have been obtained for solute Grashof numbers, $(Gr)_S = (Gr)_T$, $(Gr)_S = 3 (Gr)_T$, $(Gr)_S = 4.5 (Gr)_T$, and $(Gr)_S = 10 (Gr)_T$. From the values of the Prandtl number and the Schmidt number, we can immediately deduce that the thermal field has a smoother variation over a wider portion of the flow domain. With $(Gr)_S = (Gr)_T$ the differences in the length scales between the solutal concentration and the thermal field creates a convection pattern that is influenced more by the thermal field than the solute field. As the solutal Grashof number increases, the influence of the solutal concentration on the velocity field becomes stronger. At $(Gr)_S = 3 (Gr)_T$ the major eddy is strongly influenced by the solute field and the influence of the thermal field is largely restricted to the two corner eddies at the top right and lower left corners. The corner eddies counteract the major eddy generated by the solute field. As the solute Grashof number increases to $(Gr)_S = 4.5 (Gr)_T$, the corner eddies become weaker and at $(Gr)_S = 10 (Gr)_T$, the corner eddies all but disappear and the convection pattern is dominated by the solute field. However, it must be kept in mind that the overall convection strength is now much smaller because the solute and the thermal fields induce counteracting convective effects. This is also evident from the isotherms which are essentially determined by the conduction effect as the solute Grashof number increases and become relatively insensitive to the convection field. However, the solute field is more responsive to the changes in the convection pattern and the solutal concentration profiles mostly vary in the regions close to the two side walls.

These examples serve to illustrate the effect of various phenomena such as buoyancy and surface tension, on the fluid flow and heat transport. It may be expected that for phase change phenomena which are quite sensitive to fluid flow, heat and mass transport, convection driven by buoyancy, surface tension or forced convection, these parameters play a vital role and considering the complexity of the flow field, even for single phase flows, are not easy to predict *a priori*. Another implication is that in experimental studies, it is quite difficult to simultaneously maintain kinematic and dynamic similitude since the Grashof/Rayleigh number varies as the cube of the length scale whereas the Marangoni number is proportional to the length scale. A practical situation is governed by many more parameters, greater than 10 in number in most cases. In such cases numerical methods offer the best hope for enhancing the predictive capability.

5. FURTHER READING

More detailed information of the various aspects covered in this chapter can be found in Adamson (1990), Arpaci and Larsen (1984), Bejan (1984), Carey (1992), Chandrasekar (1961), Drazin and Reid (1981), Eckert and Drake (1987), Gebhart *et. al.* (1988), Levich (1962), Myshkis *et. al.* (1987), Platten and Legros (1984), Raithby and Hollands (1985), Rowlinson and Widom (1982), and Yih (1965).

CHAPTER IX
MODELING AND COMPUTATIONAL ISSUES IN
PHASE – CHANGE DYNAMICS

1. THE DRIVING FORCE FOR SOLIDIFICATION

In dealing with phase transformations we are often concerned with the difference in free energy between two phases at temperatures different from the equilibrium temperature. For example, if a melt is undercooled by an amount ΔT below the melting temperature, there will be a decrease in the free energy ΔG as shown in the G–T diagram below.

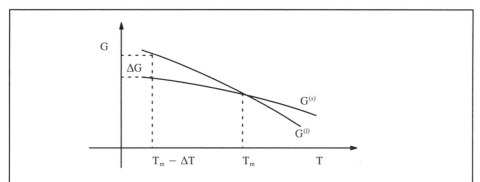

Figure 1. Free energy curve for a phase–change process indicating the decrease in free energy due to undercooling of the melt

This decrease in free energy provides the driving force for solidification. An analogous case can be made for vapor condensation.

The Gibbs free energy of the liquid and the solid at a temperature T are given by

$$G^{(l)} = H^{(l)} - TS^{(l)} \tag{1.1a}$$

$$G^{(s)} = H^{(s)} - TS^{(s)} \tag{1.1b}$$

respectively. Therefore, at the temperature T, we have,

$$\Delta G = \Delta H - T\Delta S \tag{1.2}$$

where $\Delta G = G^{(l)} - G^{(s)}$, $\Delta H = H^{(l)} - H^{(s)}$ and $\Delta S = S^{(l)} - S^{(s)}$. At the equilibrium melting temperature, T_m, the Gibbs free energy of the solid is the same as that of the liquid, i.e., $\Delta G = 0$. Consequently, $\Delta G = \Delta H_{l-s} - T_m \Delta S = 0$

$$\Rightarrow \quad \Delta S = \frac{\Delta H_{l-s}}{T_m} \tag{1.3}$$

where ΔS is the entropy of fusion. We take note of the following facts:

(1)   It is observed experimentally that the entropy of fusion is a constant quite close to R for most metals, i.e., $\Delta S \approx R = 8.3$ J/mol–K.

(2)   For small undercoolings, the difference in C_p between the solid and the liquid phases can be ignored, i.e., ΔH and ΔS are almost independent of temperature. This implies that the free energy,

$$\Delta G = \Delta H - T\Delta S \approx \Delta H_{l-s} - \frac{T}{T_m} \Delta H_{l-s} \tag{1.4}$$

i.e., for small ΔT,

$$\Delta G \approx \frac{\Delta H_{l-s} \Delta T}{T_m} \tag{1.5}$$

2. NUCLEATION IN PURE MATERIALS

If a liquid is cooled below its equilibrium melting temperature there is a driving force for solidification, namely, $\Delta G = G^{(l)} - G^{(s)}$ and it might be expected that solidification would occur spontaneously. However, this may not always happen. For example, if no impurities are present in the melt, liquid nickel can be undercooled by about 250 K without causing phase change. However, in practice, such large undercoolings are not observed because of contamination from various sources such as the walls of the container.

It is known that the phase transformation process begins by forming very small particles of the solid phase, commonly called nuclei. In the absence of impurities in the solidifying liquid, these solid nuclei must form from the solidification of the pure material and hence this kind of nucleation process is called homogeneous nucleation. Large undercoolings can be obtained only under conditions of homogeneous nucleation. However, if there are impurities in the melt, the nuclei form under conditions of heterogeneous nucleation and such nuclei form much more easily and therefore large undercoolings are not usually observed. More information can be found in Haasen (1986).

2.1 Homogeneous Nucleation

Consider a liquid of volume V, at a temperature T, which is undercooled by an amount ΔT below the melting temperature, T_m, with a Gibbs free energy of G_l. Let some of the atoms in the melt nucleate and form a small solid sphere of volume, V_s, with surface area A_s. Then the Gibbs free energy of the system changes to

$$G_2 = V_s\, g^{(s)} + (V - V_s)\, g^{(l)} + \sigma\, A_s \tag{2.1}$$

We already know that the Gibbs free energy prior to nucleation was

$$G_1 = V\, g^{(l)} \tag{2.2}$$

where $g^{(l)}$ and $g^{(s)}$ are the Gibbs free energy per unit volume of the liquid and solid phases respectively. The formation of the solid nucleus results in a free energy change of

$$\Delta G = G_2 - G_1 = -V_s \,\Delta g + \sigma \, A_s \tag{2.3}$$

where $\Delta g = g^{(l)} - g^{(s)}$. From Eq. (1.5) we can also deduce that

$$\Delta g \approx \frac{\Delta h_{l-s} \,\Delta T}{T_m} \tag{2.4}$$

where Δh_{l-s} is the latent heat of fusion per unit volume. Thus we see from Eq. (2.3) that the free energy change during nucleation comes from two sources:

(i) a lowering of the free energy due to the lower free energy of the bulk solid, and
(ii) an increase in the free energy due to the formation of the solid–liquid interface and the associated surface energy.

By considering a simple example of a spherical nucleus, we see that

$$\Delta G = -\tfrac{4}{3}\pi r^3 \Delta g + 4\pi r^2 \sigma \tag{2.5}$$

We note that the contribution from the surface energy varies as the square of the radius of the nucleus whereas the contribution from the free energy of the bulk phases varies as the cube of the radius. This implies that if the nucleus has a small radius, the surface energy dominates and the net free energy actually increases due to the formation of the solid nucleus. The newly formed nucleus is thus unstable and tends to melt back into the liquid phase. Figure 2 illustrates some aspects of the above discussion.

In Figure 2 we see that the surface energy is proportional to the square of the radius whereas the contribution due to the free energy of the bulk phases is proportional to the cube of the radius. Furthermore, for a given value of the undercooling, ΔT, there is a critical value of the nuclei radius, r_c, that yields a maximum excess of the Gibbs free energy, ΔG. Thus if the radius, $r < r_c$ then the system lowers its free energy by remelting the newly formed solid particle and if the radius, $r > r_c$ then the system lowers its free energy by continuing the growth of the solid phase from the melt. Here, r_c is the critical nucleus size at which $d(\Delta G)/dr = 0$. Unstable solid particles which tend to remelt back into the liquid phase are called clusters or embryos whereas solid particles which exceed the critical radius and can grow are referred to as nuclei.

We can also obtain an estimate of the critical radius for the case of the spherical nucleus discussed above.

$$\left.\frac{d(\Delta G)}{dr}\right|_{r_c} = 0 \quad \Rightarrow \quad r_c = \frac{2\sigma}{\Delta g_c} \tag{2.6a}$$

and

$$\Delta G_c = \frac{16\pi\sigma^3}{3(\Delta g_c)^2} \tag{2.6b}$$

Since $\Delta g = \dfrac{\Delta h_{l-s}\,\Delta T}{T_m}$, Eq. (2.4), we obtain,

$$r_c = \frac{2\sigma T_m}{\Delta h_{l-s}} \cdot \frac{1}{\Delta T} \tag{2.7a}$$

and

$$\Delta G_c = \left(\frac{16\pi\sigma^3 T_m^2}{3\Delta h_{l-s}^2}\right) \cdot \frac{1}{\Delta T^2} \tag{2.7b}$$

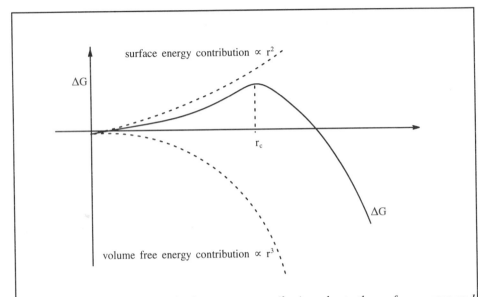

surface energy contribution $\propto r^2$

ΔG

r_c

ΔG

volume free energy contribution $\propto r^3$

Figure 2. This figure shows the free energy contributions due to the surface energy and the volume free energy due to the bulk phases present for a spherical nucleus of radius r_c. The surface energy has a positive contribution to the free energy whereas the volume free energy causes the overall free energy to decrease. Thus the free energy curve attains a maximum at the critical radius beyond which formation of the solid causes the free energy to decrease. Thus a critical size is necessary for the nucleus to grow.

An alternative approach can be devised based on the Gibbs–Thomson effect discussed earlier. If the melt is subject to a pressure P, then a spherical solid particle will be subjected to an excess pressure ΔP due to the curvature of the spherical particle, i.e., $\Delta P = \frac{2\sigma}{r}$ obtained from the Young–Laplace equation. By definition, the Gibbs free energy contains a PV term and an increase of the pressure will lead to an increase in the Gibbs free energy, i.e.,

$$\Delta G = V_s \, \Delta P = \frac{2\sigma V_s}{r} \tag{2.8}$$

In terms of unit volume, this becomes,

$$\Delta g_c = \frac{2\sigma}{r_c} \tag{2.9}$$

By definition, at the critical radius, r_c, the tendency to lower the free energy by solidification will be exactly balanced by the increase in the free energy due to surface curvature. Then, rearranging Eq. (2.9) we get

$$r_c = \frac{2\sigma}{\Delta g_c} \tag{2.10}$$

which is identical to the result obtained in Eq. (2.7). Thus, a solid sphere of radius r will have a higher free energy than the bulk solid as shown in Figure 3 and consequently will acquire a tendency to melt back as clearly presented in Figure 3. Thus surface curvature results in a decrease of the local melting temperature.

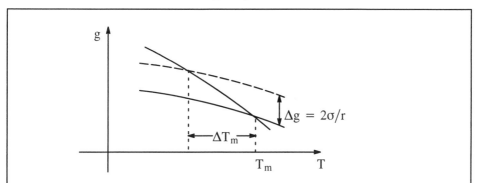

Figure 3. Free energy versus temperature curve showing the increase in free energy of the solid due to curvature of the interface and the consequent decrease in the melting point.

We have so far conducted our analysis by assuming that the nucleus is in the shape of a sphere. But this is true in general only for isotropic materials. Most materials of practical importance solidify in the shape of polyhedra, i.e., a body bounded by flat surfaces, straight edges and point vertices. We say that the solid has a crystalline structure. For example, let us consider the above analysis for a nucleus in the shape of a cube of length r. Then the Gibbs free energy,

$$\Delta G = -r^3 \Delta g + 6r^2 \sigma \tag{2.11}$$

and

$$\frac{d(\Delta G)}{dr} = 0 \quad \Rightarrow \quad r_c = \frac{4\sigma}{\Delta g_c} \quad \text{and} \quad \Delta G_c = \frac{32\sigma^3}{(\Delta g_c)^2} \tag{2.12}$$

Thus we see that $(\Delta G_c)_{\text{sphere}} \approx \frac{1}{2} (\Delta G_c)_{\text{cube}}$ which is consistent with the fact that a spherical shape minimizes the surface area for a given volume. This leads logically to the question as to why should polyhedral crystals form at all when the preferred shape is a sphere. The reason is due to anisotropy of surface energy.

It was mentioned earlier that a crystal is a polyhedral shape composed of flat faces intersecting at straight edges and point vertices. It is found that faces of different orientation have different values of the surface energy. In such a case the Gibbs criterion of minimization of surface energy acquires the following form

$$\sum_i \sigma_i A_i \quad \text{is a minimum at constant volume} \tag{2.13}$$

for the equilibrium shape of the condensed phase. Here, σ_i is the surface energy of the i'th face of the crystal and A_i is the surface area of the face.

2.2 Heterogeneous Nucleation

From the expression for the critical Gibbs free energy change, ΔG_c, it is seen that in order to make nucleation possible at small levels of undercooling (Carey 1992), ΔT_m, it is desirable to reduce the surface energy, σ. A simple way to achieve this is to have the nucleus formed in contact with a solid, such as the mould wall.

Consider a solid embryo forming in contact with a perfectly flat mould wall, with a wetting angle, θ, as shown in Figure 4.

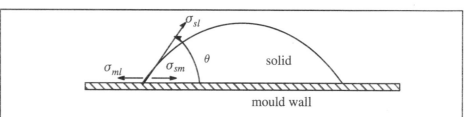

Figure 4. *Schematic of heterogeneous nucleation showing a nucleus forming in contact with the mould wall.*

Then from Young's contact condition, obtained from a static force balance, we get

$$\cos(\theta) = \frac{\sigma_{ml} - \sigma_{sm}}{\sigma_{sl}} \tag{2.14}$$

where the subscripts l, m and s designate liquid, mould and the solidified material, respectively. The formation of such an embryo will be associated with an excess free energy given by

$$\Delta G = -V_s \Delta g + A_{sl}\sigma_{sl} + A_{sm}\sigma_{sm} - A_{sm}\sigma_{ml} \tag{2.15}$$

Assuming that the embryo has the shape of a spherical cap, we get

$$\Delta G = \left\{ -\frac{4}{3}\pi r^3 \Delta g + 4\pi r^2 \sigma_{sl} \right\} f(\theta) \tag{2.16}$$

where
$$f(\theta) = \frac{(2 + \cos\theta)(1 - \cos\theta)^2}{4} \tag{2.17}$$

where $f(\theta)$ is a shape factor. Thus, the free energy for nucleation

$$(\Delta G)_{\text{heterogeneous}} = (\Delta G)_{\text{homogeneous}} \cdot f(\theta), \quad \text{with } f(\theta) \leq 1 \tag{2.18}$$

The critical radius remains the same as before, i.e.,

$$r_c = 2\sigma_{sl}/\Delta g_c \tag{2.19}$$

and the Gibbs free energy,

$$(\Delta G)_{r_c} = \frac{16\pi\sigma_{sl}^3}{3(\Delta g_c)^2} \cdot f(\theta) \tag{2.20}$$

which is lower than the corresponding value for homogeneous nucleation by a factor of $f(\theta)$. Thus it may be easily seen that the main contribution of the mould wall is to lower the activation energy barrier for nucleation by the value of the shape factor, thus reducing the level of undercooling required. On the other hand the critical radius is unaffected by the mould wall and depends on the undercooling only.

2.3 Heat Conduction with Phase Change

So far the discussion of phase change was mostly from a thermodynamic point of view. We now turn our attention to the consideration of the actual dynamics of the phase change process. In considering the dynamics of phase change, prime consideration must be given to the heat transfer mechanism which is responsible for supplying the latent heat of fusion at the phase change interface in order to drive the process (Ozisik 1968 and Crank 1984).

Consider a one dimensional phase change problem, with the liquid initially at a constant temperature of T_i. For a freezing problem, for example, T_i is greater than the fusion temperature, T_m. At time $t = 0$, the liquid surface at $x = 0$ is suddenly cooled to $T_1 < T_m$ and held at that temperature. A phase change front forms and moves into the liquid. The location of the front in time can be represented as $S(t)$. Then the heat conduction equation, in the solid and the liquid phases, with the appropriate boundary condition can be written as follows:

$$\frac{\partial T_s}{\partial t} = a_s \frac{\partial^2 T_s}{\partial x^2} \, , \qquad 0 < x < S(t) \, , \quad t > 0 \tag{2.21a}$$

$$T_s(x, t) = T_1 \quad \text{at } x = 0 \, , \quad t > 0 \tag{2.21b}$$

$$\frac{\partial T_l}{\partial t} = a_l \frac{\partial^2 T_l}{\partial x^2} \, , \qquad S(t) < x < 1 \, , \quad t > 0 \tag{2.21c}$$

$$T_l(x, t) = T_i \quad S(t) < x < 1 \, , \quad t > 0 \tag{2.21d}$$

$$\frac{\partial T_l}{\partial x} = 0 \quad \text{at } x = 1 \, , \quad t > 0 \tag{2.21e}$$

along with the coupling conditions at the solid–liquid interface given by $x = S(t)$,

$$k_s \frac{\partial T_s}{\partial x} - k_l \frac{\partial T_l}{\partial x} = \varrho_s \, \Delta h_{l-s} \frac{dS}{dt} \tag{2.22}$$

where a is the thermal diffusivity and Δh_{l-s} is the latent heat per unit mass. An important dimensionless number that is a key parameter in this problem is the ratio of the sensible heat to the latent heat defined as the Stefan number, i.e.,

$$St = \frac{C_p T_i}{\Delta h_{l-s}} \tag{2.23}$$

Although T_i is used as the temperature scale here, it is often convenient to use a characteristic temperature difference as will be the case in some of the subsequent discussions that follow.

It must be noted that although the governing equations here are the heat conduction equations with constant coefficients in each phase, the overall system is nonlinear because the boundary location, $S(t)$, is an unknown that has to be determined as part of the solution. This can be readily seen if a moving coordinate transformation is used to immobilize the moving solid–liquid interface.

To this end we define

$$\xi = \frac{x}{S(t)} \tag{2.24}$$

and

$$\eta = \frac{x - S(t)}{1 - S(t)} \tag{2.25}$$

solid:

$$\frac{\partial T_s}{\partial t} = \frac{\alpha_s}{S^2} \frac{\partial^2 T_s}{\partial \xi^2} + \frac{\alpha_s \xi}{S} \frac{dS}{dt} \frac{\partial T_s}{\partial \xi} , \qquad 0 < \xi < 1 \tag{2.26a}$$

$$T_s(x, t) = T_1 , \qquad \text{at } \xi = 0 \tag{2.26b}$$

liquid:

$$\frac{\partial T_l}{\partial t} = \frac{\alpha_l}{(1 - S)^2} \frac{\partial^2 T_l}{\partial \eta^2} + \frac{\alpha_l(1 - \eta)}{1 - S} \frac{dS}{dt} \frac{\partial T_l}{\partial \eta} , \qquad 0 < \eta < 1 \tag{2.26c}$$

$$\frac{\partial T_l}{\partial \eta} = 0 , \qquad \text{at } \eta = 1 \tag{2.26d}$$

and at the interface we have,

$$k_s \frac{\partial T_s}{\partial \xi}\bigg|_{\xi = 1} - \frac{k_l}{1 - S} \frac{dS}{dt} \frac{\partial T_l}{\partial \eta}\bigg|_{\eta = 0} = \varrho_s \, \Delta h_{l-s} \, S \frac{dS}{dt} \tag{2.27}$$

$$T_s\big|_{\xi = 1} = T_l\big|_{\eta = 0} = T_m \tag{2.28}$$

Thus these equations are clearly nonlinear. The advantage of the above transformation is that the interface is fixed in the transformed plane, but the coordinates move in time. A key feature of the solution to this problem is that $S(t) \sim \sqrt{t}$.

2.4 Heat Flow and Interface Stability for Pure Materials

In pure materials, solidification is controlled by the rate at which the latent heat of solidification can be removed and transported away from the phase change interface (Kurz and Fisher 1984). We now examine various aspects of the transport process and the interface morphology from the point of view of stability. For simplicity, we will restrict our consideration to the morphological stability of a planar interface driven by one-dimensional heat conduction.

(i) Consider the case of a solid growing with a planar interface into a superheated liquid. Then we can write the general equation governing the interface movement as

$$k_s \frac{\partial T_s}{\partial x} - k_l \frac{\partial T_l}{\partial x} = \varrho_s \, V \cdot \Delta h_{l-s} \tag{2.29}$$

where V is the velocity of the interface movement. Eq. (2.29) and Eq. (2.22) are identical except that the interface speed is represented by V to simplify the notation. Now consider a small perturbation on the planar interface as shown in Figure 5. Recalling that the interface is an isotherm, the location A, which corresponds to the peak of the perturbation now experiences a lower temperature gradient from the liquid side and a higher temperature gradient from the solid side.

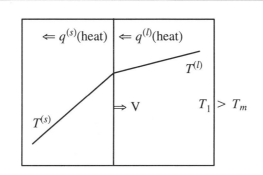

Figure 5(i). Planar interface moving into a superheated liquid showing the temperature gradients in the solid and the liquid regions.

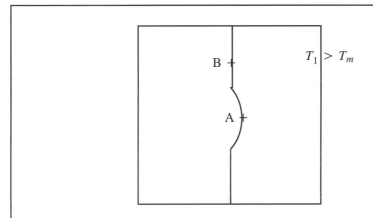

Figure 5(ii). Planar interface moving into a superheated liquid showing the effect of a small perturbation on the interface. The interface is morphologically stable.

Thus the overall heat removal rate from the interface is greater at location A than, say, at location B. Thus the growth rate of the interface at A is less and therefore the perturbation tends to melt back to a planar interface. Thus the interface is morphologically stable.

(ii) Now consider the case of a planar interface growing into an undercooled liquid. Here as shown in Figure 6 the temperature of the liquid is less than the melting temperature of the material. Thus the heat is now withdrawn from the liquid side of the interface. Now consider the effect of a small perturbation on the interface.

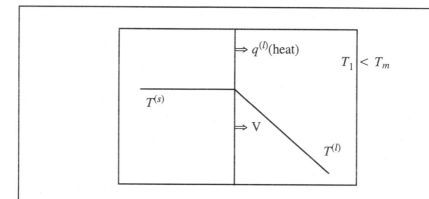

*Figure 6(i). Planar interface moving into a supercooled liquid showing the effect of a
small perturbation on the interface.*

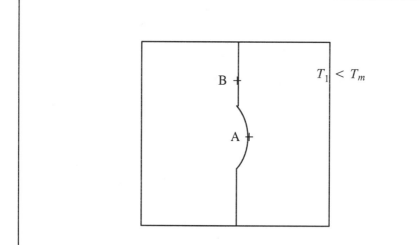

*Figure 6(ii). Planar interface moving into a supercooled liquid showing the effect of a
small perturbation on the interface. The interface is morphologically
unstable.*

Recall that the interface is an isotherm whose value is the melting temperature of the
material. Thus, the perturbation causes the local temperature gradient from the liquid side
to increase and the local rate of heat removal at location A to increase. This increases the
local interface velocity and the solidification rate. Thus the interface perturbation tends
to increase in magnitude at an increasing rate. We say that the interface experiences
morphological instability.

How does the above situation arise ? The following observations can be made to help understand the physical situation (Porter & Easterling 1992).

(i) Heat conduction away from the interface and into the solid can arise when solidification takes place from the mould walls which are cooler than the melt.

(ii) Heat conduction into the melt can arise only if the melt is undercooled below the melting temperature of the material. Such a situation arises during the initial nucleation process. As already discussed earlier, a certain degree of undercooling is required for nucleation to occur. Thus the first solid particles to be formed will grow into undercooled liquid and the latent heat of fusion will be removed by conduction through the liquid. Thus an originally spherical solid particle will develop arms in many directions. As the primary arms elongate, their surfaces also become unstable and develop secondary and tertiary arms leading to the typically observed dendrite formation.

2.5 Steady State Solution of the Planar Phase Change Problem

Consider a planar front moving under the action of a thermal field. Here only conduction effects will be considered and other modes of heat transfer will be provisionally neglected. Assume that the interface moves in the z–direction. Then we have the heat conduction equation,

$$\frac{\partial T}{\partial t} = \alpha \frac{\partial^2 T}{\partial z^2} \tag{2.30}$$

We now carry out the following non–dimensionalization procedure:

define
$$\theta \equiv \frac{C_p \, (T - T_\infty)}{\Delta h_{l-s}} \tag{2.31a}$$

and at the interface:
$$\theta \equiv St \equiv \frac{C_p(T_m - T_\infty)}{\Delta h_{l-s}} \tag{2.31b}$$

We can now immobilize the interface by transforming Eq. (2.30) and Eq. (2.31) to a frame of reference moving with the velocity, V of the interface, i.e., let $\xi = z - Vt$. Then

$$\frac{\partial \theta}{\partial t} = - V \frac{\partial \theta}{\partial \xi} \tag{2.32}$$

Then Eq. (2.30) transforms to

$$- V \frac{\partial \theta}{\partial \xi} = \alpha \frac{\partial^2 \theta}{\partial \xi^2} \tag{2.33}$$

with the boundary conditions $\quad \theta(0) = St \text{ and } \theta'(0) = - V \tag{2.34}$

We look for stationary solutions, as viewed by an observer moving with the interface, of the form

$$\theta(\xi) = St - \left[1 - \exp\left(- \frac{V \, \xi}{\alpha} \right) \right] \tag{2.35}$$

Note that

(i) V is arbitrary. There is no constraint on its permissible values.

(ii) The far field condition requires that $\theta(\xi \rightarrow \infty) = 0$. This leads to the requirement that, for a stationary solution, $St = 1$. At this value of the undercooling, the interface can move at any constant speed. If the undercooling condition is not satisfied, i.e., $St \neq 1$, then a stationary solution cannot exist.

2.6 Ivantsov Solution of the Phase Change Problem

G. P. Ivantsov, (Pelce 1988 and Kessler *et. al.* 1988) discovered a family of exact solutions for the moving boundary of the shape of a paraboloid of revolution. Assuming that surface tension can be neglected, i.e., the interface is an isotherm, exact solutions in the form of paraboloids of revolution can be written. To this end, consider a phase change interface whose shape is that of a paraboloid with tip radius, R and moving with constant velocity, V along the axis of revolution, z. Inside the paraboloid, the temperature is constant and at the phase change temperature. Outside the paraboloid, the heat is removed by the mechanism of conduction only. We now introduce the parabolic coordinates given by

$$\xi = \frac{r - z}{R} \tag{2.36a}$$

$$\eta = \frac{r + z}{R} \tag{2.36b}$$

where

$$r = \sqrt{x^2 + y^2 + z^2} \tag{2.36c}$$

and ϕ is the angle around the axis of revolution, z. Thus, the equation of the paraboloid of revolution with tip radius R is $\eta = 1$.

The stationary heat diffusion equation can be cast in the frame of reference moving with constant speed V along the axis of revolution as

$$- P \frac{1}{\xi + \eta} \left(\eta \frac{\partial T}{\partial \eta} - \xi \frac{\partial T}{\partial \xi} \right) = \frac{1}{\xi + \eta} \left(\frac{\partial}{\partial \eta} \left[\eta \frac{\partial T}{\partial \eta} \right] + \frac{\partial}{\partial \xi} \left[\xi \frac{\partial T}{\partial \xi} \right] \right) + \frac{1}{4 \eta \xi} \frac{\partial^2 T}{\partial \phi^2} \tag{2.37}$$

where $P = \dfrac{V R}{2 a}$ is the Peclet number. On the interface given by $\eta = \eta(\xi, \phi)$, the boundary conditions are:

(i) The interface temperature

$$T_i = T_m \tag{2.38}$$

(ii) Heat flux supplying the latent heat

$$P \left(\eta + \xi \frac{\partial \eta}{\partial \xi} \right) = \frac{(\mathcal{G}_s - \mathcal{G}_l) \, C_p}{\Delta h_{l-s}} \tag{2.39}$$

where $\mathcal{G} = \eta \dfrac{\partial T}{\partial \eta} - \xi \dfrac{\partial \eta}{\partial \xi} \dfrac{\partial T}{\partial \xi} - \dfrac{\eta + \xi}{4 \eta \xi} \dfrac{\partial \eta}{\partial \phi} \dfrac{\partial T}{\partial \phi}$. The solution is of the form

$$T(\eta) = T_\infty + (T_m - T_\infty) \frac{\int_\eta^\infty [\exp(-u) / u] \, du}{\int_P^\infty [\exp(-u) / u] \, du} \tag{2.40}$$

and

$$St = \frac{(T_m - T_\infty) \, C_p}{\Delta h_{l-s}} = P \exp(P) \int_P^\infty [\exp(-u) / u] \, du \tag{2.41}$$

It may be noted that the paraboloid $\eta = 1$ represents the moving boundary. The temperature field is a function of η outside the paraboloid ($\eta > 1$) and equals T_m inside the paraboloid, ($\eta < 1$).

An important feature of the Ivantsov solution is that there is a one–to–one relationship between the undercooling, St and the Peclet number, P, as is evident from Figure 7 and Eq. (2.41)

Based on the above analysis, the following statements can be made.

(i) Although steady state planar solutions do not exist except at $St = 1$, in the absence of surface tension, there exists a whole family of steady–state shape preserving paraboloid solutions, resembling the needle–crystal, at any value of the undercooling, $St < 1$.

(ii) For any given value of the Peclet number, only the product of the tip velocity and the tip radius is a constant. Thus infinitely many combinations of the tip radius and the tip velocity can exist. Specifically, as the tip radius approaches zero, the growth velocity of the tip approaches infinity.

(iii) From the above analysis, it is evident that the velocity and the length scales for the growth rate cannot be selected independently of each other. However, the velocity scale can be uniquely determined if the length scale can be fixed by any selection criteria. An obvious candidate for effecting this selection is the surface tension force between the two phases at the interface, as will be discussed next.

2.7 Inclusion of the Gibbs–Thomson Effect

We now include the effect of surface tension via the Gibbs–Thomson effect discussed earlier. Consider the growth of a needle–shaped crystal into the melt. The Ivantsov solution indicates that the sharper the crystal, the faster it will advance into the undercooled melt. However, if the curvature is high, the Gibbs–Thomson effect lowers the local melting temperature at the interface. Hence the level of the local undercooling will decrease and the interface advance will be slowed down. The actual interface velocity in regions of high curvature, such as the tip of the needle–like formation will be dictated by the balance of these two competing effects. Thus the Gibbs–Thomson effect tends to stabilize the interface morphology.

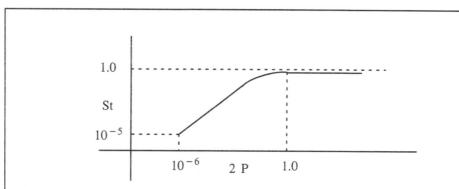

Figure 7. A log–log plot of St vs 2P showing the relationship between the undercooling and the Peclet number. As St approaches unity, the Peclet number, P approaches infinity.

3. ALLOY SOLIDIFICATION

We now turn our attention to the dynamics of solidification of alloys (Langer 1980, Kurz and Fisher 1984). Consider a dilute binary mixture. Let the mole fraction of the solute be C. Then the phase diagram can be represented as shown in Figure 8, where C_S and C_L are the mole fractions of the solute in the solid and the liquid respectively, at a given temperature. The partition coefficient, K, is the ratio C_S/C_L. For the simple idealistic cases that we consider here, the solidus and the liquidus lines are straight and thus K is independent of temperature.

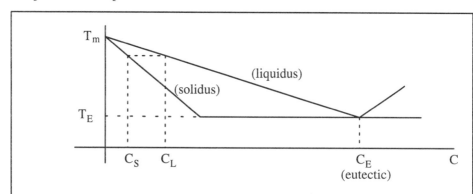

Figure 8. Phase diagram of a dilute binary mixture, in equilibrium, showing the solidus and the liquidus curves and the eutectic point. C_S and C_L are the mole fractions of the solute in the solid and the liquid respectively, at a given temperature.

If the partition coefficient, K, is less than unity, as is commonly observed, then the solute concentration in the solid is less than that in the liquid. Therefore, as solidification progresses, solute is continuously rejected at the phase change interface. The liquid progressively becomes richer in solute and referring to the phase diagram given in Figure 8, the temperature of the solidification front decreases. This in turn implies that the next layer of solid to be formed will contain a greater concentration of the solute as well. If the convection strength in the melt is low, the rejected solute at the solidification front can only be removed by diffusion which is usually a slow process compared to the solidification rate. Thus there will be a buildup of solute concentration in the liquid close to the solidification interface, leading to the formation of a concentration boundary layer ahead of the moving interface. The following diagram schematically illustrates the solute distribution as outlined below.

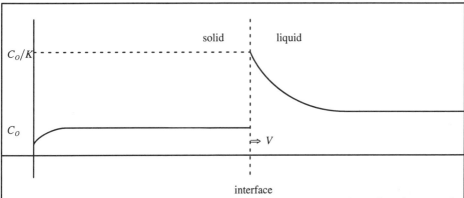

Figure 9. Schematic showing the development of a concentration boundary layer on the liquid side of the moving solidifying front. Since the partition coefficient is less than unity, solute is rejected at the interface.

3.1 Linear Stability Analysis

Our chief interest in the study of this problem is the morphological stability of the interface, which in turn impacts the microstructure of the solid and ultimately the quality of the material. Here we consider a linear stability analysis of the one–dimensional solidification problem, following the work of Mullins and Sekerka (1964). The following restrictive simplifications are made to obtain a closed form solution and thus demonstrate some of the physical phenomena involved in these systems:

(i)   Fluid convection is neglected thereby only the linear heat conduction equation is considered.

(ii) Low Prandtl number and high Schmidt number, i.e., high thermal diffusivity and low mass diffusivity

(iii) Unsteady terms are dropped

(iv) Constant material properties

(v)   Linearized phase diagram

(vi) The solidification rate is constrained to be constant

The governing equations become:

$$\frac{d^2T_k}{dx^2} + \frac{V}{\alpha_k}\frac{dT_k}{dx} = 0 \quad , \qquad k = l, s \tag{3.1}$$

$$\frac{d^2C}{dx^2} + \frac{V}{D}\frac{dC}{dx} = 0 \tag{3.2}$$

where α is the thermal diffusivity and D is the mass diffusivity. However, since $\alpha_k \gg D$ (low Pr and high Sc), the first derivative term dT_k/dx can be dropped from Eq. (3.1) so that the temperature distributions in the solid and the liquid are linear. The boundary conditions become:

(i) Conservation of energy at the interface

$$\Delta h_{l-s} V = -k_l G_l + k_s G_s \tag{3.3}$$

where $G_l =: dT_l/dx$ and $G_s = dT_s/dx$ and Δh_{l-s} is the latent heat per unit *volume*.

(ii) Conservation of solute at the interface

$$\left(C_l - C_s\right) V = -D \cdot G_c \tag{3.4}$$

where $G_c = dC_l/dx$ and the mass diffusivity in the solid is negligibly small.

(iii) Continuity of the temperature at the interface

$$T_l = T_s = T_i \tag{3.5}$$

(iv) Relation between temperature and concentration at the interface based on the linearized phase diagram,

$$T_i = T_m + M_l C_l \tag{3.6}$$

where M_l is the slope of the liquidus line in the linearized phase diagram and T_m is the melting temperature with $C = 0$. Now we fix a moving coordinate system to the interface. The temperature gradients in the solid and the liquid are linear and can be written as

$$T_l = T_i + G_l\, x \qquad \text{in the liquid} \tag{3.7a}$$

$$T_s = T_i + G_s\, x \qquad \text{in the solid} \tag{3.7b}$$

and the solute concentration becomes,

$$C = C_\infty + C_\infty\left(\frac{1-K}{K}\right)\exp\left[-\frac{V\,x}{D}\right] \tag{3.8}$$

where K is the partition coefficient and the velocity of interface advance is

$$V = \frac{1}{\Delta h_{l-s}}\,(k_s G_s - k_l G_l) \tag{3.9}$$

and the concentration gradient at the interface is

$$G_c \equiv \frac{dC}{dx}\bigg|_{x=0} = -C_\infty\frac{V}{D}\left(\frac{1-K}{K}\right) \tag{3.10}$$

Note that the temperature gradients are constants in the bulk phases.

3.2 Constitutional Supercooling

We now couple the thermal field and the concentration field according to Eq. (3.6) at the interface and then examine whether an initially planar interface will be able to maintain its planar shape under steady state conditions (Tiller *et. al.* 1953). We check whether, with a given concentration, the corresponding temperature in the liquid phase, yielded by Eq. (3.6) is higher or lower than the temperature predicted by Eq. (3.7a). If the actual temperature in the liquid phase is lower than that calculated by Eq. (3.6) then a planar interface will not be physically realizable. The condition for this is called constitutional supercooling, given by the following condition,

$$M_l \, G_C > G_l \qquad (3.11)$$

Instability due to constitutional supercooling has been observed in practice in which a planar interface assumes new morphological structures.

3.3 Linear Stability Analysis of Mullins and Sekerka

We now consider the stability analysis of the 2–D governing equations following the work of Mullins and Sekerka (1964). As before, we fix a moving coordinate system to the interface. The governing equations are

(i)   the 2–D Laplace equation for the thermal field

$$\frac{\partial^2 T_k}{\partial x^2} + \frac{\partial^2 T_k}{\partial y^2} = 0 \quad , \qquad k = l, s \qquad (3.12)$$

(ii)   the solute field

$$\frac{\partial^2 C}{\partial x^2} + \frac{\partial^2 C}{\partial y^2} + \frac{V}{D}\frac{\partial C}{\partial x} = 0 \qquad (3.13)$$

where the first derivative terms such as $\dfrac{V}{\alpha_k}\dfrac{\partial T_k}{\partial x}$ have been neglected in the heat conduction equations because the thermal diffusivity is large compared to the mass diffusivity (low *Pr* and high *Sc* case) as is the case in many practical applications. Here *V* is along the x–direction.

Now, we carry out a linear stability analysis by solving the problem for small perturbations of the interface shape,

$$x = \delta \, \sin(\omega y) \qquad (3.14)$$

where δ depends on *t* and ω. We can define the wavelength of the perturbations as $\lambda = \frac{2\pi}{\omega}$. For the condition of small perturbations to be satisfied, $\delta \ll \lambda$. Then the perturbed growth velocity in the x–direction is given by

$$v = V + \dot{\delta} \, \sin(\omega y) \qquad (3.15)$$

where *V* is the constant unperturbed growth speed and $\dot{\delta} = d\delta/dt$. We consider solutions of the following forms,

$$T_l(x, y) = T_0(x = 0) + G_l \, x + A_l \, \exp(-\omega x)\sin(\omega y) \qquad (3.16)$$

$$T_s(x, y) = T_0(x = 0) + G_s \, x + A_s \, \exp(\omega x) \, \sin(\omega y) \qquad (3.17)$$

$$C(x, y) = C_\infty(x \to \infty) + C_\infty\left(\frac{1 - K}{K}\right) \exp\left(-\frac{V\,x}{D}\right) +$$

$$B \exp(-\omega^* x) \sin(\omega y) \tag{3.18}$$

where $\omega^* = \dfrac{V}{D} + \left[\left(\dfrac{V}{D}\right)^2 + \omega^2\right]^{1/2}$ and the interface is located at $x = 0$ in the moving

frame of reference.

The boundary conditions for this problem are the same as in the previous case, i.e., Eq. (3.3–3.6). However, the boundary condition connecting the interface temperature and the concentration, Eq. (3.6), can be extended to take surface tension effects into account, by incorporating the Gibbs–Thomson effect which has already been shown to have an important effect in phase change phenomena,

$$T_i = T_m + M_l\,C_l - T_m\,\frac{\sigma}{\Delta h_{l-s}}\left(\frac{1}{R_1} + \frac{1}{R_2}\right) \tag{3.19}$$

where σ is the surface tension at the solid–liquid interface and R_1, R_2 are the principal radii of curvature of the interface, with the sign being positive for a convex crystal.

In deriving the stability criteria, a key procedure is to separate each equation into the "unperturbed parts" i.e, those terms not containing $\sin(\omega y)$ and the "perturbed part" which is linear in $\sin(\omega y)$. Let $\delta = \delta_0\,e^{f(\omega)\,t}$ where δ_0 is the initial perturbation. Then,

$$\dot{\delta}/\delta = f(\omega) \tag{3.20}$$

The interface is morphologically unstable if infinitesimal perturbations grow without bound in time i.e., if $f(\omega) > 0$.

In general, the amplification factor $f(\omega)$ contains terms due to the following phenomena:

(i)   Capillarity which has a stabilizing influence at small wavelengths

(ii)   Heat conduction which has a stabilizing effect for long wavelengths

(iii)   Solute diffusion which has a destabilizing effect for intermediate wavelengths.

It turns out that, as in the 1–D case, the parameter G_l/V controls the onset of instability. Instability is prone to appear with lower values of G_l/V. Furthermore the length scale associated with the neutral stability point, λ_i, is (Langer 1980)

$$\lambda_i \approx 2\pi\sqrt{d\delta_T} \tag{3.21}$$

which can be viewed as the morphological scale, where the thermal length scale $\delta_T = a_l/V$ and the capillary length scale $d = \sigma/\Delta h_{l-s}$.

3.4 Effect of Convection on Interface Morphology

The analysis described in Sections 3.1 to 3.3 is based on heat conduction only. In reality, of course, convection is almost always present in the melt. Thus convective effects need to be accounted for. A number of workers have made attempts to study the coupling of the solid–liquid interface morphology and convection in the melt (McFadden *et. al.* 1984, Fang *et. al.* 1985, Glicksman *et. al.* 1986, Huppert 1990 and Davis 1990). Such coupling effects have been observed in the work of Fang *et. al.* (1985) where a long circular cylinder of pure succinonitrile was heated from within by means of a heated wire along the cylinder axis to produce a melt in the annular cylinder between the wire and an outer annular cylinder of solid succinonitrile. For a sufficiently small temperature difference between the wire and the interface, the interface remains cylindrical and the fluid flow is a simple steady flow that is upward in the vicinity of the wire and downward in the vicinity of the interface. As the temperature is increased, the interface becomes distorted into a helix that travels in the vertical direction and the fluid flow becomes unstable and time–dependent.

A linear stability analysis shows that for the melting succinonitrile ($Pr = 22.8$), the onset of instability appears at a Grashof number, $Gr = 176$, with the length scale based on the distance between the surface of the hot wire and the unperturbed cylindrical interface. It is also found that when the crystal–melt interface is replaced by a rigid cylindrical wall, the critical Gr for the onset of instability is about 2150. This seems to indicate that the interface morphology and fluid convection are strongly coupled. Furthermore, the critical Gr also seems to depend on the Prandtl number. For $Pr = 0.0225$, the critical Gr is independent of whether the boundary is a rigid wall or a solid–liquid interface. Thus the fluid convection and the interface shape are strongly coupled.

In another interesting experiment by Bouissou *et. al.* (1990) revealed that the dendrite growth forms in solidification can be manipulated with the application of appropriate fluid flow. In their work, forcing the velocity field at suitable frequencies resulted in highly coherent sidebranching events, as compared to unforced dendrites where the sidebranches evolving on the main needle crystal are essentially uncorrelated, on account of their formation from noise. They also indicate the existence of a "natural" frequency of sidebranching instabilities for unforced dendrites and a resonance mechanism whereby the efficiency of external forcing is highest for a given range of frequency. This experiment appears to offer the possibility of modulating dendritic growth forms by forcing the fluid flow. This is merely one aspect of the convective effects on dendritic growth. Some experiments are aimed at investigating the possible effects on microstructures, of melt convection that invariably accompanies any real crystal growth situation. Ananth and Gill (1989 and 1991) and Canright and Davis (1991) have investigated the effects of natural convection on uniformly translating paraboloids. The former find that paraboloids of revolution are consistent solutions to the combined heat and Oseen flow equation set. This is a useful result and enables one to simply extend the Ivantsov model from the pure conduction case to the case where convection exists. So far the knowledge in this area is incomplete and more work is definitely needed. This subject will be discussed further in Section 5.5.

4. GENERAL FORMULATION FOR INTERFACE TRACKING

4.1 Literature Survey

The characteristic feature of any phase change problem is the coupling of thermal fields in the different phases with free and moving boundaries that not only separate each phase, but also dynamically evolve. The free surface and propagating phase front make such problems nonlinear. Because of the nonlinearity, only a few exact solutions have been developed, and these are limited to simple geometries and boundary and initial conditions. For most practical cases, the solutions are obtained using numerical methods. In what follows a brief survey of the existing methods for solving the phase change problems is presented. One–dimensional heat transfer with phase change was first posed mathematically by Stefan and later solved analytically by Neumann (Crank 1984, Ozisik 1980, Carslaw and Jaeger 1959). The exact solutions are limited to a number of idealized situations involving semi–infinite regions or infinite regions subject to simple boundary and initial conditions. To date, no exact solutions have been developed for phase change problems where the phase change front is a function of multiple space coordinates. The integral method, which dates back to von Karman and Pohlhausen, who used it for the approximate analysis of boundary layer equations, was applied by Goodman (1958) to solve a one–dimensional transient phase–change problems. This method provides a relatively straightforward and simple approach for approximate analysis of one–dimensional transient phase–change problems. Zien (1978) has further developed the integral method to solve the transient conduction problem with phase change, obtaining quite realistic solutions for some difficult problems.

In the area of numerical techniques, Chernouskou (1970) proposed the isotherm migration method, which interchanges the dependent variable with the space variable for one–dimensional phase change problems. Crank and Gupta (1975) were the first to apply the isotherm migration method to phase change problems in two dimensions. The advantages of this technique, in tracking a moving isotherm at the fusion temperature, are obvious, since the phase boundary is itself an isotherm, provided the phase change takes place at a constant temperature. Results predicted by this approach have been found to be satisfactory. Saitoh (1978) extended the isotherm migration method from regular geometries to arbitrary shaped, doubly connected geometries, and has applied it to solve two–dimensional problems..

The moving heat source (or the integral equation) method, originally applied by Lightfoot (1929), is based on the concept of representing the liberation (or absorption) of latent heat by a moving plane heat source (or sink) located at the solid/liquid interface. The temperature at any point will be due to the superposed contributions from this moving source (or sink). In this method, the analysis of the phase–change problem is transformed to the solution of an integral equation for the location of the liquid/solid interface. Ozisik (1978) discussed the use of a moving heat source in the conduction equations to account for the latent heat effect. He presented a mathematical development that explicitly casts the moving boundary problem into a standard heat conduction problem with a moving heat source. By doing this, the solution can immediately be written in terms of Green's function, and numerically implemented.

In general the numerical methods for solving the phase change problems can be divided into two categories: the fixed–grid approach and the transformed–grid approach. The fixed–grid approach implies that a grid is fixed in space and the interface conditions are accounted for by the definition of suitable source terms in the governing equations. In the transformed–grid approach, the governing equations and their boundary conditions are cast into a generalized curvilinear coordinate system, so the grid might adapt with the moving freezing front. A comparison between transformed grids and fixed grids on a test problem of the melting of gallium was made by Lacroix and Voller (1990). They have found that if an efficient grid generation can be achieved, the CPU usage of the transformed grid is very close to that of the fixed grid. Furthermore, a fixed grid can produce accurate predictions with the same order of mesh size as that used by a transformed grid. Thus, according to these authors, no conclusive preference can be made for either method. However, the tests covered only a limited range of grid and time resolution; more effort in this direction will be needed to further clarify this issue.

One well known method to deal with phase change problems on fixed grids is the enthalpy–porosity technique. This approach easily permits solutions for substances whose change of phase occurs over a range of temperature. Voller and Prakash (1987) and Brent *et. al.* (1988) have adopted this technique in conjunction with the SIMPLE procedure proposed by Patankar (1980) to simulate the two–dimensional convective melting of pure gallium in a rectangular cavity. The enthalpy–porosity technique was validated by comparing its results with the experimental results obtained by Gau and Viskanta (1986). A review of available implicit difference enthalpy–porosity schemes can be found in Voller (1985).

One of the drawbacks of the enthalpy formulation, particularly for the case where the change of phase occurs at a single temperature, is that a plot of temperature against time for a given node exhibits a "plateauing" tendency. This arises out of the fact that a computational grid models or represents a discrete region in space. Obviously, it requires a finite amount of time to melt a discrete region. As a consequence of a node being held fixed at a single temperature for a discrete amount of time, the effect is also felt by the surrounding nodes, causing a plateauing of the temperature. Voller and Cross (1981) have developed a smoothing technique that can be applied to a final set of numerical results. This smoothing technique has the effect of bringing the time–temperature history of a given node into better agreement with other published results. Discussion will be given later, in Section 7, in terms of some computational issues of the enthalpy formulation.

Treatments of the phase change and the associated transport process based on the fixed grid method have been developed to treat quite complicated problems. Bennon and Incropera (1987) have used a volume averaged continuum model with the SIMPLE algorithm (Patankar 1980) to investigate solidification of binary, aqueous ammonium chloride solution in a rectangular cavity. Beckermann and Viskanta (1988) have also used the same numerical method to solve the complex phase change problem. Shyy and Chen (1990a, 1991b) have combined the enthalpy formulation and a computational scheme based on an adaptive grid in curvilinear coordinates to solve the phase change problems with high Rayleigh and Marangoni numbers, and under both normal and reduced gravity conditions. With relatively finer and improved resolutions, various detailed transport phenomena and interface shapes have been reported. Recently, a turbulence model has been incorporated into this framework to predict the solidification aspects of a continuous ingot casting process (Shyy *et. al.* 1992e), yielding a highly advanced tool for analyzing transport processes during solidification, as will be discussed in Section 8.6.

Crowley (1983) applied an explicit finite difference scheme to an enthalpy formulation and coordinate transformation techniques to study crystal pulling by the Czochralski technique. She applied her model to analyze the initially steady Czochralski system with a base temperature perturbed at the melt inlet. Based on the computed response of trijunction location, she concluded that the simulated configuration is unstable to such disturbances.

A numerical solution for melting around a vertically heated cylinder incorporating the effects of natural convection was obtained by Sparrow *et. al.* (1977). They used the coordinate transformation to immobilize the solid–liquid interface and, consequently, appear to have been the first investigators to apply it to a multidimensional convection–dominated Stefan problem. However, the higher order curvature terms for the interface were neglected, limiting the validity of the method to cases where the radius of the interface varies only slightly. Furthermore, a quasi–steady assumption was invoked where the effects of interface motion were ignored; hence the method can treat the problem only with very modest phase change rate. Thompson and Szekely (1988, 1989) have applied this quasi–steady approach to predict the phase change problem in a cavity.

The above mentioned works are based on the finite volume/difference formulations. Silliman and Scriven (1980) and Cuvelier and Schulkes (1990) have used the finite element method (FEM) to predict steady viscous free surface flows. In their analysis, only liquid/gas interface is considered, and phase change is not included. Albert and O'Neill (1986) and Lynch (1982) used a finite element method with deforming elements to track a moving phase front and to model phase change problems.

Ettouney and Brown (1983) have used a finite element analysis to model the steady crystal growth process. Several variations of the methods have been developed by this group (Sackinger *et. al.* 1989, Derby and Brown 1986, Brown 1988), depending on the choice of coordinates (original or transformed), distinguished boundary condition (local heat balance or thermal equilibrium), and the method of solving the resulting system of equations (successive approximation method or Newton's method). Their numerical method was limited to the steady solidification problems. Crochet *et. al.* (1987), Dupont *et. al.* (1987) and Wouters *et. al.* (1987) investigated the oscillatory fluid flow in a horizontal Bridgman technique. The method was based on an implicit time–marching technique using FEM but was not applied to the moving solid–liquid interface problem. Other works in this area can be found in reviews of existing literature (Flemings 1974, Brown 1988, Dantzig 1989, Heinrich *et. al.* 1989, Viskanta 1989, Yao and Prusa 1989, and Ganesan and Poirier 1990). In the following both explicit and implicit interface tracking methods will be briefly reviewed and presented. Also presented in Sections 5.2 and 6 is a recent development of an interface tracking scheme based on a combined Lagrangian–Eulerian method (Shyy *et. al.* 1993c, Udaykumar and Shyy 1993b).

4.2 Governing Laws

In this Section we write down the jump conditions which are enforced at the phase–change interface (Carey 1992, Slattery 1990). Consider a pure material going through an evaporation process. Then the equations of mass continuity, momentum conservation and energy conservation apply in the bulk phases along with the appropriate jump conditions at the interface. For a liquid – gas phase change process, the interface jump conditions can be written as follows,

(i)   *mass conservation*

$$\frac{dm_g}{dt} = \varrho_g \left(- v_{Ign} + v_{In} \right) A_I = \varrho_l \left(- v_{Iln} + v_{In} \right) A_I \tag{4.1a}$$

$$\therefore \quad - \varrho_g v_{Ign} + \varrho_l v_{Iln} = (\varrho_l - \varrho_g) \, v_{In} \tag{4.1b}$$

where the subscript I designates the interface, g and l, the gas and liquid phases respectively, A the area, and n the normal direction. Here, v_{In} is the normal component of the interface velocity, v_{Ign} and v_{Iln} are the normal component of the fluid velocity on the gas side of the interface and the normal component of the fluid velocity on the liquid side of the interface, respectively; they are all defined with respect to a stationary observer.

(ii)   *momentum conservation*

In the normal direction to the interface, since $v_{Ign} \neq v_{Iln}$ there can be a jump in the normal momentum flux terms besides the normal stress terms,

$$P_l - P_g = \sigma \left(\frac{1}{R_1} + \frac{1}{R_2} \right) + \varrho_g (v_{Ign} - v_{In})^2$$
$$- \varrho_l (v_{Iln} - v_{In})^2 + 2\mu_l \frac{\partial v_{Iln}}{\partial n} - 2\mu_g \frac{\partial v_{Ign}}{\partial n} \tag{4.2}$$

where R_1 and R_2 are the principal radii of curvature of the interface measured positive on the liquid side of the interface.

In the tangential direction, the no–slip condition still holds, i.e,

$$v_{Igs} = v_{Ils} = v_{Is} \tag{4.3}$$

where s designates the tangential direction of the interface. The tangential momentum jump is given by (assuming that the liquid phase is to the left/below the interface)

$$\frac{\partial \sigma}{\partial s} = \mu_l \left(\frac{\partial v_{Iln}}{\partial s} + \frac{\partial v_{Ils}}{\partial n} \right) - \mu_g \left(\frac{\partial v_{Ign}}{\partial s} + \frac{\partial v_{Igs}}{\partial n} \right) \tag{4.4}$$

(iii)   *energy conservation*

$$\varrho_l \, (v_{lln} - v_{ln}) \, \Delta h_{l-g} = (k_l \nabla T_l - k_g \nabla T_g) \cdot \vec{n} \qquad (4.5)$$

where $\varrho_l \, (v_{lln} - v_{ln})$ is the mass flux through the interface as given by Eq. (4.1a).

We also need to know the interface temperature. For the evaporation process, recall the phase diagram, as shown in Figure 10.

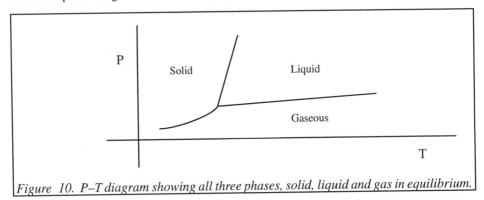

Figure 10. P–T diagram showing all three phases, solid, liquid and gas in equilibrium.

For the evaporation process, assuming thermodynamic equilibrium, we can use the Clausius–Clapeyron equation, i.e.,

$$\left.\frac{\partial P}{\partial T}\right|_I = \frac{\Delta h_{l-g}}{T \, \Delta v} \qquad (4.6)$$

where Δv is the change in specific volume that accompanies the phase change process.

For a melting process, since the solid–liquid boundary on the P–T diagram is nearly a vertical line, we can usually use a single melting temperature with a given thermodynamic pressure. The interface temperature is relatively insensitive to the local static pressure variation. However, the interface temperature is affected by the local interface curvature as was demonstrated by our discussion of the Gibbs–Thomson effect.

For an evaporation process, the interface temperature varies substantially with the local interface pressure, even with a flat interface. Furthermore, the interface curvature is usually not high enough for the Gibbs–Thomson effect to become important, except possibly during the early stages of the formation of the vapor nucleus. This is due to the fact that the density of the vapor is much smaller than that of the liquid, and hence the inertia forces are usually not high enough to sustain a locally high pressure gradient necessary to support the interface curvature.

5. MODELLING AND RELATED ISSUES

It is appropriate to briefly describe the nomenclature used in this section. In the following, x, y, and t designate the dimensional coordinates in space and time; their dimensionless counterparts are X, Y, and τ. The dependent variables u, v and T are the general symbols for the dimensional velocity components and temperature.   The corresponding dimensionless quantities are indicated by u^*, v^* and θ. The dimensional interface velocity is designated by V, and V_i is the characteristic velocity scale, and the dimensionless interface velocity is represented by V_N^*, i.e., $V_N^* = V/V_i$. In general, unless otherwise specified, the dimensionless quantities are indicated by the superscript "*".

5.1 Issues of Scaling

A striking difficulty in analyzing the materials processing problem is the presence of many dimensionless parameters, frequently numbering ten or more. In fact, because of this characteristic, the similitude approach so fruitfully used for many fluid dynamics problems is of diminished usefulness. However, the insight offered and flow regimes defined by obtaining dimensionless parameters is still very helpful. To illustrate this point, one can consider three key dimensionless parameters encountered during a phase change process. They are:

(i)   Thermal Grashof Number

$$(Gr)_T = \frac{\varrho^2 g \beta_T \Delta T l^3}{\mu^2} \tag{5.1}$$

where g is gravitational acceleration, ϱ is the characteristic density, l is the characteristic length scale of the melt, ΔT is the characteristic temperature variation in the melt, β_T is the coefficient of thermal expansion and μ is the dynamic viscosity.

(ii)   Thermal Marangoni Number

$$Ma = \frac{(\frac{d\sigma}{dT})\Delta T \, l}{\mu a_l} \tag{5.2}$$

where σ is the surface tension and a_l is the thermal diffusivity of the melt.

(iii)   Stefan Number

$$St = \frac{C_p \Delta T}{\Delta h_{l-s}} \tag{5.3}$$

where C_p is the specific heat of the melt, Δh_{l-s} is the latent heat of fusion.

Now, a quick inspection reveals that:

$$(Gr)_T \sim l^3$$

$$Ma \sim l \tag{5.4}$$

$$St \text{ is independent of } l$$

Hence, in practical terms, it is virtually impossible to conduct a laboratory experiment that can maintain the same values of all the dimensionless parameters since a change of length scale of the melt impacts different dimensionless parameters differently. No experiments have been reported that can claim that a strict similarity of the transport processes between solidification facilities of different length scales can be maintained via the change of, say, ΔT, μ and α to accommodate the variation in l. It must be observed that the Prandtl number and the Schmidt number should also be unchanged if one wants to study the transport characteristics in a truly corresponding manner.

With the aforementioned experimental difficulties in reproducing the actual operating conditions, it becomes clear that computational modeling can be immensely useful. Through such modelling, one can first assess the predictive capability via validating against carefully conducted laboratory measurements and then extend the calculations to the conditions directly relevant to actual production. However, before such a broad range extrapolation can be applied, considerable assessment and research must be carefully conducted to understand the strength and weaknesses of a given theoretical model.

As already discussed above, the importance of convection to the solidification process is currently well recognized. One of the basic problems encountered by any theoretical model is the wide disparity of length scales. Based on the dimensional analysis, several length scales can be identified in a binary eutectic system, namely, (i) capillary length $d = \sigma / \Delta h_{l-s}$ (ii) thermal transport length scale $\delta_T = \alpha_l / V_i$ where α_l is the thermal diffusivity of the melt, V_i is the characteristic growth speed, (iii) solutal transport length scale, $\delta_s = D_l / V_i$ where D_l is the mass diffusivity of the melt, and (iv) convective length scale δ_v. There is more than one way to choose a convective length scale, depending on the processing technique as well as operating conditions. For example, in Bridgman growth, the convective length scale can be determined by balancing buoyancy and shear stress effects, while in a floating zone technique, it may be deduced from the balance of Marangoni and shear stress effects. As it has become obvious by now, the overall solidification process is very complicated, involving the simultaneous presence and competition of multiple scales. For the metallic system, due to the disparity of Schmidt and Prandtl numbers, $Sc \gg 1$, $Pr \ll 1$, the following relative order of magnitudes can be identified among the length scales:

$$\mathcal{O}(d) \lessapprox \mathcal{O}(\delta_s) < \mathcal{O}(\delta_v) < \mathcal{O}(\delta_T)$$

The appropriate magnitude of the morphological length scale, λ_i, e.g., the dendritic spacing, depends on the balance resulting from these different scales and associated transport mechanisms. It should be noted that λ_i need not be the same as any of the above length scales. Its order of magnitude can be different from those already established by the capillary and transport mechanisms. The reason that this extra morphological scale can exist is due of course, to the nonlinearity of the system. A case in point is the well known morphological stability analysis performed first by Mullins and Sekerka (1964) as discussed in Section 3.3, which yields a relationship of $\lambda_i = \mathcal{O}\left(\sqrt{d\delta_T}\right)$ for an interface subject to directional solidification. However, this analysis does not consider the influence of convection. An analytical approach must address the difficulties involved with the multiple scales along with the nonlinearity of the system. In order to fully account for the various length scales that co–exist in the melt, a direct numerical simulation that accounts for the interaction among different length scales is perhaps the only viable approach. In the following, some recent developments in the numerical simulation of interface characteristics at the morphological scale, both with and without convection will be discussed along with the scaling procedure involved.

5.2 A Computational Procedure for Tracking Phase Fronts

In the following, a general computational procedure suitable for front tracking is described. This computational procedure, developed in Shyy *et. al.* (1993 c) employs a moving grid system which conforms to the interface shape. The heat conduction equation in each phase is solved in the transformed domain, viz. $\xi = \xi(x,y,t)$, $\eta = \eta(x,y,t)$. An implicit solution algorithm is employed. The interface motion and field equation solver are coupled over each iteration. The solution of the field equations in the transformed domain is straightforward and the moving grid terms are easily incorporated in the discrete, control volume formulation. The interface is tracked in a Lagrangian fashion.

Consider the non–dimensional equation for interface advance,

$$-\frac{\varrho^*}{St}V_N^* = [\frac{\partial\theta}{\partial n_l^*} - \frac{k_s}{k_l}\frac{\partial\theta}{\partial n_s^*}] \tag{5.5}$$

where V_N^* is the non–dimensional normal velocity of the interface and $(\partial\theta/\partial n^*)_{l,s}$ are the normal gradients of non–dimensional temperature, θ (normalized by ΔT, undercooling of the melt) in the liquid and solid respectively. Eq. (5.5) is non–dimensionalized based on Eq. (2.29) with ϱ_l and k_l being the scaling properties. The Stefan number, St, is defined by Eq. (5.3) and the non–dimensional velocity, V_N^*, is normalized by a velocity scale, a_l/δ_T, with δ_T being a thermal length scale. For more discussion see Eqs. (5.16) and (5.17). The local normal to the interface is given by,

$$\vec{n}^* = \frac{\nabla F}{|\nabla F|} \tag{5.6}$$

where $F = F(X,Y,\tau)$ is the curve defining the interface. X,Y, and τ are the dimensionless variables based on reference scales defined by δ_T and δ_T^2/a_l. The interface shape is defined in a piecewise fashion to facilitate handling of branched interfaces. Here a quadratic polynomial fit is performed for three successive nodal points at each point of the interface. Alternatively, a local circular arc can be used to define the interface shape as in Section 6. Thus, at the ith point on the interface we designate the curve, $Y_i = a_iX_i^2 + b_iX_i + c_i$, i.e., $F_i = Y_i - (a_iX_i^2 + b_iX_i + c_i)$ defines the interface. The a_i, b_i and c_i are determined from the known values of $(X_j,Y_j),j=i-1, i, i+1$. The local curve definition then yields the derivatives $(F_x, F_y$ and $F_{xx})$ at each point on the interface. Thus the normal and curvature at each point are obtained from,

$$\vec{n}^* = \frac{1}{|\nabla F|}(\frac{\partial F}{\partial X}\vec{i} + \frac{\partial F}{\partial Y}\vec{j}) \tag{5.7}$$

and curvature,

$$\varkappa^* = \frac{Y_{XX}}{(1 + Y_X^2)^{\frac{3}{2}}} \tag{5.8}$$

We may write Eq. (5.5) as,

$$-\frac{\varrho^*}{St} V_N^* = \frac{1}{|\nabla F|} [(F_X\theta_X + F_Y\theta_Y)_l - (F_X\theta_X + F_Y\theta_Y)_s] \tag{5.9}$$

In computing the interface normal velocity then, one seeks to obtain the derivatives F_X, F_Y and θ_X, θ_Y. The derivatives of temperature may be obtained in the transformed coordinates itself, i.e.,

$$\theta_X = \frac{(\theta Y_\xi)_\eta - (\theta Y_\eta)_\xi}{J} \tag{5.10a}$$

$$\theta_Y = \frac{(\theta X_\eta)_\xi - (\theta X_\xi)_\eta}{J} \tag{5.10b}$$

where J is the Jacobian of the coordinate transformation defined by

$$J = x_\xi y_\eta - x_\eta y_\xi \tag{5.10c}$$

F_x and F_y of course are directly available in the physical domain from the curve fit. Thus the new coordinates of the interfacial points are obtained from:

$$X^{n+1} = X^n + \frac{\partial F}{\partial X} \frac{V_N^*}{|\nabla F|} \delta\tau \tag{5.11}$$

$$Y^{n+1} = Y^n + \frac{\partial F}{\partial Y} \frac{V_N^*}{|\nabla F|} \delta\tau \tag{5.12}$$

where $\delta\tau$ is the time step size. Having obtained these new coordinates of the curve, the thermal field is solved for once again, the curve fit is performed at the interface and a fresh interface position is obtained from Eqs. (5.11) and (5.12). All these procedures are performed in a fully coupled manner involving interaction among the temperature field, interface motion and grid movement at each iteration. It may be noted that Eq. (5.5) is not an unique choice of the non–dimensional form. In fact, as will be discussed in Section 5.4, an alternative form will be used which is computationally more efficient.

5.3 Motion of Curved Fronts

The interface tracking method developed in Section 5.2 is now applied to follow the development of the phase change front for long enough times to obtain substantial curvature of an initially slightly perturbed interface. In tracking the motion of curved fronts, additional physical and computational issues enter. Physically, when the phase interface is deformed surface tension seeks to round out regions of strong curvature. This effect is expressed, for an interface in thermodynamic equilibrium (i.e. for sufficiently slow moving interfaces) by the Gibbs–Thomson formula,

$$T_{interface}|_{eq} = T_m (1 - \frac{\sigma \varkappa}{\Delta h_{l-s}}) \tag{5.13}$$

T_m is the melting temperature of a flat interface, $\varkappa$ is the local curvature of the interface. $T_{interface}|_{eq}$ is the modified melting temperature after accounting for the Gibbs–Thomson effect, discussed in Chapter VII, and in this form is applicable to an interface in kinetic equilibrium. The assumption of equilibrium requires that the kinetics at the interface be infinitely fast when compared to the time scales of motion of the interface. The interface motion is therefore limited by thermal diffusion and not by the actual attachment kinetics. Thus, surface tension σ acts to lower the temperature of regions of strong positive curvature, hence exerting a stabilising influence. In faithfully simulating the crucial role of surface tension, an accurate computation of curvature is necessary. The curvature of the front and the normal are obtained from Eqs. (5.7) and (5.8). Thus, we demand accuracy in computing the derivatives F_x, F_y, F_{xx} of the curve defining the front.

An additional mechanism of importance is related to the actual process by which solidification occurs at the moving interface. Thermodynamic equilibrium is assumed at the interface in adopting the Gibbs–Thomson condition. This is predicated on the interface motion being slow enough that the actual process of solidification—viz., that of appropriately configured and oriented clusters of liquid molecules accreting on the solid surface—is extremely rapid in comparison (infinitely fast kinetics). For non–facetted interface growth, as in metals, this is usually a reasonable approximation. However, to rigorously account for the non–equilibrium situation at the moving interface, one has to modify the Gibbs–Thomson relation to include the effect of finite–rate interfacial attachment kinetics (Chernov 1984). For continuous, or non–facetted growth usually observed in metallic crystals grown from the melt, the modification of the interfacial temperature resulting from this effect is expressed as,

$$V = d(\phi)(T_{interface} - T_{interface}|_{eq}) \qquad (5.14)$$

where V is the normal velocity of the interface, $T_{interface}$ is the actual interface temperature after accounting for the effect of kinetics and ϕ is the orientation of the normal with respect to the horizontal. Thus the interfacial attachment rate is a linear function of the temperature difference. It is noted that the crystalline anisotropy expresses itself in the attachment kinetic coefficient also. It is, however, difficult to experimentally quantify the value of $d(\phi)$ for most metals, since these are microscopic effects operating on lengths in the range of Å. Such data are available for very few materials.

5.4 A Scaling Procedure for the Conduction Driven Phase Change Problem

In this Section, we discuss appropriate procedures for scaling the governing equations involving interface movement.

(i) Inner Region

For the phase change problem , in the absence of convection ,two length scales are obtained. $\delta_T = a_l/V_i$, is the thermal diffusion length, and $d = \sigma/\Delta h_{l-s}$, is the capillary length. The capillary length, controlled by surface tension, presents a cutoff length scale in that a spherical solid nucleus below this size range will melt back. This is the range at which surface tension operates to smooth out perturbations. Linear stability analysis of a planar interface indicates that the critical wavelength for morphological stability is given by,

$$\lambda_i = O(\sqrt{d\,\delta_T})\qquad(5.15)$$

Thus the instability wavelength lies in the range of microns. Therefore three distinct scales can be defined. The only length scale that is a property of the material is the capillary length $d\,(=\sigma/\Delta h_{l-s})$. Let $\varepsilon_1 = \lambda_i/\delta_T$, the ratio of instability wavelength to the diffusion length, $\varepsilon_2 = d/\delta_T$, ratio of the capillary length to the diffusion length and $\varepsilon_3 = d/\lambda_i$, the ratio of the capillary length to the instability length. Typically, for the small undercoolings encountered, ε_1, ε_2 and ε_3 are all small quantities. Thus, when viewed from the scale of the external diffusion field the interface is planar to first order.

In practice the sole parameter that is externally controlled for solidification from a pure melt is the melt undercooling ΔT. The front velocity for a planar front is then a function of ΔT and the size of the domain. In tracking the morphological development of the interface the length scale of concern is the instability wavelength λ_i, which will be adopted as our length scale.

The following development follows Shyy *et. al.* (1993c). Let us define the region at the scale of λ_i as the inner region and that corresponding to δ_T as the outer region (here we shall take $\lambda_i = \sqrt{(d\delta_T)}$. Figure 11 illustrates the situation resulting from the differences in scale that we encounter in morphological stability phenomena. From the figure, it is clear that the temperature variations faced by the region adjoining the front, $\Delta_i T$, are of order $\Delta T \lambda_i/\delta_T$ and not ΔT and $\Delta T \lambda_i/\delta_T \ll \Delta T$. Adopting a temperature scale ΔT results in extremely slow computational development of the front due to the resulting small values of non–dimensional velocities. An appropriate choice of scaling parameters has been found to be critical for computational efficiency.

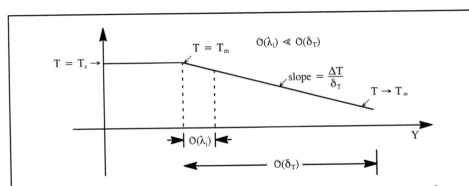

Figure 11. Illustration of temperature field and scales relevant to inner and outer regions.

Thus, with the scales decided upon above, the equation for velocity of the interface becomes,

$$\varrho_s \, \Delta h_{l-s} V_i \, V_N^* = \frac{\varrho_l \, \alpha_l \, C_p \, \Delta T}{\delta_T} \left(-\frac{\partial \theta}{\partial n_l^*} + \frac{k_s}{k_l} \frac{\partial \theta}{\partial n_s^*} \right) \tag{5.16}$$

Here Δh_{l-s} is the latent heat per unit mass, V_N^* is the dimensionless velocity, and V_i is the reference velocity scale. For $\mathcal{O}(1)$ non–dimensional front velocity, we obtain the velocity scale, $V_i = \mathcal{O}(St \, \alpha_l/\delta_T)$, where St is again the Stefan number defined as $C_p \Delta T/\Delta h_{l-s}$. This choice of the velocity scale, V_i, differs from the previous choice, mentioned in Section 5.2, by a factor of St. This difference is critical as it affects the interface speed and time step size of the computations (Shyy et. al. 1993c). With the previous choice of V_i, the non–dimensional interface velocity will be extremely slow for low St. The equation (5.16) then becomes, in non–dimensional form,

$$\varrho_s \, V_N^* = \left(-\frac{\partial \theta}{\partial n_l^*} + \frac{k_s}{k_l} \frac{\partial \theta}{\partial n_s^*} \right) \tag{5.17}$$

It is emphasized that the form of the governing equation adopted depends on the scaling procedure applied. The choice of length, time and velocity scales and the resulting governing equations have substantial implications in terms of computational efficiency. Thus the time scale of motion of the interface is,

$$\mathcal{T} = \frac{\lambda_i}{V_i} = \frac{\lambda_i \delta_T}{\alpha_l St} \tag{5.18}$$

Hence the dimensionless form of equation for heat conduction is,

$$\varepsilon_1 St \, \theta_\tau = \nabla^2 \theta \tag{5.19}$$

where $\varepsilon_1 = \lambda_i/\delta_T << 1$.

Hence, to the lowest order the inner temperature field is governed by the Laplace equation and not the unsteady diffusion equation. The diffusion equation is relevant to the interfacial processes only when $St = \mathcal{O}(1/\varepsilon_1)$ i.e., when the undercooling is large enough in comparison to $\Delta h_{l-s}/C_p$ and the size scale of instability becomes comparable to the diffusion length. However, as the undercooling increases, the length scale of instability decreases.

Defining a field of temperature difference $\theta^* = \theta - \theta_m$, the equations now reduce to,

$$\varepsilon_1 St \frac{\partial \theta^*}{\partial \tau} = \nabla^2 \theta^* \tag{5.20}$$

$$\varrho_s V_N^* = \left(-\frac{\partial \theta^*}{\partial n_l^*} + \frac{k_s}{k_l} \frac{\partial \theta^*}{\partial n_s^*} \right) \tag{5.21}$$

and,

$$\theta^*_{interface} = -\varepsilon_2 \theta_m \sigma^*(\phi) \varkappa^* + \frac{1}{d^*(\phi)} V_N^* \tag{5.22}$$

We designate $\varepsilon_2 \theta_m \sigma^*$ as σ_{eff}, the effective surface tension for our problem.

(ii) Outer region

As far as the outer region is concerned, the interface is to leading order merely a planar front. The temperature scale appropriate to the outer region is ΔT, the undercooling. Lengths scale like $\delta_T = 2a_l/V_i$, the diffusion length, where V_i is the typical velocity of the planar front as viewed from the outer scales. The outer time scale may thus be taken to be δ^2_T/a_l. Thus, the equations governing the outer field are, in dimensionless form,

$$\theta_\tau \;=\; \nabla^2\theta \tag{5.23}$$

$$\varrho^* V_N^* \;=\; St \; (-\frac{\partial\theta}{\partial n_l^*} \;+\; \frac{k_s}{k_l}\frac{\partial\theta}{\partial n_s^*}) \tag{5.24}$$

and on the interface,

$$\theta_{interface} \;=\; \theta_m \tag{5.25}$$

Conditions corresponding to the imposed undercooling apply at the boundaries of the external field. In the results to follow in Section 5.6, however, we do not compute the external field. Instead a further assumption is made that the interfacial motion is slow enough so that the external field is merely represented (in dimensional terms) by,

$$T_l \;=\; T_m \;-\; (\frac{y}{\delta_T}) \, \Delta T \tag{5.26}$$

Here the temperature relaxes to $T_m - \Delta T$ over a distance δ_T, the diffusion length. Therefore at distances y_{out} from the interface,

$$T_l|_{inner} \;=\; T_m \;-\; (\; \frac{y_{out}}{\delta_T} \;) \, \Delta T \tag{5.27}$$

Non–dimensionalizing, one obtains the temperature at y_{out} as,

$$\theta_l|_{inner} \;=\; \theta_m \;-\; Y_{out} \tag{5.28}$$

where $Y_{out} = y_{out}/\lambda_i$ according to the scales adopted. Hence this value of temperature is specified at the boundary at $Y=Y_{out}$.

5.5 Morphological Model with Convection Effect

Macrosegregation is strongly influenced by convective effects and hence the microstructural features also respond to the external solute field through the solute boundary layer. The issue of convective influence on cellular/dendritic growth is an open area of research. In what follows, we present the issues involved in the full crystal interface evolution problem with melt flow, for shapes that are not constrained to be of small distortion. The inclusion of convection calls for a rethinking of the scaling procedure for pure materials as discussed in Section 5.4. No general theoretical treatment of length scale estimation exists in the presence of convection. Ananth and Gill (1989, 1991) obtain curves of λ_i vs V_i, ie. the tip radius against translation velocity of the tip, as function of Grashof number. The selection of the tip radius is achieved with the aid of the solvability mechanism. There are several limitations to this approach. Firstly, their analysis is based on the fictitious isothermal paraboloid which has no information regarding surface tension or anisotropy. Also, Ananth and Gill's treatment is only valid when vanishingly close to the tip of the dendrite (where the Ivantsov paraboloid is a good approximation). It is unlikely that analytical results on selection criteria for the convective case will be forthcoming, when even in the case of pure conduction, this is still an open question.

A major issue then is to choose the appropriate scale for the problem to guide the numerical computation. For the pure conduction case, as already discussed, it is generally indicated that the wavelength of the morphological instability scales as $\lambda_i = \mathcal{O}\left(\sqrt{d\,\delta_T}\right)$, as mentioned before. Usually, $\delta_T \approx \mathcal{O}(\mathrm{mm})$ and $d \approx \mathcal{O}(\mathrm{\mathring{A}})$ resulting in $\lambda_i \approx \mathcal{O}(\mu)$. In our previous scaling procedure, we have adopted this length scale with the attendant difficulty that δ_T needs to be estimated to obtain λ_i. But for the pure conduction case, $\delta_T = 2a_l/V_i$, where a_l is the thermal diffusivity and V_i is the interfacial velocity, which is itself unknown. This is merely an off–shoot of the chronic problem with the Ivantsov model in that it takes the form (Pelce 1988),

$$St = \frac{\Delta T}{(\Delta h_{l-s}/C_p)} = \exp(P)\ \sqrt{(\pi P)}\ \mathrm{erfc}(\sqrt{P}) \tag{5.29}$$

where $P = \lambda_i V_i / 2a_l$ is the Peclet number, and ΔT is the undercooling at infinity. Thus a continuous family of solutions is yielded specifying the product $\lambda_i V_i$ and a selection needs to be made. This selection is induced by the surface tension and to account for this, a marginal stability criterion was theoretically derived by Langer and Muller–Krumbhaar (1978) which took the form,

$$\lambda_i^2 V_i = \frac{2a_l\,d}{\sigma^*} \tag{5.30}$$

where σ^* is a parameter depending on crystalline anisotropy. The value of σ^*, obtained from the marginal stability hypothesis was found to be remarkably close to the experimental value of 0.02 obtained by Glicksmann and Huang (1986) for succinonitrile. A firm physical basis for the above selection criterion is not established and it is at this stage a semi–empirical formula on account of σ^*. In fact, it is not confirmed that such a relationship holds for materials other than succinonitrile, e.g., in the case of pivalic Acid the relation is of doubtful validity. However, at present, Eq. (5.30) is the most widely accepted selection criterion and is useful in setting up our computational framework . It may be noted, however, that the issue of selection does not, in principle, change the quality of the solution. However, an injudicious choice of scaling parameters will result in the situation that the full movement of the interface cannot be captured economically, a fact that will only be revealed *a posteriori*. Looking at the Ivantsov paraboloid, discussed in Section 2.6, then, Eq. (5.30), implies that,

$$St = \exp(P)\ \sqrt{(\pi P)}\ \mathrm{erfc}(\sqrt{P}) \tag{5.31}$$

where St is the Stefan number defined as $\Delta T / (\Delta h_{l-s}/C_p)$. In most crystal growth cases, $P << 1$, since $\lambda_i \approx \mathcal{O}(\mu m)$, $V_i \approx \mathcal{O}(\mu m/s)$ and $a_l \approx 10^{-3}$ cm^2/s. Thus one obtains, in the limit $P \longrightarrow 0$,

$$St = \left(1 + P + \frac{P^2}{2} + ...\right)\left(1 - \frac{2}{\pi}\left(\sqrt{P} - \frac{P^3}{2} + ...\right)\right)\sqrt{\pi P} \qquad (5.32)$$

$$\text{As } P \to 0, \quad St \to \sqrt{(\pi P)} = \sqrt{\pi \lambda_i \frac{V_i}{2} a_l} \qquad (5.33)$$

Therefore,

$$\frac{\lambda_i V_i}{2 a_l} = \frac{St^2}{\pi} \qquad (5.34)$$

In order to separate the morphological length and velocity scales, λ_i and V_i, we may use Eqs. (5.30) and (5.34), to obtain,

$$\lambda_i = \frac{\pi d}{\sigma^* St^2} \qquad (5.35)$$

and

$$V_i = \frac{2 a_l St^4 \sigma^*}{\pi^2 d} \qquad (5.36)$$

Thus, the Eqs. (5.35) and (5.36) give a rough idea of the length and velocity scales assumed by a crystal growth form for given Stefan number or undercooling ΔT. Equation (5.35) is physically sensible in that as surface tension increases λ_i also increases, as the anisotropy increases λ_i decreases, the needle becoming sharper, and as the Stefan number (undercooling) decreases λ_i increases. All these behaviours are consistent. Our choice of length scale applies of course to a purely conducting situation only. We choose to extrapolate its application to the system with convection, since we are only concerned with a representative length scale. Equation (5.35) does not purport to be quantitatively accurate. A quantitative estimate of the wavelength selected in pattern–forming systems still remains a central point of investigation.

Choice of the time scale

The choice of time scale $\mathfrak{T}$ is motivated by the objective of our investigation, which is to view interfacial development. Thus, it is desirable to non–dimensionalize the interfacial motion equations so that the interface moves an $\mathcal{O}(1)$ distance in a time unit. The interfacial motion is determined by the normal velocity as follows:

$$\Delta h_{l-s} V = \Delta h_{l-s} V_i V_N^* = \mathcal{O}\left(a_l C_p \{\|\nabla\theta\| \cdot \vec{n} \, \Delta_i T / \lambda_i\}\right) \qquad (5.37)$$

where the symbol $\|\quad\|$ indicates the jump in the non–dimensional temperature gradient $\nabla\theta$ across the interface, where $\Delta_i T$ is the inner temperature scale. If ΔT is used instead, then δ_T should replace λ_i in Eq. (5.37). Before defining a time scale, it is necessary to obtain the scale $\lambda_i T$. While in the outer field an appropriate order of magnitude for the temperature field is ΔT, the imposed undercooling, the temperature variations in inner field are much smaller owing to the fact that $\lambda_i \ll \delta_T$. Figure 12 illustrates the situation. In the figure, an appropriate temperature scale in the inner region based on the slope of the global temperature profile slope of $\Delta T/\delta_T$, is $\lambda_i T = \mathcal{O}(\Delta T \lambda_i/\delta_T)$. Note that if $\delta_T = \alpha_l/V_i$ is estimated for an interface moving at (average) velocity V_i, then the ratio $\lambda_i/\delta_T = \lambda_i V_i/\alpha_l$ which is nothing but a Peclet number at the inner length scale and is $\ll 1$. The extent of the thermal length scale δ_T, depends on the type and intensity of convection present. Its value can be estimated based on an order of magnitude analysis and boundary layer treatment adopted by Garandet et al (1990). In the pure conductive case, since the procedure outlined above gives $V_i = (2\alpha_l St^4 \sigma^*/\pi^2 d)$, one gets $\delta_T = \mathcal{O}(\pi^2 d/2St^4\sigma^*)$ which for succinonitrile (SCN), and for $St = 0.1$, gives a reasonable value of approximately 1.5 cm. Also, as the Stefan number decreases, δ_T increases, which is consistent since a decrease in Stefan number implies a slower moving interface and the diffusion of latent heat is facilitated. Reverting to the question of time scale, one writes the interfacial velocity equation in normalized terms as follows,

$$V_i \, V_N^* \approx \frac{St \, \alpha_l}{\lambda_i} \| \nabla\theta \| \cdot \vec{n} \tag{5.38}$$

Now, for the Lagrangian update of interfacial points, we have,

$$\frac{dX}{d\tau} \propto V_N^* \quad , \quad \frac{dY}{d\tau} \propto V_N^* \tag{5.39}$$

Thus, for an $\mathcal{O}(1)$ development, for y–motion for example,

$$\frac{\lambda_i}{\Im} \frac{dY}{d\tau} \sim V_N^* \sim \frac{St \, \alpha_l}{\delta_T} \| \nabla\theta \| \cdot \vec{n} \tag{5.40}$$

which implies $\Im = \mathcal{O}(\lambda_i \delta_T d / \sigma^* St^3 \alpha_l)$. This gives, after substituting for λ_i,

$$\Im = \mathcal{O}(\frac{\pi \delta_T d}{\sigma^* St^3 \alpha_l}) \tag{5.41}$$

(As St decreases time scale of motion becomes larger). Thus, this is the time scale at which to view morphological evolution (Udaykumar and Shyy 1993a).

Having obtained the scales as determined above, we turn to the set of field equations to be solved. First, a fluid flow velocity scale needs to be obtained. Following the considerations involving temperature above, we estimate the velocity scale from $u_f \approx \mathcal{O}(U_{out} \lambda_i / \delta_v)$. Here U_{out} is the scale of velocity in the outer region and δ_v is the momentum boundary layer thickness. The situation is illustrated in Figure 12. u_f can be vastly different from V_i in scale. The value of u_f is connected with both the outer velocity field via U_{out} and the outer thermal field (ie. St) via λ_i.

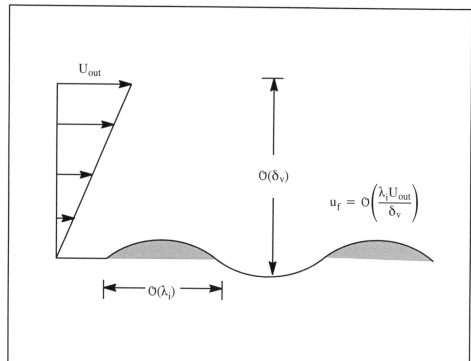

Figure 12. Illustration of velocity field assumed and the appropriate scales in the inner and outer regions for the flowfield.

Therefore, the inner field equations, after some simplification, may be written in non–dimensional form, using $\vec{w}^*$, θ^*, and P^* to represent the corresponding dimensional velocity, temperature and pressure, $\vec{w}$, T, and P, and using X, Y, τ as the nondimensional independent variables. The length, velocity, time and temperature reference scales are, respectively, λ_i, u_f, $\Im$, and $\Delta_i T = \dfrac{\Delta T \, \lambda_i}{\delta_T}$. Here θ^* is defined as $(T - T_m)/\Delta_i T$ and $\vec{w}$ is the velocity vector consisting of u and v components. The equations are as follows:

Continuity Equation:

$$\frac{\partial u^*}{\partial X} + \frac{\partial v^*}{\partial Y} = 0 \tag{5.42}$$

Momentum Equation:

$$\frac{\partial \vec{w}^*}{\partial \tau} + \left(\frac{U_{out}\delta_T}{\delta_v St^3} \cdot \frac{d\pi}{\sigma^* a_l}\right) \vec{w}^* \cdot \nabla \vec{w}^* = Pr \, \delta_T \, St \, \frac{\sigma^*}{d\pi}\left(\nabla^2\vec{w}^* - \nabla P^*\right)$$
$$- \left(\frac{\beta \, g \, d \, \pi \, \Delta T \, \delta_v}{a_l \, \sigma^* \, C_p \, U_{out} \, St^2}\right) \theta^* \, \vec{e}_j \tag{5.43}$$

where $\vec{e}_j$ is the unit vector along the vertical direction, and *Pr* is the Prandtl number.

Energy Equation:

$$\frac{\partial \theta^*}{\partial \tau} + \left(\frac{\delta_T U_{out}}{\delta_v St^3}\right)\left(\frac{d\pi}{a_l \sigma^*}\right)(\vec{w} \cdot \nabla \theta^*) = (\delta_T St)\left(\frac{\sigma^*}{d\pi}\right)\nabla^2\theta^* \tag{5.44}$$

Boundary Conditions:

   In dimensional form, one imposes boundary conditions as shown in Figure 13(a). The non–dimensional version is shown in Figure 13(b). The temperature boundary conditions at the interface are as follows.
Dimensionally,

$$T_{interface} = T_m - dT_m\varkappa \tag{5.45}$$

$\varkappa$ being the curvature. In terms of θ^*, we obtain non–dimensionally,

$$\theta^*_{interface} = -\frac{dT_m\varkappa^*}{\lambda_i} \tag{5.46}$$

and substituting for λ_i,

$$\theta^*_{interface} = -\left[\frac{St^3\sigma^{*2}T_m\delta_T}{\pi^2(\Delta h_{l-s}/C_p)d}\right]\varkappa^* \tag{5.47}$$

which is our non–dimensional form of the Gibbs–Thomson equation. These quantities are therefore required in the model presented above.

   In summary, inputs from the outer field are, (i) δ_T, thermal length scale, (ii) δ_v, viscous length scale, (iii) U_{out}, outer velocity scale and (iv) *St*, the Stefan number based on undercooling at infinity. Material parameters to be given are, (i) *d*, capillary length, (ii) σ^*, the microscopic solvability factor which is a function of anisotropy, (iii) $(\Delta h_{l-s}/C_p)$ ratio of the latent heat to specific heat, (iv) T_m, the melting temperature. Representative values of the parameters appearing in the equations above as applicable to SCN are calculated as follows. $\lambda_i = (\pi d/\sigma^* St^2) = (3.014 / St^2)$ microns For instance, *St*=0.1 implies an undercooling of 2.8 degrees at large distances from the interface and $\lambda_i \sim 300$ microns. In the experiments of Glicksman and Huang the diameter of the growth chamber is approximately 4 cm. The dendrites develop on the scale of mm or less.

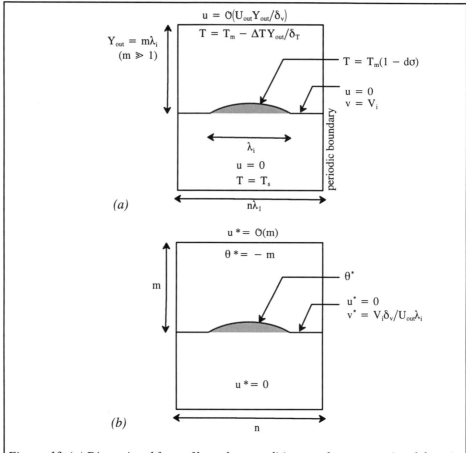

Figure 13. (a) Dimensional form of boundary conditions on the computational domain.
(b) Non–dimensional form of the boundary conditions imposed.

5.6 Examples

The effect of a modelled fluid flow field is investigated. The flow–field is imposed as a local Couette flow at each point of the interface, as shown in Figure 12. It should be noted that the 'onset' of thermosolutal convection during directional solidification was studied by Corriel et al (1980), and the onset of solutal convection alone by Hurle *et. al.* (1983). In the following, it is the impact of convection on interface development that is under study.

The magnitude of the fluid velocity at any section *x* follows the Couette profile shown. The streamlines are then set parallel to the interface at each section. This then represents a "deformable" boundary layer situation adopted by Garandet *et. al.* Thus, due to the velocity field,

$$u^{*2} + v^{*2} = u^{*2}_f = Y^2 \tag{5.48}$$

and $v/u = \tan\phi$, which determines u and v.

Periodic boundary conditions are imposed on the sides of the computational domain for the calculations presented below. The equation for energy now reads,

$$\frac{\partial\theta^*}{\partial\tau} + C_1(u^*\frac{\partial\theta^*}{\partial X} + v^*\frac{\partial\theta^*}{\partial Y}) = C_2\nabla^2\theta^* \tag{5.49}$$

We call C_1 a convection factor. In the calculations below, for cases with convection, we use $C_2 = 10$ and $C_1 = 5$.

Figure 14 shows a case with $\sigma_{eff} = 0$, see Eq. (5.22), i.e., no surface tension for the pure conduction case. As seen in Figure 14(a) rapid development of the interface with time ensues leading quickly to the formation of cusps on the interface. The imposition of periodic boundary conditions results in the formation of a finger of shorter wavelength than the dominant finger. This is explained as follows. The region marked B is one of negative curvature. The velocity at this point is slower than the surrounding portions due to the concentration of heat in the concave regions. Thus, as seen in Figure 14(b), the regions marked C and A travel faster than region B. Thus, the curvature at B becomes progressively more negative leading precipitously to cusping behavior. In the absence of surface tension no smoothing mechanism exists and the cusp deepens, ultimately stalling the calculation. It is noticed that in the conductive case perfect symmetry is maintained across the centreline of the perturbation. Also, the periodicity of the interfacial calculation is maintained well. Figure 14(b) shows that the velocity in the region B is much smaller than at region A. In fact, in the final stages, the region B may actually be remelting, although the accuracy of the calculation in the final stages may be suspect.

Figure 15(a) shows the development of the interface with time for $\sigma_{eff} = 0$, and for convection parameter $C_1 = 5$. The introduction of the flow from left to right introduces a bias in the temperature field. This bias is evident from Figure 15(a), (b) and (c). Figure 15(b) shows the interfacial shape by subtracting the mean interfacial position from the total interfacial shape. The arrows drawn on that figure show that the tip of the perturbation deviates slightly as it propagates. Thus, the finger tilts in the direction of the flow as it grows. This is also evident from Figure 15(c) which shows the normal velocity along the interface. The point of maximum velocity shifts rightward. But the effect of convection does not prevent the eventual development of cusps at the interface. Thus surface tension is still a necessity in preventing singularities on the interface.

Figure 16 represents the case with $\sigma_{eff} = 0.2$ and $C_1 = 0$, i.e. pure conduction. Periodic boundary conditions are again applied. The most striking fact, in comparison to Figure 14 is of course the absence of the second finger in this case. The action of surface tension stabilises the region B shown above and there is no tendency of the regions of negative curvature to lag behind the neighboring regions since the negatively curved parts are now warmer on account of the Gibbs–Thomson effect. This dramatically changes the picture as far as interface evolution goes.

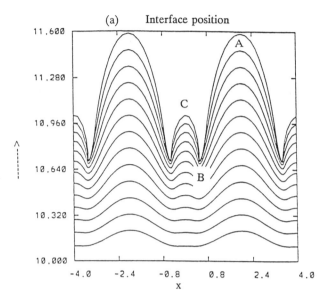

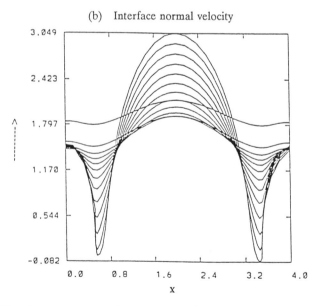

Figure 14. Development of interface with time for pure conduction case , $C_l = 0$, with no surface tension. (a) Evolution of interface shape with time. Features are formation of a small finger in addition to the main propagating one. Periodic boundary conditions imposed at the sides are well maintained. (b) Normal velocity along the interface shown at different time instants. The arrow shows the sequence in time.

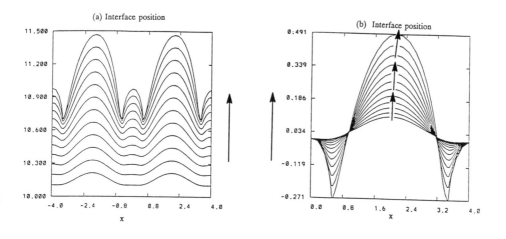

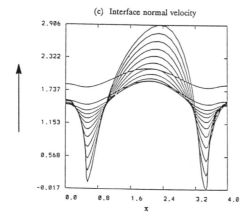

Figure 15. Development of interface with time for case with no surface tension and convection parameter $C_1 = 5$. (a) Evolution of interface shape with time. Features are formation of a small finger in addition to the main propagating one. Periodic boundary conditions imposed at the sides are well maintained. (b) Interface shape shown after subtracting mean interface position. Deviation of peak in the direction of flow is evident as time progresses. (c) Normal velocity along the interface shown at different time instants. Maximum velocity is progressively biased to the right. The arrow shows the sequence in time

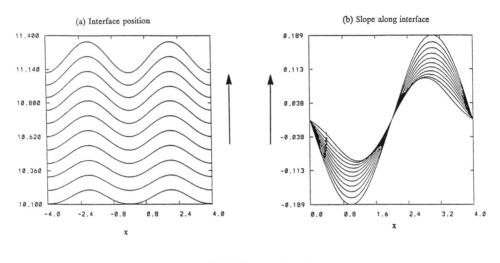

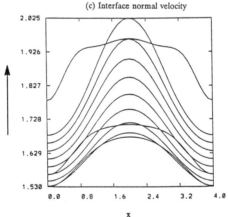

Figure 16. Development of interface with time for case with surface tension, $\sigma_{eff} = 0.2$ and $C_l = 0$. (a) Evolution of interface shape with time. The extra finger found for the zero surface tension case is noticeably missing. Periodic boundary conditions imposed at the sides are well maintained. (b) Calculated slopes along the interface. No bias is evident. (c) Normal velocity along the interface shown at different time instants. Symmetry across the centerline is observed. The arrow shows the sequence in time

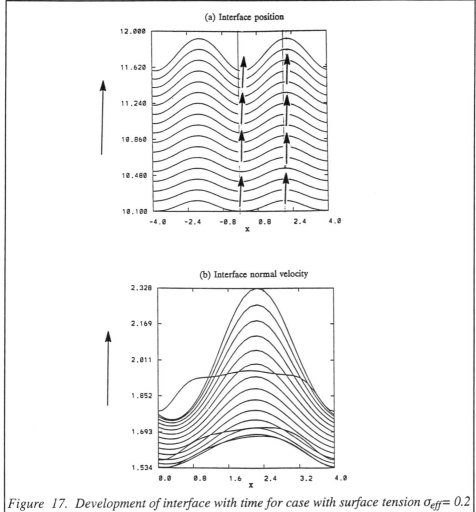

Figure 17. Development of interface with time for case with surface tension σ_{eff}= 0.2 and convection parameter C_1 = 5. (a) Evolution of interface shape with time. Interface remain smooth. Periodic boundary conditions imposed at the sides are well maintained. Interface as a whole appears to translate to the right (b) Normal velocity along the interface shown at different time instants. Maximum velocity is progressively biased to the right. The travelling wave nature of the whole motion is evident. The arrow shows the sequence in time

Figure 16(a) is intended to affirm the robustness of the calculation procedure as far a tracking interfacial properties goes. Shown in Figure 16(b) is the shape along the interface at various time steps. Good symmetry is maintained about the centerline and periodicity at the ends is perfectly enforced. Figure 16(c) shows the interfacial normal velocity and indicates the rapidly progressing unstable behavior as the tip of the perturbation travels with increasing velocity in this developing stage of the instability. More information can be found in Udaykumar and Shyy (1993a).

Figure 17 shows the motion of the interface for $\sigma_{eff} = 0.2$ and $C_1 = 5$. The distinguishing feature here is that the perturbation as a whole appears to be translating to the right, ie. a travelling wave situation results. The arrows marked on (a) display this behavior. The vertical line is drawn to show this deviation of the peaks of the perturbation. Figure 17(b) shows the interfacial velocity over one wavelength. The shift of the peak in the velocity is evident. Also noticeable from this figure is the periodicity in the interfacial velocity, which is well maintained, as seen by comparing the left and right boundaries. This figure also confirms the travelling wave nature of the instability.

In respect of the above observations, it may be noted that although the amplitudes of the perturbations are still not significant, and the value of the convection parameter is unrealistic, some results demand attention. The travelling wave nature of the instabilities in the presence of convection has been indicated also by the linear stability analyses for impure materials (Forth and Wheeler 1992). The effect of convection on fully–developed dendrites is likely to be interesting. The current numerical method is in the process of being extended to introduce more sophisticated interface tracking techniques to capture the highly nonlinear stages of the evolution. Furthermore, the velocity field above was merely a model and in fact violates the continuity equation. The velocity field will be solved also as a further refinement of the computational procedures.

6. TRACKING OF HIGHLY DISTORTED FRONTS

The interface tracking method employed in the previous sections involved a moving, boundary conforming grid (Shyy et. al. 1993c). However, when the interface becomes highly branched, as in the formation of a dendritic structure, the generation of a boundary conforming grid is a very difficult task. Furthermore, in the event of topological changes such as the merger or break–up of interfaces, the boundary fitted grid would have to be rearranged, leading to intensive logical manipulation and interpolation errors in the vicinity of the discontinuity, i.e., the interface. Thus, there is a need to "decouple" the motion of the phase front from the grid motion. To accomplish this task, an interface tracking methodology has been developed (Udaykumar and Shyy 1993b) as described below.

6.1 Background

The conditions engendered by the typical interfacial instability, which a numerical procedure is required to simulate, may be listed as follows:

(i) The interface is likely to be highly branched. There also exists the possibility of fragmentation or merger of interfaces. The generation of a body–fitted grid to conform to the interface shape (Shyy *et. al.* 1993c , Thompson *et. al.* 1985) would then be impractical.

(ii) The equations governing the instability phenomena depend critically on such features of the curve defining the interface as the local normal, and curvature. These quantities appear in the form of derivatives of the defining curve $F(X, Y, \tau) = 0$ for the interface. The first and second derivatives F_x , F_y, F_{xx}, F_{yy} need to be obtained very accurately. The calculation of the derivatives is prone to numerical inaccuracy, since these are obtained by the division of small numbers by other small numbers. Their accuracy is nevertheless of utmost importance for a successful numerical calculation.

(iii) Boundary conditions are to be imposed at the exact location of the interface separating the two phases. The interface is itself in motion. There is a discontinuity across the interface in such quantities as temperature gradient and species concentration. Thus care needs to be exercised in computing these field variables to prevent smearing of solutions. A physically conservative and geometrically consistent difference formulation is thus called for.

The above three primary restrictions influence the choice of numerical scheme for interface tracking. Additionally, a robust numerical procedure is desired, which does not, in its implementation further complicate an already formidable problem. Numerical schemes that apply to phenomena attended by moving interfaces across which some type of discontinuity exists, have been extensively researched for quite some time (Floryan and Rasmussen 1989 , Oran and Boris 1987). The literature abounds in schemes applied thus far to the unstable solidification problem. Some popular approaches are mentioned below:

(i) *Boundary element methods with simplified physics.* Conduction is the only mechanism treated in each phase. Dendritic crystalline forms, including side–branched structures have been produced (Sato *et. al.* 1988, Strain 1989). Analogous viscous fingering phenomena have also been treated by this technique (DeGregoria and Schwartz 1986). The applicability of this method with the inclusion of the fluid flow field and complicated time–dependent boundary conditions (Bouissou *et. al.* 1990 , Fabietti *et. al.* 1990) is yet to be demonstrated.

(ii) *Stochastic methods such as the random vortex method* (Aref and Tryggvason 1984), Monte Carlo method (Xiao *et. al.* 1988, 1990), and random walk method (Liang 1986) have been applied to various instability phenomena. Such calculations, while very effective in illustrating qualitative features, still need to be developed further to yield accurate values of local curvature etc., in order to apply to the solidification problem.

(iii) *Domain decomposition methods* have been extensively used (Ungar amd Brown 1984, McFadden and Coriell 1987) to study the cellular growth problem. However, the method still relies partially on domain transformation (mapping), which is inapplicable for highly contorted interfaces with multiply defined functions $F(X,Y, \tau)$ in X and Y. Glimm et al (1986) have used characteristics based methods and Chern et al (1986) applied finite element methods in conjunction with an interface tracking procedure similar in spirit to that presented in this work and a local Lagrangian grid in the vicinity of the interface. The method was applied to the Rayleigh–Taylor instability in (Chern *et. al.* 1986) . The effectiveness of a Lagrangian grid is diminished by grid skewness and entanglement problems and requires periodic reconnection of grids and consequent interpolations which may result in numerical diffusion. In their method, however, since the local grid near the interface conforms to the interface, the control volume formulation may be applied so as to essentially obviate numerical dissipation at the interface. The application is in this respect akin to the method employed by Shyy *et. al.* (1993c).

(iv) *Eulerian Methods*: The Volume of fluid (VOF) methods method in its various versions (Youngs 1984, Hirt and Nichols 1981, Chorin 1985 and Liang 1991) has emerged as a popular interface tracking procedure. The basic idea is to define a liquid fraction variable field, F, on an Eulerian grid. In a grid cell $F=1$ if the cell is completely in liquid phase, $F=0$ if the cell is empty, and $0< F<1$ if the cell contains the interface. The method accommodates naturally mergers and breakups of the interface without additional logical manipulation when such events occur. Until recently, the VOF method yielded interfaces that were not continuous, but currently this limitation is eliminated (Poo and Ashgriz 1989, Ashgriz and Poo 1991) in a modified VOF procedure. However, the operations involved in reconstructing the interface and extracting curvature values are computationally taxing. Adaptive gridding or local refinement to enhance resolution are yet to be attempted. Volume fraction based methods have been applied to the solidification problem via the enthalpy formulation by Chorin (1985). The success of the methods depends critically on the accuracy with which the interface shape, especially the curvature (Chern *et. al.* 1986, Youngs 1984, Hirt and Nichols 1981) is obtained by the interface reconstruction procedure. The method developed in this work employs a Cartesian grid to track the interfacial markers, as in the VOF technique. However, while in the latter, the interface is reconstructed, in a Lagrangian fashion, the interface location is explicitly obtained. Also, the Lagrangian nature of the interface description yields accurate values of interfacial shape directly. In a simulation, this affords the possibility of application of boundary conditions such as capillarity effects at the exact interface position instead of 'somewhere' in the computational cell. Recently, Unverdi and Tryggvason (1992) have employed an interface tracking procedure that incorporates an unstructured, moving grid with an underlying Eulerian grid. The boundary conforming unstructured grid yields flexibility in capturing distorted interfaces, while maintaining an explicitly defined interface. The solution of the field equations is performed on the stationary grid. Complex, three dimensional interfacial behaviours have been simulated. Communication between the Eulerian and Lagrangian grids involves interpolations and redistributions. The impact of such procedures in terms of numerical dissipation remains to be assessed.

(v) *Phase Variable Methods:* An interface capturing scheme under development by Sethian and coworkers (1990, 1992) relies on defining a field ϕ corresponding to a distance function over an Eulerian grid. The distance function ϕ is defined initially as the distance of a grid point from the interface and the time evolution of this function is computed over the domain. The contour corresponding to $\phi = 0$ defines the interface at any time. This function is computed throughout the domain based on a Hamilton–Jacobi formulation and advanced numerical schemes available for hyperbolic conservation law systems are applied. Highly complicated three–dimensional interfaces have been captured in this fashion. Mergers and breakups of interfaces are naturally taken care of from the evaluation of the ϕ field. However, the interface motion is computed using an integral equation form as in the case of the boundary element techniques. The resolution of the method and the application of special boundary conditions is not apparent from the results published thus far. Also, in the presence of a fluid flow field and time dependent boundary conditions, the method in its current form is subject to the same limitation as the boundary element method. Further development could provide an attractive method for interface tracking. Recently, some interesting work has been conducted by Kobayashi (1993) and also by Wheeler et. al. (1993) in simulating dendritic crystal growth. The model includes two variables; one is a phase field and the other is a temperature field. The phase field is based on the Landau–Ginzberg free energy functional, and the interface is expressed by the thin layer of the phase variable. Highly interesting results have been obtained using this model; however, since the interface is of finite thickness, the issues of feature resolution and computational efficiency have not been resolved.

(vi) *Marker–particle based methods*: The marker particle based methods are attractive on account of their simplicity and the ability to provide accurate interfacial data such as curvatures and slopes. The essential idea is to identify the interface with the aid of a string of marker particles, which are translated in Lagrangian fashion with a calculated velocity field. Several applications of this method are available in the literature (Daly 1967, 1969, Miyata 1986, Fromm 1981), and the boundary integral methods (Sato *et. al.* 1988, Strain 1989, DeGregoria and Schwartz 1986) essentially use the same concept, without a grid for field calculations. The drawbacks with marker particle methods, that have prevented their use for complicated interface shapes are mainly two–fold. Firstly, as the interface perimeter increases and the marker particles are transported, local concentrations or depletions of particles may result, impacting negatively on interface resolution and accuracy. This can be corrected by periodically redistributing (Shyy *et. al.* 1993 c) or adding/ deleting particles (Daly 1967). Numerical dissipation is expected to accompany such procedures. Secondly, difficulties with marker particle methods arise when an interface breaks up or merges with various parts of the same interface or other interfaces. The reordering of interfacial points has been thought to present formidable logical hurdles (Floryan and Rasmussen 1989, Oran and Boris 1987) . Both these problems are addressed in the work presented hereunder. An additional problem with the marker particle method is the anticipated difficulty in extending the method to the three–dimensional situation, where the neighbouring particles along the interface are no longer in sequence. However, a systematic and effective screening scheme can be devised in a logical fashion without heavily penalising the efficiency aspect of the method.

In the following a simple interface update procedure that is relatively inexpensive is presented. By using a marker–based method;

(a)   The number of data points involved in interface tracking equals the number of marker particles. No additional equations are to be solved for each cell of the domain as in the Eulerian or phase variable methods.

(b)   There is no ambiguity regarding interface location as in the VOF method. Crisp, smooth interfaces can be obtained and slopes and curvatures calculated accurately. No elaborate logical operations are involved in obtaining an updated interface shape.

Hence, Udaykumar and Shyy (1993 b) have chosen to develop a grid–supported marker–particle scheme to simulate the dynamics of interface development. In what follows, to facilitate the presentation of the methodology developed, the domains defined by the interface are termed solid and liquid, with the particular instability of a solidifying front as a working example. The applicability of the tracking procedure is, of course, not limited to any particular physical phenomenon.

6.2 Basic methodology

There are three essential elements to the combined Eulerian–Lagrangian interfacial tracking procedure, namely cells, interface particles and interfacial curves connecting the marker particles. The interfacial particles define the terminal points of the interfacial curves. Daly (1969) employed a cubic spline fit to define the interfacial curves. However, when the interface became multi–valued with respect to the dependent variable, he switched to a circular arc definition in such regions. Thus, decisions were involved in the type of curve fit to be used for a particular segment of the interface. The method therein was a standard marker–particle method in that particle depletion/ accumulation was surmounted by adding/deleting particles. In Shyy *et. al.* (1993c), piecewise parabolic approximations were used to define the interface and obtain accurate interface tracking facility, with the aid of marker redistribution based on equal arclength intervals. However, in using parabolic arcs, for multiple–valuedness such as shown in Figure 18 , the representation $Y_i = A_i \; X_i{}^2 + B_i \; X_i + C_i$ would have to be switched to the form $X_i = A_i' Y_i^2 + B_i' Y_i + C_i'$ which will involve keeping track of the turning regions .

A simple way around this situation is to use circular arc elements to obtain interfacial   segments.   Thus,   one   expresses   the   interfacial   curves   as, $(X_i - A_i)^2 + (Y_i - B_i)^2 = R_i^2$ where R_i is the radius of the osculating circle at point i and A_i and B_i are the coordinates of the centre of the osculating circle. The calculation using circles enables evaluation of normal and curvature in a very simple fashion. The determination of the direction assumed by the normal, which is required to point from the solid into the liquid, is also facilitated and will be discussed shortly. Furthermore, the use of circular arc is especially attractive when the estimation of local curvature is needed. Using circular arcs to represent the interface shape, constant curvatures will be maintained between any two consecutive marker points, which is intrinsically consistent with most of the numerical schemes for the solution of the governing field equations, such as the heat transport equation. In this manner, no curvature variation will be generated between the neighbouring marker points, meaning that the accuracy of the scheme will not be artificially inflated. The disadvantage attached to the circular arc representation is the impossibility of dealing with straight interfaces. The method of evaluation of R will then experience an infinite value. However, if required such straight sections could be approximated using a parabolic arc representation, which does not suffer from this difficulty. Or even more easily, a simple cut–off criterion can be assigned beyond which a straight line, instead of a circle is assigned for the curve in such regions.

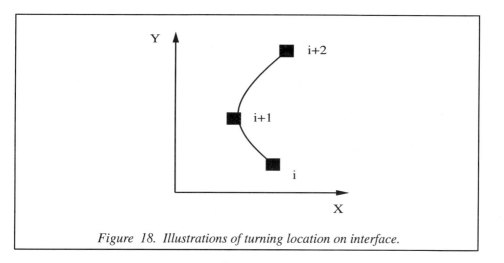

Figure 18. Illustrations of turning location on interface.

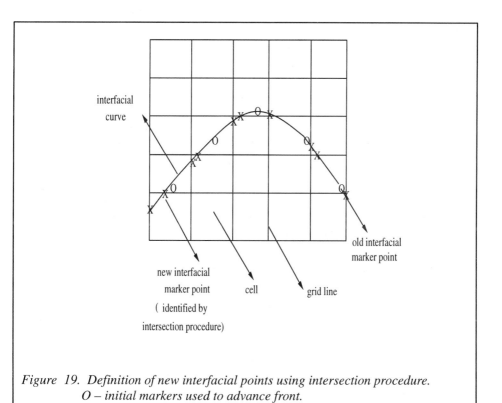

Figure 19. Definition of new interfacial points using intersection procedure.
O – initial markers used to advance front.
X – new markers resulting from redistribution via intersection procedure.

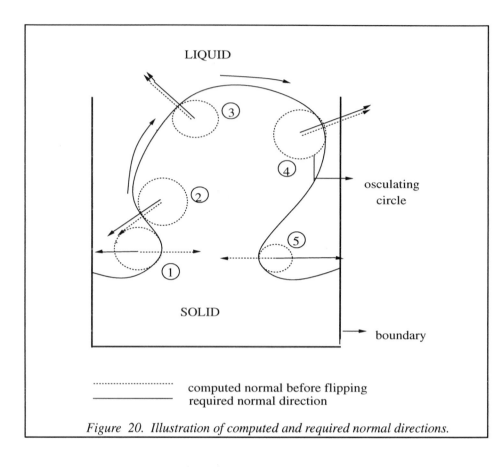

LIQUID

osculating
circle

SOLID

boundary

· · · · · · · · · · · · · · · · computed normal before flipping
_____ required normal direction

Figure 20. Illustration of computed and required normal directions.

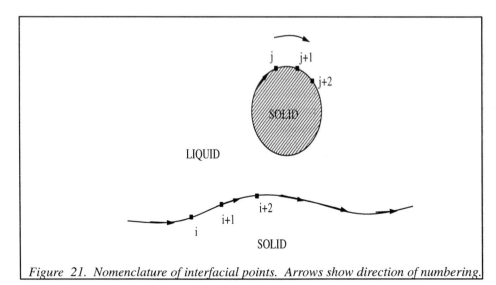

Figure 21. Nomenclature of interfacial points. Arrows show direction of numbering.

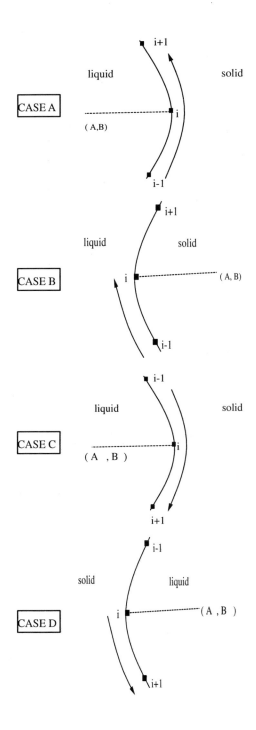

Figure 22. Illustration of conditions for switching direction of normals.

The use of cells, in line with Miyata et al (1986), offers significant advantages. The procedure employed is illustrated in Figure 19. Let the open circles in Figure 19 represent an initial distribution of interfacial points or the interfacial point locations obtained after advancing the interface at any given time step. The points are found lying within the cells. These interfacial points are connected using circular arcs as shown in Figure 19.

For three given points, such as shown in Figure 18, we obtain A_i, B_i, R_i by elimination among the relations $F(X_i, Y_i, \tau) = 0$, $F(X_{i+1}, Y_{i+1}, \tau) = 0$, and $F(X_{i-1}, Y_{i-1}, \tau) = 0$. The intersection of each segment of the interface with the grid lines is then obtained in a straightforward manner. The X's on Figure 19 represent the new locations of the interfacial points. Thus, at each step, an entirely fresh set of particles is obtained. The advantage of such a procedure is that as the interface expands, new intersection points are automatically generated, and the numerical accuracy dictated by the initial grid layout is maintained. The problem of particle rarefaction/accumulation is thus avoided. New points are automatically generated as the interface invades a particular cell. In this regard, the shape function used to link the original marker particles is of critical importance in maintaining accuracy during the particle reassignment procedure. As will be demonstrated later, the use of constant curvature concept to link the particles proves to be very satisfactory. The resolution of the interface tracking scheme is here governed by the grid spacing. However, this is not necessarily a limitation. Since the interfacial point motion is driven by the gradients of the thermal field, the maximum expected resolution is in effect controlled by the resolution of thermal gradients, i.e., the grid spacing. Thus the attainment of improved resolution by using several interfacial points per cell is merely artificial. If further resolution is necessary in any region of the interface, adaptive regridding can be performed (Shyy *et. al.* 1993c), thus improving both the interface and field variable resolution. The other advantage with locating interfacial particles on cell faces is that this facilitates the computation of interfacial velocity and the assignment of boundary conditions and control volumes corresponding to cell control points. This information is necessary anyway and thus no superfluous operations are introduced by the simple intersection procedure. Once the new interfacial points are determined and they are again connected via interfacial curve segments, the directions of the normals can be determined.

Consider the shape of the interface as shown in Figure 20. The osculating circles at different points of the interface are shown. If one were to obtain the normal to the circle at a point , one would compute,

$$\vec{n} = \frac{(X_i - A_i)\vec{i} + (Y_i - B_i)\vec{j}}{R_i} \tag{6.1}$$

The directions of the normal to the osculating circles at the sample points are shown along with the required normal direction, which is to point from the solid to the liquid. Also, at points such as 1 and 5, the curvatures would erroneously be indicated as being positive (since R is always positive), while the values expected are negative, as viewed from the solid. Therefore a decision as to the sign of the normal and curvature is required to be made. This problem is not unique to the circle representation. It is in fact more involved to resolve this issue with, say, the parabolic arc representation. The criterion that has been found to work well in determining whether to invert the sign of the normal is outlined here:

(i) Nomenclature: The interfacial points are firstly assigned numbers in sequence such that the solid lies to the right as one advances along the interface, as shown in Figure 21. When renumbering interfaces after mergers etc., this convention is adhered to.

(ii) Decision regarding inversion : Four basic types are identified and are displayed in Figure 22(a) – (d). In cases A and D, the signs of the normal and curvature are inverted. In cases B and C , the signs are retained. This is in line with the interfacial point numbering convention, whereby the solid lies always to the right as we traverse the interface. In case A for instance, the centre of the osculating circle at the point i, lies in the liquid. This is so since $X_i > A_i$. For this case, $Y_{i+1} > Y_{i-1}$. The normal therefore points from the liquid into solid and needs to be reversed. The criterion for case B is $(Y_{i+1} > Y_{i-1} , X_i < A_i)$. In such a case, the centre of the circle lies in the solid, and no reversal is necessary. Cases C and D may be argued similarly. Figure 23 shows a schematic of the definition of the arc radius at each point.

6.3 Mergers of Interfaces

Another advantage of the use of cells in the interface tracking procedure is related to the treatment of mergers. The treatment of mergers has received little attention in connection with marker–based methods, since it has been thought to involve intense logical manipulation. This is indeed the case in the absence of an underlying grid. Such difficulties are mitigated considerably when one associates with each point of the interface a cell number as detailed below.

Let us define the following data structure:

For each interfacial marker particle we record

Interfacial marker particle number : i

index of cell to which the point belongs (x and y index of cell): *cellx* (i), *celly(i).* After having found the cell number in which a particular interface point lies, that cell is declared as an interfacial cell. Conversely, with each cell, the associated point numbers are obtained. This is achieved by sorting through the interfacial cells and finding the identity of the interfacial markers that lie in the cell. Thus, we define the variables

cell number (x, y): *cellx*(i) , *celly*(i)

interfacial point number in that cell: *pointnum(cellx*(i), *celly*(i), *n*) , where *n* is the number of markers in that cell. The use of *pointnum*, which contains the integer value indicating the index of the marker particle in that cell, enables us to reduce the number of sorting operations when the interfaces interact.

There are three steps in the merger procedure, namely

(i) Detecting merger locations

(ii) Setting up for executing mergers

(iii) Picking up the new merged interface

Each of these steps will be explained below.

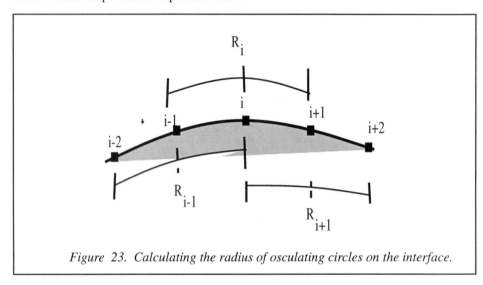

Figure 23. Calculating the radius of osculating circles on the interface.

Step 1. Detecting mergers :

Equipped with the data structures outlined above, the merger locations are detected. Consider elements of the interface shown as in Figure 24. In Figure 24, the markers (j+1 → j+4) and (i → i+2) are found in adjacent cells. This information is provided by the cellnumber values *cellx*(i) , *celly*(i) and *cellx*(j), *celly*(j) accompanying these points. Thus, to detect merger, one traverses along the interface, as in the situation shown in Figure 25, where the marker points along both interfaces are numbered along the clockwise direction.

Let it transpire that the points (j to j+n1) and (i to i+n2) fall in adjacent cells. This can be done by checking for, say, the point j+3 in Figure 24, the four cells ,i.e., top, bottom, right and left side cells. If any of these cells contains a point, then the number of that interfacial point is obtained from the value of *pointnum*. Now, however, in particular reference to point j+3, the point j+4 will lie in the bottom cell. This is merely a neighbour and not a candidate for merger with j+3. In order to avoid erroneously indicating merger in such a situation, merger is declared only when the following is true:

Let the index value of the interfacial point in the current cell be *pointnumber_current* and that in the adjacent cell be *pointnumber_adjacent*. *Pointnumber_current* and *pointnumber_adjacent* are obtained from the stored array associating each cell with the corresponding interfacial marker points, namely *pointnum.*

If (an adjacent cell contains an interfacial point) then

  If (abs(*pointnumber_adjacent – pointnumber_current)* > say 4) then

    declare merger

 else

    the *pointnumber_adjacent* is just a neighbour. No merger.

 endif

endif

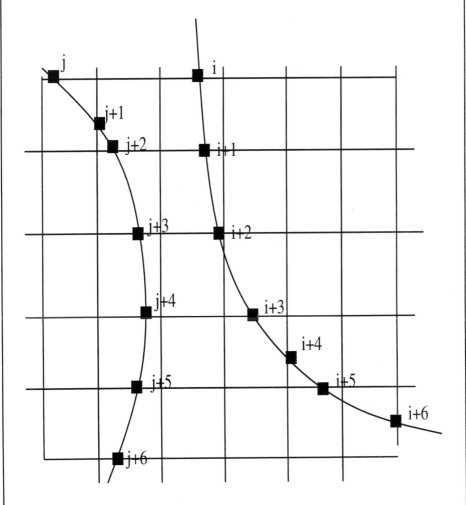

Figure 24. Illustration of interfacial point arrangement during impending merger.

The latter condition is to prevent adjacent points or very close points from being declared erroneously as merger candidates. The value 4 assigned above is chosen since we expect that each cell contains no more than two interfacial markers. If merger is detected, the merger operations are performed. Thus, in the situation depicted in Figure 25, one firstly designates the points j to j+n1 and i to i+n2 as points which have merged. Specifically, one incorporates arrays of merger points. The merger points occur in pairs, i.e., j connects to i and j+ n1 connects to i+n2. Thus declare

mergept1 (1) = j
mergept2 (1) = i
mergept1 (2) = j+n1
mergept2 (2) = i+n2

Also eliminate the points lying between i and i +n2 and j and j+n1 as this is a region that has merged.

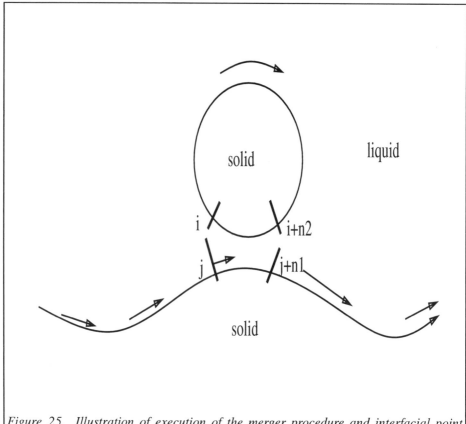

Figure 25. Illustration of execution of the merger procedure and interfacial point numbering.

Step 2. Setting up to execute merger:

However, assigning i,j and i+n2, j+n1 as merger points and joining the pairs accordingly usually leads to very sharp curvatures in that region, since, as can be envisioned by joining j and i in Figure 24, the curvature of the interface changes abruptly over one cell width. This occasions numerical difficulties associated with the large curvatures. To avoid such complications, the merger event is smoothed by smearing the merger region. This is done as follows – (i) step back *nstep* points from j, i.e., $j* = j - nstep$ (ii) step forward *nstep* points from i, i.e., $i* = i + nstep$ (iii) step back *nstep* points from i+n2, i.e., i +n2*=i+n2– *nstep* (iv) step forward *nstep* points from j+n1 , $j+n1* = j+n1 + nstep$. The choice of the forward or backward steps for smearing can be directly determined by the direction of the sequence of marker points along the interface. This smearing by *nstep* points relaxes the sharp curvature situation that may result after merger. The appropriate value of *nstep* depends on the grid spacing employed. In our calculations, for a 161×161 uniform grid, *nstep* = 8 has been found to be suitable, and is used in all calculations unless specified otherwise.

The merger procedure may now be explained with reference to the situation illustrated in Figure 26, where four circles approach each other to merge.

Firstly, the nomenclature of interfacial point numbering is to be noted. The directions of the arrows on the interfaces shown in Figure 26 indicate the direction of point numbering. The solid circles expand and merge. the merger points (after smearing) are i1,i2, ...etc. The pairs are

mer0gept1 (1) = i1,   *mergept2* (1) = j1
mergept1 (2) = i2,   *mergept2* (2) = j2
mergept1 (3) = k1,   *mergept2* (3) = l1
mergept1 (4) = k2,   *mergept2* (4) = l2

and so on. Thus, the start of a merger segment is indicated by the odd indexed merger point, while the end of the segment is given by the even indexed merger point. Therefore in the above, *mergept1* (1) is the starting point of a merging segment while *mergept1* (2) is the end point of that segment.

For each merging circle, record the start point number and end point number as shown in Figure 27. The string *istart* to *iend* forms a segment, i.e., one complete object. Thus for the object shown above, corresponding to the index *nseg* for example, the point number of the start point of the string defining the object is *istart(nseg)*. The end point of the string is *iend (nseg)*. Thus each of the four circles has a designated start point and end point value. The marker points are numbered by a simple sequence, punctuated by *istart(nseg)*, *iend(nseg)* to give the identity of each interface. Let us number the four circles 1 to 4 clockwise. Thus *istart(1)* = 1. If each circle is composed of 51 points, *iend(1)* = 51, *istart(2)* = 52, *iend(2)* = 102 etc.

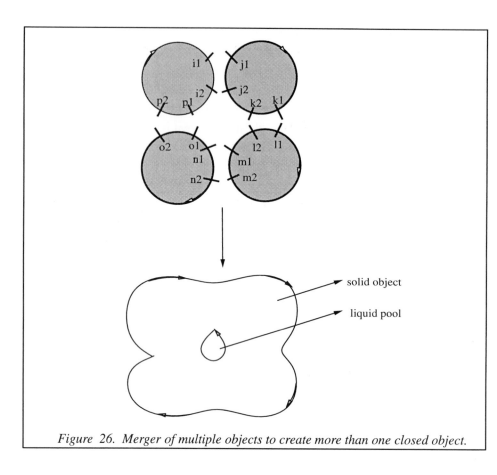

Figure 26. Merger of multiple objects to create more than one closed object.

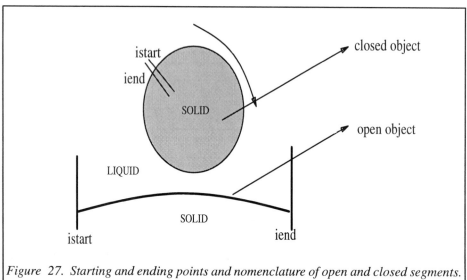

Figure 27. Starting and ending points and nomenclature of open and closed segments.

Step 3. Picking up the merged interfaces:

Now, after the merger has been declared to occur, one has to form a solid object as shown and one is left also with a liquid pool. The merger procedure has to pick up both these objects. Additionally, since for the interface defining the liquid pool, the solid lies to the right , the nomenclature direction for each object should be as shown in Figure 26. To achieve this, we proceed as follows:

Let us call the start points (*istart)*, the corresponding end points (*iend)* and the merge point pairs (*mergept1, mergept2*) *significant* points. Such *significant* points facilitate the merger procedure by reducing the number of points to be searched to ensure merger has been effected at all candidate locations. Thus, instead of sorting through each interfacial point to determine whether all the interfacial points have been taken care of during merger, one merely searches through the *significant* points, which are generally very few in number. Also, the use of *significant* points enables one to perform the connections between merging interfacial segments as described below:

Begin a startpoint loop, i.e., until all the startpoints *istart* (1 to 4) have been covered, march along the interface collecting points without deleting them from the merger process. All the points along the interface that are encountered are immediately flagged 'done' to indicate that such points have already been covered by the merger procedure . As shown in Figure 27, suppose we start from *istart(1)* = 1 . We check the point lying in front of the current point along the interface. Since the next point has not yet been covered in the merger procedure, i.e. it has not been flagged 'done', we pick up that point . We proceed in this fashion along the interface, each time checking the point ahead and flagging it as done when we pick up that point. Thus, we travel clockwise on circle 1, as depicted in Figure 26, picking up the as yet untouched points encountered. We assign the picked up points to new interfacial point arrays, say *xint_new (icount)*, *yint_new(icount)* where *icount* is a counter that records the number of points collected thus far for the new interface. We stop when we hit a *significant* point, which in this case happens to be the merge point i1, as shown in Figure 26. Stepping one point ahead of i1 will land us in the merger region. In this region, the points have already been eliminated from the interface. The only alternative then is to join this merge point to its pair.

We thus join i1 to its corresponding merge point pair j1. We now find ourselves on circle 2. We march along that curve in the direction decided upon previously, and as shown by the arrows. If the next *significant* point encountered is an end point , which is likely, then we connect the end point to the corresponding start point (a start point for a particular segment is associated with an end point and vice versa) and carry on.

If the next *significant* point is a merge point, then we check if the next point along the interface has already been eliminated from the interfacial marker array, i.e., if it is part of the merger region. If it is, then the only alternative is to connect the merge point to its pair on the other interface and continue, all the while flagging points covered as 'done'.

If not, then we continue to march along the same interface. The procedure results in the capture of the entire outer curve of the merged solid surface shown in Figure 26. The procedure stops when one encounters a point ahead that has already been covered (i.e. flagged 'done') , which in this particular case will be *istart(1)* . At this point an entire closed curve has been traced. Thus, *istart(1)* will form the start point of the new segment, whose new *nseg* value will be 1. The *icount* value resulting after the entire shape has been traced out will be the new end point value *iend(1)* of this segment. To pick up the remaining parts of the interfaces not yet covered so far, such as the pool of liquid left out in the middle, one then employs a merge point sweep, i.e., one goes to the first mergepoint that has not yet been covered in the previous procedure. Here this is the point i2 in Figure 26. Beginning now from i2 (which again is promptly flagged 'done'), one advances along the interface. The next *significant* point encountered is p1. Join to its pair o1 and continue. The procedure is carried out as described above and results in the pool of liquid. It is noted that the points are automatically picked up in the appropriate direction in agreement with the nomenclature convention by this procedure. All the necessary information regarding the post merger interface is also obtained at the end of the merger treatment. The interface undergoing merger can also be of the open type as illustrated in Figure 27. Similar considerations as detailed above apply to such interfaces. The situation with the breakup of interfaces is similarly handled. For instance, consider the situation depicted in Figure 28.

The merge point pairs are (i1,j1) and (i2,j2) . Thus, one starts at the start point of the open interval and reaches point i1. Here the points i1 to i2 have already been eliminated as merger regions. So we do not proceed beyond i1, but join it to its pair j1 and carry on. Thus the open segment has now been obtained. We then search for the left over start points. In this case the only available start point , i.e., of the original open segment has already been covered. So a mergepoint sweep is performed. The mergepoint i2 is first encountered. One travels from i2 to j2 in the direction of the point sequence on that interface, as shown by the arrow. The region beyond j2 has been eliminated already. So one merely joins j2 to i2 and thus the second segment, of closed type is obtained. Once a complete new segment is obtained, the type of segment is recorded, namely whether it is open or closed. Currently, we employ, and require, only one open interval. The segments are arranged in the order of open intervals first, followed by closed solid and liquid pools. The merger detection procedure first takes care of the first, and only, open interface segment and then the start point and merge point searches pick up all the closed objects and left over pools in succession. In what follows, we present results pertaining to some cases tried so far in applying the merger procedure.

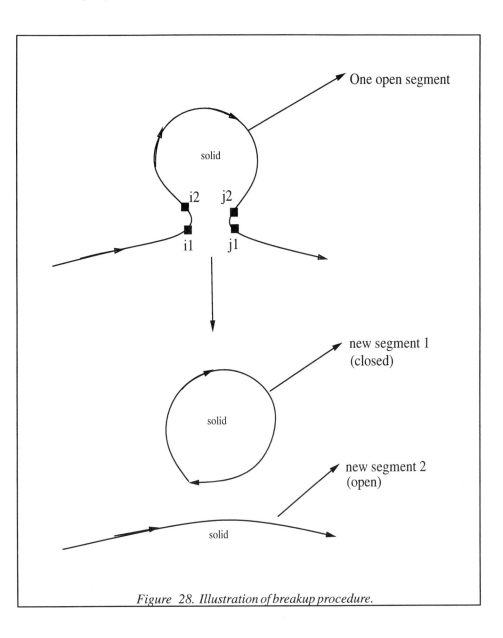

One open segment

solid

i2   j2

i1   j1

new segment 1
(closed)

solid

new segment 2
(open)

solid

Figure 28. Illustration of breakup procedure.

6.4 Summary of Merger Procedure

The merger procedure described above may be summarized as follows :

(i)   Detect merger locations.

(ii)   Set up data structures.

(iii)   Collect open interface along with its merged appendages.

(iv)   Perform start point loop for collecting closed interface segments.

(v)   Perform merge point loop to obtain remaining closed interfaces and pools.

(vi)   Redefine the interface by updating interfacial point string with newly
        obtained string of points.

(vii)   Perform the curve fit to obtain new normals and curvatures.

(viii) Translate particles to new positions in cells.

(ix)   Perform intersection procedure to obtain fresh interfacial points.

(x)   Test for merger.

Return to (i).

6.5 Results for Non–Merging Interfaces

The interface tracking method described above is first tested for accuracy . The
method is capable of handling interfaces of closed and open types. For both types periodic
conditions are enforced at the ends of the interface. In Figure 29 we follow the expansion
of a circle on a 161 x 161 grid. The initial radius of the circle is 1 and the circle expands
to a radius of 5 at the end of the calculation. The expansion occurs at a given normal
velocity $V_N^* = 1$. Figure 29(a) shows the interfacial shape at equal intervals of time. The
time step used $\Delta\tau = 0.01$, and the circle reaches a radius of 5 in 400 time steps.
Figure 29(b) shows the curvature along the interface at the same time intervals as
Figure 29(a). The interfacial curvatures are found to be constant as expected.
Figure 29(c) shows that the number of points on the interface increases from less than 80
at the start of the calculation to more than 200 at the end. Figure 29(d) shows the
development of radius with time. The superimposed crosses are the expected radii
according to the imposed linear growth rate. The agreement is excellent, the error being
in the vicinity of the machine roundoff.

The case presented in Figure 29 corresponds to constant interfacial curvature
which is naturally suited to the present circular arc interfacial representation. As will be
demonstrated in the following, other generic cases of variable curvatures can be equally
well handled by the present technique as well. However, it is observed that wiggles in the
curvature can occur at the locations where the interface is horizontal or vertical. It is also
found that in the absence of an intersection procedure, the curvature behaves very well.
Thus it appears that these small oscillations are caused by errors in evaluating the new
interfacial locations. Such behaviour of curvature is also reported in Daly (1969). Some
corrective measures have been taken in improving the curvature calculations. These are:
(i) The value of the radius R assigned to a point is an equally weighted average of the radius
values of the three points used in obtaining the interfacial curve passing through the point.
As shown in Figure 23, the points involved in calculating R_i are i, i–1 and i+1. The radius
value assigned to point i is given by, $R_i^* = (R_i + R_{i-1} + R_{i+1})/3$. Some such smoothing
of the curvature value is found to be aid in preventing the small wiggles in curvature along
the interface. The curvature, being evaluated from the second derivative of the interfacial

curve occasions more problems with accuracy than does the interface shape and slope which are found to be smooth.

(ii) It is ascertained that the points on the interface are not too close together. It was found that when interfacial points are very closely spaced significant errors and jumps in curvature values can result, mainly due to the difficulty in numerically evaluating the derivatives. One therefore runs through the interfacial points and calculates the arc lengths between two successive points. If the arclength corresponding to a particular point is less than a given fraction of the average arclength that point is deleted. Generally, very few points actually qualify for such treatment and the number of points defining the interface does not fall significantly.

As an example displaying varying interfacial curvature, Figure 30 involves growing an open interface. Here an initial perturbation of the form $Y_i = 2.0 + (1 - \cos(2\pi X_i/10))$ is applied and the velocity function for this case is $V_{Ni} = Y_i - Y_1$. The velocity function has been modeled to mimic the effect of undercooling in morphological instability (Langer 1980). The given velocity function causes the interface to bulge out and the final interface shape, in Figure 30(a) is seen to be multiple–valued with respect to x and y. Figure 30(b, c) show curvatures along the interface at different time instants. The shape of the curvature plot is also seen to change considerably with time. The curvature obtained is found to be very smooth without any wiggles even when the interface is highly deformed and the difference in curvature between the centre and the sides is significant. The number of points in this case as shown in Figure 30(d) increases from 161 (a 161 x 161 Cartesian grid was used) initially to approximately 300 finally, caused by the obvious reason that the overall arc length of the interface increases with time, resulting in increased number of intersecting points between the interface and grid lines.

As a further demonstration, Figure 31 shows the interface shape with a double wave. The curvatures are shown in Figure 31(c,d) and are found to be remarkably smooth. Figure 32 shows the development of an initial interfacial perturbation of the form, $Y_i = 3.5 + (1 - \cos(2\pi X_i/X_{out}) + \cos(8\pi X_i/X_{out}))$. The highly contorted final interface shape is well captured. The curvatures at various instants are shown in Figure 32(b–d). In this case wiggles in the curvature may be detected in the Figure 32(c,d) , although the general distribution is quite good. In this case, the wiggles appear to predominate in the regions where the curvature transitions from a positive to a negative value or vice versa. This turning of the interface occurs over a very small number of points, possibly over one or two cells only, and hence the interface definition suffers. Some further fine–tuning of the intersection procedure and/or a grid refinement procedure is perhaps necessary to smooth out the remaining wiggles.

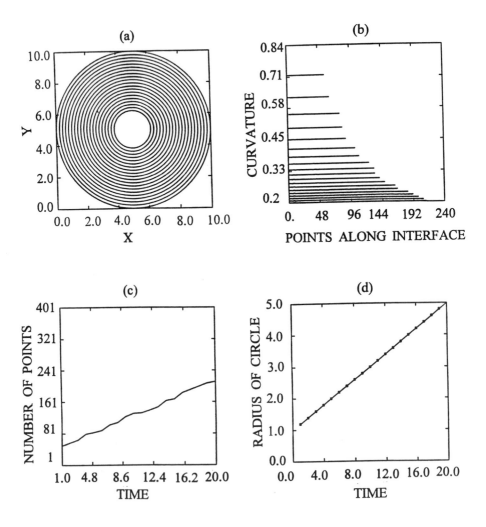

Figure 29. (a) Shapes of expanding circles at equal intervals of time. (b) Curvatures along the interface at the same time instants as in (a). (c) Plot of number of points along interface versus time. Number of points increases as circle expands. (d) Superposed computed and expected linear radius variation with time. The crosses correspond to exact values.

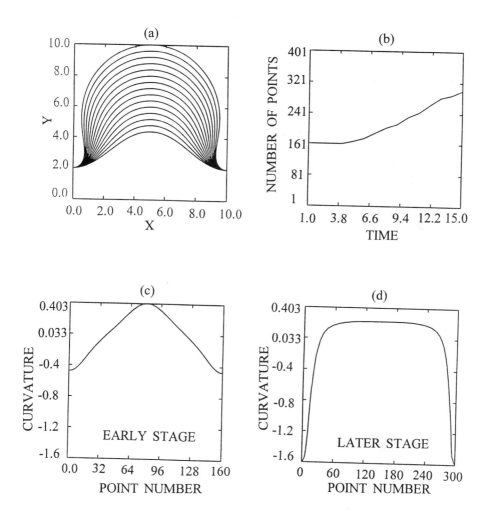

Figure 30. (a) Shape of the growing finger at equal intervals of time. (b) The number of points along the interface is shown to increase with time for the growing finger. (c) Curvature along the interface early in the development of finger. (d) Curvature along the interface in later stage of development.

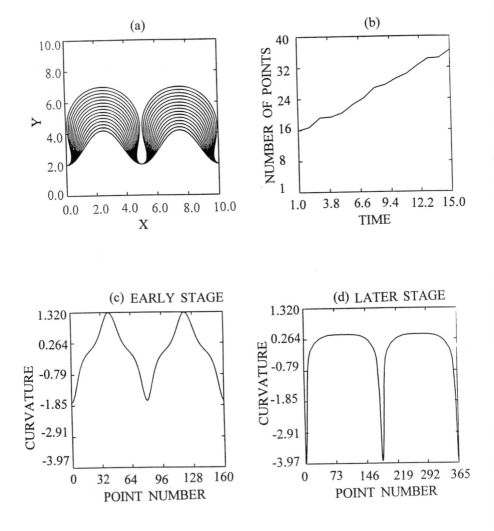

Figure 31. (a) Shape of two fingers at equal intervals of time. (b) The number of points along the interface is shown to increase with time for the growing finger. (c) Curvature along the interface early in the development of fingers. (d) Curvature along interface in later stage of development.

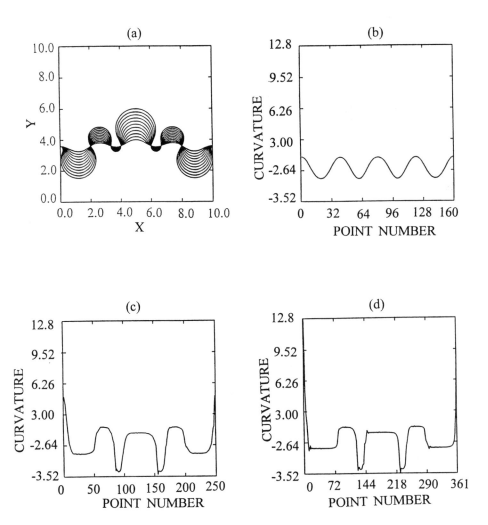

Figure 32. (a) Shape of highly deformed interface at equal intervals of time. (b) Curvature along interface in the initial stages of growth. (c) Curvature along interface late in the development of fingers. (d) Curvature along the interface at the end of the calculation.

6.6 Results for Merging Interfaces

The merger procedure was first tested to determine the effect of grid spacing and the smearing factor *nstep* in faithfully simulating the merger event. Figure 33(a–d) show the merger of two circular regions, the growth velocity being modelled as $V_N = 1$. 81×81 grid points are used in Figure 33(a), with *nstep* = 1. 81×81 grid points and *nstep*=5 are used to obtain Figure 33(b). Figure 33(c) and Figure 33(d) correspond to 161×161 grid and *nstep*=8 and *nstep*=12 respectively. The progressive improvement in the intermediate interface shapes from (a) to (d) is evident, and the final interface shape in Figure 33(d) is quite satisfactory. Employing a finer grid in itself contributes to the improved results in the merger region. Using a generous value of smearing *nstep*, leads to further improvement. However, this may lead to dissipation in the region where mergers occur. Adopting *nstep* = 8 in all further calculation, we present test cases to validate our merger/breakup procedure.

Figure 34 shows the interaction of an upward travelling (the normal velocity of all three interfaces is 1, i.e., $V_N^* = 1$) open interface with two expanding circles, in a domain of dimension 10×10 units, whose centers are at (2.5,7) and (6.5,7) and initial radii are both equal to 1. The merger procedure first captures the merger of each circle with the open interface. The resulting single open object then grows and there is a further merger which occurs between the circular bulges at around x=4.5 at a later time. Thus, multiple simultaneous mergers and successive mergers are both demonstrated in this figure.

In the next example shown in Figure 34, a protuberance is caused to grow, using a velocity function $V_N = (Y_i - Y_1)$. The fingers in the middle grow outward and expand, leading to the situation shown in Figure 34(b) where merger impends. Figure 34(c) shows the situation immediately upon merger. There is, in the centre, a pool of liquid that has been captured as desired. That the normals for the pool are directed correctly is evident from Figure 34(c–d) where the pool shrinks with time until it disappears as it should. The highly distorted final interface is well captured. Figure 34(e) to (h) show curvatures along the interface at times corresponding to the shapes shown. The symmetry of the curvatures obtained is highly satisfactory. However, for the pool of liquid, the number of points available (approx. 30) is perhaps not sufficient, for accurate computation of curvature. The resolution there may be enhanced by local grid refinement if required.

A further test of the merger procedure lies in assessing its ability to handle breakups. In Figure 35(a, b) , two circles expand and coalesce to form one single solid object (161×161 grid, *nstep* = 8). At the time instant shown in Figure 35(b) the merger is reversed, i.e., the velocity function now is $V_N^* = -1$, so that we expect to obtain two circles from the merged shape of Figure 35 (b). The breakup process is shown in Figure 35(c–d) and the final separated elements are shown in Figure 35(e). Curvature corresponding to Figure 35(d) is shown in Figure 35(f). It is evident that the merger–breakup sequence is not reversible. There is significant dissipation involved in the simulation and such damping, in the form of the value of *nstep* is essential for handling the rapid change of curvature occurring in the region of mergers.

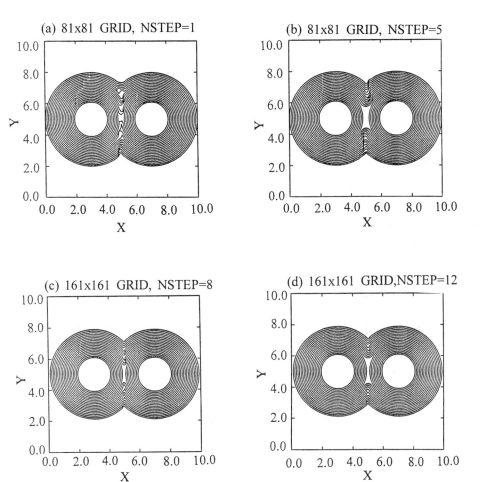

Figure 33. Mergers of circles implemented with different parameters : (a) 81 x 81 grid with smearing, nstep value of 1. (b) 81 x 81 grid with nstep = 5. (c) 161 x 161 grid with nstep = 8. (d) 161 x 161 grid with nstep = 12. The result noticeably improves with grid refinement.

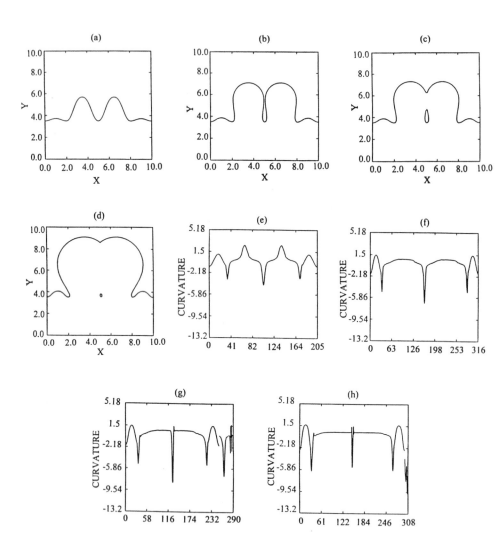

*Figure 34. The merger of two growing fingers. (a) and (b) The fingers grow and spread
under the given velocity model. (c) Merger has occurred at the center of the
domain. (d) The finger shape after merger. There is a pool of liquid left after
the merger. The finger expands, while the pool of liquid shrinks, and finally
disappears. (e) and (f) Curvatures along interface corresponding to (a) and
(b). (g) and (h) Curvatures corresponding to (c) and (d). The discontinuity
in curvature is on account of the liquid blob being formed.*

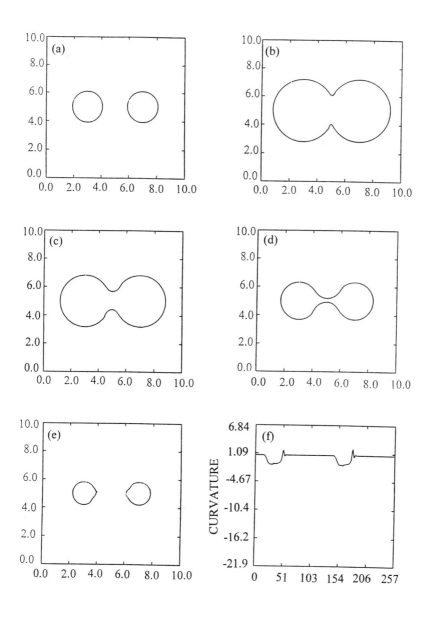

Figure 35. Case of merger of two circles and reversal of merger to cause breakup of resulting object. In (a) and (b) the two circles expand and coalesce. At that point, the process is reversed. In (c) and (d) the resulting dumbbell shaped object contracts until it breaks off into two objects i (e). Due to the nonlinearity, reversibility is not achieved as seen from the shape and the size of the object seen in (e). (f) Curvature along interface shown in (d).

The above test demonstrates the ability of the merger procedure to handle breakup phenomena.

Another example is shown in Figure 36 where an open interface merges with a growing closed object. The merger is then reversed and the separation of the two objects results. The final shape obtained subsequently, as in Figure 36(e) are very similar to the initial shapes involved in the merger phenomenon.

In the above, an interface tracking procedure has been presented. The method uses marker particles and grid cells to track interfacial points. The marker particles are connected by interfacial curves. The use of cells has facilitated the tracking of an interface that progressively increase in perimeter with time. New points are incorporated on the interface, automatically, via an intersection procedure thus circumventing the problems and corrective measures accompanying marker particle depletion/accumulation. A strategy has been developed, applying fairly simple data structures to simulate mergers and breakups of interfaces. This method is independent of the field equation solver, and can be applied to a variety of problems including the phase change problems with or without convection.

7. ENTHALPY FORMULATION FOR PHASE CHANGE PROBLEM

As already discussed in Section 5.2, besides the explicit tracking of the interface, which may be moving and irregular in shape, alternative formulations are possible (Crank 1984). For example, one can reformulate the problem so that the phase boundary is implicitly accounted for as part of the solution in the domain, without explicitly accounting for the jump conditions across the interface. The position of the interface is determined *a posteriori*, usually by interpolation. A very appropriate and convenient way of accomplishing this goal is to reformulate the energy equation utilizing the enthalpy as the dependent variable. Such an approach is called the enthalpy formulation. A major advantage of the enthalpy formulation is that it is not necessary to split the domain into separate subdomains consisting of the different phases; a fixed grid can be employed to facilitate the computations.

7.1 Basic Concepts and Implementation

We first consider an idealized phase change material, with constant thermo–physical properties, including density, and with phase change occurring at a constant, specified temperature. This implies the absence of convection, the phase change process being driven entirely by the conduction process and the release of latent heat at the phase change interface. The bench–mark problem considered here is that of conduction driven phase change in a rectangular cavity. A schematic of the problem is shown in Figure 37.

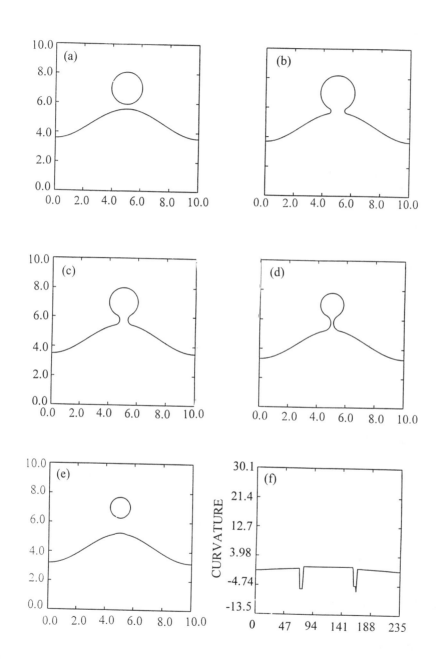

Figure 36. Case of merger of an open and a closed interface and reversal to induce breakup. (a) The initial shape of the interface. (b) Coalesced form of the composite interface. (c) and (d) the interface shapes as the breakup procedure is in progress. (e) Interfaces after breakup. (f) Curvature along interface shown in (b).

The phase–change material is initially at the phase change temperature, T_m. At time, $t = 0$, the temperature of the left wall is suddenly raised to $T_h = 1$. The objective is to study the motion of the interface in time. The interface is defined by $T = T_m = 0$. For the pure conduction problem we have, in each phase,

$$\frac{\partial T_i}{\partial t} = \alpha_i \, \nabla^2 T_i \qquad i = l, s \tag{7.1}$$

along with the jump condition

$$k_s \frac{\partial T_s}{\partial n} - k_l \frac{\partial T_l}{\partial n} = \varrho_s \, \Delta h_{l-s} \, V \tag{7.2}$$

where V is the interface velocity. For this problem, a closed form solution exists for the time–dependent location of the phase change front given by: $x = 2\beta \sqrt{\alpha t}$ where β is the root of

$$\beta \, e^{\beta^2} \, \text{erf}(\beta) = (T_h - T_m) \frac{C_p}{\Delta h_{l-s} \, \sqrt{\pi}} \tag{7.3}$$

The solution of equations (7.1) and (7.2) involves the explicit tracking of the interface in a time–dependent manner as discussed in Section 5.2. The governing equations can be cast in the weak form

$$\frac{\partial (\varrho h)}{\partial t} = \nabla \cdot (k \nabla T) \tag{7.4}$$

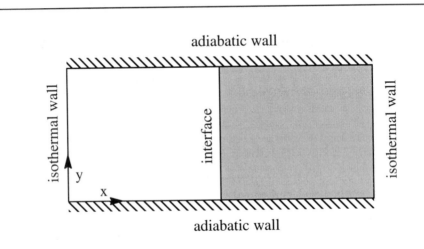

Figure 37. Schematic of the phase change problem showing the domain and boundary conditions. For the conduction driven problem, the interface is planar. For the problem with gravity driven convection included, the gravity vector is in the negative y direction. For the g–jitter problem, the initial condition is a planar interface at the center of the cavity as shown.

where h is the total enthalpy. The equation can be discretized according to the conventional control volume procedure wherein, the total enthalpy h and the temperature T can be interpreted as averaged values within a control volume. Then h can be written as

$$h = C_p T + f \, \Delta h_{l-s} \tag{7.5}$$

where f is the phase fraction of phase 1 and can be interpreted as the fractional volume of a computational cell occupied by that phase. The phase fraction of phase 1 is zero in the region occupied by phase 2 and unity in the region occupied by phase 1. The phase fraction lies between zero and unity when the control volume is undergoing phase change. Substituting Eq. (7.5) into the governing equation (7.4) we get

$$\frac{\partial(\varrho C_p T)}{\partial t} = \nabla \cdot (k \nabla T) - \Delta h_{l-s} \frac{\partial(\varrho f)}{\partial t} \tag{7.6}$$

where the latent heat now appears as a source term. This formulation is a single region formulation, wherein one set of governing equations can describe both phases and hence can more readily handle complex interface shapes, branched and multiple–valued interfaces. To close the physical model, a unique relationship between the fluid fraction f and the temperature T has to be formulated. For a pure substance undergoing isothermal phase change, the total enthalpy h, is a discontinuous function of the temperature. However, from a computational viewpoint, discontinuities are difficult to track and it is often necessary to smear the phase change over a small temperature range to attain numerical stability. Thus $T_s = T_m - \varepsilon$, $T_l = T_m + \varepsilon$ and ε is a phase change interval. The discontinuity is thus replaced by a small interval over which phase change occurs. For a pure substance, undergoing isothermal phase change, this is a purely numerical artifact and needs to be as low as possible in order to accurately model the physical system. The above relationship is substituted into Eq. (7.5) to obtain

$$f = \frac{T - T_s}{2\,\varepsilon} \tag{7.7}$$

which is used to iteratively update the phase fraction from the computed temperature field. This technique will henceforth be referred to as the T–based update method. An alternative formulation is to express the phase fraction as a function of the total enthalpy rather than the temperature, whereby a continuous relationship can be obtained via an implicit procedure. First, we invert the $h = h(T)$ relationship so that, where h_s and h_l are the enthalpy values corresponding to the T_s and the T_l temperatures respectively, $h_s = C_p T_s$ and $h_l = C_p T_l + \Delta h_{l-s}$, assuming a constant C_p. Substituting the above relationships into Eq. (7.5), we obtain,

$$f = \frac{h - h_s}{h_l - h_s} \tag{7.8}$$

for the iterative update of the fluid fraction. The motivation for the above formulation stems from the fact that the temperature is always a continuous function of the total enthalpy. In particular, it is noted that $\varepsilon = 0$ is allowable in this formulation, thereby accurately modelling phase change of a pure substance. Another motivation is the fact that in the phase change zone, the phase fraction is a rapidly varying function of the temperature causing it to be sensitive to small errors in temperature. In particular, this may result in an unphysically thick phase change zone (mushy region) due to round–off errors in the temperature calculation. Contrast this with the fact that if the latent heat is large relative to the sensible enthalpy, the phase fraction is a slowly varying function of the total enthalpy resulting in a more stable update procedure. This update procedure will henceforth be referred to as the H–based update method. The one–dimensional Stefan problem is used as the test vehicle for the comparisons. More information regarding the relative computational efficiency of the H– and T–based methods can be found in Shyy and Rao (1993). The following parameters were used: thermal diffusivity of both phases, $\alpha = 1$, Stefan number $St = 0.042$, and $\varepsilon = 10^{-5}$.

Calculations were carried out on uniform grids of 21 points, 41 points and 81 points with a time step $\Delta t = 0.01$, and with time steps, $\Delta t = 0.1$, 0.01 and 0.001 on a 81 grid. Comparisons were made with the exact solution to assess the accuracy of the two methods on the basis of the interface location. The interface location was obtained by interpolating for the $f = 0.5$ contour. Second order central differences were used for all spatial derivative terms and first order Euler backward differencing was employed for marching in time. Figure 38(a) and Figure 38(b) show that, as required, the H–based and T–based methods yield identical solutions even for coarse grids. An error distribution was formed by computing the relative error at every time instant. Figure 38(c) shows that the error distribution oscillates about a zero mean. This may be correlated with the fact that a significant part of the latent heat content in a computational cell is released when it starts to undergo phase change. Thus, the interface slows down when a computational cell just starts to change state and subsequently speeds up. The computed solution coincides with the exact solution when the interface is at the boundary between two control volumes. The amplitude of the error distribution drops with increasing spatial resolution and its frequency doubles when the grid spacing is halved. A measure of the cumulative error is obtained from the norm of the error distribution, which is the r.m.s. value of the error distribution. A log–log plot of the cumulative error versus Δx (Figure 38(d)) shows that the cumulative error varies as $(\Delta x)^{1.5}$. A calculation conducted with a time step of 0.001 yielded a solution that could not be distinguished from the solution for the time step of 0.01. Thus the solution is time accurate. More information can be found in Shyy and Rao (1993).

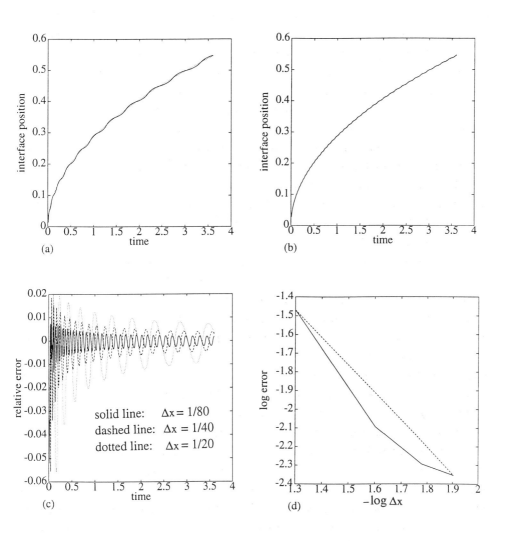

Figure 38. Solution features for the pure conduction problem with St = 0.042. (a)
Coarse grid solution (21 grid points) (b) Fine grid solution (41 grid points).
In each case the solid line designates the solution obtained using the H–based
method, the dashed line due to the T–based method and the dotted line, the
exact solution. (c) Error at every time instant for various grid sizes. The solid
line is the error with $\Delta x = 1/80$, the dashed line is the error with $\Delta x = 1/40$
and the dotted line is the error with $\Delta x = 1/20$. The error oscillates about
a zero mean and its amplitude decreases and the frequency doubles when the
grid spacing is halved (d) Cumulative error versus grid spacing for $\Delta t =$
0.01.   The solid line is the numerical result and the dotted line has an
approximate slope of 1.5.

7.2 Convection Effects

Thus far, only conduction dominated effects have been taken into account, but, as will be demonstrated in the following, convection can substantially modify the two–phase flow field. This will be demonstrated by considering, as an example, the melting of gallium in a square cavity. For this case, Gau and Viskanta (1986) have reported experimental information of interface shapes and position at specified time instants. The shape of the phase change interface is determined by the combined effects of conduction and convection driven by buoyancy forces. It will be shown that convective effects can have a substantial impact on the interface shape. A schematic of the problem is shown in Figure 37.

7.2.1 Governing equations

The governing equations in non dimensional form can be written as:

continuity :
$$\frac{\partial u^*}{\partial X} + \frac{\partial v^*}{\partial Y} = 0$$

X − momentum :
$$\frac{\partial u^*}{\partial \tau} + \frac{\partial u^{*2}}{\partial X} + \frac{\partial u^* v^*}{\partial Y} = -\frac{\partial P^*}{\partial X} + Pr\left[\frac{\partial^2 u^*}{\partial X^2} + \frac{\partial^2 u^*}{\partial Y^2}\right] + S_u^*$$

Y − momentum :
$$\frac{\partial v^*}{\partial \tau} + \frac{\partial u^* v^*}{\partial X} + \frac{\partial v^{*2}}{\partial Y} = -\frac{\partial P^*}{\partial Y} + Pr\left[\frac{\partial^2 v^*}{\partial X^2} + \frac{\partial^2 v^*}{\partial Y^2}\right] + S_v^* + Ra\ Pr\ \theta \qquad (7.9)$$

energy :
$$\frac{\partial \theta}{\partial \tau} + \frac{\partial u^* \theta}{\partial X} + \frac{\partial v^* \theta}{\partial Y} = \frac{\partial^2 \theta}{\partial X^2} + \frac{\partial^2 \theta}{\partial Y^2} - \frac{1}{St}\left[\frac{\partial f}{\partial \tau} + \frac{\partial u^* f}{\partial X} + \frac{\partial v^* f}{\partial Y}\right]$$

where Pr is the Prandtl number, St is the Stefan number, S_u and S_v are appropriate source terms to be discussed later. For isothermal problems (no physical mushy zone) the spatial derivatives in the source term of the energy equation can be neglected because f undergoes a step change at the interface. Dropping these terms also facilitates the T–based implementation. The cavity containing solid gallium is initially at the phase change temperature and the top and bottom walls of the cavity are insulated. The left wall is suddenly raised to a temperature above the melting temperature and the problem is to predict the subsequent development and shape of the phase change interface.

7.2.2 Source terms for the momentum equations

When computing the solidification process on a fixed grid, special treatment is needed to enforce the solid velocity to be zero. The approach taken here is the inclusion of Darcy–type source terms in the momentum equations. Thus the phase change material is considered to be a porous medium, with the porosity changing from zero to unity as the material melts. For pure materials, the porosity changes abruptly as the phase change occurs, but to ensure numerical stability, a continuous variation is imposed. Here, the

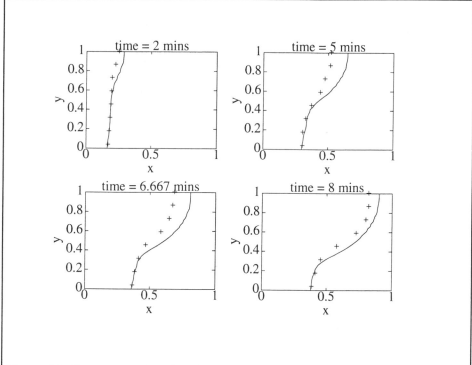

Figure 39. Comparison of predicted interface shapes obtained on a 41 × 41 grid (solid line) with experimental data (+) for the case including buoyancy driven convection .

source terms have the form: $S_u^* = - Au^*$ and $S_v^* = - Av^*$ where A varies linearly with the fluid fraction as: $A = (1 - f) \times 10^5$, where f is the fluid fraction and A should be as large as possible in the solid region and still retain numerical stability.

In order to make a comparison of the predicted interface shape, at various time instants, with the experimental results of Gau and Viskanta (1986), the calculations were carried out with the following parameters: $Ra = 2.2 \times 10^5$, $Pr = 0.0208$, $St = 0.042$. A value of $\varepsilon = 0.001$ was used in conjunction with the T–based method to ensure numerical stability. The time stepping scheme chosen was the 1st order backward Euler and the convection scheme chosen was the 1st order upwind method. All other terms were discretized by 2nd order central differences. The interface location was determined by interpolation for the $f = 0.5$ contour. Solutions were computed on a 21 × 21 uniform grid and a 41 × 41 uniform grid and a time step of 0.01 (1.43 seconds). Numerical results have been shown for t = 2 mins, 5 mins, 6.667 mins and 8 mins in Figure 39 alongside the experimental determinations of Gau and Viskanta, after translating the predicted interface to account for the undercooling present in the experiment (Lacroix 1989). Reasonable agreement has been obtained in terms of the interface shapes at the various time instants. The effect of convection is to transport the hot fluid to the top of the cavity and the interface moves faster near the top of the cavity. The shape of the interface is distorted by the convection patterns established as the calculation progresses in time. Thus, it is evident that natural convection can substantially affect the speed and the shape of the interface from the planar one yielded by the pure conduction calculations.

8. CASE STUDIES

8.1 Effect of Convection on Phase Change

The following examples elucidate the steady–state transport characteristics of solidification processes in a systematic way. With the use of the Darcy law in the momentum equations to account for the presence of the mushy zone and with the use of the enthalpy formulation, a unified set of steady–state equations governing the mass continuity, momentum and energy transport are solved for all phases. The controlling parameters, including the Rayleigh number, the Prandtl number, the Marangoni number and the Stefan number are varied to investigate their impact on the transport processes. The size of the mushy zone as well as its influence on the solidification process is also studied by varying the temperature ranges within which the mushy zone exists. A square is taken to be the domain of interest with the boundary conditions as schematically indicated in Figure 40, and calculations are carried out for a variety of Rayleigh, Marangoni and Stefan numbers. In the following, the values of T_1 and T_2 are taken as either 0.6 and 0.4 respectively, or as 0.52 and 0.48 respectively, i.e., the nondimensional temperature variation across the mushy zone is either 0.2 or 0.04. The computations presented in Sections 8.2 to 8.5 are based on the Boussinesq approximation. They complement the case studies given in Chapter VIII.

8.2 Phase Change with Natural Convection Only

We first consider the case of a pure material, i.e., the solute field, ϕ, in Figure 40 is not considered in this section.

8.2.1 $Pr = 10^3$ and $St = 0.2$

Calculations were conducted for $St = 0.2$, with $T_2 - T_1$ being the temperature scale, for four different values of the Rayleigh number from 10^4 to 10^7. Figure 41 to Figure 43 show the streamlines, enthalpy contours and phase distributions for this case. No slip conditions were imposed on all the four walls. As the Rayleigh number increases, the mushy zone becomes thinner in the top region and thicker in the lower region. The pure liquid pool increases its width but decreases its depth as Ra increases. The shape of the interface is highly affected by the convection strength. At $Ra = 10^4$, the convection effect is weaker and the enthalpy distribution is substantially affected by conduction. Hence, the mushy zone is more aligned to the vertical boundaries. The solutions are qualitatively similar to the transient ones reported earlier in Voller and Prakash (1987) and Dantzig (1989). For $Ra = 10^5$, the convection strength is more than three times that for 10^4.

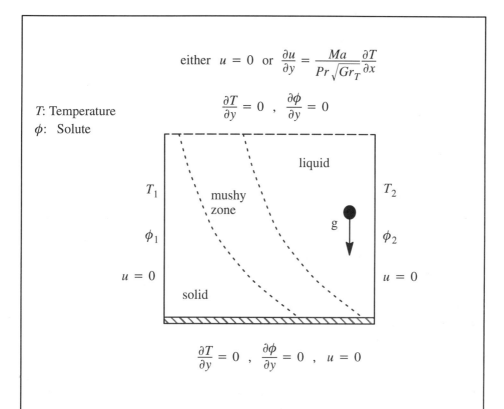

either $u = 0$ or $\dfrac{\partial u}{\partial y} = \dfrac{Ma}{Pr\sqrt{Gr_T}}\dfrac{\partial T}{\partial x}$

T: Temperature

ϕ: Solute

$\dfrac{\partial T}{\partial y} = 0$, $\dfrac{\partial \phi}{\partial y} = 0$

liquid

T_1 mushy T_2
                 zone
ϕ_1 g ϕ_2

$u = 0$ $u = 0$

solid

$\dfrac{\partial T}{\partial y} = 0$, $\dfrac{\partial \phi}{\partial y} = 0$, $u = 0$

Figure 40. Schematic of the domain illustrating the geometry. The top and bottom boundaries are adiabatic whereas the left and the right boundaries are isothermal and are held at different temperatures. The sides of the box are no–slip walls except the top boundary which can be either a no–slip boundary or a free surface. Schematically indicated are the solid and the liquid phase boundaries and the mushy zone.

Consequently, the enthalpy contours are more concentrated in the top left and lower right regions. The boundary of the mushy zone adjacent to the pure liquidus phase is highly distorted and is expected to affect the solidification dynamics and dendrite formation. As *Ra* further increases, the enthalpy contours responsively become more concentrated and the convection effect is more vigorous. Hence the mushy zone in the top left region becomes thinner. The other observable phenomenon is that at $Ra = 10^4$, convection is effectively dampened by the mushy zone; the convection cell largely conforms its boundary to that of the mushy zone. With increasing *Ra*, however, the convection becomes stronger and hence is able to penetrate more into the mushy zone. At $Ra = 10^7$, despite the mushy zone being present in half the domain, the convection cell occupies almost the whole region. More information can be found in Shyy and Chen (1990a).

   The wiggling interface of $Ra = 10^5$ appears to result from the balance of convection and release of latent heat. The wiggles exist only within a certain range of *Ra*. Similar interface characteristics have been observed by Christenson *et. al.* (1989) under different conditions. Both convection and latent heat release mechanisms are highly nonlinear and influence each other in a coupled manner. The convection strength increases as *Ra* increases and hence is able to penetrate deeper into the mushy zone, causing the phase boundaries to exert less influence on the transport characteristics. As the *Ra* number further increases, the interface boundary again is of a smoother shape. The other observable phenomenon is that with increasing *Ra*, the streamfunction shifts its peak more toward the solid phase, eventually depicting a double–peak distribution.

8.2.2 *Pr* = 10^3 *and St* = 0.2

The Prandtl number was then reduced from 10^3 to 1.49×10^{-2}. Thus, for the same *Ra*, the Grashof number increases by a factor of 6.7×10^4. Therefore, fluids of lower *Pr* exhibit much stronger nonlinear characteristics than those of higher *Pr*. This change in fluid properties causes the numerical algorithm to experience much more difficulty in yielding converged steady–state solutions. With $Pr = 1.49 \times 10^{-2}$ and $St = 0.2$, no solutions, by using either the first– or second–order upwind schemes for the convection terms, can be obtained for $Ra = 10^6$, which corresponds to $Gr = 6.7 \times 10^7$. Obviously, the source terms associated with the phase change increase the nonlinearity of the governing equations and hence make the problems more difficult to solve. For the single phase flow, steady state solutions with identical numerical treatments as well as the number of nodal points have been successfully computed for the same *Pr* (Shyy and Chen 1990).

Figure 44 to Figure 46 compare the solutions for $Pr = 1.49 \times 10^{-2}$, $St = 0.2$ and $T_2 - T_1 = 0.2$ for three different $Ra = 10^3$, 10^4 and 10^5. As in the case of $Pr = 10^3$, both the strength and size of the convection eddy increase as *Ra* increases. As expected, with almost five orders of magnitude difference in *Pr*, the enthalpy distributions, the resulting shapes of the convection cell, as well as the locations of the phase boundaries, change noticeably as *Pr* varies from 10^3 to 1.49×10^{-2}. Since the enthalpy field is under much stronger conduction influence with $Pr = 1.49 \times 10^{-2}$ than with $Pr = 10^3$, the location of the mushy zone is now closer to the middle of the domain. In this regard, it should be noted that two competing factors are at work. With the same size of the liquid pool, the lower *Pr* fluids tend to develop a stronger convection effect for the same *Ra* (which results in higher *Gr*). However, extra complexities are introduced by the phase changes; lower *Pr* also causes the enthalpy contours to be less concentrated with the same *Ra*, resulting in a smaller size of the liquid pool. Since *Gr* is dependent on the size of the liquid pool to the third power, the smaller size of the liquid pool reduces *Gr*. Hence, the overall convection strength is determined by the balance of the competing effects.

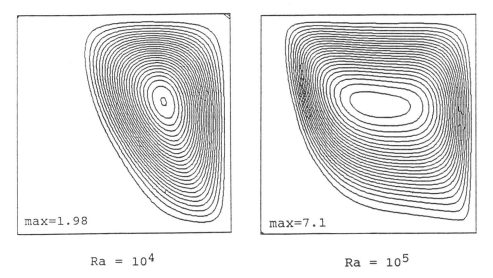

Ra = 10⁴   Ra = 10⁵

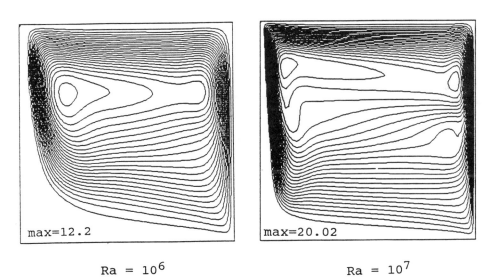

Ra = 10⁶   Ra = 10⁷

Figure 41. Streamlines of Pr = 10³, T₂ − T₁ = 0.2, and St = 0.2, temperature and induced natural convection (solid surface on top and no solute induced natural convection).

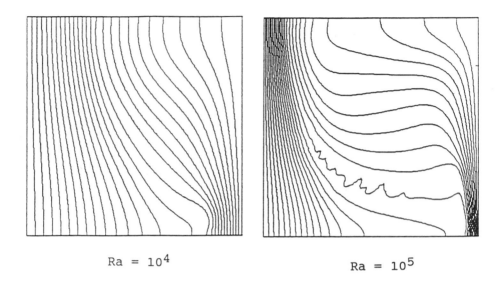

Ra = 10^4 Ra = 10^5

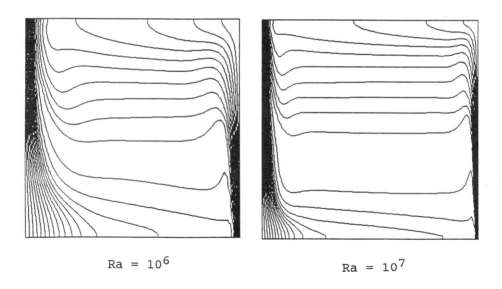

Ra = 10^6 Ra = 10^7

Figure 42. Enthalpy contours of Pr = 10^3, $T_2 - T_1$ = 0.2, and St = 0.2, temperature induced natural convection (solid surface on top and no solute induced natural convection).

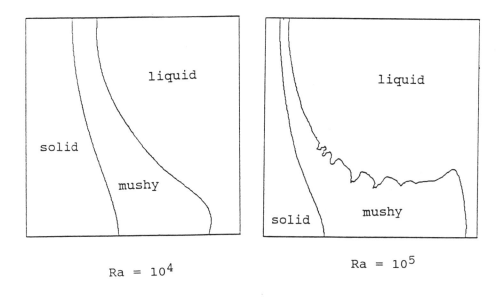

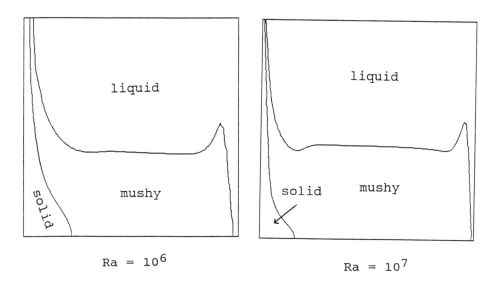

Figure 43. Phase distributions of $Pr = 10^3$, $T_2 - T_1 = 0.2$, and $St = 0.2$, temperature induced natural convection (solid surface on top and no solute induced natural convection).

max = 1.53×10^{-1}

Ra = 10^3
Gr = 6.7×10^4

max=1.36

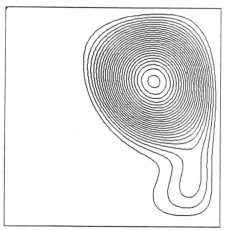

Ra = 10^4
Gr = 6.7×10^5

max=3.86

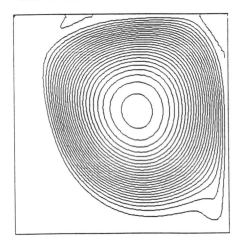

Ra = 10^5
Gr = 6.7×10^6

Figure 44. Streamlines of Pr = 1.49 $\times$ 10^{-2}, T$_2$ − T$_1$ = 0.2, and St = 0.2, temperature induced natural convection (solid surface on top and no solute induced natural convection).

Ra = 10^3
Gr = 6.7×10^4

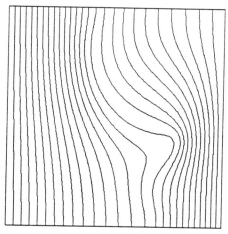

Ra = 10^4
Gr = 6.7×10^5

Ra = 10^5
Gr = 6.7×10^6

Figure 45. Enthalpy contours of Pr = 1.49 × 10⁻², T₂ – T₁ = 0.2, and St = 0.2, temperature induced natural convection (solid surface on top and no solute induced natural convection).

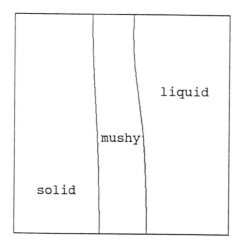

Ra = 10^3
Gr = $6.7\text{x}10^4$

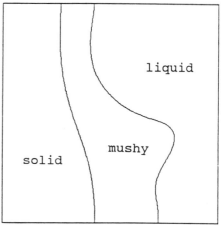

Ra = 10^4
Gr = $6.7\text{x}10^5$

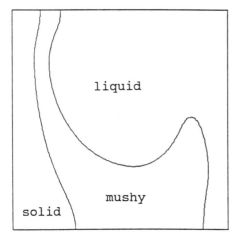

Ra = 10^5
Gr = $6.7\text{x}10^6$

*Figure 46. Phase distributions of Pr = 1.49 × 10^{-2}, $T_2 - T_1$ = 0.2, and St = 0.2,
temperature induced natural convection (solid surface on top and no solute
induced natural convection).*

8.3 Interaction of Thermocapillary and Natural Convection Flows during Phase Change

Marangoni–Rayleigh convection has a profound effect on solidification processes (Harter et. al. 1989, Nadarajah and Narayanan 1990). Examples of cases where thermocapillary convection interacts with buoyancy driven convection and phase change are given in the following. As in the previous section, $St = 0.2$ and $T_2 - T_1 = 0.2$. Two thermal Rayleigh numbers have been considered, 4.6×10^4 and 4.6×10^6. The Marangoni number varies from 0 to 8.4×10^2. Both augmenting and counteracting cases have been considered. The geometry and boundary conditions have been schematically illustrated in Figure 40.

Figure 47 compares the solutions with $Ra = 4.6 \times 10^4$, $Ma = 8.4 \times 10^3$ and two different Prandtl numbers, 1.49×10^{-2} and 10. Variation of the Prandtl number has a qualitative and quantitative difference on the transport and solidification characteristics. Major differences exist between the augmenting and counteracting cases. For the augmenting cases, the thermocapillary convection moves towards the mushy zone, which is a porous medium, rather than directly against the solid wall. The combined effect of the porous medium and Prandtl number seems to have a major effect on the resulting transport dynamics. With $Pr = 1.49 \times 10^{-2}$, it appears that the thermocapillary convection most notably adds only one more convection cell upon the buoyancy induced cell close to the free surface. For both augmenting and counteracting cases, the multiple cell pattern remains unchanged.

With $Pr = 10$, the enthalpy distribution is more responsive to the convection effect. Since surface tension is dependent on the local temperature gradient, the two–way coupling of surface tension and the temperature field is substantially stronger for high Pr material. Consequently, drastically different transport patterns can be observed as Pr varies. Figure 47 shows that with $Pr = 10$, the thermocapillary convection does not merely add one more convection cell; the Marangoni convection cell is able to merge with the buoyancy induced convection into a single one, and collectively to yield much larger influence on the thermal field. Compared to the case of $Pr = 1.49 \times 10^{-2}$, the phase boundaries of $Pr = 10$ are farther toward the cold wall with a highly distorted shape. The bulk domain is occupied by liquid melt with highly uniform temperature distributions.

In view of the large variation of convection pattern observed for the high Pr material, further assessment has been made to study the sensitivity of transport characteristics with respect to the Marangoni and Rayleigh numbers. Figure 48 shows the effect of Ma on streamfunction, enthalpy contours and phase boundaries of high Pr cases with identical Ra of 4.6×10^4, in augmenting mode. A clear evolution in convection pattern, and consequently the thermal field and phase boundaries, can be seen as Ma varies. With $Ma = 0$, the melt occupies about half the domain, and the mushy zone is located in the middle region with substantial thickness.

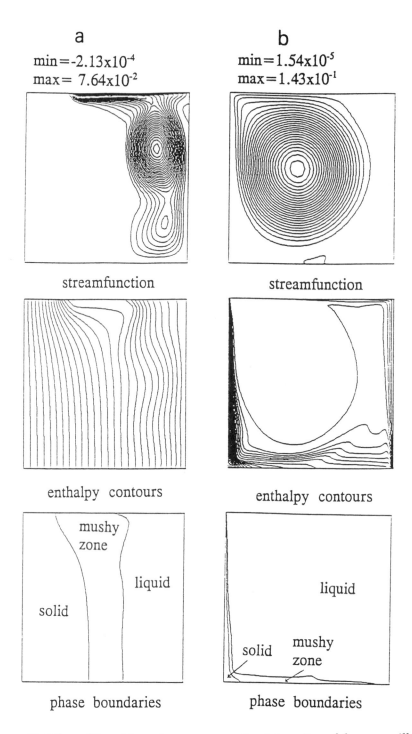

Figure 47. Effect of Prandtl number on augmenting interaction of thermocapillary and buoyancy induced flow with Ma = 8.4 × 10³, Ra = 4.6 × 10⁴ and St=0.2. (a)Pr=1.49 × 10⁻² (b)Pr=10.

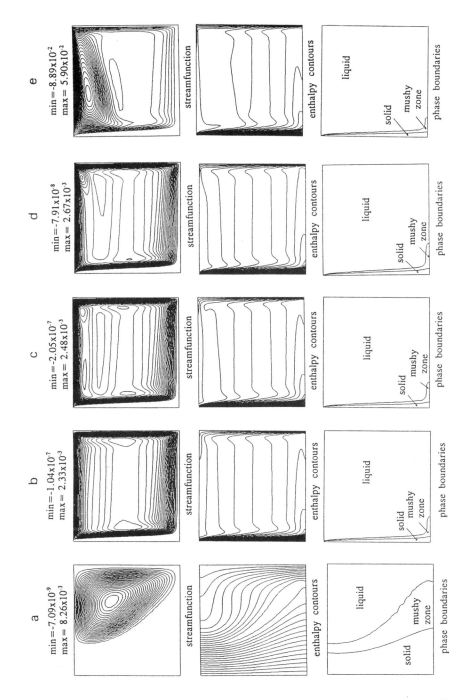

Figure 48. Effect of Marangoni number on augmenting interaction of thermocapillary and buoyancy induced flow with Pr = 10, Ra = 4.6 × 10⁴ and St=0.2. (a) Ma = 0, (b) Ma = 2.5 × 10², (c) Ma = 1.05 × 10³, (d) Ma=4.2 × 10³, (e) Ma=6.3 × 10³.

As *Ma* is increased to 2.5×10^2, the Marangoni convection and buoyancy induced convection collectively form a thin–layer pattern in the outer region of the domain. In the central region of the domain, the convection is relatively weak, as evidenced by the lack of streamfunction contours there. Furthermore, the mushy zone moves noticeably toward the cold wall and its thickness is much reduced. With $Ma = 1.05 \times 10^3$, the Marangoni convection seems strengthened, and is able to retain some of its own identity in the top free surface region. Similar multiple cell patterns have been observed experimentally. The multiple cell pattern persists at $Ma = 4.2 \times 10^3$. Overall, with *Ma* from 2.5×10^2 to 4.2×10^3, convection is pronounced only in boundary regions, the enthalpy contours and the phase boundaries are similar in character. Further increasing *Ma* to 8.4×10^3, the Marangoni convection becomes strong enough and the convection field again is of a single cell form. Moreover, unlike the case of lower *Ma*, the fluid in the central region of the melt is now of substantial convection strength. A clear transitional pattern between the multiple cell pattern (with *Ma* up to 4.2×10^3) and the single cell pattern (with $Ma = 8.4 \times 10^3$) can be seen at $Ma = 6.3 \times 10^3$. As shown in Figure 48 (e), the Marangoni convection cell grows to become substantially larger than that with $Ma = 4.2 \times 10^3$, and the buoyancy induced convection is largely limited to the region surrounding the wall and the phase interface. More information can be found in Shyy and Chen (1991a). The convective patterns depicted in Figure 48 are qualitatively in agreement with the experimental visualization obtained by Schwabe and coworkers (Schwabe 1988).

8.4 Effect of Gravity Jitter

Many devices in spacecraft contain enclosures filled with fluids. An example has been discussed in Chapter VII, Section 3, regarding the application to thermal management. In the following, we study another problem which is motivated by the need of understanding materials solidification in a much reduced but fluctuating gravitational field, the so–called g–jitter condition.

The configuration described in Section 7.2 and schematically illustrated in Figure 37 was subjected to a sinusoidal variation in the magnitude of the gravity vector:

$$g(t) \;=\; \frac{g_{max}}{2} \left[1 \;+\; \sin(\omega t) \right] \qquad (8.1)$$

The initial condition for this calculation was the conduction driven solution with a planar interface at the $x = 0.5$ location. Calculations were then conducted with an angular frequency, $\omega = 1$ and Rayleigh number, $Ra = 10^5$, based on the maximum value of the gravity vector, g_{max}. A uniform grid of 41×41 nodes was used and the time step used was 5% of the time period of the forcing function.

The following quantities are used to describe and assess the solution. q_h ; the total heat flux through the hot wall ($x = 0$), q_c ; the total heat flux through the cold wall ($x = 1$). A measure of the instantaneous interface location and shape was obtained by sampling the $f = 0.5$ contour at a total of 5 points, at every discrete time step. The coordinates of the sampled points may be described as: $(a(t),0)$, $(b(t),0.25)$, $(c(t),0.5)$, $(d(t),0.75)$, and $(e(t),1)$

where a, b, c, d and e are found by interpolating for the $f = 0.5$ contour, along $y = 0$, $y = 0.25$, $y = 0.5$, $y = 0.75$ and $y = 1$ respectively, at every time step. A schematic describing this process is shown in Figure 49.

Figure 50(a) shows the time history of the heat fluxes q_h and q_c along with their phase relationships to the forcing frequency. The heat flux through the hot wall, q_h, is mostly determined by convective effects, whereas the heat flux through the cold wall, q_c, is affected by conduction only. The quick response of convection to time variations in the body force ensures that (a) the flux q_h reaches a periodic state quickly (about 2 time periods), and (b) the phase difference between q_h and the forcing function is negligible for the chosen forcing frequency, $\omega = 1$. The heat flux q_c is determined by conduction effects and its time variation is determined by the change in boundary conditions caused by the migration of the $T = 0.5$ isotherm that constitutes the interface. Hence, q_c lags the forcing frequency by 90°. The power spectrum of q_h, and q_c (Figure 50(b)) shows the existence of superharmonics, but the energy content of the superharmonics are at least two orders of magnitude less than that of the fundamental. Thus the fundamental frequency is the dominant one. Figure 51(a) shows the time history of the interface motion at the sampled points. It is observed that the portion of the interface at the top of the cavity ($y = 1$) moves towards the cold wall ($x = 1$) as g increases in magnitude and convection strength increases, whereas, the portion of the interface near the bottom of the cavity ($y = 0$) moves towards the cold wall ($x = 0$). The spectral plots (Figure 51(b)) show that the fundamental frequency is the dominant one although higher harmonics exist.

More information of the results presented can be found in Shyy and Rao (1993). Regarding the effect of gravity level on convection, the studies by Biringen and Peltier (1990), Gebhart (1963), Venezian (1969), Wadih and Roux (1988), Shyy and Chen (1991a) and Murray et. al. (1993) can be consulted.

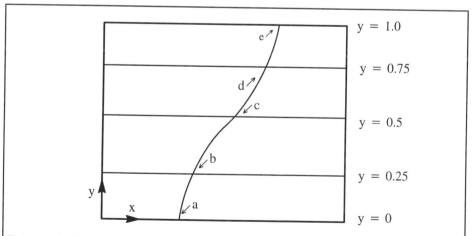

Figure 49. Schematic showing the sampled locations for the interface position at each time instant for the g–jitter problem

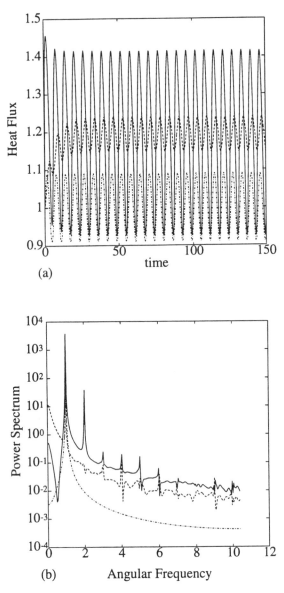

(a)

(b)   Angular Frequency

Figure 50. (a) Time history of the integrated heat fluxes for the g–jitter problem. The solid line indicates the heat flux through the heated wall and the dashed line indicates the heat flux through the cold wall. The dotted line indicates the forcing function. (b) Spectral plot of the heat flux through the walls of the cavity. The solid line indicates the heat flux through the heated wall and the dashed line indicates the heat flux through the cold wall. The dotted line indicates the forcing function.

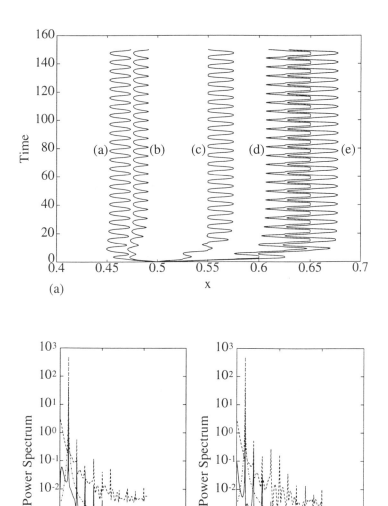

Figure 51. (a) Time history of the sampled interface motion at every time step. The letters a,b,c,d,e indicate the sampling locations as shown in Figure 49 . (b) Spectral plots of the sampled interface motion. (i) The solid line refers to location a and the dashed line to location e. (ii) The solid line refers to location b and the dashed line to location d. In both cases, the dotted line refers to the forcing frequency.

8.5 Effect of Geometry on Transport Characteristics

First, we investigate the effect of boundary perturbations on the characteristics of transport phenomena within the melt. This aspect can particularly benefit from the use of body–fitted coordinates. This case is taken from Shyy and Chen (1991b), where double diffusive convection flow is studied. The geometries and boundary conditions adopted in the present study are shown in Figure 52. Different types of boundary conditions have been employed. Dirichlet conditions are prescribed along the right and left surfaces for temperature and solutal concentration. In terms of the nondimensional quantities, both T and ϕ are set to unity along the right boundary and 0 along the left boundary. On both top and bottom surfaces, Neumann (zero gradient) conditions are assigned to both T and ϕ. The no–slip boundary condition is imposed along all four sides of the solid surfaces. No phase change takes place in this case. The irregular domain boundary demonstrates the effect of geometry on the transport and phase change characteristics.

Calculations have been performed in two different geometries as shown in Figure 52, namely, a square and an irregular domain. For the irregular domain, the right boundary is of a full period of the sinusoidal function with the magnitude equal to 5% of the width of the bottom surface–a modest variation. The grid systems employed in both domains consist of 81 x 81 nodes; the distribution is nonuniform with the spacing adjacent to all solid surfaces being 10^{-3} of the height of the domain. The mesh spacing incrementally increases from the wall to the center according to a quadratic polynomial. Besides the thermal Grashof number, the following quantities also emerge in the present problem:

Thermal Grashof Number:	$(Gr)_T = g\,\varrho^2 \beta_T \, l^3 \Delta T / \mu^2$
Solutal Grashof Number:	$(Gr)_s = g\,\varrho^2 \beta_\phi \, l^3 \Delta\phi / \mu^2$
Prandtl Number:	$Pr = \mu / \varrho\, \alpha$
Schmidt Number:	$Sc = \mu / \varrho\, D$
Lewis Number:	$Le = Sc / Pr$
Buoyancy Ratio:	$N = (Gr)_s / (Gr)_T$

$$(8.2)$$

where D is the mass diffusivity of the melt, α is the thermal diffusivity of the melt and l is the characteristic length. The Bousinnesq approximation has been adopted via the use of β_T and β_ϕ. Two sets of parameters were chosen:

(i) $Sc = 10$, $Pr = 1.49 \times 10^{-1}$, $(Gr)_T = 3.09 \times 10^6$, $N = 4.5$ and

(ii) same as (i) except $Pr = 1.49 \times 10^{-2}$

Figure 53 and Figure 54 compare the solutions obtained with the same parameters in two geometries. All the parameters, except the Prandtl number, are identical between the solutions shown in Figure 53 and Figure 54. In Figure 53, with $Pr = 1.49 \times 10^{-1}$, the convection strength is affected in a modest manner as the boundary shape changes. With $Pr = 1.49 \times 10^{-2}$, however, not only is the convection strength affected substantially by the modest boundary shape perturbation, but it also experiences a qualitative change in characteristics.

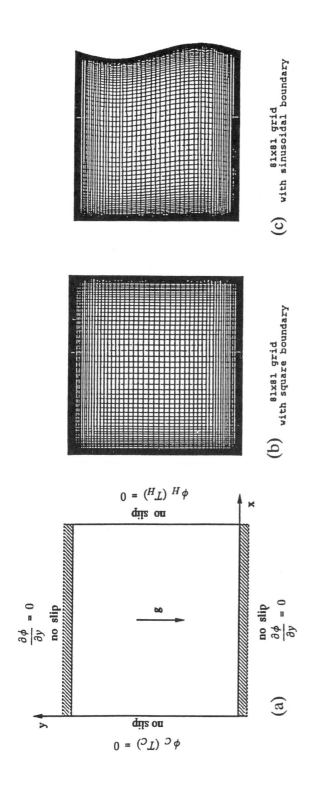

Figure 52. (a) Schematic of computational domain and boundary conditions.
(b) 81 x 81 grid with square geometry.
(c) 81 x 81 grid with sinusoidal geometry.

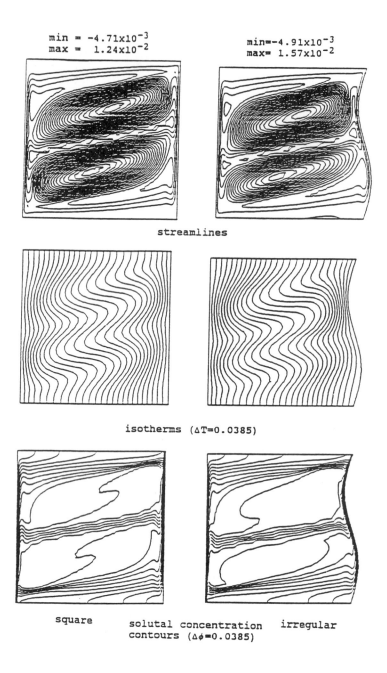

Figure 53. Effect of boundary irregularities on counteracting transport pattern
$Sc = 10$, $Pr = 1.49 \times 10^{-1}$, $(Gr)_T = 3.09 \times 10^6$, $N = 4.5$.

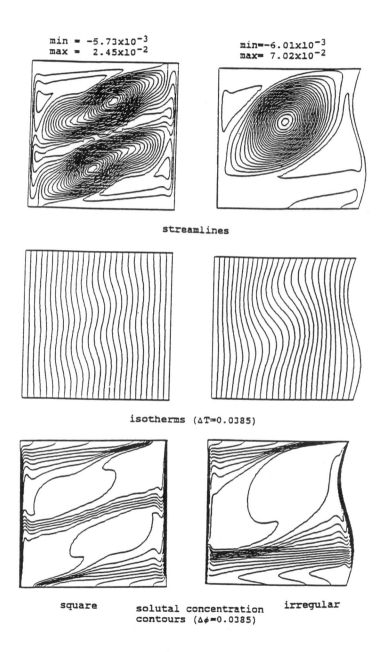

min = -5.73x10^{-3}
max = 2.45x10^{-2}

min=-6.01x10^{-3}
max= 7.02x10^{-2}

streamlines

isotherms (ΔT=0.0385)

square

solutal concentration
contours (Δφ=0.0385)

irregular

Figure 54. Effect of boundary irregularities on counteracting transport pattern,
Sc = 10, Pr = 1.49 × 10^{-2}, (Gr)$_T$ = 3.09 × 10^6, N = 4.5 .

It is noted that the dominant convection cell is largely dictated by the temperature field and the solute induced convection cells now all but disappear (Shyy and Chen 1991b). It appears that it is the Lewis number that mostly controls the sensitivity of the convection characteristics with respect to the boundary shape variations. The convection field with a higher *Le* tends to be more sensitive to geometrical perturbations.

8.6 Ingot Casting

Modern gas turbines use substantial quantities of high performance alloys. For example, titanium alloys have been extensively used in the fan and compressor sections of the power plant. However, there has long been a concern pertaining to the quality of the alloys used for these critical components. That quality concern specifically involves melt related defects. The current state of our fundamental understanding of the titanium alloys is such that even for the widely used Ti–6Al–4V, there is a lack of concrete knowledge regarding its solidification characteristics. As evidenced by the recent literature, intensive efforts have been made worldwide (Shamblen and Hunter 1989, Tripp and Mitchell 1989, Tetyukhin *et. al.* 1982, Hayakawa *et. al.* 1991, Kagawa *et. al.* 1990, Sellamuthu and Giamei 1986, Taha and Kurz 1981), to advance our understanding of the processing characteristics and their impact on the microstructural composition of these important materials.

One of the most influential process variables in controlling the cast microstructure and segregation is the movement of the solidification interface and its associated thermal and solutal gradient during the phase change process. Furthermore, the normal operating conditions of these processes are such that many transport mechanisms are present in the solidification process, including buoyancy–induced convection, surface tension–induced (Marangoni) convection, turbulence, combined conduction–radiation heat transfer, and, most critically, their interactions with the movement of the solidification front.

Physical models and numerical techniques have been developed (Shyy *et. al.* 1992e, 1993d) to perform simulations to delineate the transport phenomena and phase boundary during a Ti–6Al–4V alloy ingot casting process. The model is based on the k–ε two–equation closure (Launder and Spalding 1974) along with the concept of low Reynolds number modification (Launder and Sharma 1974) to account for the presence of dendrite branches and phase change in the mushy zone. With this new development, detailed investigation can be made to study the structure of turbulent transport during the ingot casting process, and its impact on the phase change characteristics.

A schematic illustration of an operating system is given in Figure 55(a), where the raw materials are initially melted with an external heat source in a water–cooled melting hearth before the molten metal is cast into ingots. The modeled geometry as well as conditions of the ingot casting are sketched in Figure 55(b). For the case considered, the material used is Ti–6Al–4V, with an ingot diameter of 0.432 m, and a height of the copper mold of 0.457m.

As to the convection mechanisms, both the buoyancy effect within the melt and the thermocapillary effect on the pool surface need to be included. In estimating the strength

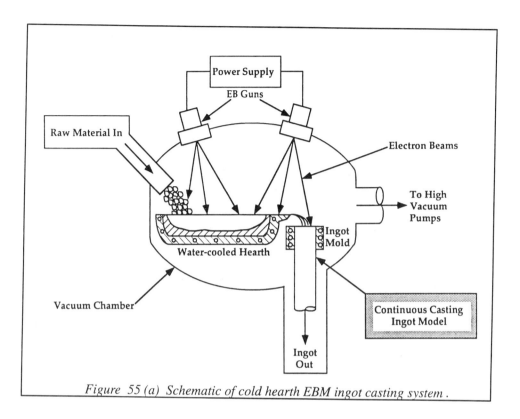

Figure 55 (a) Schematic of cold hearth EBM ingot casting system.

of buoyancy–induced convection, one can use the thermal Rayleigh number, $(Ra)_T$, as an indicator. The thermal Rayleigh number is defined as $(Ra)_T = (Gr)_T \cdot Pr$.

The thermal Grashof number is now defined without resorting to the assumption of small temperature variations. Hence, as opposed to Eq. (8.2),

$$(Gr)_T = \frac{gR^3 \varrho \; \Delta \varrho}{\mu^2} \tag{8.3}$$

where R is a characteristic length scale. Here, we choose the characteristic length R to be the ingot radius, R = 0.216 m, and the representative kinematic viscosity to be the value at the liquidus temperature, T_{liq} (1898 K), which is about 10^{-6} m²/sec. Hence, under the normal gravity level observed on earth, $Gr \sim 10^{11} \frac{\Delta \varrho}{\varrho}$ and the Prandtl number at T_{liq} is 0.165, resulting in a thermal Rayleigh number of $Ra \sim 10^{10} \frac{\Delta \varrho}{\varrho}$. For this case, the Boussinesq approximation is not employed and the property variations are completely taken into account in the computations.

It is emphasized here that although the casting material temperatures generally increase from bottom to top, their nonuniform distributions along the radial direction can still cause an unbalanced density field to produce buoyancy–induced convection. Our calculations indicated that the density imbalance $\Delta \varrho / \varrho$ could be around 10%, yielding a

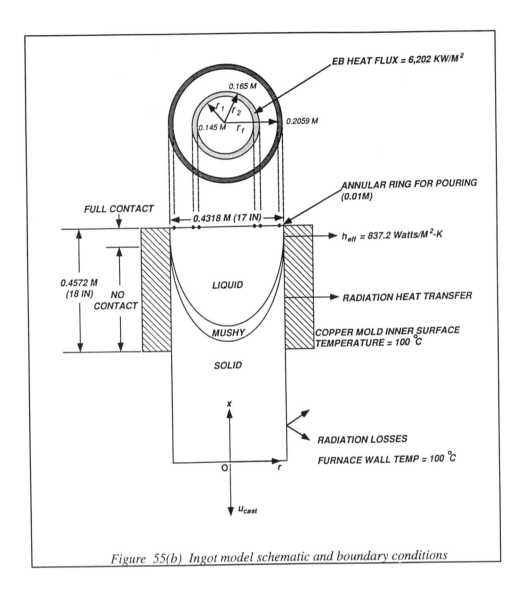

Figure 55(b) Ingot model schematic and boundary conditions

thermal Rayleigh number of the order of 10^9. Regarding the thermocapillary aspect, based on the same length scale and the appropriate material properties, it is estimated that the Marangoni number is of the order of 10^6. Hence, the flows are expected to be well within the turbulent regime.

(i) Predicted Results

The Favre–averaged Navier–Stokes equations of mass continuity, momentum, and energy transport, along with a modified k–ε two–equation turbulence closure, are the basis of our computations. We have adopted the enthalpy formulation to represent both liquid and solid phases within a unified set of equations. In terms of the momentum equations, a Darcy's law type of porous medium treatment is utilized to account for the effect of phase change on convection. Calculations have been conducted for cases that contain full contributions of both buoyancy and surface tension, and that contain only the surface tension effect. With such a systematic scrutiny, both the collective and individual effects of the two convection mechanisms can be investigated, and their interaction with solidification studied. The results presented in the following correspond to two casting speeds -2×10^{-4} m/sec (455 kg/hr) and -4×10^{-4} m/sec (910 kg/hr), and under the influence of two gravity levels, one the standard value (g_0) measured on earth (termed as earth–bound condition), and the other $10^{-5} g_0$ (termed as microgravity condition). The negative sign indicates a downward casting direction. Figure 56 shows the adaptive grid distribution generated for each case. To generate the grid, an adaptive procedure has been used. As demonstrated in Figure 56, for cases under the normal gravity level, adequate spatial resolution is needed for both enthalpy and streamfunction in the melt and mushy zone. Consequently, the grid lines do not directly correlate with the phase boundaries. Under microgravity, on the other hand, the convection field is much reduced, and hence the grid lines can more directly correspond to the characteristics of thermal field. As one can clearly observe from Figure 56, the higher the casting speed, the more downward the solidus line. The meshes around the solidification interface tend to be less orthogonal, most noticeably for the case of $u_{cast} = -4 \times 10^{-4}$ m/sec and under microgravity. This mesh distribution is formed because convection under the given condition is relatively weak, and hence the grid clustering is needed primarily for tracking the phase boundaries where the latent heat release is significant. Under such a circumstance, it is a good practice to put the emphasis on length scale resolution above mesh orthogonality.

Figure 57 presents the solutions obtained under normal gravity conditions, including isotherms and liquid fraction distribution across mushy zone. Two different casting speeds have been used for computations to investigate their impact on the resulting solidification characteristics. The model predicts a thick mean mushy zone, based on the mean temperature, as a result of the convection effect, and the zone thickness varies with the change of casting speed. Within the half domain of the cross section, the model predicts that contrarotating eddies appear due to the combined effects of surface tension and buoyancy. The liquidus line in both solutions is approximately flat, and its depth appears insensitive to the variation of the casting speed. It is noted that under the same thermal energy conditions, the mushy zone yielded by the pure conduction transport is generally thinner in order to maintain required heat flow rates by the microscopic molecular mechanism i.e., conduction, alone. With the inclusion of convection, the heat transfer rates in both the bulk melt and upper mushy zone are now enhanced by macroscopic flow motion; the degree of heat transfer enhancement depends, of course, on the local convection strength. In the bulk melt as well as the upper mushy zone where convection is vigorous, turbulence plays a dominant role in overall heat transfer, and accordingly, in these regions the temperature distribution is smeared and the overall thermal gradients are substantially reduced.

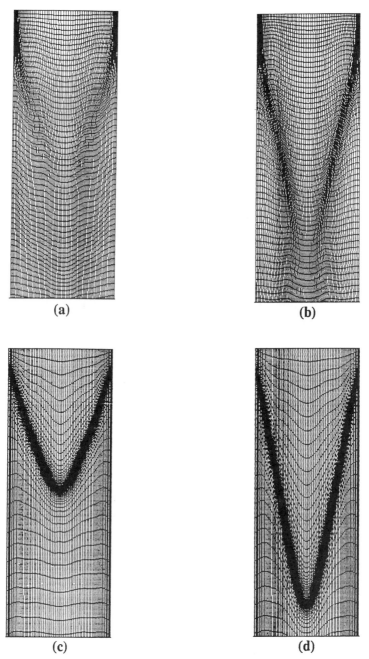

(a)   **(b)**

(c)   **(d)**

Figure 56.   Adaptive grid distribution corresponding to modelling condition (only
upper portion of the domain is shown)
(a) Normal Gravity, $u_{cast} = -2 \times 10^{-4}\,m\,s^{-1}$ (b) Normal Gravity,
$u_{cast} = -4 \times 10^{-4}\,m\,s^{-1}$ (c) Microgravity, $u_{cast} = -2 \times 10^{-4}\,m\,s^{-1}$
(d) Microgravity, $u_{cast} = -4 \times 10^{-4}\,m\,s^{-1}$.

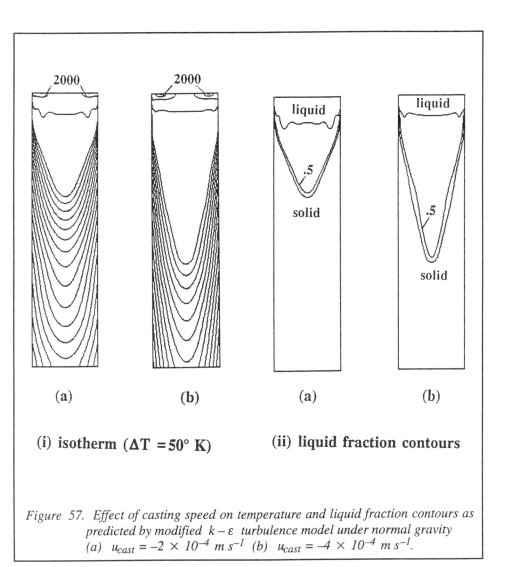

Figure 57. Effect of casting speed on temperature and liquid fraction contours as predicted by modified k – ε turbulence model under normal gravity (a) $u_{cast} = -2 \times 10^{-4}$ *m s^{-1} (b)* $u_{cast} = -4 \times 10^{-4}$ *m s^{-1}.*

Toward the solidus lines, however, the increasing presence of the solidified dendrites within the mushy zone dampens convection, and the thermal conduction again becomes the controlling mechanism. This change of dominant transport mechanisms from liquidus to solidus lines explains why in regions close to the solidus lines the temperature gradients are high, resulting in small mushy zone thicknesses encompassing large variations of the liquid fraction; while, on the other hand, close to the liquidus lines the temperature gradients are low, resulting in much increased mushy zone thicknesses there. Figure 58 shows the distribution of the corresponding streamfunction under the two casting speeds. Although the patterns of the distributions appear qualitatively the same for both speeds, it is clear that the higher the casting speed, the deeper the convection penetrates into the mushy zone.

The solutions obtained with the identical modeling condition and physical models, but under a microgravity condition with 10^{-5} times the gravity level on earth, are presented next. The gross qualitative features related to the thermal and phase change aspects are shown in Figure 60. Figure 60 shows that, under microgravity, the mushy zones are generally of much reduced thickness than those under normal gravity level. Under microgravity, the convection strength, produced virtually by the surface tension alone, is much weaker. Consequently, the liquidus line tends to have a stronger curvature, and the solidus line corresponding to the same casting speed is deeper than under normal gravity.

(ii) Experiment

A Ti–6Al–4V ingot was cast in an EBM process with the process parameters that were simplified as shown in Figure 55. The casting rate, u_{cast}, was approximately -1.8×10^{-4} m/sec. A small amount of copper was added to the ingot mold pool near the end of the casting process to mark the pool profile for subsequent ingot macroetch evaluation. The ingot top was sliced axially, and the slices were macroetched by standard nitric–hydrofluoric and ammonium bifluoride etchants. The result, shown in Figure 59, includes an outline trace of the pool profile for added clarity. Because it reduces the melting temperature, the copper probably caused some of the solid, particularly in the mushy zone, to remelt. Since there was insufficient heat available to remelt much of the solid, the profile indicated by macroetching is probably close to the solidus line. The orientation of the columnar dendritic structure that does not contain copper suggests that the macroetch profile at least approximates the solidus line.

The modified k–ε turbulence model prediction for the solidus line profile, as shown in Figure 57, has features which correlate to features observed on the macroetch profile. The "bulge" outward in the upper portion of the pool qualitatively appears in both profiles. The high curvature at the bottom of the macroetch pool profile also corresponds closely to the prediction. As predicted with both turbulence models, the pool depth is roughly proportional to the casting rate. Therefore, the pool depth predicted by the modified k–ε turbulence model is 55 cm with the -1.8×10^{-4} m/sec casting rate, which closely approximates the macroetch profile depth of 53 cm.

Comparison between the numerical and experimental results indicates that more work still remains. All of the numerical predictions have a much thinner solid layer near the top pool surface and a lower curvature at the pool bottom than observed in the macroetch profile. More efforts are required, both experimentally and theoretically, to determine the detailed thermal characteristics associated with the phase change process and to investigate their relationships with the columnar dendritic structure observed below the macroetch profile as shown in Figure 59. Issues such as the estimates of mushy zone characteristics, turbulence behavior, material properties, and process parameters need to be addressed with more quantitative information. Extensive discussion of the effect of

turbulence on the phase change process and the associated modeling issues have been
presented by Shyy *et. al.* (1993d), as will be summarized in the following.

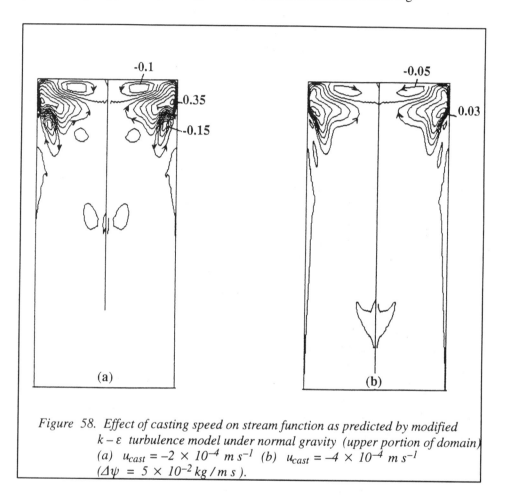

Figure 58. Effect of casting speed on stream function as predicted by modified
$k - \varepsilon$ turbulence model under normal gravity (upper portion of domain)
(a) $u_{cast} = -2 \times 10^{-4} \; m \; s^{-1}$ (b) $u_{cast} = -4 \times 10^{-4} \; m \; s^{-1}$
($\Delta \psi = 5 \times 10^{-2} \, kg \, / \, m \, s$).

(iii) Modeling Issues

Fundamentally there are two types of terms derived from turbulence that need to be
modeled. The first type is the correlation of velocity and temperature fluctuations, such
as $\overline{u' T'}$, that is commonly encountered in any turbulent heat transfer process, regardless
of whether the flow is of single– or multi–phase. In the context of the two–equation
turbulence closure model, these terms are represented by the combined eddy viscosity,
factored by a turbulent Prandtl number, and local temperature gradients. The second type
of the turbulent transport terms arises because of the presence of phase change. The
fraction of liquid which freezes locally during phase change is determined by the amount
of latent heat having been released.

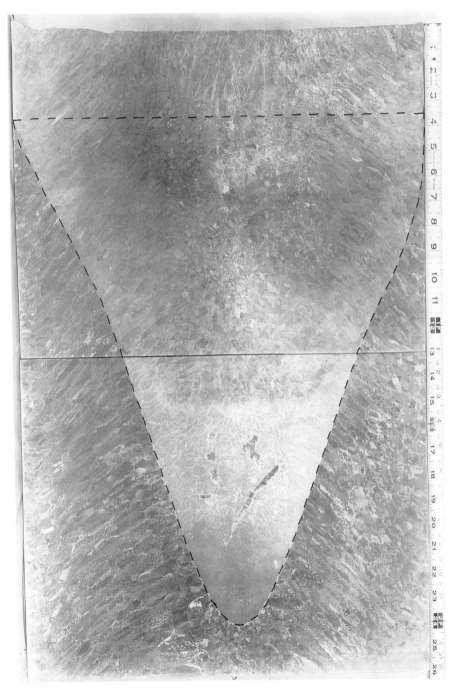

Figure 59. Macroetched axial pool profile from the ingot mold (pour entry on left)

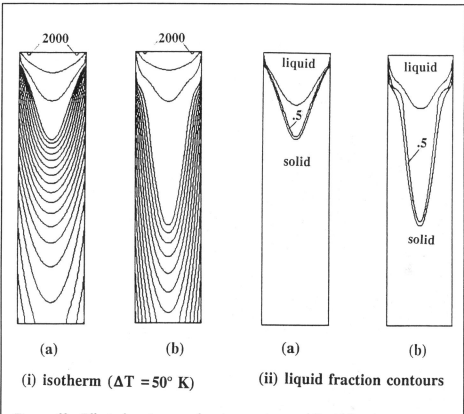

(a)　　　　(b)　　　　(a)　　　　(b)

(i) isotherm (ΔT = 50° K)　　(ii) liquid fraction contours

Figure 60. Effect of casting speed on temperature and liquid fraction contours as predicted by modified $k-\varepsilon$ turbulence model under microgravity (a) $u_{cast} = -2 \times 10^{-4}$ m s^{-1} (b) $u_{cast} = -4 \times 10^{-4}$ m s^{-1}.

This fraction can usually be estimated from the equilibrium or nonequilibrium lever rule which involves a partition ratio defined as the ratio of solid to liquid composition under the condition of equilibrium. As a simple illustration, the degree of latent heat release is assumed to vary linearly with the local temperature, i.e., $\Delta h = f \cdot \Delta h_{l-s}$, where

$$f = \frac{T - T_{sol}}{T_{liq} - T_{sol}} \quad , \quad T_{sol} < T < T_{liq} \tag{8.4}$$

Here, T represents the mean temperature of the material. Hence, terms such as $\overline{u'\Delta h'}$ can be expressed in terms of $\overline{u'T'}$, namely, $\overline{u'\Delta h'} = \alpha \frac{\overline{u'T'}}{T_{liq} - T_{sol}} \Delta h_{l-s}$ where α is an added constant to account for the correlation between the instantaneous rather than the mean value of f, as a function of the instantaneous temperature, and the instantaneous extent of latent heat release. With $\alpha = 1$, the above equation amounts to assuming that a dendrite can instantaneously form or remelt according to the local temperature level. This scenario does not appear realistic because of the finite time period required for the dendrite to solidify or melt, and hence should be viewed as the upper bound of the potential effect arising from the correlation between velocity and release of latent heat fluctuations. In reality, this correlation should be a function of local solid fraction since the time scale that a dendrite branch needs to respond to the environmental fluctuation is related to the local dendrite length scale. In the region that the solid fraction is high, the value of α must be low, and vice versa.

One can estimate the distribution of the temperature fluctuation according to the results yielded by the present turbulence model. It is noted that the correlated value of u' and T' is linked to the mean temperature gradient via a formula such as

$$\overline{u'T'} = -\frac{\mu_t}{\varrho \, \mathrm{Pr}_t} \frac{\partial T}{\partial x} \tag{8.5}$$

Based on the above equation, a correlation coefficient between u' and T' can be defined as follows $\sqrt{\overline{(u')^2}}\sqrt{\overline{(T')^2}} = \beta \overline{u'T'}$ where β is considered, for simplicity, to be a constant. Then, in accordance with the k–ε model, one can use the information of the turbulent kinetic energy k to obtain the values of $\sqrt{\overline{(u')^2}}$ by the isotropic condition, namely $\overline{(u')^2} = \frac{2}{3}k$. Hence, combining the above equations, the distribution of $\sqrt{\overline{(T')^2}}$ can be estimated. Results presented in Shyy *et. al.* (1993d) demonstrated that the location of the liquidus line is time dependent and, consequently, the mushy zone can experience a considerable variation in thickness during the casting process.

8.7 Low Pressure Conditions

For processing techniques such as chemical vapor deposition (CVD), the operating pressures are quite low, of the order of several Torr. Then the flowfield can fall into the "rarefied" regime in the sense of the Knudsen number,

$$Kn = \frac{\lambda}{l} \tag{8.6}$$

becoming appreciable (Rosner 1989). Here λ is the mean free path and l is the characteristic length scale of the transport process. If the Knudsen number is small, the continuum approach is valid, but when low pressures are encountered, the validity of the continuum hypothesis may need to be justified.

Under low pressure conditions the no–slip condition at solid boundaries may no longer be justified. Consider a straight and horizontal wall on which there is a slip velocity, U_s. Then the mean shear stress at the solid surface can be written in the form,

$$\tau = \dot{m}\left(U_s + \lambda \frac{du}{dy}\right) \tag{8.7}$$

neglecting specular reflection of the molecules from the solid surface. Equation (8.7) says that the molecules that are incident on the solid surface carry on an average, the x–momentum per molecule that prevails at a distance of the mean free path from the solid surface. Furthermore, the mass flux can be approximated as,

$$\dot{m} = \frac{1}{4}\,\omega n \bar{c} \tag{8.8}$$

where n is the number density of the molecules

$\bar{c}$ is the mean molecular speed

ω is the mass of the molecule, i.e., $n = \varrho/\omega$.

Then the expression for the wall shear stress, based on our simplified kinetic approach, becomes

$$\tau = \frac{1}{4}\,\varrho\bar{c}\left(U_s + \lambda\,\frac{du}{dy}\right) \tag{8.9}$$

Now, comparing with the continuum approach, the wall shear stress is,

$$\tau = \mu\,\frac{du}{dy} \tag{8.10}$$

and from the kinetic theory of gases we know that $\mu = \frac{1}{2}\varrho\bar{c}\lambda$. Therefore,

$$\tau = \frac{1}{2}\,\varrho\bar{c}\lambda\,\frac{du}{dy} \tag{8.11}$$

Thus we obtain the slip velocity as

$$U_s = \lambda\,\frac{du}{dy} \tag{8.12}$$

The above analysis holds for an isothermal case. If the wall is not isothermal, the slip velocity can result from the phenomenon of thermal creep,

$$U_s = \lambda\,\frac{du}{dy} + \frac{3}{4}\,\frac{\mu}{\varrho T}\,\frac{dT}{dx} \tag{8.13}$$

In this part, both the fundamental physics and the computational algorithms pertinent to the interfacial transport and its interaction with macroscopic convection–diffusion have been presented. The emphasis is heavily on the aspects related to materials processing. However, in the course of the discussion, issues such as scaling, multiple mechanisms, capillarity and gravity have been addressed. These issues are common to many other problems.

9. FURTHER READING

More detailed information of the various aspects covered here can be found in Collier (1981), Chernov (1984), Crank (1984), Haasen (1986), Hartman (1973), Kurz and Fisher (1984), Pamplin (1975), Pelce (1988), Porter and Easterling (1992), Rosenberger (1979), Rosner (1986), Szekely (1979), Tiller (1991), Walter (1987) and Williams (1985).

REFERENCES

Adamson, A.W. 1990 *Physical Chemistry of Surfaces*, 5th ed., Wiley–Interscience, New York.

Albert, M. R. and O'Neill, K. 1986 Moving Boundary–Moving Mesh Analysis of Phase Change Using Finite Element Method, *Int. J. Numer. Meths. Engrg.*, **23**, 591–607.

Ames, W.F. 1992 *Numerical Methods for Partial Differential Equations*, 3rd ed., Academic Press, New York.

Ananth, R. and Gill, W.N., 1989 Dendritic Growth of an Elliptic Paraboloid with Forced Convection in the Melt, *J. Fluid Mech.*, **208**, 575–593.

Ananth, R. and Gill, W.N., 1991 Self–Consistent Theory of Dendritic Growth with Convection, *J. Crystal Growth*, **108**, 173–189.

Anderson, D.A.,Tannehill, J.C. and Pletcher, R.H. 1984 *Computational Fluid Mechanics and Heat Transfer,* Hemisphere, New York.

Anderson, W.G. and Beam, J.E. 1992 High Temperature Capillary Pumped Loops for Thermionic Power Systems, *Proceedings of the Nineth Symposium on Space Nuclear Power Station*, 1162–1169.

Antonia, R.A. and Bilger, R.W. 1973 An Experimental Investigation of an Axisymmetic Jet in Co–flowing Air Stream, *J. Fluid Mech.*, **61**, 805–822.

Arakawa, Ch., Demuren, A.D., Rodi, W. and Schonrng, B. 1988 Application of Multigrid Methods for the Coupled and Decoupled Solutions of the Incompressible Navier–Stokes Equations, in M. Deville (ed.) *Proceedings of the seventh GAMM conference on Numerical Methods in Fluid Mechanics*, Notes on Numerical Fluid Mechanics, vol. 20, Vieweg, Braunschweig, Germany, pp 1–8.

Aref, H. and Tryggvason, G., 1984, Vortex dynamics for Passive and Active Interfaces, *Physica D,* **12**, 59–70.

Arpaci, V.S. and Larsen, P.S. 1984 *Convection Heat Transfer*, Prentice–Hall, Englewood Cliffs, NJ.

Ashgriz, N. and Poo, J. Y. 1991 FLAIR : Flux Line–Segment Model for Advection and Interface Reconstruction, *J. Comp. Phys.*, **93**, 449–468.

Babuska, I., Zienkiewicz, O.C. Gago, J. and Oliviera A. (eds.) 1986 *Accuracy Estimates and Adaptive Refinements in Finite Element Computations,* Wiley, New York.

Backman, D.G. 1990 Metal–Matrix Composites and IPM: A Modeling Perspective, *J. Metal., July Issue,* 17–20.

Bayyuk, S., Powell, K., and Van Leer, B. 1993 A Simulation Technique for Two–Dimensional Unsteady Inviscid–Flows around Arbitrarily Moving and Deforming Bodies of Arbitrary Geometry, *AIAA Paper No. 93–3391–CP*

Beckermann, C and Viskanta, R. 1988 Double–Diffusive Convection due to Melting, *Int. J. Heat Mass Transf.,* **31**, 2077–2089.

Bejan, A. 1984 *Convection Heat Transfer*, Wiley, New York.

Benek, J.A., Buning, P.G. and Steger, J.L. 1985 A 3–D Chimera Grid Embedding Technique, *AIAA Paper No. 85–1523.*

Bennon, W.D. and Incropera, F. P. 1987 A Continuum Method for Momentum, Heat and Species Transport in Binary Solid–Liquid Phase Change Systems–I. Model Formulation, *Int. J. Heat Mass Transf.,* **30**, 2161–2170.

Berge, P., Pomeau, Y. and Vidal, C. 1984 *Order within Chaos*, Wiley–Interscience, New York.

Berger, M. and Jameson, A. 1985 Automatic Adaptive Grid Refinement for the Euler Equations, *AIAA J.,* **23**, 561–568.

Bilger, R.W. 1980 Perturbation Analysis of Turbulent Non–Premixed Combustion. *Combust. Sci. Technol.,* **22**, 251–262.

Biringen, S and Peltier, L.J. 1990 Numerical Simulation of 3–D Benard Convection with Gravitational Modulation, *Phys. Fluids,* **2A**, 754–764

Birkhoff, G. and Lynch, R.E. 1984 *Numerical Solution of Elliptic Problems*, SIAM, Philadelphia, PA.

Block, J.M. 1956 Surface Tension is the Cause of Benard Cells and Surface Deformation in a Liquid Film, *Nature,* **178**, 560–562.

Blosch, E., Shyy, W. and Smith, R.W. 1993 The Role of Mass Conservation in Pressure–Based Algorithms, accepted for publication in *Numer. Heat Transf.*

Blottner, F.G. 1982 Influence of Boundary Approximations and Conditions on Finite–Difference Solutions, *J. Comput. Phys.*, **48**, 246–269.

Bogy, D.B. 1977 Drop formation in a circular liquid jet, *Ann. Rev. Fluid Mech.*, **11**, 207–228.

Boris, J.P. and Book, D.L. 1973 Flux Corrected Transport I, Shasta: An Algorithm that works, *J. Comput. Phys.*, **11**, 38–69.

Botta, E.F.F. and Veldman, A.E.P. 1982 On Local Relaxation Methods and Their Application to Convection–Diffusion Equation, *J. Comput. Phys.*, **48**, 127–149.

Bouissou, Ph., Perrin, B. and Tabeling, P., 1990 Influence of an external periodic flow on dendritic crystal growth, in *Nonlinear Evolution of Spatio–temporal Structures in Dissipative Dynamical Systems*, Eds. Busse, F.H., Kramer, L., Plenum Press, New York.

Boyce, W.E. and DiPrima, R.C. 1986 *Elementary Differential Equations and Boundary Value Problems*, 4th edition, Wiley, New York.

Braaten, M.E. 1985 Development and Evaluation of Iterative and Direct Methods for the Solution of the Equations Governing Recirculating Flows, Ph.D. Thesis, University of Minnesota, Minneapolis.

Braaten, M.E. and Shyy, W. 1986a A Study of Recirculating Flow Computation using Body–Fitted Coordinates: Consistency Aspects and Mesh Skewness, *Numer. Heat Transf.* **9**, 559–574.

Braaten, M.E. and Shyy, W. 1986b Comparison of Iterative and Direct Method for Viscous Flow Calculations in Body–fitted Coordinates, *Int. J. Numer. Meths. Fluids,* **6**, 325–349.

Braaten, M.E. and Shyy, W. 1987 A Study of Pressure Correction Methods with Multigrid for Viscous Flow Calculations in Non–Orthogonal Curvilinear Coordinates, *Numer. Heat Transf.*, **11**, 417–442.

Brandt, A. 1977 Multi–Level Adaptive Solutions to Boundary Value Problems, *Math. Comput.*, **31**, 333–390.

Brent, A.D., Voller, V. R. and Reid, K. J. 1988 Enthalpy–Porosity Technique for Modeling Convection–Diffusion Phase Change: Application to the Melting of a Pure Metal, *Numer. Heat Transf.*, **13**, 297–318.

Briggs, W.L. 1987 *A Multigrid Tutorial*, SIAM, Philadelphia, PA.

Brown, R.A. 1988 Theory of Transport Processes in Single Crystal Growth from the Melt, *A.I.Ch.E. J.,* **34**, 881–911.

Bruneau, C.H, and Jouran, C. 1990   An Efficient Scheme for Solving Steady Incompressible Navier–Stokes Equations, *J. Comput. Phys.*, **89**, 389–413.

Butler, T.D., Cloutman, L.D., Duckowicz, J.K. and Ramshaw, J.D. 1981 Multi–Dimensional Numerical Simulation of Reactive Flow in Internal Combustion Engine, *Prog. Energy Combust. Sci.*, **7**, 293–315.

Canright, D. and Davis, S.H. 1991 Buoyancy Effects of a Growing, Isolated Dendrite, *J. Crystal Growth*, **114**, 173–189.

Carey, V.P. 1992 *Liquid–Vapor Phase–Change Phenomena*, Hemisphere, Washington, DC.

Carslaw, H.S. and Jaeger, J. C. 1959 *Conduction of Heat in Solids*, London, Oxford University Press, Oxford, UK.

Chalmers, D.R., Fredly, J.E., Ku, J. and Kroliczek, E.J. 1988 Design of a Two–Phase Capillary Pumped Flight Experiment, *Paper SAE–88–1086.*

Chan, Y.T., Gibeling, H.J. and Grubin, H.L. 1988 Numerical Simulation of Czochralski Silicon Growth, *J. Appl. Phys.*, **64**, 1425–1439.

Chandrasekar, S. 1961 *Hydrodynamic and Hydromagnetic Stability*, Oxford University Press, Oxford, UK.

Chang, P.Y. and Shyy, W. 1991a Adaptive Grid Computation of Three–Dimensional Natural Convection in Horizontal High Pressure Mercury Lamps, *Int. J. Numer. Meths. Fluids,* **12**, 143–160.

Chang, P.Y and Shyy, W. 1991b Three–Dimensional Heat Transfer and Fluid Flow in the Modern Discharge Lamp, *Int. J. Heat Mass Transf.,* **34**, 1811–1822.

Chang, P.Y., Shyy, W. and Dakin, J.T. 1990 A Study of Three–Dimensional Natural Convection in High Pressure Mercury Lamps, Part I: Parametric Variations with Horizontal Mounting, *Int. J. Heat Mass Transf.,* **33**, 483–493.

Chernov, A.A. 1984 *Modern Crystallography III: Crystal Growth*, Springer–Verlag, New York.

Chen, C.P., Jiang, Y., Kim, Y.M. and Shang, H.M. 1991 A Computer Code for Multiphase All–Speed Transient Flows in Complex Geometries, NASA Contract Report NAG8–092.

Chen, J.Y. and Kollman, W. 1990 Chemical Models for PDF Modeling of Hydrogen–Air Nonpremixed Turbulent Flames, *Combust. Flame*, **79**, 75–99.

Chen, T.H. and Goss, L.P. 1989 Flame Lifting and Flame/Flow Interaction of Jet Diffusion Flames, *AIAA Paper No. 89–0156*.

Chen, T.H., Goss, L.P., Talley, D. and Mikolaitis, D. 1989 Stabilization Zone Structure in Jet Diffusion Flames from Liftoff to Blowout, *AIAA Paper No. 89–0153*.

Chen, M.–H., Hsu, C.–C. and Shyy, W. 1991a Assessment of TVD Schemes for Inviscid and Turbulent Flow Computation, *Int. J. Numer. Meths. Fluids*, **12**, 161–177.

Chen, M.–H., Shyy, W., Sun, C.–S., and Liang, S.–J. 1991b Assessment of Several Linear Multigrid Solvers for Transport Problems, *Proceedings of Seventh Interational Conference on Numerical Methods in Laminar and Turbulence Flow*, **vol. 7**, C. Taylor, J. H. Chin and G. M. Homsy (eds), Part 2, 1283–1294, Pineridge press, Swansea, U.K.

Chen, Y.S. 1988 Viscous Flow Computations using a Second–Order Upwind Differencing Scheme, *AIAA Paper 88–0417*.

Chen, Y.S. 1989 Compressible and Incompressible Flow Computations with a Pressure Based Method, *AIAA Paper 89–0286*.

Chern, I.–L., Glimm, J., McBryan, O., Plohr, B. and Yaniv, S. 1986 Front Tracking for Gas Dynamics, *J. Comp. Phys.*, **62**, 83–110.

Chernousko, F. L. 1970 Solution of Non–Linear Heat Conduction Problems in Media with Phase Change, *Int. Chem. Engrg*, **10**, 42–48.

Chesshire, G. and Henshaw, W.D. 1990 Composite Overlapping Meshes for the solution of Partial Differential Equations, *J. Comput. Phys.*, **90**, 1–64.

Chyu, W.J., Rimlinger, M.J. and Shih, T. I.–P. 1993 Effects of Bleed–Hole Geometry and Plenum Pressure on Three–Dimensional Shock–Wave/Boundary–Layer/Bleed Interactions, *AIAA Paper No. 93–3259*.

Chorin, A. J. 1985 Curvature and Solidification, *J. Comp. Phys.*, **58**, 472–490.

Christensen, M.S., Bennon, W.D. and Incropera, F.P. 1989 Solidification of an Aqueous Ammonium Chloride Solution in a Rectangular Cavity—II. Comparison of Predicted and Measured Results. *Int. J. Heat Mass Transf.*, **32**, 69–79.

Collier, J.G. 1981 *Convection Boiling and Condensation*, 2nd edition, McGraw–Hill, New York.

Concus, P. and Golub, G. 1973   Use of Fast Direct Methods for the Efficient Numerical Solution of Nonseparable Elliptic Equations, *SIAM J. Numer. Anal.*, **10**, 1103–1120.

Connell, S. and Holmes, D.G. 1993 A 3D Unstructured Adaptive Multigrid Scheme for the Euler Equations, *AIAA Paper No. 93–3339–CP.*

Conte, S.D. and de Boor, C. 1980 *Elementary Numerical Analysis*, 3rd ed., McGraw–Hill, New York.

Correa, S.M. and Gulati, A. 1992 Measurements and Modeling of Bluff Body Stabilized Flame, *Combust. Flame*, **69**, 195–213.

Correa, S.M., and Shyy, W. 1987 Computational Models and Methods for continuous Gaseous Turbulent Combustiion, *Prog. Energy Combust. Sci.*, **13**, 249–292.

Correa, S.M., Drake, M.C., Pitz, R.W. and Shyy, W. 1985   Prediction and Measurement of a Non–Equilibrium Turbulent Diffusion Flame, *Twentieth Symposium (International) on Combustion,* The Combustion Institute, Pittsburgh, PA, 337–343.

Corriel, S.R., Cordes, M.R., Boettinger, W.J. and Sekerka, R.F. 1980 Convective and Interfacial Instabilities during Unidirectional Solidification of a Binary Alloy, *J. Crystal Growth*, **49**, 13–28.

Crank, J. 1984 *Free and Moving Boundary Problems,* Oxford University Press. Oxford, UK.

Crank, J. and Gupta, R. S. 1975 Isotherm Migration Method in Two–Dimensions, *Int. J. Heat Mass Transf.*, **18**, 1101–1106.

Crochet, M. J., Geyling, F. T. and van Schaftingen, J. J. 1987 Numerical Simulation of the Horizontal Bridgman Growth. Part 1: Two–Dimensional Flow, *Int. J. Numer. Meths. Fluids*, **7**, 27–49.

Crowley, A. B. 1983 Mathematical Modeling of Heat Flow in Czochralski Crystal Growth, *IMA J. Appl. Math.*, **30**, 173–189.

Cuvelier, C. and Schulkes, R. M. S. M. 1990 Some Numerical Methods for the Computation of Capillary Free Boundaries Governed by the Navier–Stokes Equations, *SIAM Review*, **32**, 355–423.

Dahlquist, G. 1963 A Special Stability Problem for Linear Multistep Methods, *BIT*, **3**, 27–43.

Dahlquist,G. and Bjorck, A. 1974 *Numerical Methods*, Prentice Hall, Englewood Cliffs, NJ.

Dakin J.T. and Shyy, W 1989 The Prediction of Convective and Addictive Demixing in Vertical Metal Halide Discharge Lamps. *J. Electrochem. Soc.* **136**, 1210–1215.

Daly, B.J. 1967 Numerical Study of two–fluid Rayleigh Taylor Instability, *Phys. Fluids*, **10**, 297–307.

Daly, B.J. 1969 A Technique for Including Surface Tension Effects in Hydrodynamic Calculations, *J. Comp. Phys.*, **4**, 97–117.

Dantzig, J.A. 1989 Modeling Liquid–Solid Phase Change with Melt Convection, *Int. J. Numer. Meths. Engrg.* **28**, 1769–1785.

Dasgupta, A., Li, Z., Shih, T. I.–P., Kundu, K. and Deur, J.M. 1993 Computations of Spray, Fuel–Air Mixing, and Combustion in a Lean–Premixed–Prevaporized Combustor, *AIAA Paper No. 93–2069*.

Davies, C.B. and Venkatapathy, E. 1992 Application of a Solution Adaptive Grid Scheme to Complex Three–Dimensional Flows, *AIAA J.*, **30**, 2227–2233.

Davis, D.L. and Thompson, H.D. 1992 The Impact of Computational Zone Interfacing on Calculated Scramjet Performance, *AIAA Paper No. 92–0390*.

Davis, S.H. 1987 Thermocapillary Instabilities, *Ann. Rev. Fluid Mech.*, **19**, 403–435.

Davis, S.H. 1990 Hydrodynamic Interactions in Directional Solidification, *J. Fluid Mech.*, **212**, 241–262.

deGennes, P.G. 1985 Wetting: Statics and Dynamics, *Rev. Modern Phys.*, **57**, 827–863.

DeGregoria, A. J. and Schwartz, L. W., 1986, A Boundary–Integral Method for Two–phase Displacement in Hele–Shaw Cells, *J. Fluid Mech.*, **164**, 383–400.

Derby, J. J. and Brown, R. A. 1986 Thermal–Capillary Analysis of Czochralski and Liquid Encapsulated Czochralski Crystal Growth, *J. Crystal Growth*, **74**, 605–624.

DeVahl Davis, G. and Jones, I.P. 1983 Natural Convection in a Square Cavity, *Int. J. Numer. Meths. Fluids*, **3**, 227–248.

DeVahl Davis, G. and Mallinson, G.D. 1976 An Evaluation of Upwind and Central Difference Approximations by a study of Recirculating Flow, *Comput. Fluids*, **4**, 29–43.

Demirdzic, I., Gosman, A.D., Issa, R.I. and Peric, M. 1987 A Calculation Procedure for Turbulent Flow in Complex Geometries, *Comput. Fluids*, **15**, 251–273.

Drake, M.C., Pitz, R.W. and Lapp, M. 1986a Laser Measurements on Nonpremixed Hydrogen–Air Flames for Assessment of Turbulent Combustion Models, *AIAA J.*, **24**, 905–917.

Drake, M.C., Pitz, R.W. and Shyy, W. 1986b Conserved Scalar Probability Density Functions in a Turbulent Jet Diffusion Flame. *J. Fluid Mech.*, **171**, 24–51.

Drake, M.C., Correa, S.M., Pitz, R.W., Shyy, W. and Fenimore, C.P. 1987 Superequilibrium and Thermal Nitri–Oxide Formation in Turbulent Diffusion Flames. *Combust. Flame,* **69**, 347–365.

Drazin, P.G. and Reid, W.H. 1981 *Hydrodynamic Stability*, Cambridge University Press, Cambridge, UK.

Dupont, S., Marchal, J. M., Crochet, M. J. and Geyling, F. T. 1987 Numerical Simulation of the Horizontal Bridgman Growth. Part 2: Three–Dimensional Flow, *Int. J. Numer. Meths. Fluids*, **7**, 49–67.

Dussan V, .E.B. 1979 On the Spreading of Liquid on Solid Surfaces: Static and Dynamic Contact Lines, *Ann. Rev. Fluid Mech.*, **11**, 371–400.

Dwyer, H.A., Kee, R.J. and Sanders, B.R. 1980 Adaptive Grid Method for Problems in Fluid Mechanics and Heat Transfer, *AIAA J.*, **18**, 1205–1212.

Dyson, D.C., 1978 The Energy Principle in the Stability of Interfaces, *Prog. Surf. Memb. Sci.*, **12**, 479–564.

Eckbreth, A.C. 1988 *Laser Diagnostics for Combustion Temperature and Species*, Abacus Press, Cambridge, MA.

Eckert, E.R.G. and Drake, R.M.,Jr. 1987 *Analysis of Heat and Mass Transfer*, Hemisphere, Washington, D.C.

Eick, J.D., Good, R.J., and Neumann, A.V. 1975 Thermodynamics of Contact Angles, II. Rough Solid Surfaces, *J. Coll. Int. Sci.*, **53**, 235–248.

Eiseman, P.R. 1988 Adaptive Grid Generation, *Comput. Meths. Appl. Mech. Engrg.*, **64**, 475–489.

Elenbass, W. 1951 *The High Pressure Mercury Vapor Discharge,* North–Holland, Amsterdam, The Netherlands.

Elenbaas, W. 1972 *Light Source.* Crane, Russak & Company, Inc., New York.

Engquist, B. and Majda, A. 1977 Absorbing Boundary Conditions for the Numerical Simulation of Waves, *Math. Comput.,***31**, 629–651.

Engquist, B., Lotstedt, P. and Sjorgreen, B. 1989 Nonlinear Filters for Efficient Shock Computations. *Math. Comp.,* **52**, 509–537.

Ettouney H. M. and Brown, R. A. 1983 Finite–Element Methods for Steady Solidification Problems, *J. Comput. Phys.*, **49**, 118–150.

Fabietti, L. M., Seetharaman, V. and Trivedi, R. 1990 The Development of Solidification Microstructures in the Presence of Lateral Constraints, *Metall. Trans. A,* **21**, 1299–1310.

Fang, Q.T., Glicksman, M.E., Coriell, S.R., McFadden, G.B. and Boisvert, R.F. 1985 Convective Influence on the Stability of a Cylindrical Solid–Liquid Interface, *J. Fluid Mech.*, **151**, 121–140.

Flemings, M.C. 1974 *Solidification Processing.* McGraw–Hill, New York.

Fletcher, C.A.J. 1988 Computational Techniques for Fluid Dynamics, 2 Volumes, Springer–Verlag, New York.

Floryan J. M. and Rasmussen,H. 1989 Numerical Methods for Viscous Flows with Moving Boundaries, *Appl. Mech. Rev.,* **42**, No. 12, 323–341.

Forsythe, G.E. and Wasow, W.R. 1960 *Finite–Difference Methods for Partial Differential Equations*, Wiley, New York.

Forth, S.A. and Wheeler, A.A. 1992 Coupled Convective and Morphological Instability in a Simple Model of the Solidification of a Binary Alloy Including a Shear Flow, *J. Fluid Mech.*, **236**, 61 – 94.

Fromm, J. 1981 Finite Difference Computation of the Capillary Jet, Free Surface Problem, in *Lecture Notes in Physics,* **238**, Springer Verlag, New York.

Ganesan, S. and Poirier, D.R. 1990 Conservation of Mass and Momentum for the Flow Interdendritic Liquid during Solidification, *Metall. Trans. B* **21**, 173–181.

Garandet, J.P., Duffar, T. and Favier, J.J. 1990 On the Scaling Analysis of the Solute Boundary Layer in an Idealized Growth Configuration, *J. Crystal Growth*, **106**, 437–444.

Garandet, J.P., Duffar, T. and Favier, J.J. 1990 On the Scaling Analysis of the Solute Boundary Layer in an Idealized Growth Configuration, *J. Crystal Growth*, **106**, 437–444.

Gartling, D.K. 1990 A Test Problem for Outflow Boundary Conditions – Flow Over a Backward Facing Step, *Int. J. Numer. Meths. Fluids.*, **11**, 953–967.

Gau, C. and Viskanta, R. 1986 Melting and Solidification of a Pure Metal on a Vertical Wall, *ASME J. Heat Transf.*, **108**, 174–181.

Gear, C. W. 1971 *Numerical Initial Value Problems in Ordinary Differential Equations*, Prentice–Hall, Englewood Cliffs, NJ.

Gebhart, B. 1963 Random Convection under Conditions of Weightlessness, *AIAA J.*, **1**, 380–383

Gebhart, B., Jaluria, Y., Mahajan, R.L. and Sammakia, B. 1988 *Buoyancy–Induced Flows and Transport*, Hemisphere, Washington, D.C.

Ghia, U., Ghia, K.N. and Shin, C.T. 1982 High–Re Solution for Incompressible Flow Using the Navier–Stokes Equations and a Multigrid Method, *J. Comput. Phys.*, **48**, 387–411.

Gibbs, J. 1931 *The Collected Works of J. Willard Gibbs*, **1**, 219 – 274, Longmans, New York.

Glicksman, M.E., Coriell, S.R. and McFadden, G.B. 1986 Interaction of Flows with the Crystal–Melt Interface, *Ann. Rev. Fluid Mech.* **18,** 307–335.

Glicksman, M.E. and Huang, S.–C. 1986 Fundamentals of Dendritic Solidification, I and II, *Acta. Metall.*, **29**, 701–734.

Glimm, J., McBryan, O., Melnikoff, R. and Sharp, D.H. 1986 Front Tracking Applied to Rayleigh–Taylor Instability, *SIAM J. Sci. Stat. Comput.*, **7**, 230–251.

Gnoffo, P. 1983 A Vectorized Finite–Volume Adaptive Grid Algorithm Applied to Planetary Entry Flowfields, *AIAA J.*, **9**, 1249–1254.

Godunov, S.K. and Ryabenkii, V.S. 1987 *Difference Schemes*, North–Holland, Amsterdam, The Netherlands

Golub, G.H. and van Loan, C.F. 1989 *Matrix Computations*, 2nd ed.,The Johns Hopkins University Press, Baltimore, MD.

Goodman, T. R. 1958 The Heat Balance Integral and Its Application to Problems Involving a Change of Phase, *ASME J. Heat Transf.*, **80**, 335–342.

Gosman, A.D. and Ideriah, F.J.K. 1976 TEACH–T: A General Computer Program for Two–dimensional, Turbulent Recirculating Flow. Imperial College, Department of Mechanical Engineering Report (unnumbered), London, UK.

Goss, L.P. and Switzer, G.L. 1986 Laser Optics/Combustion Diagnostics. *AFWAL–TR–2023*, Wright Patterson Air Force Base, OH.

Greenspan, D. and Casulli, V. 1988 *Numerical Analysis for Applied Mathematics, Science, and Engineering*, Addison–Wesley, Reading, MA.

Gresho, P. 1991 Incompressible Fluid Dynamics: Some Fundamental Formulation Issues, *Ann. Rev. Fluid Mech.*, **23**, 413–453.

Gresho, P. and Lee, R. 1981 Don't Suppress the Wiggles — They are Telling You Something, *Comput. and Fluids*, **9**, 223–255.

Guckenheimer, J. and Holmes, P. 1983 *Nonlinear Oscillations, Dynamical Systems, and Bifurcations of Vector Fields*, Springer–Verlag, New York.

Guggenheim, E.A. 1957 *Thermodynamics*, Wiley–Interscience, New York.

Haasen, P 1986 *Physical Metallurgy*, 2nd edition, Cambridge University Press, Cambridge, UK.

Hackbusch, W. 1985 *Multi–Grid Methods and Applications*, Springer–Verlag, New York.

Hageman, L.A. and Young, D.M. 1981 *Applied Iterative Methods*, Academic Press, New York.

Hairer, E., Norsett, S.P. and Wanner, G. 1987 *Solving Ordinary Differential Equations I*, Springer–Verlag, New York.

Hairer, E. and Wanner, G. 1991 *Solving Ordinary Differential Equations II*, Springer–Verlag, New York.

Han, T., Humphrey, J.A.C. and Launder, B.E. 1981 A Comparison of Hybrid and Quadratic–Upstream Differencing in High Reynolds Number Elliptic Flows, *Comput. Meths. Appl. Mech. Engrg.*, **29**, 81–95.

Harlow, F.H. and Welch, J.E. 1965 Numerical Calculation of Time–Dependent Viscous Incompressible Flow of Fluid with Free Surface, *Phys. Fluids*, **8**, 2182–2189.

Harten, A. 1983 High Resolution Schemes for Hyperbolic Conservation Laws, *J. Comput. Phys.*, **49**, 357–393.

Harten, A. 1984 On a Class of High Resolution Total Variation Diminishing Schemes, *SIAM J. Numer. Analysis*, **21**, 1 – 23.

Harten, A. and Osher, S. 1987 Uniformly High–Order Accurate Nonoscillatory Schemes, *SIAM J. Numer. Analysis*, **24**, 2, 279–309.

Harter, W.E., Zhao, A.X. and Narayanan, R. 1989 Low Gravity Interfacial Instabilities in Liquid Encapsulated Crystal Growth, *Mater. Sci. Forum*, **50**, 205–212.

Hartman, P. (ed.) 1973 *Crystal Growth: An Introduction*, North–Holland, Amsterdam, The Netherlands.

Hatsopoulos, G.N., and Keenan, J.H. 1965 *Principles of General Thermodynamics*, Wiley, New York.

Hayakawa, H., Fukada, N., Udagawa, T., Koizumi, H.G. and Fukuyama, T. 1991 Solidification Structure and Segregation in Cast Ingots of Titanium Alloy Produced by Vacuum Arc Consumable Electrode Method, *ISIJ International*, **31**, 775–784.

Hayase, T., Humphrey, J.A.C. and Greif, R. 1992 A Consistently Formulated QUICK Scheme for Fast and Stable Convergence using Finite–Volume Iterative Calculation Procedures, *J. Comput. Phys.*, **98**, 108–118.

Hawken, D.F., Gottlieb, J.J. and Hansen, J.S. 1991 Review of Some Adaptive Node–Movement Techniques in Finite Element and Finite Difference Solutions of Partial Differential Equations, *J. Comput. Phys.*, **95**, 254–302.

Heinrich, J. C., Felicelli, S., Nadapukar, P. and Poirier, D. R. 1989 Thermosolutal Convection during Dendritic Solidification of Alloys, Part II: Nonlinear Convection, *Metallurgical Transaction* B **20B**, 883–891.

Heitor, M.V. and Whitelaw, J.M. 1986 Velocity, Temperature and Species Characteristics of the Flow in a Gas–Turbine Combustor, *Combust. Flame,* **64**, 1–32.

Henrici, P. 1962 *Discrete Variable Methods in Ordinary Differential Equations*, Wiley, New York.

Henshaw, W.D. and Chesshire, G. 1987 Multigrid on Composite Meshes, *SIAM J. Sci. Stat. Comput.*, **8**, 914–923.

Hinatsu, M. and Ferziger, J.H. 1991 Numerical Computation of Unsteady Incompressible Flow in Complex Geometry Using a Composite Multigrid Technique, *Int. J. Numer. Meths. Fluids*, **13**, 971–997.

Hirsch, C. 1990 *Numerical Computation of Internal and External Flows*, Wiley, New York, 2 volumes.

Hirschfelder, J.O., Curtiss, C.F. and Bird, R.B. 1954 *Molecular Theory of Gases and Liquids*, Wiley, New York.

Hirt, C. W. and Nichols, B. D. 1981 Volume of Fluid (VOF) Method for the Dynamics of Free Boundaries, *J. Comp. Phys.*, **39**, 201–225.

Holmes, D.G. and Conner, S. 1989 Solution of the 2–D Navier–Stokes Equations on Unstructured Adaptive Grids, *AIAA 9th Computational Fluid Dynamics Conference, AIAA Paper No. 89–1932*.

Homsy, G.M. 1987 Viscous Fingering in Porous Media, *Ann. Rev. Fluid Mech.*, **19**, 271–312.

Hortmann, M., Peric, M. and Scheurer, G. 1990 Finite Volume Multigrid Prediction of Laminar Natural Convection: Benchmark Problems, *Int. J. Numer. Meths. Fluids*, **11**, 189–207.

Hsu, C.C. and Lee, C.L. 1991 On a Zonal Method for Transonic Turbulent Flow Past a Wing Fuselage Configuration, Proceedings of the International Congress, **1**, 279–287, Melbourne, Australia.

Huang, P.G., Launder, B.E. and Leschziner, M.A. 1985 Discretization of Nonlinear Convection Processes: A Broad–Range Comparison of four Schemes, *Comput. Meths. Appl. Mech. Engrg.*, **48**, 1–24.

Huang, S.–C. and Glicksman, M.E. 1981 Fundamentals of Dendritic Solidification –I. Steady–State Tip Growth, *Acta Metall.*, **29**, 701–715, and II. Development of Side–Branch Structure, *Acta Metall.*, **29**, 717–734

Huppert, H.E. 1990 The Fluid Mechanics of Solidification, *J. Fluid Mech.*, **212**, 209–240

Hurle, D.T.J., Jakeman, E. and Wheeler, A.A. 1983 Hydrodynamics Stability of the Melt during Solidification of a Binary Alloy, *Phys. Fluids*, **26**, 624–626.

Isaacson, E. and Keller, H.B. 1966 *Analysis of Numerical Methods,* Wiley, New York.

Issa, R.I. 1986 Solution of the Implicitly Discretized Fluid Flow Equations by Operator Splitting, *J. Comput. Phys.*, **62**, 40–65.

Ivantsov, G.P. 1947 Temperature Field around Spheroidal, Cylindrical and Circular Crystal Growing in a Supercooled Melt, Translated from *Daklady Akad. Nauk.*, **58**, 567–569.

Iverach, D., Basden, K.S. and Kirov, N.Y. 1972 Formation of Nitri–Oxide in Fuel–Lean and Fuel–Rich Flames, *Fourteenth Symposium (Int,l.) on Combustion,* 767–775, The Combustion Institute, Pittsburgh, PA.

Jameson, A., Schmidt, W. and Turkel, E. 1981 Numerical Simulation of the Euler Equations by Finite Volume Methods using Runge–Kutta Time Stepping Schemes, *AIAA Paper No. 81–1259*.

Janicka, J. and Kollman, W. 1979 A Two–Variable Formalism for the treatment of Chemical Reactions in Turbulent H_2–Air Diffusion Flames, *Seventeenth Symp. (Intl.) on Combustion*, 421–431, The Combustion Institute, Pittsburgh, PA.

Jansons, K. M. 1986 Moving Contact Lines at Non–zero Capillary Numbers, *J. Fluid Mech.*, **167**, 393–407.

Jeng, Y.N. and Chen, J.C. 1992a Geometric Conservation Law of the Finite–Volume Method for the SIMPLER Algorithm and a Proposed Upwind Scheme, *Numer. Heat Transf.*, **22B**, 211–234.

Jeng, Y.N. and Chen, J.C. 1992b Error Analysis of the Finite Volume Method for a Model Steady Convection Equation, *J. Comput. Phys.*, **100**, 64–76.

Jiang, Y., Chen, C.P. and Tucker, P.K. 1990 Multigrid Solution of Unsteady Navier–Stokes Equations using a Pressure Method, *AIAA Paper 90–1522*.

Jin, G. and Braza, M. 1993 A Nonreflecting Outlet Boundary Condition for Incompressible Unsteady Navier–Stokes Calculations, *J. Comput. Phys.*, **107**, 239–253

Johnson, R.E. Jr. 1959 Conflicts Between Gibbsian Thermodynamics and Recent Treatments of Interfacial Energies in Solid–Liquid–Vapor Systems, *J. Phys. Chem.*, **63**, 1655–1658.

Jones, W.P. and Priddin, C.H. 1979 Prediction of the Flow Field and Local Gas Composition in Gas Turbine Combustor, *Seventeenth Symp. (Int'l) on Combustion*, 399–409, Combustion Institute, Pittsburgh, PA.

Jones, W.P. and Toral, H. 1983 Temperatures and Compositions Measurements in a Research Gas Turbine Combustor Chamber, *Combust. Sci. Tech.*, **31**, 249–275.

Jones, W.P. and Whitelaw, J.H. 1984 Modeling and Measurements in Turbulent Combustor. *Twentieth Symp. (Intl.) on Combustion*, 233–249, The Combustion Institute, Pittsburgh, PA.

Joos, F. and Simm, B. 1987 Comparison of the Performance of a Reverse Flow Annular Combustion Chamber under Low and High Pressure Condition, *AGARD Propulsion and Energetics 70th Symp.*, AGARD CP No. 442.

Joshi, D.H. and Vanka, S.P. 1991 Multigrid Calculation Procedure for Internal Flows in Complex Geometries, *Numer. Heat Transf.*, **20**, 61–80.

Kagawa, A., Hirata, M. and Sakamoto, Y. 1990 Solute Partition on Solidification of Nickel–Base Ternary Alloys, *J. Mater. Sci.*, **25**, 5063–5069.

Kao, K.–H., Liou, M.–S. and Chow, C. 1993 Grid Adaptation using Chimera Composite Overlapping Meshes, *AIAA Paper No. 93–3389–CP.*

Karki, K.C. and Patankar, S.V. 1988 Calculation Procedure for Viscous Incompressible Flows in Complex Geometries, *Numer. Heat Transf.*, **14**, 295–307.

Kaviany, M. 1991 *Principles of Heat Transfer in Porous Media*, Springer–Verlag, New York

Kelkar, K.M. and Choudary, D. 1992 A Numerical Method for the Computation of Flow and Scalar Transport using Non–Orthogonal Boundary–Fitted Coordinates, in J.L.S. Chen and K. Vafai (eds.) *Modern Developments in Numerical Simulation of Flow and Heat Transfer*, ASME HTD– **194**, 1–9, New York.

Keller, H.B. 1971 A New Difference Scheme for Parabolic Problems, *Numerical Solution of Partial Differential Equations, Vol 2 (B. Hubbard , ed.)*, Academic Press, New York.

Kent, J.H. and Bilger, R.W. 1976 The Prediction of Turbulent Diffusion Flame Fields and Nitric Oxide Formation. Sixteenth Symp. (Int'l) on Combustion, 1643–1656, The Combustion Institute, Pittsburgh, PA.

Kessler, D.A., Koplik, J. and Levine, H. 1988 Pattern Selection in Fingered Growth Phenomena, *Advances in Physics*, **37**, 255–339.

Kenty, C. 1938 On Convection Currents in High Pressure Mercury Arcs, *J. Appl. Phys.* **50**, 53–66.

Kenworthy, M.J., Correa, S.M. and Burrus, D.L. 1983 Aerothermal Modeling, *NASA CR–168296.*

Kestin, J. 1979 *A Course in Thermodynamics*, 2 vols., revised printing, Hemisphere, Washington, D.C.

Kim, J., Moin, P. and Moser, R. 1987 Turbulence Statistics in Fully Developed Flow at Low Reynolds Number, *J. Fluid Mech.*, **177**, 133–166.

Kobayashi, M.H. and Pereira, J.C.F. 1991 Calculation of Incompressible Laminar Flows on a Non–staggered, Non–orthogonal grid, *Numer. Heat Transf.*, **19**, 243–262.

Kobayashi, R. 1993 Modeling and Numerical Simulations of Dendritic Crystal Growth, *Physica D*, **63**, 410–423.

Kroliczeck, E.J., Ku, J. and Oliendorf, S. 1984 Design, Development and Test of a Capillary Pumped Loop Heat Pipe, *AIAA Paper No. 84–1720.*

Ku, J., Kroliczek, Taylor, W.J. and Mcintosh, R. 1986a Functional and Performance Test of Two Capillary Pumped Loop Engineering Model, *AIAA Paper No. 86–1248*.

Ku, J., Krolzek, E.J., Putler, D. Schweickart, R.B. and Mcintosh, R 1986b Capillary Pumped Loop GAS and Hitchhiker Flight Experiment, *AIAA Paper No. 86–1249*.

Kuo, K.K. 1986 *Principles of Combustion*, Wiley–Interscience, New York.

Kurz, W. and Fisher, D.J. 1984 *Fundamentals of Solidification*, Trans. Tech. SA, Switzerland.

Kurokawa, J., and Nagahara, H. 1986 Flow Characteristics in Spiral Casing of Water Turbines, *IAHR Symposium*, **2**, Paper No. 62, Montreal, Canada.

Lacroix, M. 1989 Computation of Heat Transfer During Melting of a Pure Substance from an Isothermal Wall, *Numer. Heat Transf.*, part B, **15**, 191–210.

Lacroix, M. and Voller, V. R. 1990 Finite Difference Solutions of Solidification Phase Change Problems: Transformed versus Fixed Grids, *Numer. Heat Transf.*, part B, **17**, 25–41.

Lamb, H. 1945 *Hydrodynamics*, 6th ed., Dover, New York, Art. 58, I, Art. 194 and 195, III. sec. 265–275.

Lambert, J.D. 1973 *Computational Methods in Ordinary Differential Equations*, Wiley, New York.

Lan, C.W. and Kou, S. 1990 Thermocapillary Flow and Melt/Solid Interface in Floating Zone Growth under Microgravity, *J. Crystal Growth*, **102**, 1043–1058.

Langer, J.S. 1980 Instabilities and Pattern Formation in Crystal Growth, *Rev. Modern Phys.*, **52**, 1–28.

Langer, J.S. and Muller–Krumbhaar, H. 1978 Theory of Dendritic Growth, Parts I,II and III, *Acta. Metall.*, **28**, 1681–1708.

Lapidus, L. and Pinder, G.F. 1982 *Numerical Solution of Partial Differential Equations in Science and Engineering*, Wiley, New York.

Launder, B.E. and Sharma, B.I. 1974 Application of Energy Dissipation Model of Turbulence to the Calculation of Flow near a Spinning Disc, *Lett. Heat Mass Transf.*, **1**, 131–138.

Launder, B.E. and Spalding, D.B. 1974 The Numerical Computation of Turbulent Flow, *Comput. Meths. Appl. Mech. Engrg.*, **3**, 269–289.

Lax, P.D. 1973 *Hyperbolic Systems of Conservation Laws and the Mathematical Theory of Shock Waves*, SIAM, Philadelphia, PA.

Lee, D. and Tsuei, Y.M. 1992 A Formula for Estimation of Truncation Errors of Convection Terms in a Curvilinear Coordinate System, *J. Comput. Phys.*, **98**, 90–100.

Lee, D., Yeh, C.L., Tsuei, Y.M., Jiang, W.T. and Chung, Y.L. 1990 Numerical Simulations of Gas Turbine Combustor Flows, *AIAA Paper No. 90–2305*.

Lefebvre, A.H. 1989 *Atomization and Sprays*, Hemisphere, New York.

Leonard, B.P. 1979 A Stable and Accurate Convective Modelling Procedure Based on Quadratic Upstream Interpolation, *Comput. Meths. Appl. Mech. Engrg.*, **19**, 59–98.

Leschziner, M.A. and Dimitriadis, K.P. 1989 Computation of Three–Dimensional Turbulent Flow in Non–Orthogonal Junctions by a Branch–Coupling Method, *Comput. Fluids*, **17**, 371–396.

Levich, V.G. 1962 *Physicochemical Hydrodynamics*, Prentice–Hall, Englewood Cliffs, NJ.

LeVeque, R.J. 1990 *Numerical Methods for Conservation Laws*, Birkhäuser, Boston, MA.

Liang, P.Y., 1991, Numerical Method for Calculation of Surface Tension Flows in Arbitrary Grids, *AIAA Journal,* **29**, No. 2, 161–167.

Liang, S. 1986 Random–Walk Simulation of Flows in Hele–Shaw Cells, *Phys. Rev. A,* **33**, No. 4, 2663–2674.

Libby, P.A. and Williams, F.A. (eds) 1980 *Turbulent Reacting Flows,* Springer–Verlag, New York.

Lien, F.–S. and Leschziner, M.A. 1991 Multigrid Convergence Acceleration for Complex Flow including Turbulence, in W. Hackbusch and U. Trottenberg (eds.), *Multigrid Method III*, 277–288, Birkhauser Verlag.

Liew, S.W., Bray, K.M. and Moss, J.B. 1981 A Flamelet Model of Turbulent Non–Premixed Combustion, *Combust. Sci. Technol.,* **27**, 69–73.

Liew, S.W., Bray, K.M. and Moss, J. B. 1984 A Stretched Laminar Flamelet Model of Turbulent Non–Premixed combustion, *Combust. Flame,* **56**, 199–213.

Lightfoot, N. M. H. 1929 The Solidification of Molten Steel, *London Mathematical Society Proceedings*, **31**, 97–116.

Liou, M.–S., Van Leer, B. and Shuen, J.–S. 1990 Splitting of Inviscid Fluxes for Real Gases, *J. Comput. Phys.*, **87**, 1–24

Liou, M.–S. 1992 On a New Class of Flux Splittings, *13th Intern. Conf. Numer. Meth. Fluid Dynamics,* Rome, Italy.

Liou, M.–S. and Steffen, C.J. 1993 A New Flux Splitting Scheme, *J. Comp. Phys.*, **107**, 23–39

Lovin, J.K. and Lubkowitz, A.W. 1969 User's Manual for RAVFAC (a Radiation View Factor Computer Program), *NASA CR–61321.*

Lowke, J.J. 1979 Calculated Properties of Vertical Arcs Stabilized by Natural Convection. *J. Appl. Phys.*, **50**, 147–157.

Lumley, J.L. (ed.) 1990 *Whither Turbulence ? Turbulence at the Crossroads*, Lecture Notes in Physics, vol. **357**, Springer–Verlag, New York.

Lynch, D. R. 1982 Unified Approach to Simulation on Deforming Elements with Application to Phase Change Problem, *J. Comput. Phys.*, **47**, 387–441.

MacCormack, R.W. and Baldwin, B.S. 1975 A Numerical Method for solving the Navier–Stokes Equations with Application to Shock–Boundary Layer Interactions, *AIAA Paper No. 75–1.*

Majumdar, S. 1988 Role of Underrelaxation in Momentum Interpolation for Calculation of Flow with Nonstaggered Grids, *Numer. Heat Transf.*, **13**, 125–132.

Masri, A.R., Dibble, R.W. and Barlow, R.S. 1992 The Structrue of Turbulent Nonpremixed Flames of Methanol over a Range of Mixing Rates, *Combust. Flame,* **89**, 167–185.

Mavriplis, D.J. and Martinelli, L. 1991 Multigrid Solution of Compressible Turbulent Flow on Unstructured Meshes using a Two–Equation Model, NASA Langley Research Center, *ICASE Report 91–11.*

McFadden, G. B. and Coriell, S. R., 1987, Non–Planar Interface Morphologies During Unidirectional Solidification, *J. Crystal Growth*, **84**, 371–388.

McFadden, G.B., Coriell, S.R., Boisvert, R.F., Glicksman, M.E. and Fang, Q.T. 1984 Morphological Stability in the Presence of Fluid Motion in the Melt, *Met. Trans.*, **15A**, 2117–2124.

McGuirk, J.J. and Palma, J.M.L.M. 1992 Calculations of the Dilution System in an Annular Gas Turbine Combustor, *AIAA J.*, **30**, 963–972.

McKay, T.D. and McCay, M.H. 1993 Measured and Predicted Effects of Gravity Level on Directional Dendritic Solidification of $NH_4Cl–H_2O$, *Microgravity Sci. Technol.*, **6**, 2–12.

Melton, J., Enomoto, F. and Berger, M. 1993 Three–Dimensional Automatic Cartesian Grid Generation for Euler Flows, *AIAA Paper No. 93–3386–CP.*

Merkle, C.L. and Choi, Y.–H. 1987 Computation of Low Speed Flow with Heat Addition, *AIAA J.*, **25**, 831–838.

Merkle, C.L., Venkateswaran, S. and Buelow, P.E.O. 1992 The Relationship between Pressure–Based and Density–Based Algorithms, *AIAA Paper No. 92–0425.*

Miller, J.A., Kee, R.J., Smooke, M.D. and Grear, J.F. 1984 The Computation of the Structure and Extinction Limit of a Methane–Air Stagnation Point Diffusion Flame. *Western States/CI Pap. 84–20.*

Milne–Thompson, L.M., 1979 *Theoretical Hydrodynamics*, MacMillan, London.

Minkowycz, W.J., Sparrow, E.M., Schneider G.E. and Pletcher, R.H. (eds) 1988 *Handbook of Numerical Heat Transfer*, Wiley, New York.

Mitchell, A.R. and Griffiths, D.F. 1980 *The Finite Difference Method in Partial Differential Equations,* Wiley, New York.

Miyata, H. 1986 Finite Difference Simulation of Breaking Waves, *J. Comp. Phys.*, **65**, 179–214.

Mongia, H.C. *et al.*, 1979 Combustor Design Criteria Validation. USATL–TR–78–55A, 55B, 55C.

Moon, Y. and Liou, M.–S. 1989 Conservative Treatment of Boundary Interfaces for Overlaid Grids and Multi–level Grid Adaptations, *AIAA Paper No. 89–1980.*

Morse, P.M. 1969 *Thermal Physics*, 2nd ed., Benjamin, New York.

Mullins, W.W. and Sekerka, R.F. 1964 Stability of a Planar Interface During Solidification of a Dilute Binary Alloy, *J. Appl. Phys.*, **35**, 444–451.

Murray, B.T., Coriell, S.R., McFadden, G.B. and Wheeler, A.A. 1993 The Effect of Gravitational Modulation on Convection in Vertical Bridgman Growth, *Microgravity Sci. Technol.*, **6**, 70–73.

Myshkis, A.D., Babskii, V.G., Kopachevskii, N.D., Slobozhanin, L.A. and Tyuptsov, A.D. 1987 *Low–Gravity Fluid Mechanics,* Springer–Verlag, New York.

Nadarajah, A. and Narayanan, R. 1990 Comparison between Morphological and Rayleigh–Marangoni Instabilities in D. Meinkoehn and H. Haken (eds), *Dissipative Structures in Transport Processes and Combustion*, 215–228, Springer–Verlag, New York.

Neumann, A.V.,and Good, R.J., 1972 Thermodynamics of Contact Angles, I. Heterogeneous Solid Surfaces, *J. Coll. Int. Sci.*, **38**, 341–358.

Ni, R.H. 1982 A Multiple Grid Scheme for Solving Euler equations, *AIAA J.*, **20**, 1565–1571.

Nield, D.A. 1958 Surface Tension and Buoyancy Effects in Cellular Convection, *J. Fluid Mech.*, **4**, 489–493.

Oliger, J. and Sundstrom, A. 1978 Theoretical and Practical Aspects of some Initial Boundary Value Problems in Fluid Dynamics, *SIAM J. Appl. Math.*, **35**, 419–446.

Oran, E.S. and Boris, J.P. 1981 Detailed Modeling of Combustion System, *Prog. Energy Combust. Sci. 7, 1–72.*

Oran E.S. and Boris, J. P. 1987 *Numerical Simulation of Reactive Flow*, Elsevier, New York.

Osher, S. 1984 Riemann Solver, the Entropy Condition and Difference Approximations, *SIAM J. Numer. Anal.*, **21**, 217–235.

Osher, S. and Chakravarthy, S.R. 1983 Upwind Schemes and Boundary Conditions with Applications to Euler Equations in General Coordinates, *J. Comput. Phys.*, **50**, 447–481.

Ostrach, S. 1982 Low–Gravity Fluid Flows, *Ann. Rev. Fluid Mech.*, **14**, 313–345.

Ostrach, S. 1983 Fluid Mechanics in Crystal Growth, *J. Fluids Engrg.*, **105**, 5 – 20.

Ozisik, M.N. 1968 *Boundary Value Problems of Heat Conduction*, International Textbook Co., Scranton, PA.

Ozisik, M. N. 1978 A Note on the General Formulation of Phase Change Problems as Heat Conduction Problems with a Moving Heat Source, *ASME J. Heat Transf.*, **100**, 370–377.

Ozisik, M. N. 1980 *Heat Conduction*, Wiley, New York.

Pamplin, B. (ed.) 1975 *Crsytal Growth*, Pergamon, New York.

Patankar, S.V. 1980 *Numerical Heat Transfer and Fluid Flow,* Hemisphere, Washington DC.

Patankar, S.V. 1988 Recent Developments in Computational Heat Transfer, *ASME J. Heat Transf.*, **110**, 1037–1045.

Patankar, S.V. and Spalding, D.B. 1972 A Calculation Procedure for Heat, Mass and Momentum Transfer in Three–Dimensional Parabolic Flows, *Int. J. Heat Mass Transf.*, **15**, 1787–1806.

Peaceman, D.W. and Rachford, H.H. 1955 The Numerical Solution of Parabolic and Elliptic Differential Equations, *SIAM J.,* **3**, 28–41.

Pearson, J.R.A. 1958 On Convection Cells Induced by Surface Tension, *J. Fluid Mech.,* **4**, 489–500.

Pelce, P. 1988 *Dynamics of Curved Fronts*, in *Perspectives in Physics*, ed. P. Pelce, Academic Press, New York.

Pember, R., Bell, J. Collela, P., Crutchfield, W. and Welcome, M. 1993 A Three–Dimensional Adaptive Cartesian Grid Algorithm for Inviscid Compressible Flow, *AIAA Paper No. 93–3385–CP.*

Penner, S.S., Wang, C.P. and Bahadori, M.Y. 1984 Laser Diagnostics Applied to Combustion System. Twentieth S*ymp.(Int'l) on Combustion*, 1149–1176, The Combustion Institute, Pittsburgh, PA.

Peric, M. 1985 A Finite Volume Method for the Prediction of Three–Dimensional Fluid Flow in Complex Ducts, Ph.D. Thesis, University of London.

Perng, C.Y. and Street, R.L. 1991 A Coupled Multigrid–Domain Splitting Technique for Simulating Incompressible Flows in Geometrically Complex Domains, *Intl J. Numer. Meths. Fluids*, **13**, 269–286.

Peters, N. 1984 Laminar Diffusion Flamelet Models in Non–Premixed Turbulent Combustion. *Prog. Energy and Combus. Sci,* **10**, 319–339.

Peters, N. and Rogg, B. (eds) 1993 *Reduced Kinetic Mechanisms for Applications in Combustion Systems*, Lecture Notes in Physics, m 15, Springer–Verlag, New York.

Peyret, R. and Taylor, T.D. 1983 *Computational Methods for Fluid Flow*, Springer, New York.

Phillips, R.E. and Schmidt, F.W. 1984 Multigrid Techniques for Numerical Solution of the Diffusion Equation, *Numer. Heat transf.*, **7**, 251–168.

Pitts, E., 1976 The Stability of a Meniscus Joining a Vertical Rod to a Bath of Liquid, *J. Fluid Mech.*, **76**, 641–651.

Platten, J.K. and Legros, J.C. 1984 *Convection in Liquids*, Springer–Verlag, New York.

Poo, J. Y. and Ashgriz, 1989, N., A Computational Method for Determining Curvatures, *J. Comp. Phys.*, **84**, 484–491.

Pope, S.B. 1985 PDF Methods for Turbulent Reacting Flow. *Prog. Energy and Combust. Sci.* **11**, 119–192.

Porter, D.A. and Easterling, K.E. 1992 *Phase Transformations in Metals and Alloys*, 2nd edition, Chapman and Hall, London, UK.

Prandtl, L. and Tietjens, O.G. 1957 *Fundamentals of Hydro– and Aeromechanics*, Dover, New York.

Priddin, C.H. and Coupland, J. 1986 Impact of Numerical Methods on Gas Turbine Combustor Design and Development. In *Calculation of Turbulent Reacting Flows.* (Edited by R.M.C. So, J.H. Whitelaw and H.C. Mongia) 335–348, New York.

Probstein, R.F. 1989 *Physicochemical Hydrodynamics*, Butterworths, Boston, MA.

Rai, M.M. 1986 a A Conservative Treatment of Zonal Boundaries for Euler Equation Calculations, *J. Comput. Phys.*, **62**, 472–503.

Rai, M.M. 1986 b An Implicit Conservative, Zonal–Boundary Scheme for Euler Equation Calculations, *Comp. Fluids*, **14**, 295–319.

Raithby, G.D. 1976 A Critical Evaluation of Upstream Differencing Applied to Problems Involving Fluid Flow, *Comput. Meths. Appl. Mech. Engrg.*, **9**, 75–103.

Raithby, G.D. and Schneider, G.E. 1979 Numerical Solution of Problems in Incompressible Fluid Flow: Treatment of Velocity–Pressure Coupling, *Numer. Heat Transf.*, **2**, 417–440.

Raithby, G.D. and Hollands, K.G.T. 1985 Natural Convection, in *Handbook of Heat Transfer*, (eds) W.M. Rohsenow, J.P. Hartnett, E.N. Ganic, 6.1–6.93, McGraw–Hill, New York.

Rayleigh, Lord 1964 *Scientific Papers*, Dover, New York.

Reggio, M., Trepanier, J. and Camarero, R. 1990 A Composite grid Approach for the Euler Equations, *Int. J. Numer. Meths. Fluids*, **10**, 161–178.

Reif, F. 1965 *Fundamentals of Statistical and Thermal Physics,* McGraw–Hill, New York.

Rhie, C.M. 1986 A Pressure Based Navier Stokes Solver using the Multigrid method, *AIAA Paper No. 86–0207.*

Rhie, C.M. and Chow, W.L. 1983 Numerical Study of the Turbulent Flow past an Airfoil with Trailing Edge Separation, *AIAA J.*, **21**, 1525–1532.

Rimlinger, M.J., Shih, T. I.–P. and Chyu, W.J. 1992 Three–Dimensional Shock–Wave / Boundary–Layer Interactions with Bleed Through a Circular Hole, *AIAA Paper No. 92–3084.*

Richtmyer, R.D. and Morton, K.W. 1967 *Difference Methods for Initial–Value Problems*, 2nd ed., Wiley, New York.

Roache, P.J. 1972 *Computational Fluid Dynamics*, Hermosa Publishers, Albuquerque, NM.

Rodi, W. Majumdar, S. and Schonung, B. 1989 Finite Volume Methods for Two–Dimensional Incompressible Flows with Complex Boundaries, *Comput. Meths. Appl. Mech. Engrg.*, **75**, 369–392.

Roe, P.L. 1981 Approximate Riemann Solver, Parameter Vectors and Difference Scheme, *J. Comput. Phys.*, **43**, 357–372

Roe, P.L. 1985 Some Contribution to the Modelling of Discontinuous Flows, in E. Engquist et al. (eds.), *Large Scale Computations in Fluid Mechanics, Part 2, Lectures in Applied Mathematics,* **22**, American Mathematical Society, New York.

Roe, P.L. 1986 Characteristic–Based Schemes for Euler Equations, *Ann. Rev. Fluid Mech.*, **18**, 337–356.

Rosenberger, F. 1979 *Fundamentals of Crsytal Growth I,* Springer–Verlag, New York.

Rosner, D.E. 1986 *Transport Processes in Chemically Reacting Flow Systems,* Butterworths, Boston, MA.

Rosner, D.E. 1989 Side–wall Gas Creep and Thermal Stress Convection in Macrogravity Experiments on Film Growth by Vapor Transport, *Phys. Fluids* A, **1**, 1761–1763.

Rowlinson, J.S. and Widom, B.L. 1982 *Molecular Theory of Capillarity,* Oxford University Press, Oxford, UK.

Sackinger, P. A., Brown, R. A. and Derby, J. J. 1989 A Finite Element Method for Analysis of Fluid Flow, Heat Transfer and Free Surface in Czochralski Crystal Growth, *Int. J. Numer. Meths. Fluids*, **9**, 453–492.

Saffman, P.G. and Taylor, G.I. 1958 The Penetration of a Fluid into a Porous Medium or Hele–Shaw Cell Containing a More Viscous Liquid, *Proc. Royal Soc. Lond., Ser. A,* **245.**, 312 – 329.

Saitoh, T. 1978 Numerical Method for Multi–Dimensional Freezing Problems in Arbitrary Domains, *ASME J. Heat Transf.*, **100**, 294–299.

Sathyamurthy, P. and Patankar, S.V. 1990 A Multilevel Correction Base Coupled Point Solution Procedure for Fluid Flow Problems, in *Recent Advances and Applications in Computational Fluid Mechanics*, O. Baysal (ed), ASME, New York, 13–22.

Sato, Y., Goldbeck–Wood, G. and Muller–Krumbhaar, H. 1988 Numerical Simulation of Dendritic Growth, *Phys. Rev. A,* **38**, 2148–2157.

Schneider, G.E. and Zedan, M. 1981 A Modified Strongly Implicit Procedure for the Numerical Solution of Field Problems, *Numer. Heat Transf.*, **4**, 1–19.

Schawbe, D. 1988 Surface–Tension Driven Fluid Flow in Crystal Growth Melts, *Crystals,* **11**, 75–112, Springer–Verlag, New York.

Schwarz, H.R. 1989 *Numerical Analysis*, Wiley, New York.

Scriven, L.E. and Sterling, C.V. 1964 On Cellular Convection Driven by Surface Tension Gradients: Effects of Mean Surface Tension and Surface Viscosity, *J. Fluid Mech.*, **19**, 321–340.

Sekerka, R.F. 1973 Morphological Stability, in P. Hartman (ed.) *Crystal Growth: An Introduction*, vol. 1, 403–443, North–Holland, Amsterdam, The Netherlands

Sellamuthu, R. and Giamei, A.F. 1986 Measurement of Segregation and Distribution Coefficients in MAR–M200 and Hafnium–Modified MAR–M200 Superalloys., *Metall. Tran.,* **17A**, 419–428.

Sethian, J. A. 1990 Numerical Algorithms for Propagating Interfaces: Hamilton–Jacobi Equations and Conservation Laws, *J. Diff. Geom.* **31**, 131–161.

Sethian, J. A. and Strain, J., 1992, Crystal Growth and Dendritic Solidification, *J. Comp. Phys.,* **98**, No. 2, 231–253.

Seydel, R. 1988 *From Equilibrium to Chaos,* Elsevier, New York.

Shamblen, C.E. and Hunter, G.B. 1989 Titanium Base Alloys Clean Melt Process Development, in L.W. Lherbier and J.T. Cordy (eds), *Proceedings of the 1989 Vacuum Metallurgy Conference, Iron and Steel Spciety, Inc., Warrendale, PA, 3–11.*

Shih, T. I.–P., Bailey, R.T., Nguyen, H.L. and Roelke, R.J. 1990 GRID2D/3D – A Computer Program for Generating Grid Systems in Complex–Shaped Two – and Three–Dimensional Spatial Domains, *NASA TM 102453*.

Shuen, J.–S., Chen, K.–H. and Choi, T. 1992 A Time –Accurate Algorithm for Chemical Non–Equilibrium Viscous Flows at all Speeds, *AIAA Paper No. 92–3639*.

Shyy, W. 1984 a Determination of Relaxation Factors for high Cell Peclet Number Flow Simulation, *Comput. Meths. Appl. Mech. Engrg.* **43**, 221–230.

Shyy, W. 1984 b A Note on Assessing Finite Difference Procedures for Large Peclet/Reynolds Number Flow Calculation, in J.J.H. Miller (ed), Boundary and Interior Layers: Computational and Asymptotic Methods, Vol. **3**., 303–308, Boole Press, Dublin, Ireland.

Shyy, W. 1985 a A Study of Finite Difference Approximations to Steady–state, Convention Dominated Flow Problems, *J. Comput. Phys.,* **57**, 415–438.

Shyy, W. 1985 b Numerical Outflow Boundary Condition for Navier–Stokes Flow Calculations by a Line Iterative Method, *AIAA J.,* **23**, 1847–1848.

Shyy, W. 1985 c A Numerical Study of Annular Dump Diffuser Flows, *Comput. Meths. Appl. Mech. Engr.,* **53**, 47–65.

Shyy, W. 1986 An Adaptive Grid Method for Navier–Stokes Flow Computation 2: Grid Addition, *Appl. Numer. Math,* **2**, 9–19.

Shyy, W. 1987 a An Adaptive Grid Method for Navier–Stokes Flow Computation, *Appl. Math. Comput.,* **21**, 201– 219.

Shyy, W. 1987 b Effects of Open Boundary Condition on Incompressible Navier–Stokes Flow Computation: Numerical Experiments, *Numer. Heat Transf.,* **12**, 157–178.

Shyy, W. 1988 Computation of Complex Fluid Flows using an Adaptive Grid Method, *Int. J. Numer. Meths. Fluids.,* **8**, 475–489.

Shyy, W. 1989 A Unified Pressure Correction Algorithm for Computing Complex Fluid Flows. *Recent Advances in Computational Fluid Mechanics* (C.C. Chao, S.A. Orzag and W. Shyy, eds.), Lecture Notes in Engineering, Springer–Verlag, New York, **43**, 135–147.

Shyy, W. 1991 a Structure of an Adaptive Grid Computational Method from Viewpoint of Dynamic Chaos, *Appl. Numer. Math.,* **7**, 263–285.

Shyy, W. 1991 b Structure of an Adaptive Grid Computational Method from Viewpoint of Dynamic Chaos, Part II: Grid Addition and Probability Distribution, *Appl. Numer. Math.*, **7**, 523–545.

Shyy, W. and Braaten, M.E. 1986 CONCERT–Cartesian or Natural Coordinates for Elliptic Reacting Turbulent Flows: A Package of Two–Dimensional and Three–Dimensional Computer Code, *GE Technical Information Series Report No. 86CRD187,* Schenectady,NY.

Shyy, W. and Braaten, M.E. 1988 Adaptive Grid Computation for Inviscid Compressible Flows using a Pressure Correction Method, Proceedings of the AIAA/ASME/SIAM/APS First National Fluid Dynamics Congress, *AIAA–CP 888,* 112–120.

Shyy, W. and Chang, Y. 1990a A Study of Three–Dimensional Natural Convection in High Pressure Mercury Lamps, Part II: Wall Temperature Profiles and Inclination Angles, *Int. J. Heat Mass Transf.,* **33**, 495–506.

Shyy, W. and Chang, Y. 1990b Effects of Convection and Electric Field on Thermofluid Transport in Horizontal High Pressure Mercury Arcs, *J. Appl. Phys.,* **67**, 1712–1719.

Shyy, W. and Chen, M.–H. 1990a Steady–State Natural Convection with Phase Change, *Int. J. Heat Mass Transf.,* **33**, 2545–2563.

Shyy, W. and Chen, M.–H. 1990b Effect of Prandtl Number on Buoyancy–Induced Transport Process with and without Solidification, *Int. J. Heat Mass Transf.,* **33**, 2565–2578.

Shyy, W. and Chen, M.–H. 1991a Interaction of Thermocapillary and Natural Convection Flows during Solidification: Normal and Reduced Gravity Conditions, *J. Crystal Growth,* **108**, 247–261.

Shyy, W. and Chen, M.–H. 1991b Double–Diffusive Flow in Enclosures, *Physics of Fluids A,* **3**, 2592–2607.

Shyy, W. and Chen, M.–H. 1993 A Study of Buoyancy Induced and Thermocapillary Flow of Molten Alloy, *Comput. Meths. Appl. Mech. Engrg.*, **105**, 333–358.

Shyy, W. and Dakin, J.T. 1988 Three–Dimensional Netural Convection in a High Pressure Mercury Discharge Lamp, *Int. Comm. Heat Mass Transf.,* **15**, 51–58.

Shyy, W. and Gingrich, W.K. 1992 Transient Two–Phase Heat Transfer in a Capillary–Pumped–Loop Reservoir for Spacecraft Thermal Management, to be published.

Shyy, W. and Rao, M.M. 1992 Convection Treatment for High Rayleigh Number Laminar Natural Convention Calculation, *Numer. Heat Transf.*, **22B**, 367–374.

Shyy, W. and Rao, M.M. 1993 Enthalpy Based Formulations for Phase–Change Problems with Application to G–Jitter, *AIAA Paper No. 93–2831,* also accepted for publication in *Microgravity Sci Technol.*

Shyy, W. and Sun, C.–S. 1993 Development of a Pressure–Correction / Staggered Grid Based Multigrid Solver for Incompressible Recirculating Flows, *Comput. Fluids.*, **22**, 51–76.

Shyy, W. and Thakur, S.S. 1993 Development of a Controlled Variation Scheme in a Sequential Solver for Recirculating Flows, to be published.

Shyy, W. and Vu, T.C. 1991 On the Adoption of Velocity Variable and Grid System for Fluid Flow Computation in Curvilinear Coordinates, *J. Comput. Phys.*, **92**, 82–105.

Shyy, W., Correa, S.M. and Braaten, M.E. 1988 Computation of Flow in a Gas Turbine Combustor, *Combust. Sci. and Tech,* **58**, 97–117.

Shyy, W., Tong, S.S. and Correa, S.M. 1985 Numerical Recirculating Flow Calculation Using a Body–Fitted Coordinate System, *Numer. Heat Transf.,* **8**, 99–113.

Shyy, W., Braaten, M.E. and Burrus, D.L. 1989 Study of Three–Dimensional Gas Turbine Combustor Flow, *Int. J. Heat Mass Transf,* **32**, 1155–1164.

Shyy, W., Gingrich, W.K., Krotiuk, W.J., Fredley, J.E. and Chalmers, D.R. 1991 Modeling of Two–Phase Thermocapillary Flow in a Spacecraft Thermal Control, *Microgravity Flows – 91,* A. Hashemi, B.N. Antar, I. Tanasawa(eds)., ASME, 15–20.

Shyy, W., Gingrich, W.K. and Gebhart, B. 1992a Adaptive Grid Solution for Buoyancy Induced Flow in Vertical Slots, *Numer. Heat Transf.*, **22A**, 51–70.

Shyy, W., Thakur, S. and Wright, J. 1992b Second–Order Upwind and Central Difference Schemes for Recirculating Flow Computation, *AIAA J.,* **30**, 923–932.

Shyy, W., Chen, M.–H., Mittal, R. and Udaykumar, H.S. 1992c On the Suppression of Numerical Oscillations using a Nolinear Filter, *J. Comput. Phys.*, **102**, 49–62.

Shyy, W., Chen, M.–H. and Sun, C.–S. 1992d Pressure–Based Multigrid Algorithm for Flow at All Speeds, *AIAA 30th Aerospace Sciences Meeting, AIAA Paper No. 92–0548,* also *AIAA J,* **30**, 2660–2669.

Shyy, W., Pang, Y., Hunter, G.B., Wei, D.Y. and Chen, M–H. 1992e Modeling of Turbulent Transport and Solidification during Continuous Ingot Casting, *Int. J. Heat Mass Transf.*, **35**, 1229–1245.

Shyy, W., Gingrich, W.K. and Gebhart, B. 1992f Adaptive Grid Solution for Buoyancy Induced Flow in Vertical Slots, *Numer. Heat Transf.*, **22**, 51–70.

Shyy, W., Udaykumar, H.S., and Liang, S.–J. 1992 g A Study of Meniscus Formation with Application to Edge–Defined Fibre Growth Process, *AIAA 27th Thermophysics Conference, AIAA Paper No. 92–2903* also accepted for publication in *Phys. Fluids*

Shyy, W., Sun, C.–S., Chen, M.–H. and Chang, K.C. 1993a Multigrid Computation for Turbulent Recirculating Flows in Complex Geometries, *Numer. Heat Transf.*, **23A**, 79–98.

Shyy, W., Wright, J.A. and Liu, J. 1993b A Multilevel Composite Grid Method for Fluid Flow Computations, *AIAA Paper No. 93–0768.*

Shyy, W., Udaykumar, H.S., and Liang, S.–J. 1993 c An Interface Tracking Method Applied to Morphological Evolution During Phase Change, *Int. J. Heat Mass Transf.,* **36**, 1833–1844.

Shyy, W, Pang, Y., Hunter, G.B., Wei, D. and Chen M.–H. 1993d Effect of Turbulent Heat Transfer on Continuous Ingot Solidification, *ASME J. of Engineering Material and Technology,* **115**, 8–16.

Shyy, W., Thakur, S.S. and Udaykumar, H.S. 1993e A High Accuracy Sequential Solver for Simulation and Active Control of a Longitudinal Combustion Instability, accepted for publication in *Computer Systems in Engineering*

Silliman, W. J. and Scriven, L. E. 1980 Separating Flow near a Static Contact Line: Slip at a Wall and Shape of a Free Surface, *J. Comput. Phys.*, **34**, 287–313.

Slattery, J.C. 1990 *Interfacial Transport Phenomena*, Springer–Verlag, New York.

Smith, G.D. 1985 *Numerical Solution of Partial Differential Equations: Finite Difference Methods, 3rd ed.*, Oxford University Press, Oxford, UK.

Smoller, J. 1983 *Shock Waves and Reaction–Diffusion Equations*, Springer–Verlag, New York.

Smooke, M.D. (ed) 1991 *Reduced Kinetic Mechanisms and Asymptotic Approximations for Methane–Air Flames*, Lecture Notes in Physics, Vol. 384, Springer–Verlag, New York

Spalding, D.B. 1955 *Some Fundamentals of Combustion*, Butterworths, London, UK.

Spalding, D.B. 1970 Mixing and Chemical Reaction in Steady Confined Turbulent Flame. *Thirteenth Symp. (Intl) Combustion.* 649–658, The Combustion Institute, Pittsburgh, PA.

Spalding, D.B. 1972 A Novel Finite Difference Formulation for Differential Expressions involving Both First and Second Derivatives, *Int. J. Numer. Meths. Engrg*, **4**, 551–559.

Sparrow, E. M., Patankar, S. V. and Ramadhyani, S. 1977 Analysis of Modeling in the Presence of Natural Convection in the Melt Region, *ASME J. Heat Transf.*, **99**, 520–526.

Steger, J.L. and Benek, J.A. 1987 On the Use of Composite Grid Schemes in Computational Aerodynamics, *Comput. Meths. Appl. Mech. Engrg.*, **64**, 301–320.

Steger, J. and Warming, R.F. 1981 Flux–Vector Splitting of the Inviscid Gasdynamics Equations with Applications to Finite–Difference Methods, *J. Comput. Phys.*. **40**, 263–293.

Stoer, J. and Bulirsch, R. 1980 *Introduction to Numerical Analysis*, Springer–Verlag, New York.

Stone, H.L. 1968 .Iterative Solution of Implicit Approximation of Multidimensional Partial Differential Equations, *SIAM J. Numer. Anal.*, **5**, 530–558.

Strain, J. 1989 A Boundary Integral Approach to Unstable Solidification, *J. Comp. Physics*, **85**, 342–389.

Strang, G. 1968 On the Construction and Comparison of Difference Schemes, *SIAM J. Numer. Anal.*, **5**, No. 3, 506–517.

Stuben, K. and Trottenberg, U. 1982 Multigrid Methods: Fundamental Algorithms, Model Problem Analysis and Applications, in W. Hackbusch and U. Trottenberg (eds.) *Multigrid Methods, Lecture Nores in Mathematics*, **960**, 1–176, Springer–Verlag, Berlin.

Switzer, G.L., Goss, L.P., Trump, D.D., Reeves, C.M., Stutrud, J.S., Bradley, R.P. and Roquemore, W.M. 1985 CARS Measurements in the Near–Wake Region of an Axisymmetric Bluff–Body combustor, *AIAA Paper No. 85–1106*.

Szekely, J. 1979 *Fluid Flow Phenomena in Metals Processing*, Academic Press, New York.

Tabeling, P., Zocchi, G. and Libchaber, A. 1987 An experimental study of the Saffman–Taylor instability, *J. Fluid Mech.*, **177**, 67–82.

Taha, M.A. and Kurz, W. 1981 About Microsegregation of Nickel Base Superalloys, *Z. Metallkde.*, **72**, 546–549.

Tatarchenko, V.A., and Brenner, E.A. 1980 Crystallization Stability During Capillary Shaping, I and II, *J. Crystal Growth*, **50**, 33–50.

Tetyukhin, V.V., Denisov, Y.P., Dubrovina, N.T., Trubin, A.N. and Sarelyev, V.V. 1982 in *Titanium and Titanium Alloy,* ed by J.C. Williams and A.F. Belov, Plenum Press, New York, 141.

Thakur, S.S. and Shyy, W. 1992 Unsteady, One–Dimensional Gas Dynamics Computations Using a TVD Type Sequential solver, *AIAA Paper No. 92–3640*.

Thakur, S.S. and Shyy, W. 1993a Development of High Accuracy Convection Schemes for Sequential Solvers, *Numer. Heat Transf.*, **23B**, 175–199.

Thakur, S.S. and Shyy, W. 1993b Some Implementational Issues of Convection Schemes for Finite Volume Formulations, *Numer. Heat Transf.*, part B, **24**, 31–55.

Thomas, P.D., Ettouney, H.M., and Brown, R. A., 1986 A Thermal–Capillary Mechanism for a Growth Rate Limit in Edge–Defined Film–Fed Growth Rate Limit Silicon Sheets, *J. Crystal Growth*, **76**, 339–351.

Thomas, P.D. and Lombard, C.K. 1979 Geometric Conservation Law and its Application to Flow Computations on Moving Grids, *AIAA J.*, **17**, 1030–1037.

Thompson, C.F., Leaf, G.K. and Vanka, S.P 1988 Application of a Multigrid method to a Buoyancy induced Flow Problem, S.F. McCormick (ed.), *Multigrid Methods*, Lecture Notes in Pure and Applied Mathematics, Dekker, Inc, New York, 101–110.

Thompson, J.F. 1985 A Survey of Dynamically Adaptive Grids in the Numerical Solution of Partial Differential Equations, *Appl. Numer. Math.*, **1**, 3–27.

Thompson, J.F., Warsi, Z. U. A., and Mastin, C.W. 1985 *Numerical Grid Generation,* Elsevier, New York.

Thompson, M.C. and Ferziger, J.H. 1989 An Adaptive Multigrid Technique for the Incompressible Navier Stokes Equations, *J. Comput. Phys.*, **82**, 94–121.

Thompson, M.E. and Szekely, J. 1988 Mathematical and Physical Modeling of Double–Diffusive Convection of Aqueous Solution Crystallizing at a Vertical Wall, *J. Fluid. Mech.,* **187**, 409–433.

Thompson, M. E. and Szekely, J. 1989 The Transient Behavior of Weldpools with a Deformed Free Surface, *Int. J. Heat Mass Transf.*, **32**, 1007–1019.

Tiller, W.A. 1991 *The Science of Crystallization*, 2 volumes, Cambridge University Press, Cambridge, UK.

Tiller, W.A. Jackson, K.A., Rutter, J.W. and Chalmers, B. 1953 The Redistribution of Solute Atoms During the Solidification of Metals, *Acta. Metall.*, **1**, 428–437.

Tirmizi, S. H. and Gill, W. N. 1987 Effect of Natural Convection on Growth Velocity and Morphology of Dendritic Ice Crystals, *J. Crystal Growth,* **85**, 488–502.

Tripp, D.W. and Mitchell, A. 1989 Segregation and Hot–Topping of Titanium Alloy Ingots, in Lherbier, L.W. and Cordy, J.T. (eds), *Proceedings of the 1989 Vacuum Metallurgy Conference,* Iron and Steel Society, Inc., PA, 83–89.

Tu, J.Y. and Fuchs, L. 1992 Overlapping Grids and Multigrid Methods for Three–Dimensional Unsteady Flow Calculations in IC Engines, *Int. J. Numer. Meths. Fluids*, **15**, 693–714.

Udaykumar, H.S. and Shyy, W. 1993a Modelling Solidificaiton Processes at Morphological Scales, *ASME National Heat Transfer Conference*, Atlanta, GA.

Udaykumar, H.S., and Shyy, W. 1993b Development of a Grid–Supported Marker Particle Scheme for Interface Tracking, *AIAA 11th Computational Fluid Dynamics Conference*, AIAA Paper No. 93–3384.

Ungar, L. H. and Brown, R. A. 1984 Cellular Interface Morphologies in Directional Solidification. The One Sided Model, *Phys. Rev. B,* **29**, No. 3, 1367–1380.

Unverdi, S.O. and Tryggvason, G. 1992 A Front Tracking Method for Viscous, Incompressible, Multidimensional Flows, *J. Comp. Phys.,* **100**, 25–37.

Turner, J.S. 1973 *Buoyancy Effects in Fluids*, Cambridge University Press, Cambridge, UK.

Van Doormaal, J.P., Raithby, G.D. and McDonald, B.M. 1987 The Segregated Approach to Predicting Viscous Compressible Fluid Flows, *ASME J. Turbomachinery,* **109**, 268–277.

Van Leer, B. 1974 Towards the Ultimate Conservative Differencing Scheme II. Monotonicity and Conservation Combined in a Second Order Scheme, *J. Comput. Phys.*, **14**, 361–370.

Van Leer, B. 1979 Towards the Ultimate Conservative Difference Scheme V. A Second Order Sequel to Godunov's Method, *J. Comput. Phys.*, **32**, 101–136

Vanka, S.P. 1985 Block Implicit Calculation of Steady Turbulent Recirculating Flows, *Int. J. Heat Mass Transf.*, **28**, 2093–2103.

Vanka, S.P. 1986 Block–Implicit Multigrid Solution of Navier–Stokes Equations in Primitive Variables, *J. Comput. Phys.*, **65**, 138–158.

Vanka, S.P. 1987 Second–Order Upwind Differencing in a Recirculating Flow, *AIAA J.*, **25**, 1435–1441.

Varga, R.S. 1962 *Matrix Iterative Analysis*, Prentice–Hall, Englewood Cliffs, NJ.

Venezian, G. 1969 Effect of Modulation on the Onset of Thermal Convection, *J. Fluid Mech.*, **35**, 243–254.

Viskanta, R. 1989 Heat Transfer during Melting and Solidification of Metals, *ASME J. Heat Transf.*, **110**, 1205–1219.

Voller, V.R. 1985 Implicit Finite–Difference Solutions of the Enthalpy Formulation of Stefan Problems, *IMA J. Numerical Anal.*, **5**, 201–214.

Voller, V.R. and Cross, M. 1981 Accurate Solutions of Moving Boundary Problems Using the Enthalpy Method, *Int. J. Heat Mass Transf.*, **24**, 545–556.

Voller, V.R. and Prakash, C. 1987 A Fixed Grid Numerical Modeling Methodology for Convection–Diffusion Mushy Region Phase–Change Problem, *Int. J. Heat Mass Transf.*, **30**, 1709–1719.

Vu, T.C., and Shyy, W. 1988 Navier–Stokes Computation of Radial Inflow Turbine Distributor, *ASME J. Fluids Engrg.*, **110**, 29–32.

Vu, T.C., and Shyy, W. 1990 Navier–Stokes Flow Analysis for Hydraulic Turbine Draft Tubes, *ASME J. Fluids Engrg.*, **112**, 199–204.

Wachspress, E.L. 1966 *Iterative Solutions of Elliptic Systems*, Prentice–Hall, Englewood Cliffs, NJ.

Wadiah, M. and Roux, B. 1988 Natural Convection in a Long Vertical Cylinder under Gravity Modulation, *J. Fluid Mech.*, **193**, 391–415.

Walter, H.U. (ed.) 1987 *Fluid Science and Materials Science in Space*, Springer–Verlag, New York.

Warming, R.F. and Beam, R.M. 1976 Upwind Second–Order Difference Schemes and Applications in Aerodynamic Flows, *AIAA J.*, **14**, 1241–1249.

Waymouth, J.F. 1971 *Electric Discharge Lamps*, MIT Press, Cambridge, MA.

Weatherhill, N., Hassan, O., Marchant, M. and Marcum, D. 1993 Adaptive Inviscid Flow Solutions for Aerospace Geometries on Efficiently Generated Unstructured Tetrahedral Meshes, *AIAA Paper No. 93–3390–CP*.

Wheeler, A.A., Murray, B.T. and Schaefer, R.J. 1993 Computation of Dendrites using a Phase Field Model, *Physica D.*, **66**, 243–262

Whitham, G.B. 1974 *Linear and Nonlinear Waves*, Wiley, New York.

Williams, F. A. 1985 *Combustion Theory, 2nd edition*, Benjamin–Cummings, Menlo Park, California.

Withington, J.P., Shuen, J.S. and Yang, V. 1991 A Time Accurate, Implicit Method for Chemically Reacting Flows at all Mach Numbers, *AIAA Paper No. 91–0581*.

Woodward, P.R. and Collela, P. 1984 a The Numerical Simulation of Two–Dimensional Flow with Strong Shocks, *J. Comput. Phys.*, **54**, 115–173.

Woodward, P.R. and Collela, P. 1984 b The Piecewise Parabolic Method (PPM) for Gas Dynamical Calculations, *J. Comput. Phys.*, **54**, 174–201.

Wouters, P., van Schaftingen, J. J., Crochet, M. J. and Geyling, F. T. 1987 Numerical Simulation of the Horizontal Bridgman Growth. Part 3: Calculation of the Interface, *Int. J. Numer. Meths. Fluids*, **7**, 131–153.

Wright, J. 1993 A Pressure–Based Composite Grid Method for Complex Fluid Flows, Ph.D. Thesis, University of Florida.

Wright, J. and Shyy, W. 1992 A Pressure–Based Composite Grid Method for the Incompressible Navier–Stokes Equation. *28th AIAA/SAE/ASME/ASEE Joint Propulsion Conference,* Paper No. 92–3641, also *J. Comput. Phys.*, **107**, 225–238

Xiao, R.–F., Alexander, J. I. D., and Rosenberger, F. 1988 Morphological Evolution of Growing Crystals: A Monte Carlo Simulation, *Phys. Rev A*, **38**, 2447–2456

Xiao, R.–F., Alexander, J. I. D., and Rosenberger, F. 1990 Growth Morphology with Anisotropic Surface Kinetics, *J. Crystal Growth*, **100**, 313–329.

Yang, H.Q and Yang, K.T. 1988 Buoyant Flow Calculations with Non–orthogonal Curvilinear Coordinates for Vertical and Horizontal Parallelpiped Enclosures, *Int. J. Numer. Meths. Engrg.* **25**, 331–345.

Yang, H.Q., Habchi, S.D. and Przekwas, A.–J. 1992 A General Strong Conservation Formulation of Navier–Stokes Equations in Non–Orthogonal Curvilinear Coordinates, *AIAA Paper No. 92–0187.*

Yang, S.C. 1990 Numerical Simulation of Transonic Turbulent Flows Over a Complex Configuration, Ph.D. Thesis, University of Florida.

Yao, L.S. and Prusa, J. 1989 Melting and Freezing. In *Advances in Heat Transfer* (Edited by J.P. Hartnett and T.F. Irvine), **19**, 1–95, Academic Press, New York.

Yee, H.C. 1986 Linearized Form of Implicit TVD Schemes for the Multidimensional Euler and Navier–Stokes Equations, *Comput. & Math. with Appl.*, **12A**, 413–432.

Yee, H.C. 1987a Upwind and Symmetric Shock–Capturing Schemes, *NASA TM* 89464.

Yee, H.C. 1987b Construction of Explicit and Implicit Symmetric TVD Schemes and Their Applications, *J. Comput. Phys.* **68**, 151–179.

Yee, H.C. and Warming, R.F. 1985 Implicit Total Variation Diminishing Schemes (TVD) for Steady State calculations, *J. Comput. Phys.*, **57**, 327–360.

Yih, C.–S. 1965 *Dynamics of Nonhomogenous Fluids,* Macmillan, New York.

Youngs, D. L., 1984, Time Dependent Multimaterial Flow With Large Fluid Distortion, in *Numerical Methods for Fluid Dynamics,* Eds. Morton, K. W. and Baines, M. J., Academic, New York, 273–285.

Young, D.M. 1971 *Iterative Solution of Large Linear Systems*, Academic Press, New York.

Young, D.M. and Gregory, R.T. 1988 *A Survey of Numerical Mathematics*, Two Volumes, Dover, New York.

Young, N.O., Goldstein, J.S. and Block, M.J. 1959 The Motion of Bubbles in a Vertical Temperature Gradient, *J. Fluid Mech.*, **6**, 350–356.

Zalesak, S.T. 1979 Fully Multidimensional Flux Correct Transport Algorithms for Fluid, *J. Comput. Phys.*, **31**, 335–362.

Zebib, A, Homsy, G.M. and Meiburg, E. 1985 High Marangoni Number Convection in a Square Cavity, *Phys. Fluids*, **28**, 3467–3476.

Zien, T.F. 1978 Integral Solutions of Ablation Problems with Time–Dependent Heat Flux, *AIAA J.*, **16**, 1287–1295.

Zollweg, R.J. 1978 Convection in Vertical High–Pressure Mercury Arcs., *J. Appl. Phys.*, **49**, 1077–1091.

INDEX